SPACEFLIGHT

SPACEFLIGHT

THE COMPLETE STORY, FROM **SPUTNIK** TO **CURIOSITY**

GILES SPARROW

CONTENTS

SECOND EDITION

DK DELHI

SENIOR ART EDITOR Ira Sharma
PROJECT EDITOR Priyanka Kharbanda
ART EDITORS Mansi Agrawal, Priyanka Bansal
ASSISTANT EDITORS Vatsal Verma, Kathakali Banerjee
DTP DESIGNERS Anita Yadav, Ashok Kumar
PICTURE RESEARCHER Nishwan Rasool
PICTURE RESEARCH MANAGER Taiyaba Khatoon
PRE-PRODUCTION MANAGER Balwant Singh
PRODUCTION MANAGER Pankaj Sharma
MANAGING EDITOR Kingshuk Ghoshal
MANAGING ART EDITOR Govind Mittal

DK LONDON

SENIOR EDITOR Peter Frances
SENIOR ART EDITOR Ina Stradins
PROJECT EDITOR Miezan van Zyl
JACKET DESIGNER Surabhi Wadhwa-Gandhi
JACKET DESIGN DEVELOPMENT MANAGER Sophia MTT
PRODUCER, PRE-PRODUCTION David Almond
PRODUCER Mary Slater
MANAGING EDITOR Angeles Gavira Guerrero
MANAGING ART EDITOR Michael Duffy
ASSOCIATE PUBLISHING DIRECTOR Liz Wheeler
ART DIRECTOR Karen Self
DESIGN DIRECTOR Philip Ormerod
PUBLISHING DIRECTOR Jonathan Metcalf
EDITORIAL CONSULTANT Anthony Brown
INDEXER Vanessa Bird

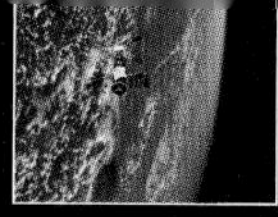

AFTER APOLLO 168

WORKING IN SPACE 186

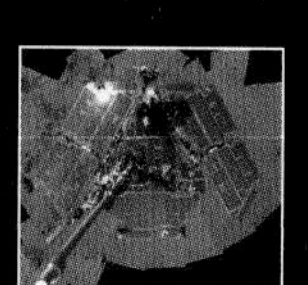

SATELLITES AND SPACEPROBES 242

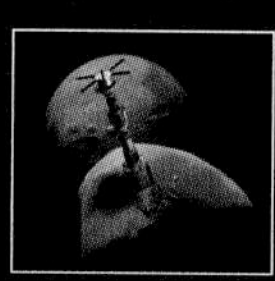

INTO THE FUTURE 286

FIRST EDITION

SENIOR EDITOR Peter Frances
DTP DESIGNER Laragh Kedwell
PRODUCTION Elizabeth Warman
JACKET DESIGNERS Lee Ellwood, Duncan Turner

MANAGING EDITOR Sarah Larter
SENIOR MANAGING ART EDITOR Philip Ormerod
PUBLISHING MANAGER Liz Wheeler
PUBLISHER Jonathan Metcalf
ART DIRECTOR Bryn Walls

EDITORIAL CONSULTANT Carole Stott
INDEXER Hilary Bird

MP3 DESIGN AND EDITORIAL SERVICES

PROJECT EDITOR David Preston
ART EDITOR Jerry Udall
DESIGNERS Gadi Farfour, Peter Laws, Colin Tilley-Loughrey
ASSISTANT EDITORS Joe Booker, Emma Northam
PICTURE RESEARCHERS Ieva Augustaityte, Marian Pullen
ILLUSTRATORS Julian Baum, Peter Bull, Hugh Johnson, Ian Palmer, Chris Taylor, Adrian Thompson, Terry Pastor (The Art Agency)
CONSULTANT Robin Scagell
SPECIAL PHOTOGRAPHY Gary Ombler

This edition published in 2019
First published in Great Britain in 2007 by
Dorling Kindersley Limited
80 Strand, London, WC2R 0RL

A Penguin Random House Company
10 9 8 7 6 5 4 3 2 1
001-310755-May/2019

A CIP catalogue record for this book is available from the British Library.
ISBN: 978-0-2413-4679-2

Printed and bound in Malaysia

A WORLD OF IDEAS:
SEE ALL THERE IS TO KNOW

www.dk.com

60 years and counting

By Buzz Aldrin

The Apollo 11 moonwalk was the first time that humans had set foot on another world. It was the symbolic highlight of the whole Apollo programme, but the credit for it should be shared widely. That first landing would not have been possible without the astronauts who flew the earlier exploratory missions. For example, before we could establish that a landing was possible at all, we had to send two especially hazardous missions, Apollo 8 and Apollo 10, into orbit around the Moon. In fact, it was not until Apollo 13 came close to ending in disaster that we realized just how much danger the astronauts on those earlier missions had been in. Yet, when I talk to anyone about the Apollo programme, one of the first questions they usually ask is, "How many people walked on the Moon?" I have to give them the answer – 12. However, it should not be forgotten that, in all, 24 men left Earth orbit (a feat that has not been repeated since) and went to the Moon.

As well as the other astronauts, we also owe a great debt to the many thousands of people who were involved on the ground. Since leaving NASA, I have tried to continue to play an active part in the future of space exploration, and some of the pioneering engineers who worked on the Apollo programme continue to inspire me in that work.

In one way or another, I have now been involved in the manned exploration of space for more than 40 years – that is to say, for almost the entire duration of the Space Age and a large part of the period covered by this book.

It is generally accepted that the Space Age began when the Soviet Union sent the satellite Sputnik 1 into orbit on 4 October 1957. I have to be honest and say that at the time Sputnik did not make a great impression on me. I have my reasons, though. At the time, the United States was in the thick of the Cold War with the Soviet Union. As a pilot in the United States Air Force, I was stationed in West Germany, where I was training to fly tactical fighters to send nuclear weapons into the Soviet Union. In the event that a nuclear exchange did break out, it was unlikely that I would have a base to return to. That was a sobering reality for a 28-year-old with a young family. When Sputnik went into space and sent back its radio signal, it seemed to me little more than a stunt.

It was different when Yuri Gagarin became the first person to fly in space, in April 1961. By then, my own circumstances had changed. I was halfway through a doctoral thesis on piloting techniques for space rendezvous. In those early years of the Space Age, one thing quickly led to

"Walking on the Moon was **a piece of cake.** It was easy. But getting to the Moon was anything but easy."

FIRST STEPS ON A NEW WORLD
This photograph of me was taken by Neil Armstrong in the Sea of Tranquillity. I'm leaning forward slightly, to avoid overbalancing under the weight of my backpack, and smudges of moondust can be seen on my shins because I misjudged the distance between the bottom step of the ladder on the Lunar Module and the surface. But on the whole, it was easy to walk around on the Moon. Those first steps attracted a lot of attention – understandably, because of the symbolism of the event – but from a technical point of view this was one of the easiest parts of the mission.

another. Less than a month after Gagarin's return, Alan Shepard made his suborbital flight, becoming the first American in space. And just 20 days after that, President Kennedy made his famous speech committing our country to send a man to the Moon – and bring him safely back to Earth – before the end of the decade.

All this time, I was drawing closer to taking an active part in the space programme. By the end of 1962, I had finished my thesis, and in October of the following year I accepted an invitation from the Head of the Astrounaut Office, Deke Slayton, to join the astronaut programme. The tensions between East and West remained high, but it is also undoubtedly true that the Cold War gave a great impetus to the speed of development in spaceflight. As we pursued our own Mercury, Gemini, and then Apollo programmes, we knew that the Soviets were training their own people to fly in space and that they had some extremely capable engineers, men such as their Chief Designer, Sergei Korolev.

As we embarked on the Gemini programme, I was able to put my earlier studies of space rendezvous to use. It was clear that to make a landing on the Moon we would need spacecraft made up of modules that could separate from one another and then link up again, either in orbit around the Earth or around the Moon. I was able to contribute to the development of the techniques that were used to do this. I was also able to help find new ways of training for spacewalks. After Michael Collins's spacewalk on Gemini 10, it was decided that we should experiment with underwater training. This is now a staple part of astronaut preparation, but it was new and untried back then. I had done some scuba diving before joining NASA, so it was an environment in which I was already at home. I was able to put both my rendezvous expertise and my underwater training into practice when I flew on Gemini 12 with Jim Lovell in 1966. During that mission, we performed docking manoeuvres and I made three spacewalks.

After Gemini came Apollo. Of course, an enormous amount has already been said and written about that programme. In retrospect, Apollo now seems remarkable to me not only for its boldness and its ultimate success but also for how much we accomplished in a short time (there were just eight years between the announcement of our intention to go the Moon and the first

A LIFE IN FLIGHT
Flying aircraft and flying spacecraft are very different experiences. In space, the pilot usually has less control and relies more on computers and help from the ground. I began my own flying career in the Air Force and flew in combat in the Korean War (above). In all, I spent 290 hours in space, about five of them outside the Gemini 12 spacecraft (below).

"I remember this fleeting thought from the surface of the Moon: the two of us, Neil and I, are **further away** than two humans have ever been before, not just in distance but in what we have to do to get back, and yet there are **more people paying attention** to what we are doing now than have ever paid attention to other people before."

THE NEXT STEPS
When I took this photograph of my own footprint, I had little idea that it would become a symbol of human exploration in space. It is now 50 years since we left the Moon. Although it seems that it will be at least ten years before another astronaut leaves their mark there, it is still encouraging to know there is now a plan to go back.

landing) and for how adaptable we showed ourselves to be. For example, we recovered very quickly from the awful fire on Apollo 1. Apollo 8 also comes to mind again. This was the first manned mission into lunar orbit, so it was a big step forwards but it was one that we had to make sooner than we had originally planned because we suspected that the Soviets were gaining ground on us and were about to attempt their own circumnavigation of the Moon.

When we look back over humanity's first 60 years in space, it is important that we not only celebrate what we have done but also properly understand the past, seeing clearly where mistakes have been made, so that we can plot the right course for the future.

Engendering a spirit of adaptability similar to the one we had on Apollo is something that we should be doing as we look ahead. NASA's two main projects over the last four decades have been the Space Shuttle and the International Space Station. Both are technically marvellous but they are highly ambitious and incredibly complex. Despite some successes, they have not lived up to all their expectations. One way to make ourselves more adaptable in future is to have several projects running along parallel paths. The main US project for the future is Orion. We need to develop the best possible solution for this programme, because it will ultimately provide us with the technology to return to the Moon, scheduled for the end of the next decade.

In my view, spaceflight should not be an exclusive preserve of professional astronauts. I would like to see as many people as possible become involved. Ever since the Shuttle programme, there has been a widening of the net for crew selection, something that should continue and be extended. We should support opportunities for so-called "space tourists" or, as I would prefer to call them, star flyers or star travellers. This can be done by making a concerted effort to educate children about space and make it appealing to them. Education is, after all, the key to our future. I hope that this book will help to inspire a future generation of astronauts and, in some way, help to make the next 60 years of human endeavour in space as rich, exciting, revealing, and successful as the first 60.

Buzz Aldrin

JANUARY 1940: ROCKET MEN
The American rocket pioneer Robert Goddard (foreground) and his team of engineers inspect one of their newly created rockets in their workshop at Roswell, New Mexico.

ROCKET DREAMERS

FOR CENTURIES, dreams of travel into space had been the preserve of fantasists, satirists, and the occasional speculative scientist, but the 20th century changed all that. Rockets were for the first time recognized as the only practical means of space travel, while a series of design advances transformed them from outsized fireworks to intercontinental ballistic missiles. The genealogy of the space age is a complex one. Nineteenth-century novelists influenced rocket theorists such as Konstantin Tsiolkovskii in Russia, and later Robert Goddard in America and Hermann Oberth in Germany. Each made his own unique contributions to rocket science, but there were many parallel discoveries, too. In turn, these pioneers influenced another generation – most notably in Russia and Germany, where the rocket would grow from an experimental plaything to a weapon of war. In the aftermath of the Second World War, rival superpowers scrambled to adapt the secrets of German rocketry to their own purposes. But throughout all this, the engineers and scientists behind the rockets retained the dream of spaceflight.

The earliest rockets

Before spaceflight, there was the rocket – at first little more than a novelty, but later a powerful weapon of war whose importance waxed and waned over the centuries, nevertheless playing a pivotal role in several military conflicts around the world.

HERO'S ENGINE
An early reaction motor was designed in the first century CE by Greek-Egyptian scientist Hero of Alexandria. Heat applied from below boiled water in a spherical vessel, and steam spouting from the nozzles caused the sphere to spin on its axis.

Though it took some time for scientists to realize it, a rocket is a "reaction motor" operated by the simple principle of action and reaction – as fuel exhaust escapes from the rocket in one direction, the rocket is pushed in the other. The basic requirement for any rocket, then, is a propellant that can be stored in a relatively stable state, but which expands violently when required. And until the 20th century, only one thing fitted the bill: a black powder, also known as gunpowder – made up of charcoal, sulphur, and saltpetre (potassium nitrate) – that exploded when a flame was applied.

History does not record the invention of this early weapon of mass destruction, but it is thought to have originated in Song Dynasty China around the middle of the 11th century. The use of exploding powder as a propellant would have followed naturally, probably a result of seeing containers flung across the room by the force of its explosions. By 1232, self-propelled "flying bombs" were being used to defend the Chinese city of Kai'feng from the advancing Mongol army of Genghis Khan. These early rockets would have been almost as dangerous for the defenders as the attackers, since whenever any part of their flimsy paper or card wrapping burnt through to form another exhaust, they might shoot off in another direction, setting fires wherever they landed.

The new weapons, however, were not enough to save China from conquest, and by 1241 the Mongols were expanding their empire to the west, using rockets themselves in battles across Eastern Europe. And with them came the secret of black powder – the recipe was first written down around 1250 by English scholar Roger Bacon (though he disguised it in code, wary of its potential as a weapon). While the Mongol threat disintegrated in a series of internal disputes, the knowledge of black powder spread rapidly. In 1288, Arab forces were using rockets in an attack on the Spanish city of Valencia, and by 1405 rockets were a familiar part of the medieval war machine, depicted by German engineer Konrad Kyeser in his military manual *Bellifortis*.

While China now entered a more peaceful era under the Ming Dynasty, the evolution of rockets in Europe continued apace. Kyeser's rockets were already mounted on the top of long stabilizing rods, which could be placed in a gutter-like launcher to allow for basic targeting. Seventeenth-century Polish author Casimirus Siemienowicz illustrated a variety of rockets that are sometimes strikingly similar to the designs of today, with long tubular bodies, stabilizing fins, and even multiple stages (see p.23). In 1715, Russian Tsar Peter the Great's plans for a new capital at St. Petersburg included an enormous factory for

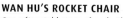

CONGREVE ROCKETS
The rockets designed by William Congreve (top) featured a number of important innovations, including stabilizing fins around their bases. Later Congreve designs were fired from copper launching tubes, which helped to direct their flight and reduced the risk of misfires during launch.

WAN HU'S ROCKET CHAIR
One often-told story of early Chinese rocketry is the legend of Wan Hu, a Ming-Dynasty official who supposedly flew into space using a chair supported on 47 rockets.

ROCKETS AT WAR
One key element in the popularity of rockets was their portability. Congreve rockets could be moved, set up on their launching frames, and fired with relative ease, even from small unstable platforms such as boats. This painting depicts an attack by the British on the American fleet during the War of 1812.

the mass production of rockets. But in truth, the rocket was on the verge of obsolescence, thanks to the increased range and accuracy of artillery.

The metal-clad rocket

The invention that "saved" the rocket in the 18th century once again emerged from the East – this time from India. Around the mid-1700s, Hyder Ali, Sultan of Mysore, ordered the construction of rockets sheathed in iron, not card or paper. Because the heavier cladding directed the rocket's exhaust more efficiently and would not burn through, the new rockets had a vastly increased range of more than 800m (½ mile), despite their greater weight. By the late 18th century, as the British grip on India tightened, Ali's son Tippu Sultan put his father's invention to good use in the sieges of Seringapatam (1792 and 1798), though rockets were not enough to save him from eventual defeat. In fact, they ultimately helped his enemy, since captured rockets shipped back to Britain probably inspired William Congreve, working at the Royal Arsenal, to develop a more advanced model. Congreve gave his rockets a payload, or cargo, for the first time, mounting a separate charge of black powder in the rocket's nose, where it would explode on impact – the first warhead. He also invented an improved launching platform and came to realize that a ring of five smaller exhaust nozzles gave a rocket much more stability than a single outlet.

ADVANCED ROCKETRY
This illustration shows armourers constructing a rocket to the plans of Casimirus Siemienowicz. In his time, the study of ballistics, allowing flight paths and targets to be calculated, was advancing rapidly.

The 19th century saw a series of modifications to Congreve's design. In 1807, Englishman Henry Trengrouse devised a rocket that could carry a line to a ship in difficulties, which soon became an important part of coastguard equipment in Britain and beyond. Shortly after this came the invention of the rocket-powered harpoon and the signal flare. The most important advance, though, came in 1844 when another Englishman, William Hale, tilted the exhaust nozzles of his designs, causing the rockets to rotate around their long axes and fly with greater stability. This meant that the clumsy stabilizing stick could at last be eliminated.

TECHNOLOGY

LAWS OF MOTION AND GRAVITY

Until the late 17th century, philosophers were resigned to the idea that the Universe was driven by supernatural, or at best inscrutable, forces, with little relationship between the behaviour of objects on Earth and those above. This view began to change around 1609, when Johannes Kepler finally swept aside old notions of heavenly spheres and celestial clockwork, replacing them with laws of planetary motion that could accurately describe (though not explain) the orbits of the planets. However, it was the English scientist Isaac Newton (left) who finally showed how the movement of the planets, and of objects on Earth, could all be explained through three simple laws of motion, and the effects of a force he called gravity that was produced by any object with a substantial mass.

WEIGHTLESS BLUNDER
Jules Verne's travellers experienced weightlessness only as they crossed the region where the Earth and Moon's gravitational pull were balanced. In reality, though, since they and their ship were travelling at constant speed, they would have been weightless immediately after launch.

Space visionaries

Writers have fantasized about journeys beyond Earth since classical times, but the industrial revolution of the 19th century, coupled with advances in scientific knowledge, gave rise to a wave of speculative fiction that would inspire later generations to make space travel a reality.

The Roman poet and satirist Lucian of Samosata is widely acknowledged as the world's first science-fiction writer. His *True History*, written around 150 CE, is a tale of travellers carried into space and eventually to the Moon on a giant water spout. Lucian, however, was principally writing a fantasy, at a time when the rigours of space travel were purely matters of guesswork. Later literature throws up similar tales, such as the proposal by the 17th-century English bishop Francis Godwin for a lunar expedition in a carriage pulled by geese; a more thoughtful fantasy comes from no less an authority than Johannes Kepler, the astronomer who finally worked out the laws of planetary orbital motion and clinched the case for a Sun-centred Solar System. In his *Somnium* (Dream) of 1634, Kepler tells the tale of an expedition to the Moon, recognizing that it would involve a traumatic launch and travel beyond Earth's atmosphere and that conditions in space were dangerous, with fierce radiation from the Sun.

JULES VERNE
In a prolific career, Verne wrote 54 novels and numerous short stories. He frequently returned to the theme of exploration in unknown environments and to making predictions of the world of the future.

BULLET TO THE MOON
Although this illustration for From the Earth to the Moon *implies some kind of engine driving the moonship, Verne never wrote of such a device.*

Verne and Wells

However, the true colossi of science fiction belong to the 19th century. French author Jules Verne wrote a series of adventure novels on scientific themes, of which the most influential was *From the Earth to the Moon* (1865). Verne made a serious attempt to address the problem of launching a spacecraft towards the Moon, opting to propel his heroes and their moonship from a giant cannon, the *Columbiad*. His understanding of the laws of physics was somewhat awry, though: he didn't realize that the ship's occupants would be crushed by the sudden acceleration as the shell was fired (something that Kepler had taken into account); and he misinterpreted the influence of gravity (see caption, above left). In contrast, Herbert George Wells made little or no attempt at realism in his story of lunar flight. In *The First Men in the Moon*, written a generation later in 1901, he solved

EARLY EDITIONS
From the Earth to the Moon *(left) and its sequel,* Around the Moon *(1870), were the bestsellers of their age, published in many languages and rarely out of print.*

the problem of space travel by having one of his characters invent a material, cavorite, that blocks out the effects of gravity. What's more, when the spacecraft's cavorite coating is complete, it simply shoots into the sky without any apparent propulsion. However, while Verne's story was focused on the human characters and the voyage to the Moon itself, Wells's imagination can be forgiven, since in his case space travel was largely a means to an end. Once on the Moon, the story turns into a political allegory as the travellers encounter a strange alien society. (In his even more influential *The War of the Worlds* (1898), Wells's Martians seem to use a "space cannon" to fire their ships towards Earth.)

Both Verne's and Wells's errors would have been obvious to anyone well-versed in the physics of the time, but a generation later educated people were still criticizing the pioneers of rocketry and claiming that space travel was impossible – at a time when it was on the verge of becoming a reality. The old geocentric instincts died hard, and there seems to be something about the laws of motion and gravitation that makes them particularly susceptible to misunderstanding.

BIOGRAPHY

NIKOLAI KIBALCHICH

Rocket-powered spaceflight was suggested more than once in the 19th century, but the strangest case must surely be the tale of Nikolai Kibalchich (1853–81). In the 1880s, Russia was an absolute monarchy ruled by the Tsar and his noblemen, and the lower classes were seething with the resentment that would eventually lead to the revolution of 1917. Kibalchich, an engineer by training, became involved in a revolutionary group and made grenades used to kill Tsar Alexander II in 1881. While in prison awaiting execution for his role in the assassination, Kibalchich sketched a design for a rocket-powered passenger platform similar to that shown here. He sent it to the government, but after his death it languished in the archives until its rediscovery in 1917.

Despite these problems, Wells, Verne, and others that followed them helped shape the imaginations of a generation of scientists. Space travel no longer seemed a fantastical dream, but an attainable goal. At least in some circles, one could now discuss the issues seriously and without risk of mockery.

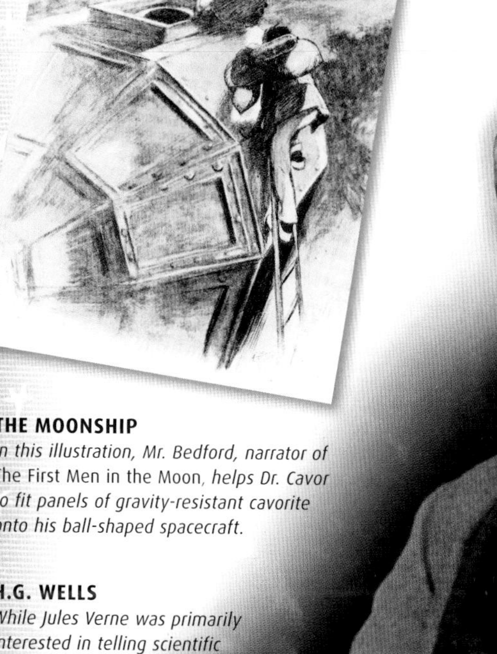

THE MOONSHIP
In this illustration, Mr. Bedford, narrator of The First Men in the Moon, *helps Dr. Cavor to fit panels of gravity-resistant cavorite onto his ball-shaped spacecraft.*

H.G. WELLS
While Jules Verne was primarily interested in telling scientific adventure stories, H.G. Wells was a more political writer, frequently using his novels about the future to promote his socialist beliefs.

Rocket prophet

Konstantin Tsiolkovskii did more than any other single person to make space travel a reality, developing many of the techniques and principles still used in rocketry today. Yet he spent most of his life as an obscure provincial teacher, winning recognition only in his old age.

While Verne and Wells had made the subject of space exploration acceptable, it was left to other talents to make it achievable, and the greatest of these was undoubtedly Konstantin Eduardovich Tsiolkovskii. Born in 1857 in the Russian town of Izhevskoye, Tsiolkovskii overcame a childhood illness that left him almost entirely deaf, to become one of the greatest practical scientific thinkers of his time.

> "From the moment of using **rocket devices** a new great era will begin in astronomy."
>
> **Konstantin E. Tsiolkovskii, 1896**

Fascinated from an early age with airships and balloons, Tsiolkovskii came to believe that a sealed ship of this type might be suitable for space travel. By this time, most scientists had realized that space beyond Earth was a vacuum – balloon-borne experiments had shown that air pressure fell rapidly at high altitudes. Since most propulsion systems relied on the presence of a medium to push against, they would be useless in a vacuum, so how could a spacecraft be propelled and steered?

A high-speed launch from a Verne-style cannon was out of the question. Tsiolkovskii carried out experiments that showed living creatures could survive an acceleration of up to 60m/s (200ft/s) per second (roughly six times the acceleration due to Earth's gravity, or 6*g*), but not much more. He also worked out the Earth's escape velocity, the speed required to launch an object from the Earth's surface so that it could never be pulled back by the planet's gravity. This turned out to be about 11.2km (7 miles) per second. Clearly any attempt to reach such a speed in a near-instantaneous burst of acceleration would result in a spacecraft's occupants being crushed to death.

The solution, Tsiolkovskii suggested, was a self-contained rocket or "reaction engine", which could produce steady acceleration inside or outside the atmosphere, eventually reaching speeds where a spacecraft could remain in orbit – its tendency to move away from the Earth perfectly balanced against the force of the planet's gravity – or even break free of the Earth altogether to travel across interplanetary space. Tsiolkovskii was not actually the first person to suggest rockets as a means of space travel (that honour goes to the 17th-century French author Savinien Cyrano de Bergerac), but he was the first to treat the idea seriously, publishing a number of detailed scientific papers that reached fruition in *The Exploration of Cosmic Space by Means of Reaction Devices* (1903).

Among Tsiolkovskii's breakthroughs was the realization that multi-stage rockets would be more efficient than single-stage ones (see p.23). He was also the first to show how steering vanes, used to deflect the exhaust, could control a rocket's direction in a vacuum. Tsiolkovskii was certain that liquid fuels would be needed to reach space, since black powder (see p.12) was too weak a propellant, and combusted only by reacting with oxygen in the atmosphere. A truly self-contained rocket would have to carry not only fuel, but also a chemical oxidant, in tanks onboard, and the best choice would be liquid hydrogen fuel burning with liquid oxygen. At the time, however, these substances were impossible to manufacture in large amounts.

THE INVENTOR'S WORKSHOP
While working as a teacher in the town of Kaluga, Tsiolkovskii produced numerous models to demonstrate his ideas, but never attempted to launch a rocket himself.

GRAND OLD MAN
Tsiolkovskii spent much of his life in relative obscurity. It was only after the 1917 Russian Revolution that his ideas, such as liquid rocket designs (right), received widespread recognition. When he died, in 1935, he was acknowledged as the pioneer of a new science.

TECHNOLOGY

HOW ROCKETS WORK

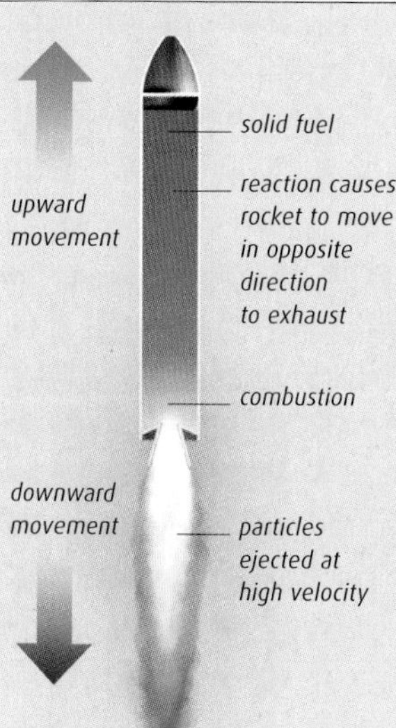

Rockets rely on the principle of conservation of momentum – the immutable law that says that unless an external force is applied, the momentum of a system (its mass multiplied by its velocity, or speed in a particular direction) must stay the same. So when the exploding fuel inside a typical rocket forces gas out of the exhaust at high speed, the rocket itself must move in the opposite direction in order for the overall momentum to stay the same. Because the rocket has far more mass than the exhaust gas, it accelerates far more slowly than the exhaust. Once it has left the Earth's atmosphere, the rocket can work equally well, if not better, in a vacuum.

New World pioneer

Although Konstantin Tsiolkovskii is rightly acknowledged as the founder of modern rocketry, many of his ideas were unknown to contemporaries such as Robert Goddard, the American physics lecturer who, in 1926, heralded a revolution with his launch of the first liquid-fuelled rocket.

Born in Worcester, Massachusetts, in 1882, Goddard was fascinated by physics from an early age, though his interest in spaceflight was not ignited until 1898, when he read H.G. Wells's *The War of the Worlds*. Lagging behind his schoolmates due to constant illness, Goddard nevertheless pursued his studies, which took him as far as a research fellowship at Princeton before a near-fatal bout of tuberculosis in 1913 forced a return to Worcester. Once recovered, he took a teaching post at nearby Clark University, where he had studied for his doctorate and where he would remain for the next 20 years.

Goddard realized the potential efficiency of liquid fuels in 1909 (independently of Tsiolkovskii), but his convalescence at Worcester must have inspired him to act, since it was shortly after this, in 1914, that he began to register patents for rocket designs, including a multi-stage vehicle and a liquid-fuelled rocket. Unlike Tsiolkovskii's theoretical mix of liquid hydrogen and liquid oxygen, Goddard's combination of gasoline for fuel and liquid nitrous oxide for oxidant (see panel, opposite) was practical, using the technology of the day. Once back at work, he began developing working engines, initially funding his own experiments, but later with backing from the Smithsonian Institute and, once the United States had become involved in the First World War, the Army. Experiments using small charges of solid fuel led him to discover the optimum design of rocket nozzle – a shape first proposed by Swedish engineer Gustav de Laval in 1890 for use in steam engines. Although Goddard demonstrated an early form of bazooka, the war ended before he could put many of his theories into practice.

In 1919, Goddard summarized his work so far in his book *A Method of Reaching Extreme Altitudes*. For many outside Russia, this was the first serious

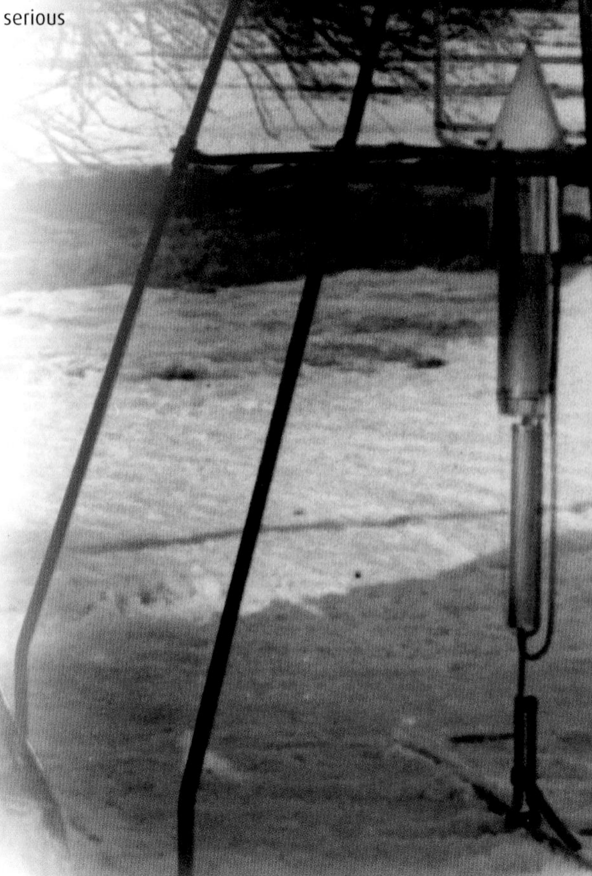

ROBERT AND NELL
Goddard proudly displays the first liquid-fuelled rocket on its ladder-like support structure. The rocket is actually "back-to-front" compared with later designs – the fuel and oxidant tanks sit at the bottom, capped by a protective cone and linked by parallel pipes to a combustion chamber at the top.

BIGGER AND BETTER
Goddard's rockets developed rapidly, from early tabletop models (above), through Nell, to larger, advanced designs, such as his P-series of the early 1940s (right). By this time, components were arranged in a more familiar order, with fuel tanks, topped with an aerodynamic nosecone, sitting on top of the combustion chamber.

"... the dream of yesterday is **the hope of today** and the reality of tomorrow."

Robert Goddard, 1904

WORK AT ROSWELL
By 1940, Goddard was working on more advanced rocket designs at his Roswell laboratories. This design was one of the first to use turbopumps to force fuel into the combustion chamber at a high rate.

SPACEFLIGHT ADVOCATE
The most famous portrait of Goddard dates from 1924 and shows him at the blackboard at Clark University, discussing the possible use of rockets to reach the Moon. Unfortunately, Goddard's promotion of such ideas led to merciless teasing by the press.

proposal for space travel they had encountered, and Goddard had to endure a great deal of scorn, often from journalists who delighted in attacking his ignorance of basic physics when, in fact, they were revealing their own. One *New York Times* article of January 1920 was particularly withering, though the paper eventually saw fit to issue a retraction in 1969, on the day after the Apollo 11 Moon landing.

Nell takes flight

Despite all this, Goddard persevered, and on 16 March 1926, he saw his liquid-fuelled rocket, nicknamed Nell, take flight for the first time. The flight lasted only two and a half seconds and reached a height of 13m (41ft), but the principle had been proved.

By the late 1920s, the regular launches from Worcester were attracting a lot of attention. Goddard wanted more privacy, and through his friendship with aviator Charles Lindbergh (the first man to fly across the Atlantic), he attracted funding from financier Daniel Guggenheim that allowed him to relocate to Roswell, New Mexico. Here, he continued to improve his rockets until his death in 1945, and also worked on experimental aircraft for the US Navy. His attempts to recapture Army interest were met with indifference, though his work found a more appreciative audience in Europe – Germany even attempted to plant spies among his researchers.

TECHNOLOGY

LIQUID-FUELLED FLIGHT

Liquid fuels are far more efficient than the black powder used in rockets before Goddard's time, but they have a number of inherent risks, since the chemicals they use are sometimes highly unstable and difficult to manufacture or store. While the fuels will frequently react with oxygen in the air, in order to be self-contained a rocket must carry a chemical oxidant onboard. As shown here, fuel and oxidant are carried in separate tanks, and travel through a network of pipes to reach the combustion chamber, where they may either react spontaneously, or require a spark in order to explode. Although modern solid-fuel rockets are far more efficient than their black-powder ancestors, liquid-fuel designs retain one key advantage – the rate of burn can be throttled up and down, and even stopped and restarted later.

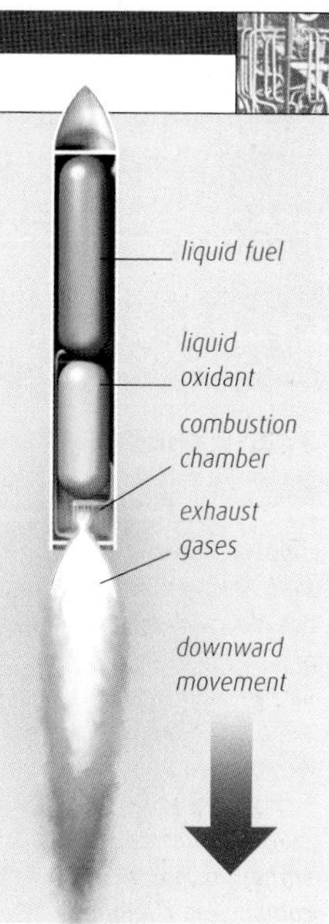

The dream takes shape

While much of the groundwork for modern rocketry was laid in Russia and America, it was in Germany that the idea of space travel really took hold. In the early 20th century, visionaries planted the seeds that would ultimately give rise to the rocketry programmes of the Second World War.

LIFE ON THE MOVE
Born in the Austro-Hungarian city of Hermannstadt (now Sibiu, Romania) in 1894, Oberth moved to Munich for his medical training, and remained in Germany after the old empire was partitioned in 1918. He spent brief periods of his later life in Austria, Italy, and the United States, but eventually retired to Germany in 1962. He died in 1989, at the age of 95.

Germany's equivalent of Goddard and Tsiolkovskii was without doubt Hermann Oberth. While Goddard had been inspired by Wells, Oberth's interest in space travel was fired by reading Verne's *From the Earth to the Moon* at the age of 11, and within a few years he was making his own model rockets. Though he initially trained to be a doctor at his father's behest, harsh experience as a medic in the trenches of the First World War prompted him to follow his interest in physics. However, he failed to gain a doctorate from the University of Heidelberg, largely because his dissertation explored physiological and medical aspects of rocket travel in which his physicist supervisor had no grounding. Rather than revise and resubmit, Oberth instead financed its publication as a book, *The Rocket into Interplanetary Space* (1923). Just as had happened with Goddard in America, his public promotion of such apparently outlandish ideas soon turned Oberth into something of a celebrity. However, while Goddard was openly mocked for his comparatively modest suggestions, Oberth's far more visionary proposals were welcomed with open arms, partly thanks to bestselling popularizations of the work by authors such as Max Valier and Willy Ley. Even though Oberth had certainly developed most of his ideas, such as liquid-fuelled rockets and multiple stages, independently of either Goddard or Tsiolkovskii, the fact that he had written to the isolated American in 1920 to request a copy of his early papers was enough to arouse some suspicion and envy across the Atlantic – for the rest of his life, Goddard referred to Oberth as "that German".

Throughout the 1920s, Oberth's fame and popularity increased, and in 1929 he published a revised and expanded version of his work, *Ways to Spaceflight*, that attracted even more attention.

OBERTH'S WORK
Hermann Oberth's book The Rocket into Interplanetary Space *(above) inspired engineers such as the members of the VfR. In this photograph (right), Oberth, in the dark overalls, is standing by the rocket intended for launch at the premiere of* Frau im Mond.

ROCKETS AND CINEMA
Frau im Mond *was only a moderate success at the cinema – largely because it was silent at a time when "talkies" were becoming increasingly popular. Ironically, this silent movie can claim credit for inventing the launch countdown, added to the script by Fritz Lang in order to increase tension.*

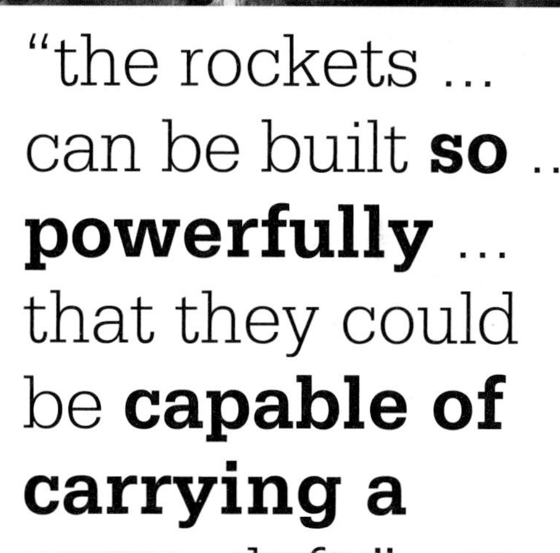

IMAGINARY JOURNEY
Although it offered the first realistic depiction of spaceflight, Lang's film was not the first to touch on the idea of a trip to the Moon – that honour goes to French pioneer Georges Méliès's rather more light-hearted Le Voyage dans la Lune *of 1902 (right).*

FRITZ LANG'S VISION
Lang's lunar voyage is a tale of jealousy and mistrust among the crew of an expedition visiting the Moon in search of gold discovered by the astronomer Professor Manfeldt.

> "the rockets ... can be built **so** ... **powerfully** ... that they could be **capable of carrying a man** aloft."
>
> **Hermann Oberth, 1923**

Rockets, society, and rocket societies

Why was battle-scarred Germany so open to the opportunities of the rocket when the victorious and more prosperous United States was not? In reality it was just another aspect of the scientific and cultural blossoming that briefly occurred under the Weimar Republic in the 1920s, freed from the repressive conservatism of the old kaiser. The German expressionist films of the time were another, and, not surprisingly, the two eventually came together.

In 1929 the film director Fritz Lang recruited Oberth and Willy Ley to act as consultants for his ambitious new project *Frau im Mond* (*The Woman in the Moon*), which was to be the first serious film about space travel. The film borrowed heavily from Oberth's ideas and popularized an image of the rocket that has persisted to this day. It also proved that while Oberth was a fine theorist, he was no engineer: persuaded by Lang that a rocket launch would be the perfect publicity stunt to open the film, Oberth and his colleagues laboured for months to build Germany's first liquid-fuelled rocket, but they met with little success, and an explosion during testing cost Oberth the sight in one eye.

Nevertheless, later in 1929 Oberth did manage to test-fire a liquid-fuelled rocket engine called *Kegeldüse* in the laboratory. By this time, he had begun to collaborate with a young and enthusiastic engineer called Wernher von Braun. The two had met through the VfR space and rocketry society (see p.22). Now back on familiar territory as a theoretician and the elder statesman for a new generation of rocketeers, Oberth was finally able to see some of his dreams become reality, as the VfR made a series of successful rocket launches. Throughout the later development of the German missile programme, and the US space programme that followed it, he would remain a peripheral, though influential, figure.

HISTORY FOCUS

ROCKET CARS

Another manifestation of the German obsession with rockets were the rocket-propelled vehicles of the late 1920s. Developed by manufacturing magnate Fritz von Opel in collaboration with powder-rocket maker Friedrich Wilhelm Sander and Austrian space enthusiast and author Max Valier, the Opel-RAK series of cars, aircraft, and even railway carriages began with the RAK 1 (left) driven by Kurt Volkhart to a top speed of 75kph (47mph) on 15 March 1928. The much-improved RAK 2, powered by 24 separate rockets, reached 230kph (143mph) just two months later. Although they were mainly intended as publicity stunts, Valier in particular seems to have been keen to develop the idea further. Tragically, he was killed by shrapnel when the liquid-fuelled engine he planned to fit on the RAK 7 car exploded in his laboratory.

Rocket societies

The 1920s and 1930s saw the formation of a number of rocket societies – clubs where like-minded physicists and engineers collaborated to develop new and more powerful types of rocket.

Most of the world's early rocket societies started out as groups of keen amateur enthusiasts, such as the American and British Interplanetary Societies (established in 1930 and 1933), and the astronautical section of the French Astronomical Society (established in 1927, and for which the term astronautics was first coined). But two in particular caught the attention of their respective governments.

The VfR

Germany's *Verein für Raumschiffahrt* or VfR (Society for Space Travel) was founded in 1927 at Breslaw (now Wroclaw in Poland) by Johannes Winkler, an engineer at aircraft manufacturer Junkers. The authors Max Valier and Willy Ley were early members, and numbers soon swelled to 500 people, including influential figures such as Hermann Oberth, Eugen Sänger, Arthur Rudolph, and a young student called Wernher von Braun.

By February 1931, Winkler was able to launch Europe's first liquid-fuelled rocket, the HW-1, from Dressau. He used a powerful combination of liquid methane and liquid oxygen in his rocket, which was able to reach altitudes of 500m (1,600ft). Over the following months, VfR members conducted a series of increasingly ambitious launches from their rocket airfield near Berlin, using a design conceived by

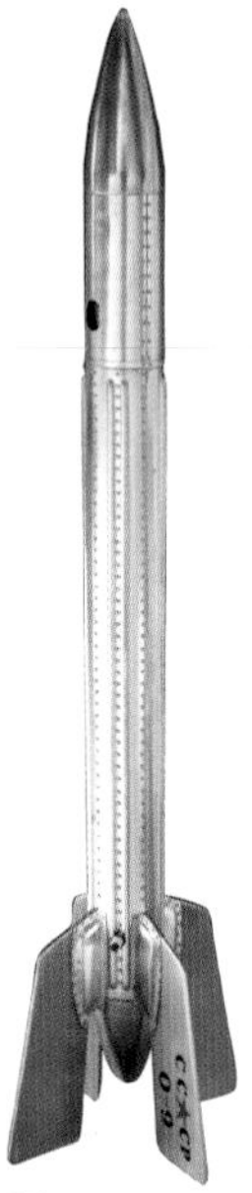

GIRD 09
Yefremov's early rocket design used liquid oxygen to burn a petroleum gel fuel. This hybrid design achieved far better performance than the GIRD-X. The first launch reached 400m (1,300ft), and later ones reached 1,500m (5,000ft).

YOUNG ENTHUSIAST
The young Wernher von Braun (right) carries an HW-series rocket at the VfR's Raketenflugplatz *(rocket airfield) outside Berlin.*

LAUNCHING GIRD 09
Nikolai Yefremov contends with a dangerous leak of liquid oxygen during one of GIRD's several attempts to launch its hybrid-fuel rocket in August 1933.

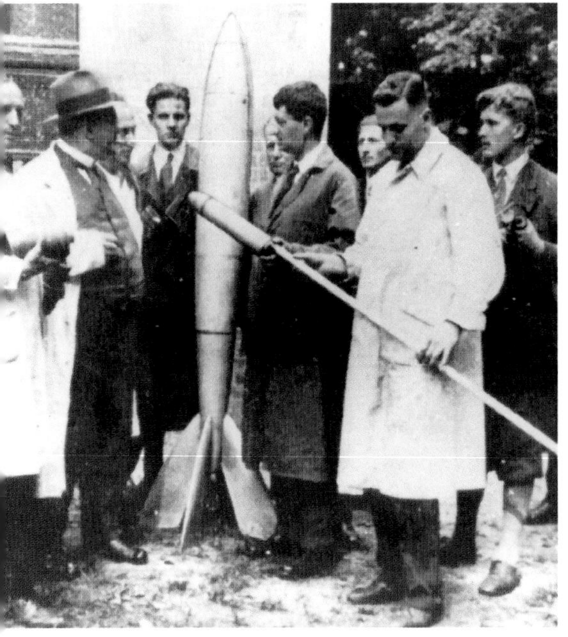

GERMAN ROCKETEERS
VfR members including Rudolf Nebel (far left), Hermann Oberth (centre), Klaus Riedel (right of centre in light coat), and Wernher von Braun (far right) are seen here with Oberth's Frau im Mond *rocket. Riedel is holding a Mirak rocket.*

Rudolf Nebel and built by Klaus Riedel. These Mirak rockets were soon able to reach altitudes of more than 1km (⅔ mile).

In 1932, the VfR invited Captain Walter Dornberger of the German Army to view a demonstration. The test launch ended in failure, but Dornberger was sufficiently impressed to offer the group funding – if they would keep their work secret and concentrate on military applications. The Army was particularly interested in rocket weapons because their development was one of the few areas not strictly regulated under the Treaty of Versailles. The VfR eventually turned down Dornberger's offer, but fierce arguments about whether or not to accept it nearly tore them apart. Within a year, the Nazi Party had seized power, and one of their early measures was to outlaw civilian rocket experiments. Von Braun and several other VfR members were soon lured to work for the Army under Dornberger, while many of the others retreated from practical research back into the realms of theory.

GIRD

In the Soviet Union, the VfR's equivalent was the Group for the Study of Reactive Motion (GIRD), founded from the merger of two earlier rocket clubs in 1931. It had many local branches, but the most important were those in Moscow and Leningrad (MosGIRD and LenGIRD respectively). MosGIRD was largely instigated by Friedrich Tsander, an enthusiastic advocate of spaceflight. Many of its members were to play influential roles in the Soviet space programme, most importantly Sergei Korolev and Mikhail Tikhonravov. LenGIRD, meanwhile, included Valentin Glushko among its members.

By August 1933, MosGIRD had launched the GIRD 09, based on a hybrid semi-solid rocket engine (which combined fuel and an oxidizer to produce combustion gases and thrust) designed by Tikhonravov and Nikolai Yefremov. In November of that year, the Soviet Union's first true liquid-fuelled rocket, Tsander's GIRD-X, took flight to a height of 80m (250ft), powered by alcohol and liquid oxygen. Tsander, however, did not live to see it – he had died that March from typhus, with Korolev taking his place as GIRD's nominal leader.

While the VfR was an independent civilian group, GIRD never had the same degree of autonomy because most of its members were already working for the Soviet state on rocket-related research projects. As GIRD's activities drew increasing attention from the state, the organization was absorbed into the Red Army under Field Marshal Tukachevsky. Here it was merged with the Gas Dynamics Laboratory (GDL) of Leningrad, creating the Jet Propulsion Research Institute (RNII). The RNII was an organization riven with factional infighting: former GIRD members were frequently in conflict with each other and ex-GDL staff. The RNII's director was more concerned with jet propulsion than with rocketry, and Glushko had work on liquid-oxygen rockets cancelled in favour of his own nitric acid systems. When the paranoid Stalin turned on Tukachevsky at the beginning of the Great Purges of 1937–38, the consequences for the former GIRD members would prove terrible.

GIRD MEMBERS
Proud Soviet space enthusiasts surround their GIRD-X liquid-fuelled rocket prior to its launch in November 1933. A young Sergei Korolev is standing to the right of the rocket.

TECHNOLOGY

PRINCIPLES OF ROCKET STAGING

All the early rocket pioneers soon hit upon the idea of multi-stage rockets (which Tsiolkovskii called rocket trains). The major problem faced by any rocket is the sheer weight of fuel that it must carry if it is to generate sufficient thrust at launch to get moving against Earth's gravity. As the rocket picks up speed and starts to burn down the fuel supply, it makes no sense to carry the huge volume and weight of a mostly empty fuel tank along for the ride. Instead, it's simpler to split the rocket into separate elements, each with its own fuel tanks and engines. The initial stage is by far the largest, and may be supplemented by booster rockets to produce even more thrust during launch, but once these have burned out, they can be jettisoned and a smaller upper stage can take over its job, accelerating the suddenly lighter rocket at a much faster rate.

payload
4 Once in orbit, the 2nd stage falls away
3 The 2nd stage takes the payload into orbit
payload
2 While the 1st stage cuts out, detaches, and drops away, the 2nd-stage engine fires
2nd stage
1 A multi-stage rocket launches with a burn from its 1st-stage engines
1st stage

The birth of the missile

At first, Walter Dornberger struggled to convince the German Army that rockets could be a practical weapon. But with a rocket team assembled from the cream of the VfR, he would ultimately build the first missiles.

TESTING THE A4
An A4 rocket is launched in 1943. The first two A4 launches, in June and August 1942, crashed due to guidance failures, but in October that year a third test rocket reached a height of 85km (53 miles).

The VfR chose the right man when they approached Walter Dornberger about potential funding in 1932. A veteran of the First World War who had studied physics for several years, he was in charge of a small weapons-testing facility at West Kummersdorf. Dornberger was a strong advocate of the idea that rockets could be used as ballistic missiles - burning their engines to the peak of their flight path, and then descending on a trajectory similar to that of any other projectile. The first person to suggest such an application had been French rocket enthusiast Robert Esnault-Pelterie in the late 1920s, but he had been unable to interest the French military in the proposal.

Wernher von Braun was among those who argued that the VfR should accept funding to work on military applications. The society's members were as a whole far more interested in spaceflight than in missiles, but there were sharp divisions between those who wanted nothing to do with warfare and those who saw military funding as a means to finally get some serious support for their work. While the VfR ultimately rejected the Army offer, it was inevitable that some of its leading lights would follow the money.

The fruits of their effort were a number of increasingly ambitious rocket designs, the A-series. While the A1 never got off the drawing board let alone the launch pad, von Braun himself was able to launch a pair of A2 rockets, christened Max and Moritz, from the island of Borkum in December 1934. These rockets burned an ethyl alcohol/liquid oxygen propellant mix, and incorporated an important new feature - a spinning gyroscope in their mid-sections. The weight of this spinning mass helped to stabilize the entire rocket, ensuring it maintained a steady flightpath that reached a peak at around 2,000m (6,600ft).

"This ... is **the first day of a new era** ... that of space travel."

Walter Dornberger, on the first successful A4 launch, 3 October 1942

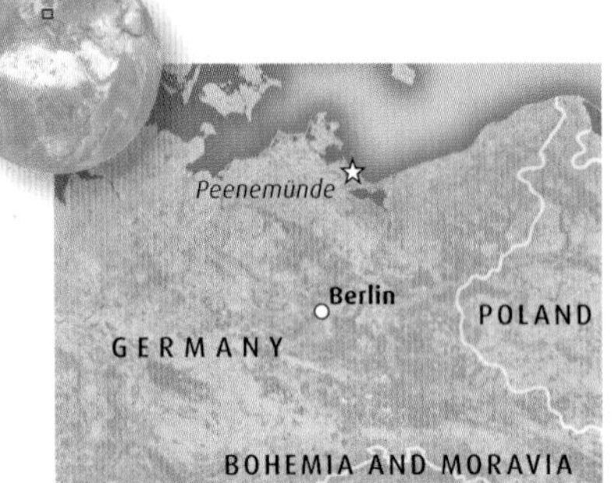

BALTIC HIDEAWAY
The peninsula of Peenemünde, on Germany's northeastern coast, was suggested by Wernher von Braun's mother when he mentioned the need for an isolated site.

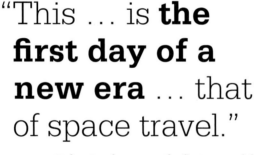

TEST STAND
Preparations for a test at Peenemünde reveal the true scale of history's first large rocket.

HIGH-LEVEL VISIT
The rocket team frequently welcomed high-ranking visitors to Peenemünde (though Hitler visited them only once at Kummersdorf, and seemed unimpressed). Here, Admiral Dönitz and his entourage are seen during an inspection in May 1943. Wernher von Braun is on the right, wearing a dark civilian suit.

This success, and the development of static test engines with far greater power than the A2's motors, convinced the Army to put more money into rockets. As von Braun's team outgrew Kummersdorf, a new base was established at Peenemünde on the Baltic coast. Over the next few years, it would grow to incorporate test ranges, test stands for measuring engine performance, and factories for rocket assembly.

LENGTH	14m (46ft)
MAXIMUM DIAMETER	1.7m (66in)
TOTAL MASS	12,870kgf (28,373lbf)
UNFUELLED MASS	4,008kg (8,836lb)
ENGINES	1 x A4
LIFT-OFF THRUST	25,000kg (55,125lb)
MANUFACTURER	Mittelwerk

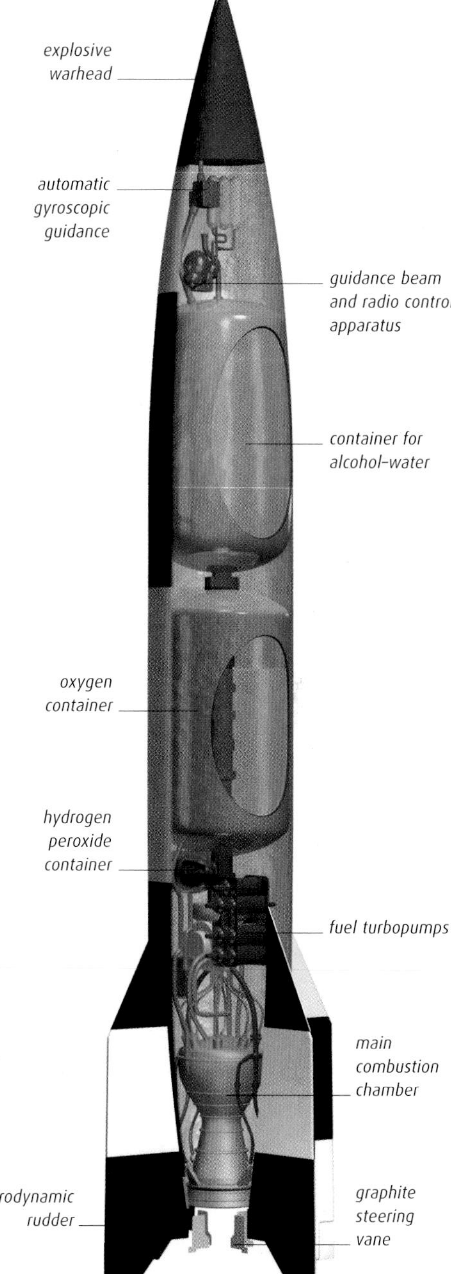

THE FIRST MISSILE
The A4 is the ancestor of all other liquid-fuelled rockets, including most modern ballistic missiles and space launch vehicles. To this day, rocket technology still follows the same principles set out by the A4 team.

Building the A4

By the summer of 1936, Nazi Germany was preparing for war. A new test rocket, the A3, was in development, but the pressure to produce a practical missile for warfare was growing. The result was an outline for the A4, a massive rocket that would scale up all their existing systems. Dornberger suggested that it should be able to carry a 100-kg (220-lb) explosive warhead over more than 260km (160 miles).

Tests of the A3 began in late 1937. This was a much more powerful rocket than its predecessors, developing 1,500kg (3,300lb) of thrust at launch and capable of burning for up to 45 seconds. A small tank of liquid nitrogen, heated so it evaporated into high-pressure gas, was used to force propellant into the engine at high speed. The A3 also incorporated a guidance system of gyroscopes and accelerometers. Although this had practical problems, the theory was sound – the first use of a technique still in use today.

The A4's gestation was a long one – and fortunately for the Allies, its debut was partly delayed by Germany's own military hierarchy. The success of the Blitzkrieg attacks of 1939 and 1940 made the Army so convinced of its superiority that it decided to cut back on funding to Dornberger's project. But building an engine with a thrust of 25,000kg (55,125lb) was also a massive technical challenge. The main problem for the scientists was how to force fuel into the combustion chamber at a fast enough rate to produce the required thrust. The solution was a high-speed turbopump, powered by the fierce chemical reaction of hydrogen peroxide and potassium permanganate. In the meantime, other systems were tested on a new research rocket, called the A5. By the time the first A4 was successfully launched, on 3 October 1942, Germany's hopes for a swift end to the war had been dashed and much was expected of the new weapon, which was soon to also bear a new name.

TESTING KIT
This box of electronics was used for pre-launch tests on A4 rockets. Generally the launch sequence was triggered from inside an armoured vehicle or blockhouse close to the launch site, with all other personnel kept back at a safe distance.

PEENEMUNDE RAID
Their suspicions aroused by aerial photographs and Polish resistance reports, the Allies launched air raids against Peenemünde in August 1943. The most devastating, Operation Hydra, involved 596 RAF bombers.

GUIDANCE GYROSCOPES
The later A-series rockets used several gyroscopes and accelerometers to calculate the rocket's precise trajectory, speed, and distance travelled, so that it could be guided using steering vanes that deflected the exhaust.

The missile goes to war

Although the first successful A4 launches occurred in late 1942, it took almost two more years before the missile was ready to enter service. By that time, it had acquired a more infamous pseudonym – the V-2.

MOBILE LAUNCHER
One reason for delays to the V-2 entering service was the need to develop a transport system for it. The rockets had to be fuelled and transported from Nordhausen to forward units via the rail network before they came within range of the enemy cities.

HIT AND RUN
A standard V-2 battery would consist of about 30 vehicles – Meillerwagens, mobile generators, fire trucks, tow trucks, and troop transporters. They became proficient at moving into forest clearings or tree-lined lanes, setting up, and launching within just a few hours.

As the A4 edged closer to mass production, it was increasingly clear that Peenemünde was not well-suited to large-scale manufacturing or perhaps even continued testing. The devastating Allied raids of August 1943 proved the final straw – among the casualties was Walter Thiel, the ingenious designer of the A4's high-performance engine. The decision was made to move much of the flight testing to Blizna in southern Poland, while large-scale manufacture of the missiles would begin in the huge Mittelwerk tunnel complex under the Kohnstein mountain near the town of Nordhausen in central Germany. The main focus of those remaining at Peenemünde would be research and development.

At the insistence of Joseph Goebbels's propaganda ministry, the A4 was henceforth to be called the Vergeltungswaffe-2 (Reprisal Weapon 2), or V-2 for short. The name V-1 had been given to the Luftwaffe's jet-propelled "flying bomb", tested alongside the V-2 at Peenemünde and also manufactured at the Mittelwerk. To provide a slave workforce for the huge underground factories, concentration camps at nearby Mittelbau-Dora were expanded. By the end of 1943, thousands of prisoners were being worked to death in appalling conditions beneath the mountain.

Even with these massive resources, production did not go smoothly – von Braun's team at Peenemünde was still refining the rocket, and every modification resulted in lost time and lost lives among the labourers. Meanwhile, construction of infrastructure for the new weapons was under way. While the V-1 flying bombs required long concrete "ski ramps" to assist their takeoff, the V-2 was designed for rapid deployment on mobile launchers. Nevertheless, the German military began planning a number of bunkers along the French coast that it intended to use for V-2 launches, but the Allied air forces subjected these to such heavy bombing that they were eventually abandoned.

By mid-1944, shortly after the D-day landings that started the liberation of mainland Europe, the Allies had a good idea of what was coming – parts of crashed V-2s had been smuggled to London from both Poland and Sweden, and aerial reconnaissance missions had given intelligence analysts an insight of the rocket's dimensions and capabilities.

SCENE OF DEVASTATION
When a V-2 struck the corner of Smithfield Market in central London, early on the morning of 8 March 1945, it killed 110 people. However, this was one of the last missiles to reach Britain – within weeks the V-2 launchers had been driven back out of range.

"... they travelled **faster than the speed of sound** ... The first you knew was the explosion."

Eyewitness to a V-2 attack, London, 1944

Terror from the skies

The first V-1s began to fall on London in June 1944, but at least the distinctive engine sound from the flying bombs offered the civilians below some advance warning. When the V-2 was finally deployed in early September, its silent approach had a far greater psychological effect. As a ballistic missile, the V-2's engines only fired until it reached the peak of its trajectory. It then fell almost noiselessly towards its target. The guidance systems were too primitive to make the V-2 anything other than a blunt instrument, but when it did strike a populated area, the effect could be devastating: 567 people were killed when one struck a cinema in Antwerp, and 160 died in a strike on southeast London. The difficulty in targeting meant that most of the missiles were aimed at major cities rather than smaller towns – between them, London and Antwerp endured some 90 per cent of more than 3,000 attacks by the rockets. Each V-2 carried a warhead with more than a tonne of explosives, and its final approach, dropping from an altitude of around 100km (60 miles) at up to four times the speed of sound, left no possibility for interception or defence.

By the time the V-weapons were deployed, though, the German Army was in retreat, and firing them from their *Meillerwagen* launch platforms became a hit-and-run operation. A practised V-2 crew could set up, launch a missile, and depart within two hours – usually moving too quickly for the launcher to be spotted and destroyed on the ground. Ultimately the V-2 attacks were only brought to an end as the German Army was driven into retreat and the target cities fell out of range. The bombardment finally ceased in March 1945.

Final attack

More than 6,000 V-2s were produced in total, but they were far from reliable weapons, and early models had a tendency to disintegrate in mid-flight that was not corrected until late 1944. Despite inflicting heavy casualties, the missiles never came close to turning the tide of the war – by the time the V-2 entered service it had truly become the Nazi regime's Reprisal Weapon, a final psychological attack on the Allies at home, even as their armies were sealing Germany's own fate on mainland Europe.

UNDER THE MOUNTAIN
Colour photographs from the Nazi era grimly capture the reality of life and work beneath the mountain at Nordhausen. Up to 10,000 forced labourers worked underground in the factory at its peak, helping to produce not only the V-2, but also the V-1. Many of the workers died of pneumonia in the cold, damp conditions.

The winners take all

The V-2 attacks revealed Germany's massive lead in rocket development, and as wartime alliance turned to Cold-War rivalry, the United States and the Soviet Union were racing to plunder the German rocket programme.

ALLEN DULLES
The US end of Operation Overcast/Paperclip was driven by Allen W. Dulles of US Army Intelligence, later director of the CIA under President Eisenhower.

Although their intervention had come too late to affect the outcome of the war, von Braun's missiles had proved one thing beyond doubt – the rocket was once again a formidable weapon: fast, almost silent on its final approach, and difficult to intercept. Clearly, ballistic missiles would have a major role to play in future warfare, and as the leaders of the democratic West and communist East contemplated a situation in which their rival spheres of influence would clash around the world, getting hold of superior German technology became a high priority.

In February 1945, Captain Robert Staver arrived in Europe under instructions to track down the V-2 masterminds and bring them into US custody. A month later, Colonel Holger Toftoy, head of Army Ordnance Intelligence in Paris, was ordered to track down as many intact V-2s as possible, for later use in a testing programme. For both men, it was a race against time – many of the Germany's key missile sites lay directly in the path of the advancing Red Army. Fortunately for the Americans, though, von Braun and his team had plans of their own.

DORA ARTEFACTS
The Nazis kept paperwork on all aspects of the rocket programme, ranging from construction contracts to special camp banknotes.

Willing captives

When the Soviet armies swept into Peenemünde on 5 May 1945, they found that the birds had flown – the missile range and factories had been abandoned in mid-February, and explosives deliberately set to destroy as much as possible. Von Braun and his team had moved first to Nordhausen, close to the Mittelwerk factory, and closer to the advancing US troops. On 19 March, orders came from Berlin that all records of German experimental programmes were to be destroyed – instead, von Braun had 14 tonnes of material spirited away in the night and hidden in a cave for later recovery.

Since August 1944, the V-2 team had been under the command of SS General Hans Kammler, a former concentration camp commandant who apparently planned to use them to barter for his life. In late March he had the entire team shipped south to Bavaria. During April, Kammler disappeared – perhaps assassinated, perhaps fleeing for his life. When the 44th Infantry Division finally reached the village on 2 May, von Braun and his team were able to surrender to their captors of choice.

The soldiers may not have immediately realized the value of their prize, but Robert Staver certainly did. Nordhausen had been captured on 11 April, with huge stores of missile components, but no sign of the scientists or their paperwork. Another problem was that both Nordhausen and Garmisch-Partenkirchen, where the scientists were being questioned, were due to be handed over to Soviet administration in June. In Paris, Toftoy saw the need to move fast. He despatched a team to Nordhausen to collect the parts for 100 V-2s, and

"TAKE US TO IKE!"
Leaders of the V-2 team after capture in Bavaria. Von Braun had broken his arm in a car crash a few weeks earlier. Dornberger is on the left in the hat.

ENTRANCE TO UNDERGROUND FACTORY
At Kohnstein the US troops discovered a huge complex carved out of the soft gypsum rock. In the early years of the Second World War the complex was used as a store for fuel and poison gas. Most prisoners who died at Mittelwerk had perished in the first few months of heavy labour.

THE PRODUCTION LINE
The assembly lines ran through two parallel tunnels, A and B, each 1.9km (1¼ miles) long. Missiles were placed on carriages and rolled along the railway tracks that connected the many halls in Tunnel B. The final hall, over 15m (50ft) high, allowed missiles to stand vertically for testing.

STORES OF FINISHED PARTS
Tunnel A was used for transporting parts and equipment into and around the factory, while the numerous shorter cross-tunnels were used for storage. The production-line manufacture of the V-2s meant that large stores of completed parts were kept on site.

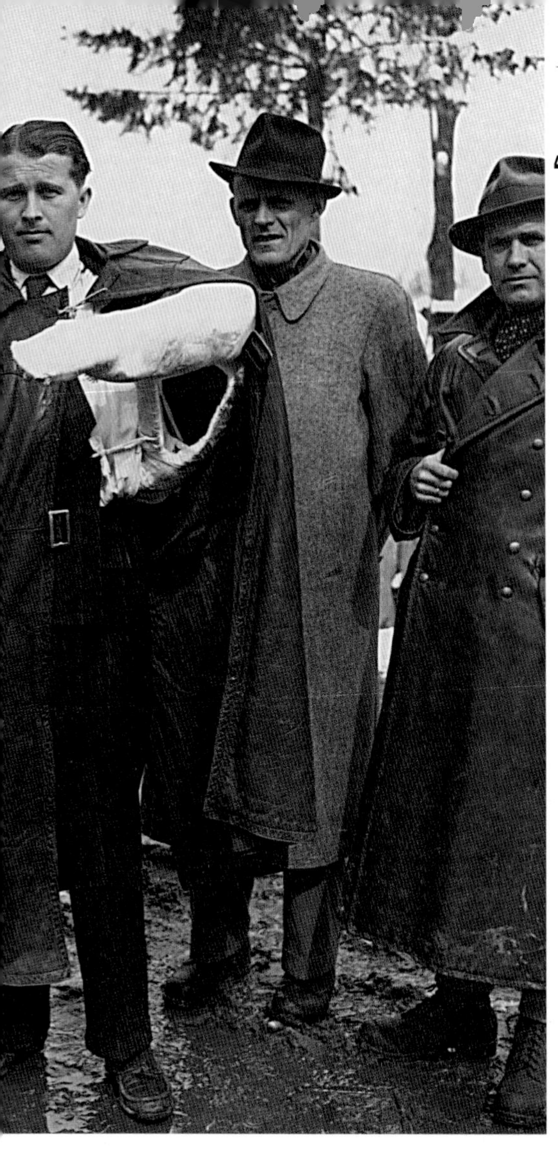

"Germany has lost the war, but let us not forget that it was our team that first succeeded in **reaching outer space** ..."

Wernher von Braun, referring to the V-2 reaching an altitude of 100km (60 miles), 1945

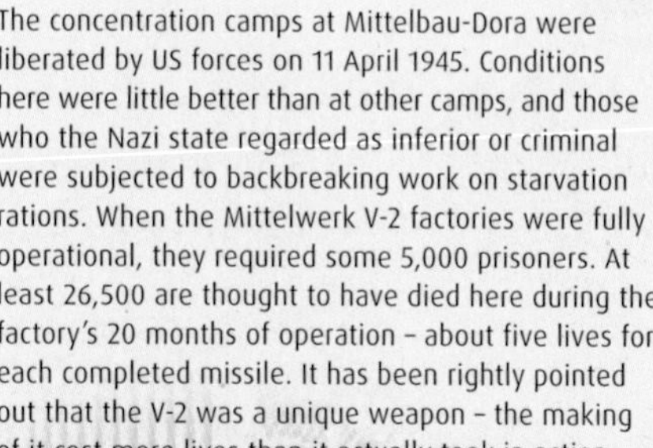

HISTORY FOCUS

MITTELBAU-DORA

The concentration camps at Mittelbau-Dora were liberated by US forces on 11 April 1945. Conditions here were little better than at other camps, and those who the Nazi state regarded as inferior or criminal were subjected to backbreaking work on starvation rations. When the Mittelwerk V-2 factories were fully operational, they required some 5,000 prisoners. At least 26,500 are thought to have died here during the factory's 20 months of operation – about five lives for each completed missile. It has been rightly pointed out that the V-2 was a unique weapon – the making of it cost more lives than it actually took in action.

ship them back to the US zone. Staver also based himself at Nordhausen, where he tracked down a number of senior members of the rocket team that had not gone south and located the hidden cache of documents. As the handover deadline neared, Staver and Toftoy arranged a mass evacuation of Peenemünde staff and their families to the US zone.

On 19 July, Operation Overcast (later Paperclip) was given the go-ahead by Washington. Staver and Toftoy were authorized to offer the Germans six-month contracts for work in the US. Most took some persuading – their families would have to remain behind in Germany – and the future beyond those six months was uncertain. The first contracts were finally signed on 12 September, and within days, the initial wave of scientists was flying out of Europe.

Although most of the rocket team had eluded them, the Soviets still had a large share of the plunder. Peenemünde and Nordhausen were now theirs, and a few key scientists, notably Helmut Gröttrup of the guidance and control team, also cast their lot to the east. On balance, then, the superpowers were fairly evenly matched as they began their post-war missile race.

OPERATION BACKFIRE

In the summer and autumn of 1945, the British Army conducted a number of test firings of V-2s (left) captured during their advance through the low countries. Launches took place at Cuxhaven in northern Germany, and rocket team scientists (above) were frequently flown in from their camp in the south to assist.

Von Braun: model American

The most important figure on the American side of the Space Race was to be a naturalized German whose intimate involvement in the V-2 programme would continue to raise awkward questions even as he became a renowned national figure.

THE V-2 ARRIVES
Captured German rockets were transported to the United States aboard aircraft carriers such as USS Midway, *before being transferred to Fort Bliss or direct to White Sands.*

THE ROCKET TEAM
This group photo shows 105 of the 116 German rocket experts at White Sands Proving Ground in 1946. Von Braun is in the front row, right of centre, with one hand in his pocket.

Wernher von Braun arrived at Fort Bliss, Texas, in October 1945. He was one of an initial group of seven German scientists brought into the United States as part of Operation Overcast. During nearly five months of evaluation in Germany, he had been interviewed by American rocket experts such as Tsien Hsue-Shen, prepared reports for the US Army on the status of the German rocket programme, and helped the Americans round up the rest of his team. If there had ever been any doubt among the US military that von Braun was the man they needed in their new missile programme, his willing cooperation dispelled it.

Von Braun was born in 1912 at Wirsitz in the German province of Posen (now part of Poland), the second son of a noble family. Fascinated by astronomy from an early age, his interest in rockets was fired by reading Hermann Oberth's influential work. At the age of 12, he attempted to imitate the exploits of Wan Hu (see p.12) by strapping rockets to the back of a trolley and piloting it along the Berlin streets. Keen to master Oberth's theories, the young von Braun applied himself to the study of physics and maths, eventually enrolling at the Technical High School of Charlottenburg aged 18. It was here that he joined the VfR and demonstrated his talent for practical rocket engineering.

Von Braun's Nazi past was a constant source of controversy throughout his American career – even today it continues to cast a shadow over the early days of the US space programme. It is all but impossible to judge in hindsight, but his primary motivation seems always to have been an obsession with the rocket's use in space travel. He joined the Nazi party in 1937 and was made a lieutenant of the infamous SS in 1940, but he always argued that these were politically necessary manoeuvres if he was to keep working on his rockets; that he never paid more than lip service to Nazi ideals; and that his SS position was purely for show – he claimed only to have worn the uniform once. However, von Braun was certainly well aware of the concentration camps around Mittelwerk and of the foreign forced labour used at Peenemünde. Whether his involvement in a missile programme ultimately launched against civilian targets was a war crime in itself is still a subject for debate, particularly as Dornberger was prosecuted after the war for his role in the project. However, Dornberger's support for the Nazi Party was always far more explicit, and even he was imprisoned for only two years before also travelling to the United States.

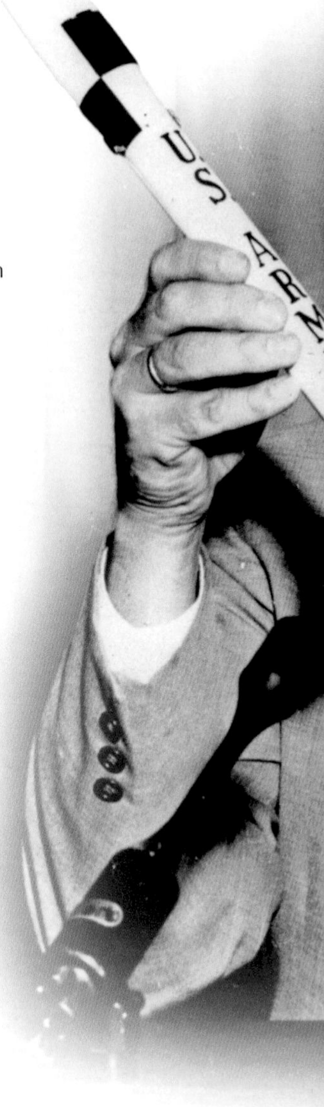

The life of Fort Bliss

Throughout 1946, the Americans reassembled the nucleus of von Braun's old team at Fort Bliss. The documentation spirited away from Nordhausen arrived, and dozens of captured missiles and tonnes of parts from the Mittelwerk factories were shipped

ROCKET SALESMAN
Von Braun soon proved himself as influential an advocate of rocketry and space exploration in the media as he was within the corridors of power. Here, he brandishes a model of the Redstone rocket at a press conference. In the 1950s, he would cement his place in the American public consciousness by collaborating with Walt Disney on a series of popular television documentaries.

LAUNCHING BUMPER
One of the Fort Bliss team's early projects was Bumper, a basic test of a two-stage rocket design consisting of a small WAC Corporal missile mounted on top of a V-2 first stage. The Bumper launches set a series of altitude records, topping out at 393km (244 miles). These missions, and the White Sands V-2 launches, were among the first to carry scientific instruments to the edge of space.

to America. The chief role for the group was to advise on the reconstruction and test-firing of the captured V-2s at the nearby White Sands range in the New Mexico desert and to help interpret the wealth of paperwork containing the secrets of the German rocket programme. By this time, there was another important factor driving the development of ballistic missiles – the devastating explosions over Japanese cities that had finally ended the Second World War in the Pacific had also heralded a new era of atomic warfare. The Soviets would now be racing to develop their own nuclear weapons, and while the Hiroshima and Nagasaki bombs had been dropped from aircraft, it was clear that if the United States was to maintain its military superiority, the delivery system of choice would be a continent-spanning supersonic missile armed with a nuclear warhead. Indeed, the desire of the superpowers to demonstrate superior missile technology became a key factor during the early decades of the Space Age.

COLLIER'S MAGAZINE
Between 1952 and 1954, von Braun, Willy Ley, and Hans Haber (an expert on the effects of space travel on the human body) explained their vision for the future of space exploration in the pages of Collier's *magazine.*

The months spent at Fort Bliss were frustrating ones – work on reconstructing and testing the V-2s was painfully slow, not helped by a slender budget and occasional bureaucratic hold-ups caused, for example, by FBI director J. Edgar Hoover's fears that the rocket team was a security risk. The Germans leapt at any chance for more stimulating activity, such as writing reports and articles, and later overseeing the flight of scientific instruments, mice, and even a few small monkeys onboard the V-2s.

Von Braun, as ever, was keen to promote the idea of spaceflight, and began to voice his opinions to anyone who would listen – both within the military and, increasingly, in the popular media. And so the American public began its complex relationship with Wernher von Braun – one that would see him pilloried as a closet Nazi in the satirical movie *Dr. Strangelove*, yet fêted as the genius who ultimately put Americans on the Moon.

HISTORY FOCUS

LEGALIZING THE ROCKET TEAM

As the Germans were joined by their families and became settled in America, Operation Paperclip faced a bureaucratic dilemma. Since the rocket team had been flown directly into the United States without visas, it was impossible to officially acknowledge their presence in the country. The eventual solution was somewhat farcical – the Germans were driven in buses across the Mexican border to the town of Juárez, where they applied for US entry visas before being readmitted to the country. It wasn't until 1955, however, that many took citizenship at a ceremony in Huntsville (below).

The Chief Designer

While Wernher von Braun soon became a familiar figure in the US media, his main rival carried out his greatest work beneath a cloak of anonymity. Sergei Pavlovich Korolev, known as the Chief Designer of Spacecraft, was a talented engineer who gave the Soviet Union a lead in space which it maintained until his death.

YOUNG KOROLEV
Korolev's passion for aviation was supposedly triggered by watching an air display as a boy in 1913. Up to and during the Second World War, his principal work was on aircraft designs.

The contrast between the two men who would ultimately drive opposing sides in the Space Race could be traced back to their birth. Von Braun was a scion of German nobility, Korolev the son of a poor craftsman, born in Zhytomyr, Ukraine, in 1907. His parents separated when he was three, though his mother told the young boy that his father had died. Raised largely by his grandparents, Korolev proved himself a good student with a head for mathematics. When his mother married an electrical engineer in 1916, his new stepfather introduced him to practical engineering. Shortly afterwards, on the eve of the Russian Revolution, the family moved to Odessa, where Korolev pursued a growing interest in aircraft.

In 1923, he joined a local Ukrainian aviation society, where he gained first-hand flying experience and was also subjected to communist indoctrination (though he would not join the Communist Party until 1952). The new Soviet government extended its reach into every aspect of public and private life, and Korolev would later learn to his cost just what happened when personal and political rivalries collided. In 1925, he moved with his family to Moscow, where he studied at the Bauman Technical School before working in an aviation design bureau. It was during this period that he got to know rocketry expert and space enthusiast Friedrich Tsander, and he became one of the founder members of MosGIRD in 1931 (see p.23).

RESEARCH ROCKET
An elegant Soviet R-2 awaits launch on the test range at Kasputin Yar in southern Russia. Derived from an undeveloped German modification of the V-2, it first flew in 1950. The R-2 was longer and slimmer than the V-2/R-1 design, had a more powerful engine, and incorporated two pods along its hull, which could carry scientific instruments and animal passengers to high altitudes.

COMPLEX PERSONALITY
A charismatic but demanding figure to those who worked under him, Korolev was driven by dreams of manned spaceflight. However, his experiences in the Gulag made him slow to trust, and he had little time for those he considered liars, which led to many personality clashes as he manoeuvred his way through the Soviet political system.

Rise, fall, and rise

By 1933, the Soviet government had decided to formalize the study of rocket propulsion, and Korolev was appointed Deputy Chief of the

YOUNG DAREDEVIL
In 1929, Korolev (left) and Savva Lyushin (centre) designed a rocket-powered glider, the SK-9, which Korolev copiloted from Moscow to Koktebel in Ukraine. Here, the famous pilot K.K. Artseulov (right) inspects the glider.

AVIATION SCHOOL
Korolev (second left) is seen here during his time at Moscow's Bauman Technical School, where one of his advisers was the aircraft designer Andrei Tupolev.

new Jet Propulsion Research Institute (RNII), working on the development of missiles and rocket-powered aircraft. It was here that he first met Valentin Glushko (see panel, right), who was to be a lifelong rival. The 1930s saw the peak of Soviet leader Joseph Stalin's reign of terror, and in 1937–38 a wave of purges of suspected saboteurs, spies, and traitors swept through the country. Many people seized the chance to advance their careers or settle old scores by informing on anything that could be seen as remotely subversive, and Korolev's iconoclastic personality made him a target. The knock on the door came on 22 June 1938. Only later did Korolev learn that Glushko had been his chief accuser.

At first Korolev was sent to the harsh environment of a Siberian Gulag, but with the outbreak of the war expertise like his was too valuable to waste, even in a paranoid society like Stalin's Russia. By 1942, he was working in a specialist Moscow prison for scientists, ironically under Glushko's leadership. Here Korolev worked on rocket-assisted fighter aircraft. His loyalty to the Soviet Union duly demonstrated, he was finally released in June 1944 (though he would wait until 1957 for a pardon).

BIOGRAPHY

VALENTIN GLUSHKO

Ukrainian-born like Korolev, Valentin Petrovich Glushko (1908–89) developed a complex rivalry with his colleague, which saw them forced to work together throughout the heyday of the Space Race. His enthusiasm for space travel fired by reading the novels of Jules Verne as a boy, Glushko studied physics at Leningrad (modern St. Petersburg), working at the Gas Dynamics Laboratory (GDL) and joining the Leningrad branch of GIRD in 1931. The merger between GIRD and the GDL first brought him into contact and conflict with Korolev, as they worked together for the RNII. Glushko was arrested during the purges of 1938, and seemingly turned informant under intense pressure. In 1946, he was made Chief Designer of OKB 456, in charge of rocket engine development.

Within months, Korolev found himself commissioned as a colonel in the Red Army and flown to Soviet-zone Germany to work under General Lev Gaidukov on the recovery and reconstruction of V-2 related material. One story recounts how he and another officer attempted to gatecrash a British Operation Backfire launch at Cuxhaven, only to be reduced to watching from outside the perimeter fence.

Returning to Russia, Korolev was put in charge of a design team at Scientific Research Institute 88 (NII-88). The institute's main role was to develop ballistic missiles, but the post allowed him to work on technology that could later be used for space travel. Early programmes involved replication of the German V-2 (called the R-1) and improvements in the form of the R-2 and R-3. However, the limitations of V-2- based designs were becoming clear, and Korolev had ambitious plans for the future.

TESTS AT GORODOMLYA
A number of German V-2 scientists working for the Soviet Union, led by Helmut Gröttrup, were transported to Russia in 1946 and mostly settled on Gorodomlya Island, north of Moscow. Here they worked on various projects for Korolev and others. Engine test stands, wind tunnels, and other equipment were shipped in, but the main role of the Germans was to train a new generation of Soviet rocket scientists.

Missiles and rocket planes

The late 1940s saw a missile race in which both the United States and the Soviet Union strove to develop more powerful and longer-range missiles, and even experimental planes, from captured German technology.

X-1 ROCKET PLANE
Bell Aircraft's rocket-propelled X-1 was the first aircraft to travel through the sound barrier and to near the edge of space (see panel, below, and over).

In early 1950, the German rocket team at Fort Bliss began its relocation to Huntsville, Alabama, where wartime munitions factories were to be transformed into a new centre for the development and manufacturing of long-range ballistic missiles. A project called Hermes-C1, previously under slow development at General Electric, was reassigned to the new base at Redstone Arsenal. The intention was that the new missile, later known simply as the Redstone, should be capable of delivering a nuclear warhead over 320km (200 miles). Although the warhead was much heavier than that carried by the V-2, the range was similar and the Germans saw little challenge in the task. They effectively re-engineered the V-2 with a range of improvements, some of which had first been suggested back at Peenemünde. By mid-1953, the first Redstones were being launched from a little-known test range at Cape Canaveral, Florida. Elsewhere in the United States, other groups were also developing new applications of rocket technology. The Jet Propulsion Laboratory (JPL), part of the California Institute of Technology, was working on short-range tactical missiles, often powered by solid propellants much more energetic than black powder. The Navy was developing a series of research rockets called Viking, capable of carrying cameras and experimental payloads high into the atmosphere. And the National Advisory Committee for Aeronautics (NACA) was working on a variety of bizarre experimental ideas, often drawing on schemes developed in Germany during the last years of the Second World War. These ultimately became the first X-planes, prototypes that tested technologies which later became commonplace. The first of these strange aircraft was the Bell X-1, a highly successful rocket-propelled plane (see panel, right, and over).

The Soviet programme continued to be led by Korolev. In his role at NII-88, and from 1946 as Chief Designer at the head of his own experimental design bureau, OKB-1, he was the driving force behind the Soviet Long-Range Ballistic Missile (LRBM) programme. The first fruits of this project were derived from German technology – the R-1 missile, which entered service in 1950, was a more or less direct copy of the V-2, and the R-2, with its extended range and payload, still copied essential elements. However, the next major missile project, the R-3, required a huge leap in capability and a completely new missile design. Korolev established an informal Council of Chief Designers, at which the heads of the six OKB bureaus would meet and collaborate on ambitious projects such as this. Valentin Glushko in particular played a crucial part – he led the design team at OKB-456, working on more powerful and reliable liquid-fuelled engines.

Many of Korolev's early experiments had involved solid-fuel rockets, but his experience with the captured V-2s brought him to the realization that

ON THE PAD
The sheer size of the R-7 called for a new type of launch pad at the Tyuratam complex (see p.60). The entire assembly had to be held in place by a series of sloping support gantries that lifted away from the rocket during launch. Beneath the engines, a flame pit caught the exhaust and diverted it away through escape channels so it did not harm the rocket.

NAVAL RIVAL
Von Braun's team were not the only branch of the US military working on rocket development. The US Navy had its own programme, Viking, under development, though these were sounding rockets intended for scientific purposes.

HISTORY FOCUS

THROUGH THE SOUND BARRIER

While the engineers at Huntsville and JPL worked on a new generation of rocket-propelled missiles, the National Advisory Committee for Aeronautics had other plans. Rockets had potential as powerful engines for a new generation of fighter aircraft, even though jets were still very new technology. In the closing months of the war, Germany had considered a rocket-propelled variant of its Me 262 jet fighter, but this never went beyond a prototype stage. Nevertheless, in 1946, NACA set up an experimental aircraft programme that would produce a wide range of so-called X-planes in the coming decades. The first of these, the stubby XS-1 (later just X-1) was manufactured by Bell Aircraft Inc. Piloted by Chuck Yeager (right), it became the first aircraft to break the sound barrier (see over).

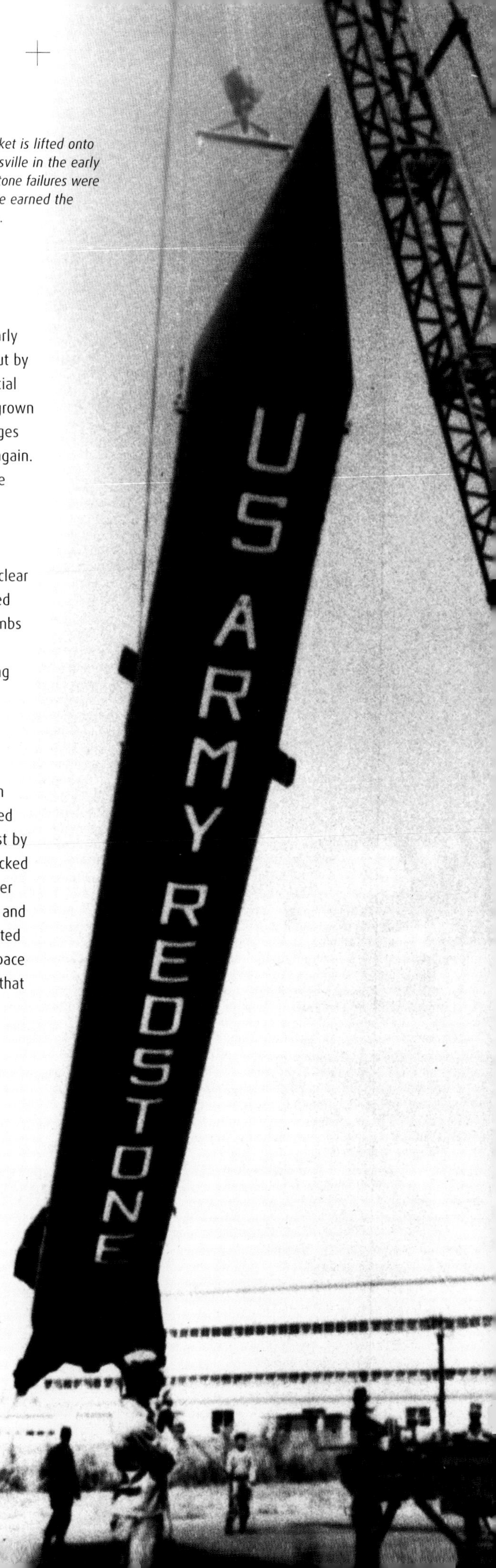

OLD RELIABLE
An early Redstone rocket is lifted onto the test stand at Huntsville in the early 1950s. In service, Redstone failures were so rare that the missile earned the nickname Old Reliable.

liquid fuel was the key to long-range missiles. The captured German scientists were also regularly quizzed during the rocket's long development, but by the early 1950s, they were no longer such a crucial part of the Soviet rocket programme, as home-grown expertise, so devastated in Stalin's paranoid purges of the 1930s and 1940s, began to flourish once again. Before long, most of the captured Germans were allowed to return to their homeland.

Missile delivery

With the successful testing of the first Soviet nuclear weapons in 1949, the pressure for a missile-based delivery system grew. The early Soviet atom bombs were heavier than their US equivalents, and it was clear that the only practical way of deploying them would be with a very powerful rocket – an Intercontinental Ballistic Missile (ICBM). Work on the R-3 was abandoned in order to concentrate on the more ambitious missile.

Korolev and his team ultimately came up with the design officially designated as R-7 (nicknamed Semyorka or Little Seven) and known in the West by the codename "Sapwood". Rather than using stacked stages, each firing in series, the R-7 used a cluster of booster rockets, grouped around a central core and all firing at once. The idea had first been suggested by Mikhail Tikhonravov in 1948 for a proposed space launcher, and fittingly this was to be the rocket that would make spaceflight a reality.

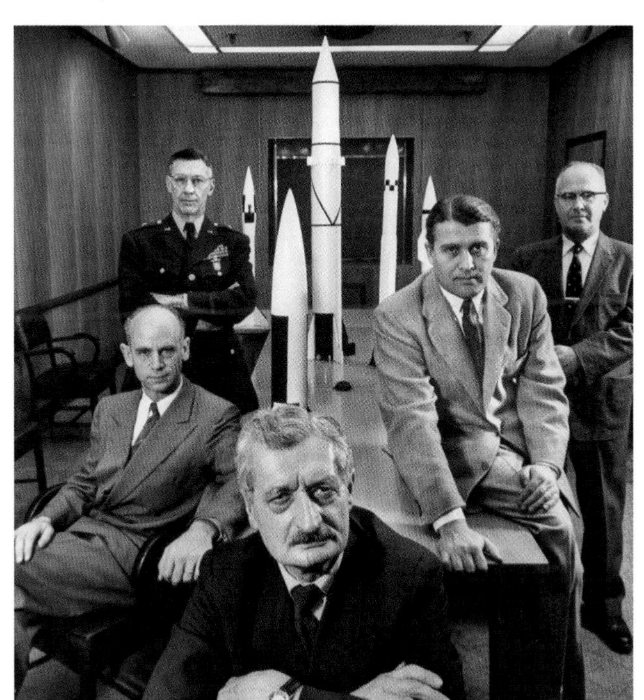

THE ROCKETEERS
This famous picture shows prominent figures working at Huntsville in the early 1950s – left to right are guidance expert Ernst Stuhlinger, Major General Holger Toftoy, Hermann Oberth, Wernher von Braun, and jet aircraft pioneer Robert Lusser.

EXPERIENCE

THE FIRST SUPERSONIC FLIGHT

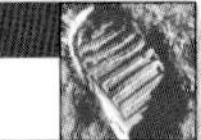

TOP GUN
Yeager became a pilot just a year after enlisting in the USAF as a mechanic in 1941.

Pushing the envelope

Chuck Yeager's supersonic flight aboard the Bell X-1 was a prelude to the Space Age – an audacious mission with an unpredictable outcome that would shape the coming age of jet aviation. In five hair-raising minutes, Yeager would pass through the sound barrier for the first time and learn what lay beyond.

The Bell X-1 had been ordered by the US Army Air Force (as it then was) and the National Advisory Committee for Aeronautics (NACA, the forerunner of NASA), in March 1946. In the aftermath of the Second World War, the full potential of jet fighters was becoming apparent, but there were vital questions that needed to be answered about what would happen when an aircraft went through the sound barrier. For this reason, Bell Aircraft was asked to manufacture three experimental aircraft to be powered by Reaction Motors' new XLR-11 rocket engine. After initial testing at Bell, the aircraft were handed over to the newly formed US Air Force for taking up to, and through, the sound barrier.

"Priorities were, get the airplane **above Mach one** as soon as you can, and don't kill yourself, and **don't embarrass the Air Force**."

Chuck Yeager, looking back on his epic flight

GLAMOROUS GLENNIS
Initially the X-1 was the XS-1 (EXperimental, Supersonic), but this was soon shortened by NACA. Yeager, however, named the aircraft after his wife.

LAUNCH TO LANDING
Taken to high altitude on a B-29 carrier aircraft, Yeager boarded the X-1 by lowering himself through the bomb-bay doors of the B-29. Once in the X-1's cockpit, he had difficulty closing the hatch – he had cracked two ribs falling from a horse a few days before, but kept quiet about it. With the hatch secured, the B-29 dropped its cargo like a bomb, and seconds later, Yeager triggered the rocket engines. After five historic minutes, he glided back to a landing at Muroc Field, California (now Edwards Air Force Base).

3 - 2 - 1 - DROP
A rare colour photograph shows the X-1 in its distinctive orange livery, moments after release from the carrier aircraft.

"If you want to grow old as a pilot, you've got to **know when to push it**, and when to back off."

Chuck Yeager reflects on the special skills required by a test pilot

On 14 October 1947, Yeager climbed aboard *Glamorous Glennis* and fired the engines to attempt to break the sound barrier. The flight's communications transcript records the exchanges between Yeager, the project engineer, Captain Jackie Ridley, and the Drop Pilot and project manager, Major Robert Cardenas:

Robert Cardenas: 8-0-0. Here is your countdown: 10 - 9 - 8 - 7 - 6 - 5 - 4 - 3 - 2 - 1 - Drop.

Chuck Yeager: Firing [rocket chamber] four. Four fired okay. Will fire two. Two on. Will cut off four. Four off. Will fire three. Three burning now. Will shut off two and fire one. One on. Will fire two again. Two on ... will fire three again. Three on. **Acceleration good.** Have had mild buffet – usual instability. Say, Ridley, make a note here. Elevator effectiveness regained.

Jackie Ridley: Roger, noted.

CY: Ridley – make another note. There's something wrong with this Machmeter. It's gone screwy!

JR: If it is, we'll fix it – personally, I think you're seeing things!

Yeager had, in fact, passed through the sound barrier while his instruments indicated a speed of Mach 0.96, leaving the turbulence behind as he did so. Seconds later, and still climbing, he shut off the rocket engines with 30 per cent of their fuel remaining, and prepared for the glide home.

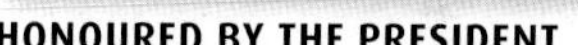

HONOURED BY THE PRESIDENT
Yeager was later honoured by President Truman, along with Bell Aircraft's Larry Bell and John Starck of NACA, at a White House reception. He remained a public figure and American hero throughout the 1950s, and is seen here with Truman's successor, Dwight D. Eisenhower.

REMEMBERED ON FILM
Yeager's supersonic flight plays a pivotal role in Tom Wolfe's book The Right Stuff, *an account of relations between the test pilot community and the Mercury Seven astronauts, later filmed with Sam Shepard (above) as Yeager.*

Reaching for orbit

While the missile programmes of the rival superpowers moved steadily forwards in the 1950s, enthusiasts on both sides were convinced that the same technology could be used for the exploration of space. The announcement of a forthcoming International Geophysical Year (IGY) of scientific study would finally spur their superiors into action.

A MOUSE IN SPACE?
Delegates at a 1954 New York symposium inspect a model of Dr. Fred Singer's MOUSE (Minimum Orbital Unmanned Satellite of Earth), a US project proposed in the early planning stages of the IGY but later shelved.

By the early 1950s, many in the scientific community felt that space travel was an idea whose time had come. Ideas of launching a person into space still seemed outlandish, but the prospects for putting some kind of artificial satellite into orbit seemed promising. Hawkish thinkers and military powers on both sides also saw the advantages. Space might become an important new front in the Cold War and a vital strategic element if the Cold War ever became "hot".

Struggling into space

In the United States, the main obstacles were political – space exploration was not only off the agenda, it was still not even seen as a practical possibility. When presented with a preliminary plan to send a small satellite into orbit in time for the IGY of 1957–58, President Harry S. Truman dismissed it as "hooey".

But the space advocates were not deterred, and they set about appealing directly to the public. Coordination was key, and in 1951 Willy Ley organized a Symposium on Space Travel, bringing together enthusiasts and engineers ranging from the Huntsville missile team to the amateurs of the American Rocket Society. Soon memos were exchanged about the best way to sell the dream of space travel to America. Cynics wryly commented that the US would commit itself only if the Soviets threatened to get there first.

By mid-1954, things began to move, as von Braun was approached by the Office of Naval Research to take part in a summit of leading scientists that would examine the various satellite proposals already on the drawing board and look for a way forward. The result was a proposal for Project Orbiter, a cheap launch system based on the Redstone rocket, topped with a second stage powered by cheap solid rockets. Von Braun was sure it could put a satellite weighing a few kilograms into orbit during the IGY.

But in the spring of 1955, with work on Project Orbiter already advancing, a series of political setbacks saw von Braun excluded from US plans for space. One attack came from scientists involved in the IGY, who dismissed the Orbiter plan as inelegant and argued that the first US satellite should be an all-American affair. These objections might have been overcome, were it not for a turf war between the military branches.

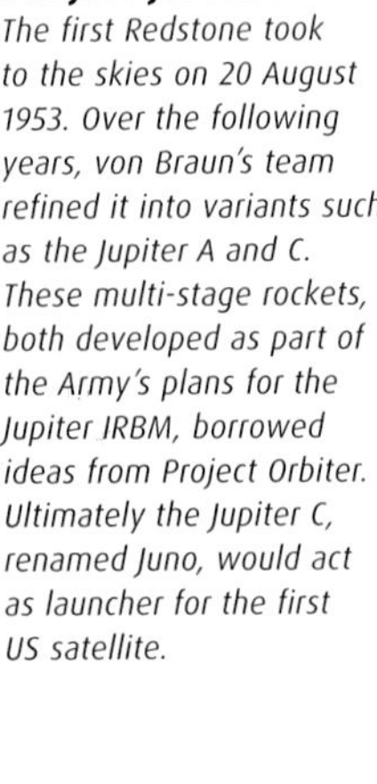

PROJECT JUPITER
The first Redstone took to the skies on 20 August 1953. Over the following years, von Braun's team refined it into variants such as the Jupiter A and C. These multi-stage rockets, both developed as part of the Army's plans for the Jupiter IRBM, borrowed ideas from Project Orbiter. Ultimately the Jupiter C, renamed Juno, would act as launcher for the first US satellite.

TECHNOLOGY

THE FIRST RETURN FROM SPACE

Within a week of the decision to abandon Project Orbiter, the US Defense Department threw a lifeline to Huntsville when it authorized the Army's development of the Jupiter Intermediate-Range Ballistic Missile (IRBM). Jupiter missiles would follow a trajectory that left and re-entered Earth's atmosphere, so permission was given for a programme of Redstone-based test vehicles that could take Jupiter components such as warhead nose cones into space and test how they survived the heat of re-entry. The resulting Jupiter A and C rockets were very similar to the original Project Orbiter. The first object to successfully return to Earth from space was a model Jupiter nose cone, launched on 8 August 1957.

TESTING REDSTONE
Exhaust billows from the base of a Redstone rocket during a static test-firing of the engines. The test stand at Redstone Arsenal allowed the Huntsville engineers to measure the amount of thrust their rocket engines could generate. With the Army unwilling to fund large building projects at the site, the Germans were forced to build the stand on a tight budget, effectively raiding the petty cash.

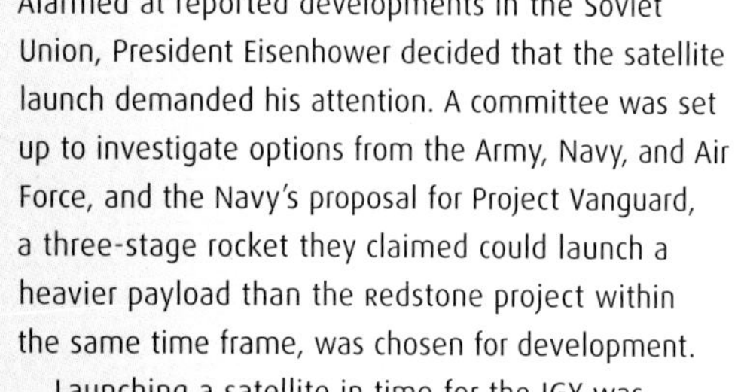

Alarmed at reported developments in the Soviet Union, President Eisenhower decided that the satellite launch demanded his attention. A committee was set up to investigate options from the Army, Navy, and Air Force, and the Navy's proposal for Project Vanguard, a three-stage rocket they claimed could launch a heavier payload than the Redstone project within the same time frame, was chosen for development.

Launching a satellite in time for the IGY was finally a US government priority, but the Huntsville team had been relegated to the sidelines.

Planning Object D

Meanwhile in the Soviet Union, Korolev was also playing politics. Various satellite projects had been floated since the end of the Second World War, but the first signs of an official commitment to space exploration came in March 1954, with the announcement of Soviet participation in the IGY. In August 1955, a satellite launch was approved, and the Third Commission on Spaceflight, chaired by Mstislav Keldysh, was set up to coordinate the Soviet space programme. But uncertainties over the programme's status lingered, until in January 1956, Korolev seized the opportunity of a visit by Soviet premier Nikita Khrushchev (to inspect the R-7 under construction) to show him Mikhail Tikhonravov's work on a satellite laboratory known as Object D. At first, Khrushchev showed little interest – until Korolev pointed out how, with the R-7 nearing completion, the Soviet Union would be able to launch a satellite far larger than anything the Americans could hope for. Enticed by the chance to humiliate the United States, Khrushchev gave the project his backing. Tikhonravov's team were transferred to Korolev's OKB-1, and the project became a top priority. The Space Race was about to begin.

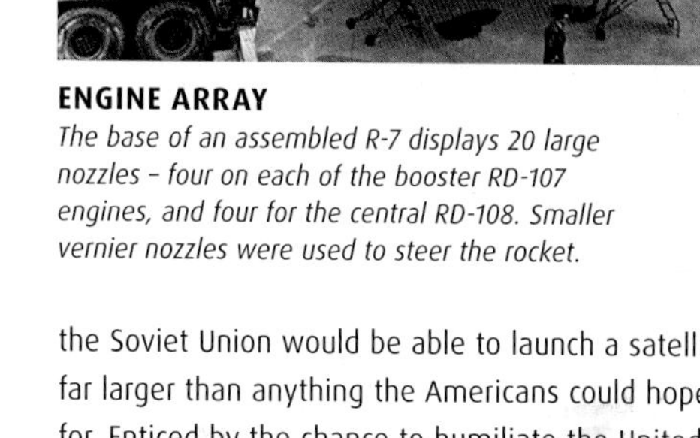

ENGINE ARRAY
The base of an assembled R-7 displays 20 large nozzles – four on each of the booster RD-107 engines, and four for the central RD-108. Smaller vernier nozzles were used to steer the rocket.

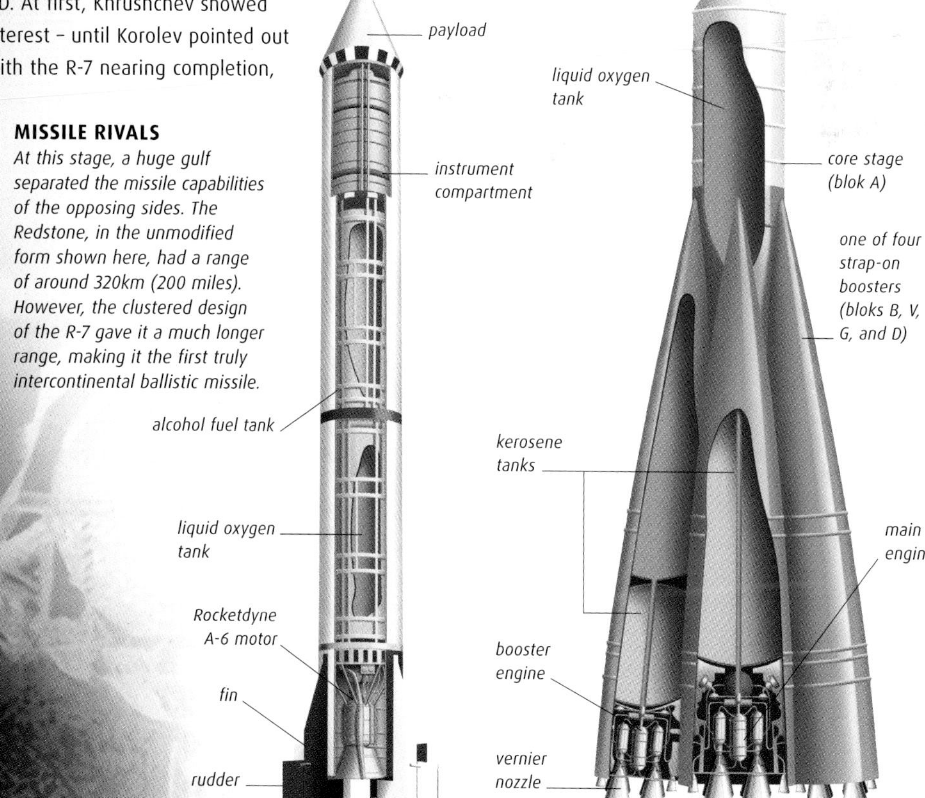

MISSILE RIVALS
At this stage, a huge gulf separated the missile capabilities of the opposing sides. The Redstone, in the unmodified form shown here, had a range of around 320km (200 miles). However, the clustered design of the R-7 gave it a much longer range, making it the first truly intercontinental ballistic missile.

JANUARY 1959: WINDS OF CHANGE
A technician examines a boilerplate Mercury capsule prior to testing its performance in the huge wind tunnel operated by NASA's Space Task Group at Langley, Virginia.

THE DAWN OF THE SPACE AGE

BY 1957, both the Soviet Union and the United States had the technology to put a satellite in space – the question of who would be the first to do it was largely one of political willpower. Korolev's team had to deal with a hasty satellite redesign in order to have something ready to launch, while von Braun's group were sidelined – and the replacement naval rocket team faced problems of their own. However, no one could have predicted the seismic effect that the first space launches had on the imagination of the general public. Americans were shaken out of their complacent belief in a technologically backward Soviet Union, while Soviet Premier Nikita Khrushchev soon realized the propaganda value of his nation's new-found space superiority.

The late 1950s and early 1960s, then, saw the beginnings of a Space Race, as the rival superpowers attempted ever more daring technological feats in the effort to claim new firsts. Most daring of all were the efforts to launch the first manned spacecraft – programmes that would create a new breed of hero for a modern age: the spaceman.

1955 1956 **1957** 1958 1959 1960 1961 1962 1963 1964 1965 1966 1967 1968 1969 1970 1971 1972 1973 1974 1975 1976 1977 1978 1979 1980 1981 1982 1983 1984 1985 1986 1987 1988 1989 1990 1991 1992 1993 1994 1995 1996 1997 1998 1999 2000 2001 2002 2003 2004 2005 2006 2007 2008 2009 2010 2011 2012 2013 2014 2015 2016 2017 2018 2019 2020

25 January 1957
The Soviet government orders that plans be made for the launch of a basic satellite following delays to the design of Object D, initially intended for the first launch.

15 May 1957
The first test-firing of a Soviet R-7 two-stage ICBM, core of the eventual Sputnik launch vehicle, is unsuccessful.

21 August 1957
A Soviet R-7 ICBM (codenamed Sapwood by Western intelligence) launches successfully from Tyuratam.

4 October 1957
A modified R-7 launch vehicle lifts off from Tyuratam carrying Sputnik 1. The Soviet news agency TASS announces the successful launch to the world.

25 October 1957
Sputnik's batteries run down and its radio signal ceases.

4 January 1958
Sputnik 1 burns up during atmospheric re-entry.

Red star in orbit

A new era of exploration began on 4 October 1957, as Sputnik 1 began its transmissions from high above the Earth. And the beeps from the satellite's radio were also a clear signal of Soviet space superiority.

But Sputnik 1, an 83-kg (184-lb) steel ball some 58cm (23in) across, and containing little more than a basic radio transmitter, was still a long way from Tikhonravov's ambitious Object D orbital laboratory. Just what had happened to derail Soviet plans for a more spectacular entry into the Space Age?

The birth of Sputnik

Ultimately, plans to develop Object D in time for a 1957 launch had fallen victim to the same problems encountered time and again in the early days of the space programme – bureaucracy and politics. Despite a promising start and direct approval from the Soviet head of state, the satellite had fallen behind schedule due to the unwillingness of other institutions to cooperate with the mysterious requirements of OKB-1. To take just one example, there were repeated delays to the delivery of high-quality silicon that the Department of Chemistry was supposed to be supplying for Object D's solar cells. By September 1956, it was clear that the project was in trouble, and Mstislav Keldysh of the Commission on Spaceflight said as much at a meeting of the Academy of Sciences. Korolev knew there was now a risk that the United States might achieve a launch while the Soviet satellite was still under construction. Against Keldysh's wishes, he began development of a much smaller and more achievable satellite, which he named Prosteshii Sputnik – Project Sputnik (PS), or "Companion".

Early in 1957, Korolev revealed his contingency measures to the Soviet leadership. The plan required two small PS models and a new "shroud" to cap the R-7 and protect the satellite on its way to orbit. The revised project was approved and made OKB-1's top priority (contingent, of course, on the successful launch of the R-7) in late January.

Testing the R-7

While work on the new PS satellites got under way, pressure was also growing on the R-7 project. In May 1957, the first missile arrived at the Tyuratam launch complex (see p.59). But a launch attempt on 15 May ended in failure 100 seconds into the flight, as one of the strap-on boosters caught fire and exploded due to a fuel leak. Modifications were made to prevent a repeat performance, but other problems appeared in the second and third tests. Finally, on 21 August, an R-7 made it nearly all the way to its target zone on the remote Kamchatka Peninsula before its warhead disintegrated. Another launch on 7 September ended in the same way – there was clearly work to be done before the R-7 could be a dependable ICBM, but its lower stages were now seen as reliable enough to attempt a satellite launch.

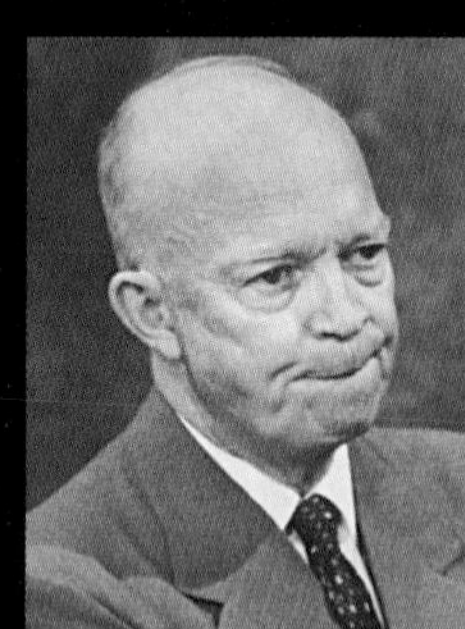

GETTING THE MESSAGE
Most Americans waking on the morning of 5 October 1957 were alarmed to learn that a Soviet satellite had flown over their heads at least four times as they slept. For President Dwight D. Eisenhower, it was a shocking sign of Soviet technological parity – and potential superiority.

DOUBLE CELEBRATION
The Sputnik launch (above) not only seemed to show Soviet superiority in rocketry, it was also the first time the R-7 missile had performed flawlessly – another reason for Soviet premier Nikita Khrushchev (right) to be happy with the results.

"... the new socialist society turns even the **most daring** of man's **dreams** into a reality."

Official TASS press statement, 4 October 1957

SPUTNIK IN ORBIT
As it separated from the shroud and the core rocket stage, Sputnik released four antennae, between 2.4 and 2.9m (94 and 114in) long. The "beeps" they broadcast carried data about pressure and temperature encoded in their duration.

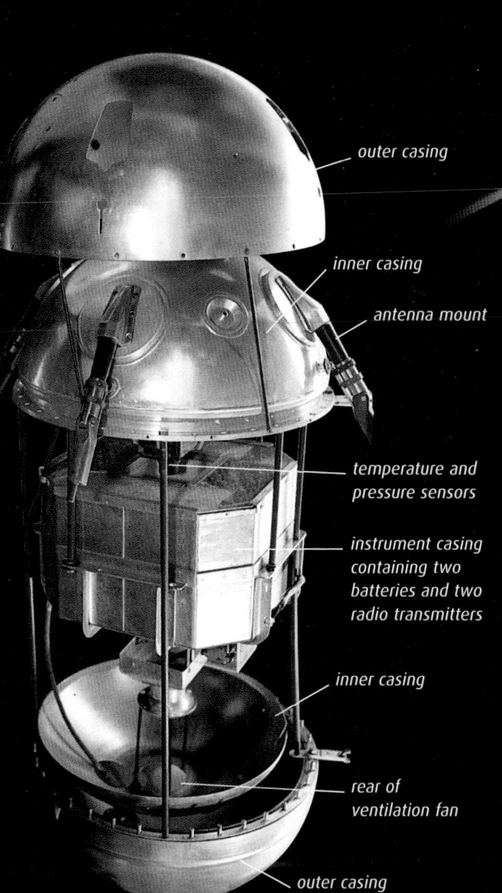

INSIDE SPUTNIK 1
Although basic, every aspect of the first satellite had a scientific purpose. The ball-like shape would aid calculations of atmospheric drag (which eventually pulled the satellite to its doom three months later), while the internal sphere tested a concept used in later pressurized spacecraft.

Countdown to launch

On 20 September, the State Commission met to approve the launch, initially setting the date for 6 October. But rumours soon spread of an American scientific paper, due to be presented on that day at an IGY conference in Washington, entitled "Satellite in Orbit". Although Soviet intelligence was confident that the US was only planning a suborbital launch (see over), Korolev did not want to miss his date with history by a whisker, and the launch was duly moved forward two days.

At 10:28pm Moscow Time on 4 October 1957, a modified R-7 carrying the PS-1 satellite fired its engines to rise clear of the launch pad at Tyuratam. The ascent went flawlessly, as did the separation of Sputnik 1, as the little satellite became known, from the central rocket stage. However, by the time Sputnik had unfolded its antennae and begun to transmit, it had passed beyond the range of Soviet receivers. There was an anxious 90-minute wait for its next passage across the sky, when a steady stream of beeps confirmed Sputnik was in a stable orbit ranging from 215 to 939km (133 to 583 miles) above the Earth. The Space Age had begun.

LIFE
THE SATELLITE
WHY REDS GOT IT FIRST
WHAT HAPPENS NEXT
GOV. FURCOLO WRITES
TALE OF ZANY CAMPAIGN
U.S. SCIENTISTS PLOT ORBIT
OCTOBER 21, 1957 25 CENTS

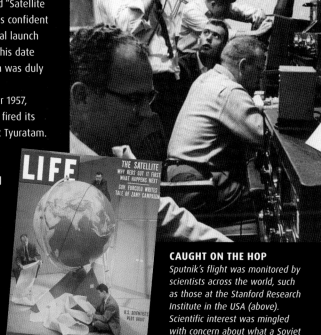

CAUGHT ON THE HOP
Sputnik's flight was monitored by scientists across the world, such as those at the Stanford Research Institute in the USA (above). Scientific interest was mingled with concern about what a Soviet presence in space could mean.

1 May 1957
A Vanguard first and third stage launch a dummy satellite on a suborbital path.

4 October 1957
News of the successful launch of Sputnik 1 spreads around the world.

23 October 1957
The Vanguard TV-2 rocket, with functioning first and second stages and an inert third stage, launches successfully. In view of the Sputnik 1 launch, the decision is made to attempt a satellite launch as part of the TV-3 test flight.

8 November 1957
Von Braun's Army-based Project Orbiter is restarted, under orders to launch a satellite within 90 days.

6 December 1957
The US Navy Vanguard rocket TV-3 fails on launch from Cape Canaveral.

The US responds

Caught completely off-guard by the Soviet launch of Sputnik 1, the United States rushed to respond with a satellite launch of their own. But as poor decisions came back to haunt them, the launch ended not in triumph, but in humiliation.

News that the Soviet Union had launched a satellite first reached the US on the evening of 4 October 1957. Scientists at the IGY meeting in Washington queued to congratulate the gleeful Soviet delegation. At Huntsville, Wernher von Braun was entertaining a group of VIPs including the new Secretary of Defense and the Army's Chief of Staff when the news broke. After proclaiming, "Today, man has taken his first step towards Mars," von Braun turned to lobbying his visitors on the importance of restarting his Redstone-based launch vehicle programme (see p.48).

In the White House there was consternation, as much over the intelligence failure as over Sputnik 1 itself. For years, the CIA had dismissed Soviet space promises as propaganda and the idea that a Russian satellite could outperform US efforts as fantasy. The level of surprise was well expressed by James M. Gavin, Army Chief of Research and Development, who likened Sputnik to a "technological Pearl Harbor" – the US had lost a seemingly one-horse race.

While the inquests would rumble on, the immediate question was how to respond. Eisenhower faced a difficult decision – he felt that the missile gap, as it was known, was largely a public misconception, fed by a need for secrecy about America's missile programmes and its intelligence-gathering capabilities. The Redstone rockets, now under the control of the Army Ballistic Missile Agency (ABMA), were increasingly reliable, while the longer-range Atlas and Titan ICBM programmes of the Air Force and Navy were progressing well. Meanwhile, U-2 spy planes provided a good idea of the state of the R-7, or Sapwood, as it was known in the West.

Nevertheless, the American public was nervous, and a swift response was called for. On 9 October, Eisenhower congratulated the Soviet Union for its launch of what he called a "small ball" into space. He insisted that US plans would not change and announced a test of the Vanguard rocket in December. The implication that the vehicle would have been ready anyway was misleading – in reality, the Vanguard team, run by John P. Hagen of the Naval Research Laboratory, knew that rushing to launch on Eisenhower's timetable was a dangerous gamble.

Vanguard stumbles

The US Naval research rocket that had beaten off the Redstone challenge to become the official US satellite launcher was a hybrid of existing and new rocket stages (see panel, right). In theory, the use of tried-and-tested components should have made

LAUNCH PAD FIASCO
After winning the battle with von Braun's US Army team to be the first to launch a satellite, the US Navy's Vanguard rocket explodes on the launch pad, watched on television by millions of dismayed Americans.

UNDER NEW MANAGEMENT
With the establishment of the Army Ballistic Missile Agency (ABMA) at Redstone in 1956, von Braun's team came under the command of Major General John B. Medaris (on the left in this picture with von Braun).

STANDING READY
By mid-1957, the Huntsville team had successfully used their Redstone-based Jupiter C rocket to reach space (see p.38). Medaris ordered that several of the missiles should be held back for a potential satellite launch, but the Department of the Army was so concerned that the ABMA might "accidentally" launch a satellite before Vanguard that they insisted the upper stages be disabled.

NASA IS BORN
On 21 November 1957, members of the US Satellite Committee for the IGY wrote to several influential figures recommending the formation of a civilian agency to manage the space programme. Within a year, Eisenhower had established NASA, with T. Keith Glennan (right) as Administrator and Hugh Dryden (left) as his deputy.

the project easier, but from the beginning Vanguard suffered from neglect – largely because it was a scientific rather than military project, with a smaller budget to match. For example, the main contractor, which had previously built the Viking rocket, transferred its experienced staff to more lucrative work on the Titan ICBM. At the time of Eisenhower's announcement, the Vanguard components had not even been test fired together – a successful maiden launch for the complete vehicle would be a major accomplishment in itself (and something not achieved by either the V-2, Redstone, or R-7). But after a test in October saw the first and second stages function well, Washington insisted that the next launch should attempt to put a US satellite in orbit. When the Soviets compounded American worries with the launch of Sputnik 2 (see over), pressure on the Vanguard team could only increase.

"Flopnik"

On 6 December, the world's press gathered at Cape Canaveral for the launch of Vanguard TV-3. The 23-m (76-ft) vehicle carried a truly "small ball" in its nose cone – a 1.8-kg (4-lb) satelllite about the size of a grapefruit. The launch had already been postponed twice on preceding days due to poor weather, but on the third day all seemed fine, and the countdown went as planned, watched by millions of television viewers across the country. At 11:44am, the rocket's engines fired, and it slowly began to lift off the launch pad, rising to an altitude of 1.3m (4ft) before, two seconds after launch, a huge explosion tore the first stage apart, sending the upper sections toppling into the flames. To rub salt in the wound, the satellite itself escaped the inferno, rolling free and transmitting its beeps to anyone with a radio. America's first step into space had gone up in flames – drastic action would be needed to restore US dignity.

TECHNOLOGY

THE VANGUARD LAUNCHER

At 23m (76ft) tall but just 1.14m (46in) across at its broadest, the Vanguard was the epitome of a tall, elegant rocket, certainly when compared with the more squat appearance of the Redstone missile and its variants. The first stage, powered by a mix of liquid oxygen and kerosene, was designed by General Electric based on the successful Viking rocket. The second stage used an engine similar to those used on the Navy's Aerobee sounding rockets, burning a mix of nitric acid and unsymmetrical dimethylhydrazine (UDMH). This stage contained the rocket's guidance systems – with no fins in the design, the rocket's attitude was controlled by tilting the exhaust nozzles on gimbals. The upper stage, a solid-propellant rocket built by the Grand Central Rocket Company, was set spinning during separation in order to stabilize it.

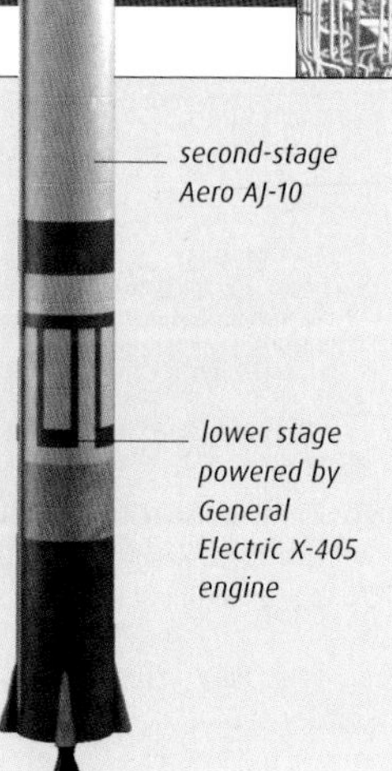

12 October 1957
After the success of Sputnik 1, Khrushchev orders Korolev's satellite and launch teams to quickly build and launch a much larger satellite, that would be capable of carrying a dog.

3 November 1957
Sputnik 2 is launched from Tyuratam carrying the dog Laika onboard. She dies from stress and overheating within a few hours following a system failure.

7 November 1957
Sputnik 2's transmitters fail, leaving Laika's fate a mystery.

10 November 1957
For many years, this was the official date of Laika's death, when her food and oxygen would have run out.

14 April 1958
Sputnik 2 falls out of orbit, re-entering the Earth's atmosphere.

Laika's travels

Within a month of their initial satellite launch, the Soviet Union again stunned the world with the launch of a much larger Sputnik, this time carrying a passenger – the first living creature in orbit.

Basking in the success of Sputnik 1, the OKB-1 team prepared for a well-earned break in early October 1957 – only to find that Nikita Khrushchev had other ideas. The Soviet premier, his interest in space fired by the reaction in the Western press, telephoned Sergei Korolev on 12 October to congratulate him in person and ask what he planned to do next. With Object D still not ready and a spare R-7 (Sputnik 1's backup) at Tyuratam, there was a clear opportunity to press home Soviet superiority with the launch of a heavier satellite. Knowing that they still had some equipment in stock from earlier test launches that had carried dogs to the edge of the atmosphere and returned many of them unharmed, Korolev suggested that the first launch of an animal into orbit would offer both scientific and propaganda advantages.

DOGS IN SPACE
Dogs played the same role in early Soviet space exploration that primates played in the US. Here, Korolev is pictured with one of two dogs that successfully reached an altitude of 100km (60 miles) aboard an R-1D rocket in July 1954.

And so Korolev rapidly recalled his team from holiday to work on the construction of Sputnik 2. Often working from Korolev's sketches rather than properly drafted plans, OKB-1 was under orders from Khrushchev to launch the satellite within a month. In retrospect, their achievement was remarkable, and it seems likely that Korolev had in fact made some early plans during the development of Sputnik 1.

Building the satellite

Sputnik 2 provided the first opportunity to get real scientific data from orbit. The basic spacecraft design was conical. At the narrow end, an array of sensors was fitted to measure high-energy radiation from the Sun. Behind these sat a pressurized sphere – the core of the original PS-2 satellite, now fitted with more sophisticated radio transmitters for returning data from orbit. At the base of the craft was a cylindrical, pressurized module within which the unfortunate canine would travel.

Korolev's team had already developed sensors to monitor the health of dogs during rocket flights, but a longer flight would create new challenges. Insulation would be needed to maintain a steady temperature despite the intense variations outside the satellite. Food, water, and air would have to be regulated, and the carbon dioxide breathed out by the animal would have to be filtered from the air

PREPARED FOR LAUNCH
Sputnik 2 was cramped, but still had room for Laika to lie down. She was strapped in two days before launch, so that she could become used to the conditions.

HISTORY FOCUS

A SOVIET TRAIL ACROSS THE SKY

One of Sputnik 2's major achievements was to disprove a vocal minority who had claimed that Sputnik 1 was nothing but a Soviet hoax. Since the 1920s, the Western world had been led to believe that the Soviet Union was technologically backward and incapable of innovation. Even when the first Soviet nuclear weapons were tested, many assumed that they were reliant on technology stolen from the West. For the average Western citizen, the arrival of Sputnik 1 was a traumatic event, so it was little wonder some people were in denial. Sputnik 2, however, was large enough to be visible to the naked eye as it moved across the night sky. The satellite itself did not have a source of light onboard – it shone only by reflecting sunlight – but long-exposure photographs such as this one provided physical evidence of a new object in Earth's sky.

LAIKA LIVES ON
Laika's image became an icon of the early Space Age, used on stamps, postcards, and even cigarette packets in the Soviet Union and beyond.

before it built up in poisonous quantities. There was also the delicate question of how to handle the dog's physical waste in weightless conditions.

A dog's life

Laika (her name means Barker) was one of ten animals considered for the flight. All the dogs used were light-coloured mongrel bitches – light so they would be seen by onboard cameras, mongrels because they were considered more robust, and bitches to simplify the design of "nappies" used to handle body waste. Laika was found wandering the streets of Moscow and trained to tolerate increasingly confined spaces prior to launch. She was ultimately selected for the mission by Oleg Gazenko, an expert in the developing area of space medicine.

Sputnik 2 took to the skies on 3 November 1957, and within hours the Soviet Union was trumpeting its latest coup. The satellite weighed 508kg (1,117lb) and entered orbit still attached to the exhausted 6,800-kg (15,000-lb) core stage of its R-7 rocket. State press agency TASS reported that Laika was adjusting well to the pressures of spaceflight, but that since it would be impossible to return her to Earth, the spacecraft carried a system to painlessly put her to sleep after ten days in orbit, before her food, water, or air supplies ran low. Regular updates followed, but nothing more was heard after a radio failure on 7 November. To the outside world, it was generally assumed that Laika had died of natural causes about one week into her flight.

It took 45 years and the collapse of the Soviet Union for the true story of Laika's fate to emerge. While the launch itself was a success, and Laika's heart rate rapidly fell back to normal levels, it seems that the satellite and core rocket stage were actually intended to separate. When the separation failed, it tore away some of the protective insulation, and also caused a malfunction in the thermal control system. Temperatures in the capsule rose rapidly to 40°C (104°F), and Laika's heart rate began to rise again. She died around six hours into the flight, from a combination of heat and stress.

Laika was one of the few animal casualties of the space programme, and the only one to have been sent into space with no hope of return. It is almost certain that, if time had allowed, Korolev would have attempted to design some form of re-entry capsule – not only could this have saved the dog, it could also have provided more useful scientific information. The flight made Laika the most famous dog in history, but in 1998 a reflective Oleg Gazenko said, "The more time passes, the more I am sorry about it. We did not learn enough from the mission to justify the death of the dog."

radiation and particle detectors

original Sputnik PS-2 pressurized core containing radio transmitter

pressurized cabin containing life-support systems

INSIDE SPUTNIK 2
Sputnik 2 had similar dimensions to Object D, allowing it to fit into the standard R-7 nose cone. This model has had the outer shielding removed.

8 November 1957
Following the failure of the Navy's Vanguard rocket, Project Orbiter is formally restarted, under orders to launch a satellite within 90 days.

31 January 1958
A Juno 1 rocket successfully puts the Explorer 1 satellite into orbit.

5 March 1958
An attempted launch of Explorer 2 fails when the upper rocket stage of its launch vehicle misfires.

17 March 1958
A US Navy Vanguard rocket finally succeeds in placing a small satellite, known as Vanguard 1, into orbit.

26 March 1958
Explorer 3 is successfully launched with another Juno I rocket.

23 May 1958
Explorer 1's batteries fail, and it ceases transmission of information, though it remains in orbit for another 12 years.

26 July 1958
Explorer 4 is successfully launched.

24 August 1958
The launch of Explorer 5 fails due to a problem with booster rocket separation.

The US reaches orbit

With Sputnik 2 capturing headlines around the world, the US government finally allowed Army planning for a spaceshot to resume. In the wake of Vanguard, Explorer 1 would become the first American satellite.

The same evening that news of Sputnik 1 had broken, Secretary of Defense Neil McElroy, who happened to be visiting Redstone Arsenal, had called Wernher von Braun's bluff – if the Army team were allowed to restart work toward a satellite launch, how soon could they have something in orbit? Sixty days, said the optimistic von Braun, while his more cautious commander, John B. Medaris, thought that 90 days was more feasible.

Nevertheless, it took the launch of Sputnik 2, and the realization that the Soviet space programme was not just a one-off publicity stunt, to force a decision in Washington. On 8 November, the Huntsville team was given approval to launch a satellite, with several provisos. The most important of these was the nature of the satellite – Eisenhower, in particular, still had reservations about using a military launcher in the space programme, and so it was decreed that the satellite must carry an instrument package to do some real science (and just perhaps to help them leapfrog the Russians).

While the launcher would be a modified Jupiter C (the refined Redstone that the team had completed and tested earlier in the year), the satellite would have to be designed from scratch. William H. Pickering of the Jet Propulsion Laboratory (JPL) in California (at the time a development facility for short-range missiles) successfully lobbied Medaris to allow his team to build the satellite, while the Huntsville group concentrated on readying the launch vehicle. It was to be the first of many satellites that would be developed at the laboratory. In order to simplify the design, the satellite, soon christened Explorer 1, was modified from one of JPL's Baby

1ST U.S. SATELLITE LAUNCHED, ORBITS IN SPACE WITH SPUTNIK

NEW MOON IS LOFTED TO SKY-RIDING PATH BY JUPITER-C MISSILE

DEFENSE NO VOTE ISSUE, NIXON SAYS

PROBERS LINK HEAD OF UNION TO MOBSTERS

BIPARTISAN FRONT URGED

Aid Pact Will Flood Egypt With Russians

IKE INFORMED OF SATELLITE

GOOD NEWS AT LAST
The surprise news that the US finally had a satellite in orbit briefly helped to calm the insecurity of the American public.

CELEBRATING SUCCESS
William Pickering, James Van Allen, and Wernher von Braun triumphantly display a full-sized model of Explorer 1 to a packed Washington press conference after the successful launch.

SATELLITE AND LAUNCH VEHICLE
The drum (with a vertical black stripe) linking Explorer 1 to the Redstone first stage contained two rings of Baby Sergeant rockets. The outer ring of 11 fired first, then the other three blasted free, carrying Explorer 1 into orbit.

JUNO BLASTS OFF
A pillar of fire shoots from the Redstone first stage as Explorer 1 begins its historic ascent into space. The Juno launcher seems short and stubby compared with the flawed elegance of Vanguard.

IN THE BLOCKHOUSE
Tense engineers monitor Explorer 1's progress under the supervision of Kurt Debus (left), director of the Launch Operations Center and another member of the V-2 team.

Sergeant solid-rocket missiles, allowing it to propel itself into orbit. Only the front half contained scientific instruments (see below). The satellite was linked to the lower-stage Jupiter C by a ring of modified Baby Sergeant rockets in a drum. These fired in two groups and separated in mid-flight, effectively acting as two intermediate stages. As in Vanguard, the upper stages were set spinning during launch and ascent in order to keep the vehicle stable.

Orbital Explorer

While the Vanguard team had to work under the pressure of a public deadline, preparations for Explorer 1 were mostly secret. The modified Jupiter C, renamed Juno to distance it from the military programme, was flown in to Cape Canaveral on 20 December, while the satellite joined it in late January 1958. Two launch attempts were abandoned, but high winds abated on 31 January, and Juno's main engines fired at 10:48pm local time. Despite the secrecy, thousands had gathered nearby to watch.

Von Braun, Pickering, and others saw the launch from the War Room at the Pentagon. As it turned out, Explorer 1 entered a higher orbit than expected, 2,520km (1,575 miles) from Earth at its highest point. This made it the most distant object launched so far, but the longer orbit also meant a delay before signals were picked up to confirm the satellite was safely in orbit. By the time von Braun, Pickering, and James Van Allen (see panel, below) arrived at the National Academy of Sciences for a news conference in the early hours of 1 February, the press had been alerted and they received a hero's welcome.

INSIDE EXPLORER 1
The rear of Explorer 1 was a miniature version of JPL's Sergeant rocket. The front contained three scientific instruments – a temperature sensor, a microphone system to detect the sound of tiny meteorites hitting the satellite, and the particle detector package (effectively an oversized Geiger counter). There was also a radio transmitter to send data back to Earth.

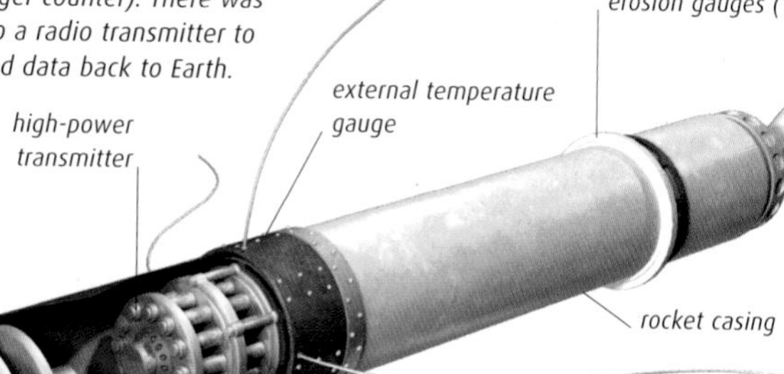

BIOGRAPHY

JAMES VAN ALLEN

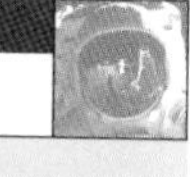

William Pickering handed the task of designing Explorer 1's scientific instruments to James Van Allen (1914–2006), a scientist who had been investigating the upper atmosphere and near-Earth space since the 1940s. He had been involved in launching particle detectors and other instruments on V-2 and Bumper rockets from White Sands and was one of the strongest advocates of the project to launch a satellite during the IGY. The reward for his work was scientific immortality – Explorer 1 discovered the presence of radiation belts beyond the Earth's atmosphere, today known as the Van Allen Belts.

15 May 1958
The launch of the Sputnik 3 orbiting laboratory emphasizes the power and dominance of Soviet launch vehicles.

18 December 1958
The experimental SCORE satellite is launched by an Atlas B rocket, broadcasting a pre-recorded message from President Eisenhower.

28 February 1959
The US launches an experimental spy satellite, Discoverer 1, intended to take photographs from space and then return a capsule of film safely to Earth.

11 August 1960
At the end of the Discoverer 13 mission, a film capsule is returned to Earth from orbit for the first time.

12 August 1960
NASA launches the experimental Echo 1 radio-reflector satellite.

4 October 1960
The US Army launches Courier 1B, the first satellite with the ability to receive and retransmit signals from the ground.

10 July 1962
Telstar, ancestor of the modern communications satellite, is launched. Television signals broadcast from space are received in the United States and Europe.

Early satellites

The opening exchanges of the Space Race were followed by a period of experimentation as both sides attempted to orbit increasingly ambitious satellites and establish the potential future uses of space.

The late 1950s and early 1960s saw many firsts: the first tests of communication via satellite, the first attempts at Earth observation and using satellites for intelligence, and the first orbital laboratories.

Solar power

Although the Vanguard launcher was ultimately doomed by the more reliable Juno and its successors, it did achieve one significant milestone. On 17 March 1958, it successfully put the Vanguard TV-4 satellite into orbit. This small "grapefruit", identical to the one the US had failed to launch the previous December, was the first spacecraft that used solar cells to generate energy.

Tikhonravov's Object D satellite, a far more impressive use of solar power, finally made it to orbit as Sputnik 3 on 15 May 1958 (a failed launch in April meant that a backup was used). One of the craft's many innovations was a data recorder that would store instrument readings on magnetic tape and play them back when the satellite came within range of Soviet receiver stations. Although the recorder did not work properly, limiting the data returned from the satellite, the principle would be widely used later. A two-year hiatus in Sputnik launches followed, before they resumed in 1960 (later Sputniks were in fact tests for the manned Vostok missions).

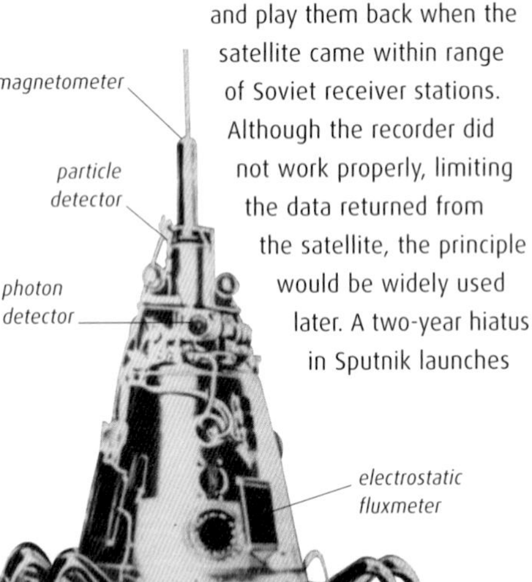

SPUTNIK 3
The third Sputnik to be launched weighed a massive 1,327kg (2,920lb) – far more than the US could hope to launch at the time. It carried a variety of instruments to study the environment around the Earth.

PROJECT ECHO
One of the earliest satellites launched by NASA, Echo 1 inflated after launch into a sphere 30m (100ft) across. When the satelloon, as it was called, was visible, signals bounced off it could be received at great distances.

Eyes in the sky

Since the US and Soviet space programmes were both driven by military priorities, the achievement of spaceflight was soon followed by the first attempts at surveillance. Space-based cameras would be less vulnerable to attack from the ground than spy planes (a point brought home when a US U-2 was shot down over the USSR in May 1960). Since the technology to send pictures back from space electronically was in its infancy (see p.52), the best way to capture high-resolution images was on photographic film that could then be returned to Earth. To test this, the US launched a series of Discoverer satellites (also known by the name CORONA), of which Discoverer 13 was the first to return its film to Earth, dropping it into the atmosphere on 11 August 1960, to be retrieved in mid-descent by a C-119 transport plane. Meanwhile, the Samos programme experimented with electronic image transmission – developing its photographs in orbit before sending them back to Earth.

TRACKING THE SATELLITES
The arrival of the Space Age left the superpowers scrambling for ways of tracking their satellites around the globe. The first large radio telescope, at Jodrell Bank, England, found a new purpose tracking launches. Soon, the United States had installed tracking stations around the world, while the Soviet Union had equipped a fleet of tracking ships that could stay in contact with a spacecraft throughout its orbit.

Extraterrestrial relays

Communications was another area in which satellite technology needed proving. A large satellite in orbit could act as a reflector for radio signals (tested by the Echo project, which began in May 1960). A satellite could also broadcast a pre-recorded signal across a wide area (as with SCORE, which transmitted a Christmas message from President Eisenhower to the world in 1958). A more direct precursor of today's communication satellites was the US Army's Courier 1B, launched in October 1960. The world's first active repeater satellite, it recorded signals sent from the ground and then retransmitted them.

TECHNOLOGY

SATELLITE DESIGN

Getting complex machines to work in space requires a number of innovations. One of the worst problems is the temperature difference between sunlight and shade, which can cause components to expand, contract, and eventually fracture. Silvered insulation can help by deflecting much of the Sun's radiation, while highly conductive heat pipes can carry heat from hot areas to cooler ones. Delicate electronics must be cushioned with internal filling against the stress of launch and must also be robust enough to cope with flying through a blizzard of radiation and charged particles. Most satellites and probes also need stabilization and steering systems.

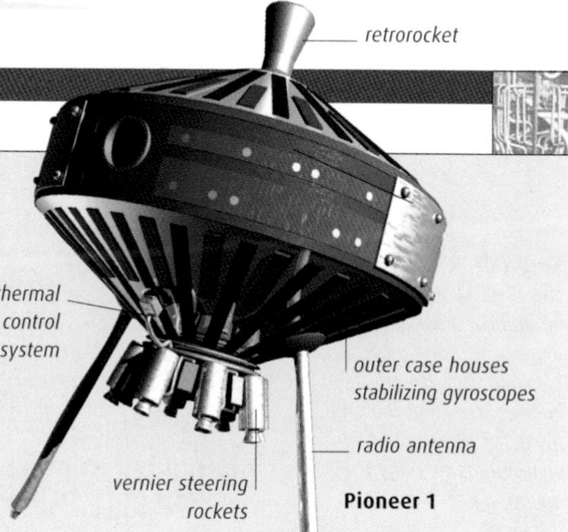

Pioneer 1

17 August 1958
An unnamed US Air Force spaceprobe intended to orbit the Moon is destroyed within two minutes of launch.

11 October 1958
Pioneer 1 fails to escape Earth orbit en route to the Moon.

2 January 1959
The Soviet Union launches Luna 1, also known as Mechta. It misses the Moon but becomes the first artificial object to leave Earth orbit.

3 March 1959
Following three launch failures, Pioneer 4 escapes Earth orbit but misses its flyby of the Moon by 60,000km (36,000 miles) and does not return any data.

12 September 1959
Luna 2 is launched.

14 September 1959
Luna 2 crashes successfully into the Moon.

4 October 1959
Luna 3 is launched.

7 October 1959
Luna 3 travels behind the Moon and returns the first images of the lunar far side.

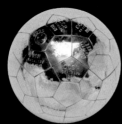

CARGO TO THE MOON
Both Luna 1 and Luna 2 carried a payload of small symbolic objects onboard, including "footballs" of pentagonal pennants designed to explode on impact and scatter their emblems over a wide area.

First to reach the Moon

The Earth's natural satellite, the Moon, sits on our cosmic doorstep and was an obvious first stop in the exploration of the Solar System. But in the 1950s, reaching it was still a major technological challenge.

Considering the Soviet Union's clear lead in the Space Race at the start of 1958, it is somewhat surprising that history's first moonshots were in fact launched from Cape Canaveral. The first attempt, in August 1958, was part of the United States' programme for the IGY. It made use of a Thor Able-I rocket – the Air Force's new Thor IRBM with the second and third stages of a Navy Vanguard bolted on top of it. Unfortunately, the first stage failed just 77 seconds into the flight.

The following month, the newly formed NASA took over the lunar programme and immediately attempted a repeat of the earlier launch. This probe, named Pioneer 1, got a little further than its unnamed sibling, but a programming error meant that it still did not achieve sufficient speed to escape the Earth's gravity and reach the Moon – instead it arced 113,584km (70,549 miles) above the Earth before falling back to crash in the Pacific Ocean.

Further attempts at lofting a probe towards the Moon also met with failure. Pioneer 2 fell swiftly back to Earth when one of its rocket stages failed, and Pioneer 3, which used the Army's Juno II launch vehicle (a modified version of the finished Jupiter IRBM), suffered a similar fate. At least onboard radiation detectors on Pioneers 1 and 3 helped to define the extent of the Van Allen Belts.

The Lunik probes

The Soviet Union might have appeared inactive during these American launch attempts, but there was a lot going on in secret. The State Commission had approved a series of Moon probes in March 1958, but with little compromise on the size and

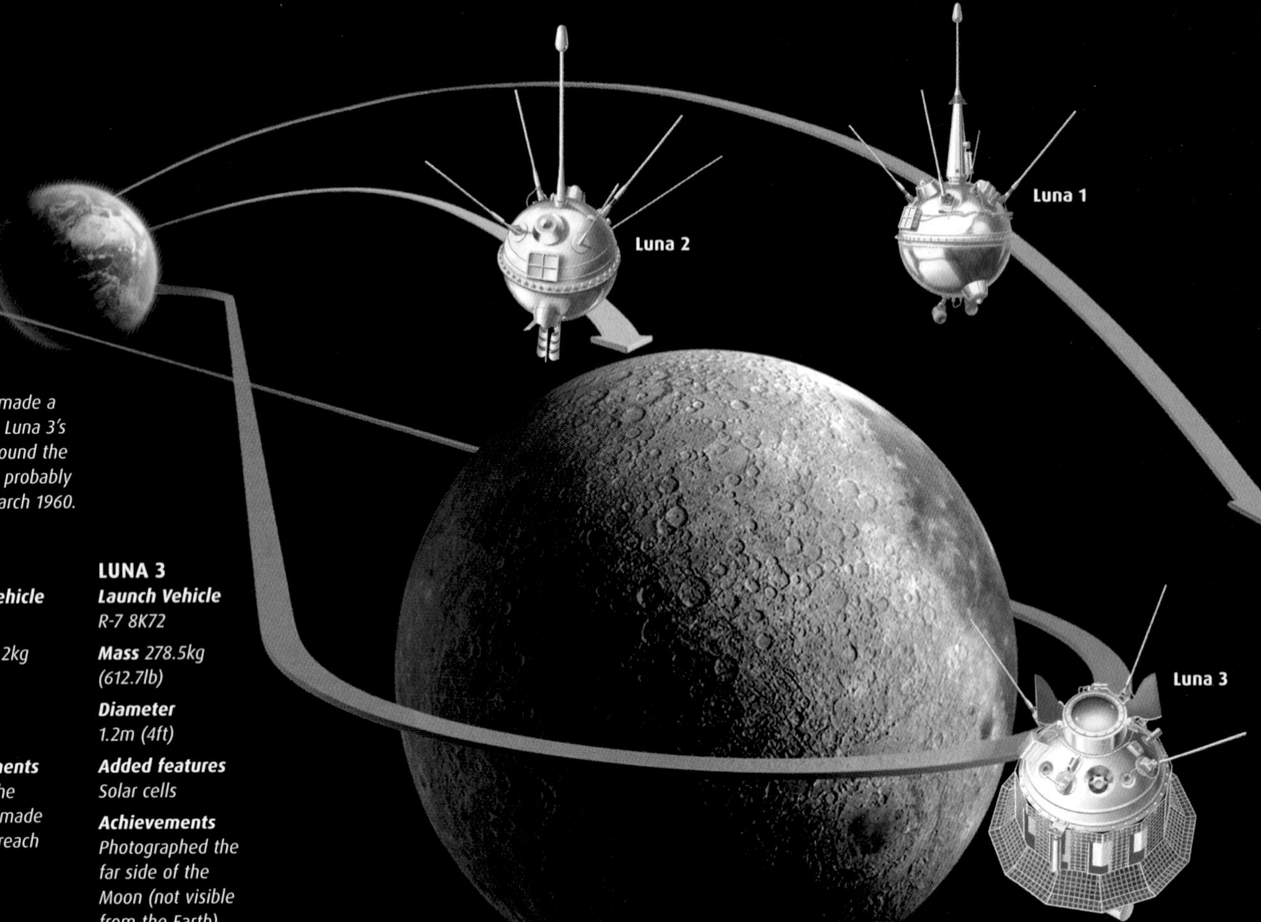

ROUTES TO THE MOON
While the first two Luna probes made a simple flyby and a crashlanding, Luna 3's orbit swung it in a figure of 8 around the Moon and back towards Earth. It probably re-entered the atmosphere in March 1960.

LUNA 1
Launch Vehicle
R-7 8K72
Mass *361kg (795.8lb)*
Diameter
0.9m (3ft)
Achievements
Failed to impact with the Moon but became the first spacecraft to fall into orbit around the Sun

LUNA 2
Launch Vehicle
R-7 8K72
Mass *390.2kg (854.4lb)*
Diameter
0.9m (3ft)
Achievements
Became the first man-made object to reach the Moon

LUNA 3
Launch Vehicle
R-7 8K72
Mass *278.5kg (612.7lb)*
Diameter
1.2m (4ft)
Added features
Solar cells
Achievements
Photographed the far side of the Moon (not visible from the Earth)

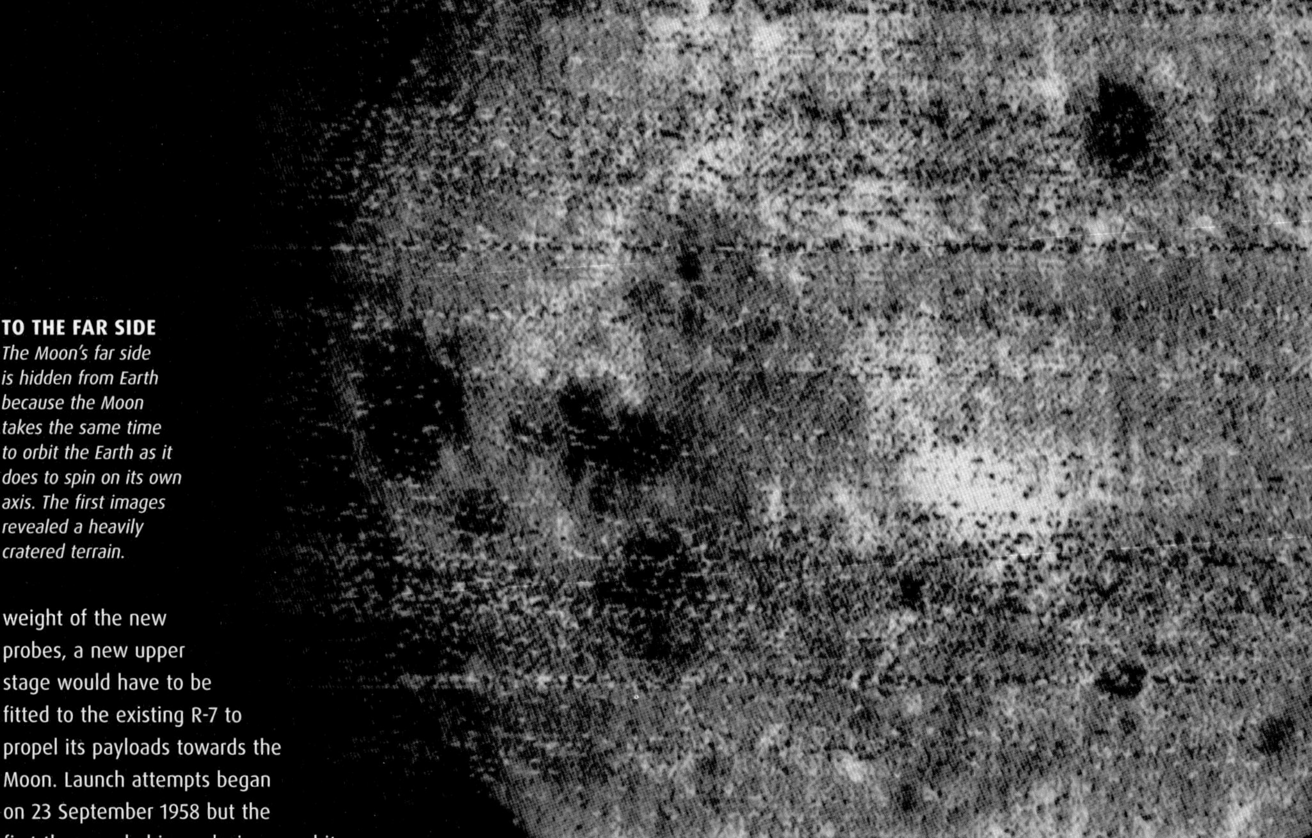

TO THE FAR SIDE
The Moon's far side is hidden from Earth because the Moon takes the same time to orbit the Earth as it does to spin on its own axis. The first images revealed a heavily cratered terrain.

weight of the new probes, a new upper stage would have to be fitted to the existing R-7 to propel its payloads towards the Moon. Launch attempts began on 23 September 1958 but the first three ended in explosions, and it was not until 2 January 1959 that Luna 1, also known as Mechta (Dream) and popularly nicknamed Lunik, broke free of the Earth's gravity and sailed past the Moon at a distance of 5,995km (3,723 miles), some 34 hours after launch. The spacecraft had been intended to crash directly into the Moon, but a control failure saw the mission change to a not-so-close flyby. However, the Soviet Union was still able to announce another triumph – and its achievement became all the more apparent in March, when Pioneer 4, the first semi-successful US moon probe, missed its target by 60,000km (36,000 miles) on its way into interplanetary space.

Throughout the remainder of 1959, the Soviets made the Moon their own. Despite further problems with the R-7 rocket, they successfully crashed Luna 2 into the Moon on 14 September. Venturing beyond the Van Allen Belts, this probe also discovered the solar wind – the stream of particles constantly blowing away from the Sun.

One final coup for 1959 came on 7 October, when Luna 3 successfully swung past the Moon and used an ingenious electronic camera system (see panel, right) to send back the first images of the permanently hidden lunar far side.

UNDER CONSTRUCTION
The Luna probes were all manufactured at Korolev's OKB-1 design bureau. Luna 3, seen here, was the most complex of the early designs, incorporating a camera and additional solar cells to power it.

TECHNOLOGY

IMAGING THE MOON'S DARK SIDE

The Yenisey-2 "phototelevision" system carried onboard Luna 3 was the most sophisticated camera sent into space up to that time – a flying photo booth and fax machine combined. As the Moon's sunlit far side came into view, a light-sensitive cell triggered the exposure of its temperature- and radiation-resistant photographic film. These were automatically processed onboard, and once the probe was back in contact with Earth, a second radio command triggered the transmission system. This involved a cathode-ray tube that shone light through the film onto a photoelectric sensor that produced a signal proportionate to the transparency of the film. As the cathode-ray tube scanned back and forth across each picture, the varying signal sent back to Earth allowed the reconstruction of 17 low-resolution, but nevertheless ground-breaking, images of the mountainous far side of the Moon.

1955 1956 1957 1958 1959 **1960** 1961 1962 1963 1964 1965 1966 1967 1968 1969 1970 1971 1972 1973 1974 1975 1976 1977 1978 1979 1980 1981 1982 1983 1984 1985 1986 1987 1988 1989 1990 1991 1992 1993 1994 1995 1996 1997 1998 1999 2000 2001 2002 2003 2004 2005 2006 2007 2008 2009 2010 2011 2012 2013 2014 2015 2016 2017 2018 2019 2020

11 March 1960
Pioneer 5 becomes the first spacecraft deliberately launched into interplanetary space.

October 1960
Two attempted Soviet missions to Mars, timed to coincide with Soviet premier Nikita Khrushchev's visit to the United States, fail during launch.

4 February 1961
The first Soviet attempt to reach Venus fails to leave Earth orbit and is concealed with the codename Sputnik 7.

12 February 1961
Venera 1 is successfully launched toward Venus, but contact is lost after just a week.

22 July 1962
The Venus probe Mariner 1 fails when its Atlas-Agena B launch vehicle veers off course.

27 August 1962
Mariner 2 is successfully launched towards Venus.

1 November 1962
The Soviets finally launch a spaceprobe, Mars 1, toward Mars.

14 December 1962
Mariner 2 executes the first successful flyby of Venus.

21 March 1963
Mars 1 loses contact with Earth.

5 November 1963
Mariner 3 fails to escape Earth orbit.

2 April 1964
The Soviet Zond 1 probe is launched towards Venus but once again contact is lost during its journey.

28 November 1964
Mariner 4 is successfully launched toward Mars.

14 July 1965
Mariner 4 makes the first successful flyby of Mars.

On to the planets

The next great targets for the racing space powers were the Earth's neighbouring planets, Venus and Mars – but they would soon learn that each presented its own unique challenges.

The Soviet Union already had an advantage in this new heat of the Space Race – the same powerful launch vehicles that allowed it to put larger satellites into orbit also allowed it to launch smaller payloads further away from the Earth and, with the addition of extra rocket stages, out of Earth's gravity altogether.

Unlike orbital or even lunar launches, however, sending spacecraft to the planets relies on the calendar. As Venus, Earth, and Mars each orbit the Sun at a different rate, the distance between them is constantly changing, and the only practical period to launch a spaceprobe is around the time of their closest approaches to the Earth, typically within a brief launch window of perhaps just a few weeks.

First attempts

Although both the US and USSR's first forays into interplanetary space had been accidental, the result of missed flybys of the Moon, the end of the 1950s saw NASA turn its Pioneer programme into a series of deliberate missions to investigate conditions away from Earth (see panel, opposite).

Meanwhile, Sergei Korolev was given approval to develop a series of planetary probes in late 1959, with the intention of sending a probe to Mars during the close approach late the following year. As the launch window approached, it became clear that it would coincide with a visit by Soviet premier Khrushchev to the United States – the ideal opportunity for another Soviet spectacular.

But for the first time, luck was against the Soviets – in mid-October 1960, two attempts to launch probes on a Martian flyby mission ended in rocket failures. These were hushed up, but rumours leaked out, along with tales of a third, disastrous rocket explosion that was for a long time seen as another failed attempt to reach Mars. The reality was rather different and for a long time remained one of the Soviet missile programme's darkest secrets (see p.64).

A few months later, around a close approach of Venus, the Soviets launched the first of their Venera probes. An initial failure was again covered up – it never left Earth orbit after its upper rocket stage failed – but the second attempt was on its way to Venus and trumpeted as another Soviet triumph when communications, embarrassingly, went dead.

INCOMING DATA
Scientists at the Jet Propulsion Laboratory in Pasadena, builders of the Mariner probe series, pose with a stream of Venusian data in front of a chart showing Mariner 2's progress.

PLANETARY VISIONS
Until the 1960s, most people's ideas of other planets were inspired by the paintings of the visionary American artist Chesley Bonestell. His most famous work, Saturn Viewed From Titan, *is shown here.*

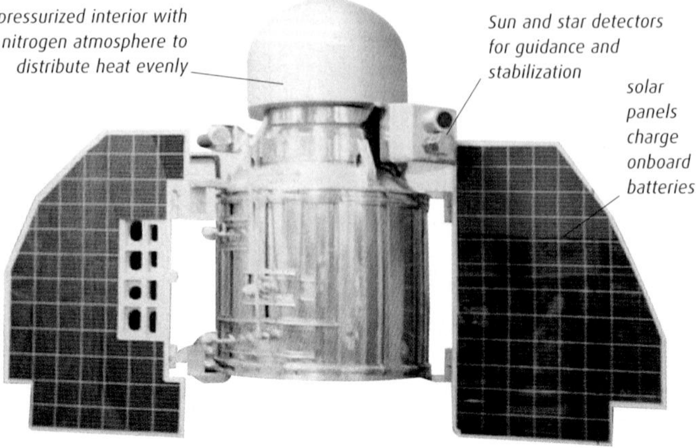

VENERA 1
Far more sophisticated than any previous spaceprobe, Venera 1 stood 2m (6½ft) tall, weighed some 643.5kg (1,416lb), and incorporated a variety of scientific instruments.

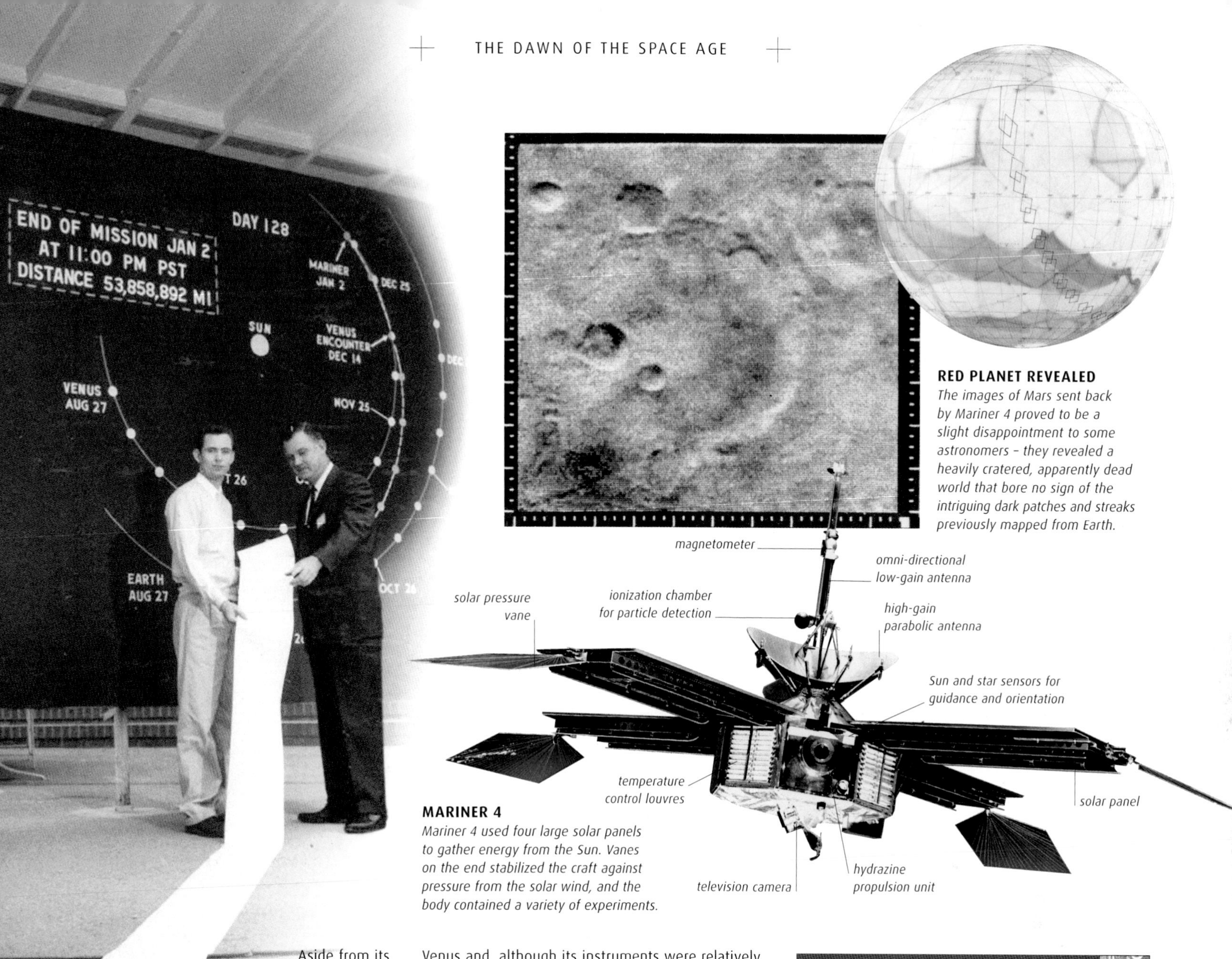

RED PLANET REVEALED
The images of Mars sent back by Mariner 4 proved to be a slight disappointment to some astronomers – they revealed a heavily cratered, apparently dead world that bore no sign of the intriguing dark patches and streaks previously mapped from Earth.

MARINER 4
Mariner 4 used four large solar panels to gather energy from the Sun. Vanes on the end stabilized the craft against pressure from the solar wind, and the body contained a variety of experiments.

Aside from its Pioneer probes, NASA was lagging behind Soviet efforts, and missed the 1960 and 1961 launch windows. By the following year, however, the Americans were ready to launch the first wave of their own armada to the planets – the Mariner probes. Both sides followed a deliberate policy of building probes in pairs so that two could be sent in the same launch window. But while the Soviets found this just doubled their failure rate, for NASA the story was "second time lucky".

So when Venus-bound Mariner 1 was destroyed after its launch vehicle (based on the Atlas ICBM) veered off course, Mariner 2 was ready to go in a matter of weeks – and this time, finally, everything went right for an interplanetary probe. Three-and-a-half months later, Mariner 2 made the first flyby of Venus and, although its instruments were relatively simple, it revealed that the planet has a baking surface, a dense, choking atmosphere, and a rotation period (243 Earth days) longer than its year.

At the next close approach of Mars, history repeated itself. Mariner 3 failed to leave Earth orbit after its protective shroud jammed, but on 28 November 1964, Mariner 4 launched perfectly. After a seven-month flight, the probe flew past Mars on 14 July 1965, coming within 10,000km (6,000 miles) of the Red Planet. It sent back 21 precious pictures of the surface and revealed that Mars lacks a magnetic field and had a much thinner atmosphere than expected.

Throughout the 1960s, the Mariner probes continued to deliver for NASA – Mariner 5 would fly past Venus in 1967, and Mariners 6 and 7 made close flybys of Mars in July and August 1969. The Soviets, meanwhile, had to endure a long learning curve before their Venera probes began to produce results (see pp.266–67).

TECHNOLOGY

PROBING INTERPLANETARY SPACE

Following the success of Explorer 1 (shown here), NASA began work on the Pioneer series of probes, launched throughout the course of the 1960s. The first of these were relatively simple probes that targeted the Moon, but in March 1960 Pioneer 5 was deliberately sent into orbit around the Sun between the orbits of Earth and Venus. For the first time, this gave NASA experience in communicating with a spacecraft over interplanetary distances, and the probe paved the way for Pioneers 6 through 9. These more sophisticated probes, launched between December 1965 and November 1968, were designed to operate for up to 180 days each, but they actually operated for several years, allowing scientists to monitor conditions across the inner Solar System.

1961
The French CNES space agency is established.

26 April 1962
Ariel 1, a US-built satellite carrying British experiments, is launched.

1964
The British government approves development of the Black Arrow satellite launcher.

26 November 1965
France successfully launches the A-1 Astérix satellite, powered by a Diamant A rocket.

6 December 1965
A NASA Scout rocket launches the French FR 1 scientific satellite.

28 June 1969
The first test launch of Britain's Black Arrow ends in failure.

10 March 1970
Following development of the launch centre at Kourou, French Guiana, France launches the first Diamant B rocket.

29 July 1971
Britain announces the cancellation of Black Arrow.

28 October 1971
The last flight of Black Arrow successfully orbits Britain's Prospero satellite.

4 April 1972
A Soviet rocket launches the French-built SRET satellite.

Britain and France in space

As the Space Race between the superpowers gathered pace, other countries began to recognize the importance of an independent launch capability. First among these were Britain and France, but the two space programmes would develop in very different directions.

Britain and France both inherited small parts of the German V-2 legacy at the end of the Second World War, and enthusiastic engineers (many of whom had started out in rocket societies) were soon at work replicating and learning from the technology. As with the superpower space programmes, these projects were staffed by space enthusiasts, even if they were driven by military priorities.

Hopes unfunded

In 1954, Britain began work on its own intermediate-range ballistic missile, the Blue Streak, with assistance from Australia (Woomera in the Australian outback was to be used as a testing range). Despite early successful tests, spiralling costs and doubts about Blue Streak's military effectiveness eventually led to the cancellation of the programme in 1960.

But Blue Streak lived on in other forms for another decade. Britain at first tried to interest Canada and Australia in a collaboration to build a three-stage launch vehicle called Black Prince, with Blue Streak as its first stage. When no deal could be reached, the missile became a crucial element in the ill-fated European ELDO project (see p.229). In parallel with Blue Streak, Britain had also been developing a smaller sounding (or research) rocket, Black Knight – initially to test warhead designs for use on Blue Streak. Uniquely, Black Knight utilized high test peroxide (HTP), a concentrated form of hydrogen peroxide, as an oxidizer. First used in some German experimental rocket engines, HTP's violently combustible nature allowed the design of Black Knight's Gamma engines to be greatly simplified.

By the 1960s, Britain led the world in HTP propulsion, and in 1964 the Royal Aircraft Establishment proposed building a small, cheap satellite launch vehicle based largely around the existing Black Knight technology. Black Arrow, as it became known, was approved by the British government, but suffered from repeated delays to its funding. Finally given the go-ahead, test launches began

AUSTRALIA-BOUND
Although the Blue Streak missiles were manufactured in Britain, the lack of a suitable launch site meant that they had to be shipped around the world from the Spadeadam Rocket Establishment in Cumbria (shown here) to the Australian missile range at Woomera for testing. Transport delays to both Blue Streak and Black Arrow added to the woes of the British space effort.

DESERT LAUNCH PAD
A Blue Streak missile is shown here ready for launch at Woomera. Despite successful tests, the British government abandoned the project in 1960, to the annoyance of their Australian partners.

in 1969, but a troubled start to the programme meant that, with astounding lack of foresight, the government cancelled the project in July 1971, on the eve of its greatest success (see panel, below). In the future, it seemed, Britain would be content to rely on small American rockets to launch its satellites. But soon British satellites themselves would be a thing of the past, and from the 1970s, Britain's space effort would generally extend no further than its limited involvement in the new European Space Agency, ESA.

France enters space

While Britain's rocket programme ultimately failed due to a lack of political vision, the French effort thrived on Gallic self-assurance. In the late 1940s, French scientists, helped by a few of their captured German counterparts, formed plans to turn the A9, von Braun's planned successor to the V-2, into reality. While these ideas proved overly ambitious, they still led to the successful Veronique series of sounding rockets, first launched in 1950. In parallel with these efforts, France developed a series of ballistic missile prototypes, each named after a precious stone.

In 1961, France's newly created space agency, the Centre National d'Etudes Spatiales (CNES), decided to build a satellite launch vehicle. Diamant, as it became known, was evolutionary rather than revolutionary – a tried and tested two-stage Saphir missile with a new third stage added. The first Diamant A launch took place at Hammaguir, Algeria, in November 1965 and was an unqualified success, launching a small satellite, the A-1 or Astérix, into orbit. This made France only the third nation on Earth to launch its own satellite. Diamant launches continued alongside French involvement in the ELDO project, but in the late 1960s France also developed the improved Diamant B launcher and a new launch complex at Kourou in French Guiana. Diamant B first launched from Kourou in March 1970, and the programme continued until 1975. When France ultimately abandoned its national launch vehicle, it was only to take the lead role in ESA's more ambitious Ariane project.

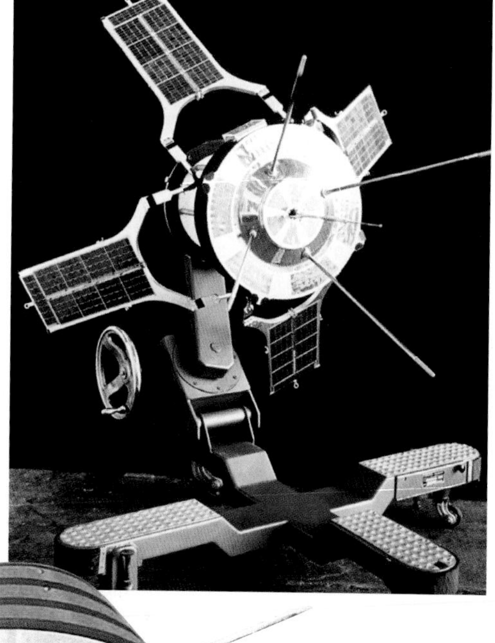

ASTÉRIX AND DIAPASON
The first French satellite, Astérix (left), was little more than an orbiting radio transmitter. More advanced was Diapason (above), which incorporated a modified transmitter for measuring the speed of the satellite and therefore the Earth's varying gravity.

DESERT LIFT-OFF
A Diamant A rocket blasts off from the original French launch site at Hammaguir in the Algerian desert in 1965.

THE ARIEL SERIES
Britain's first satellites were the Ariel series, launched by NASA on Delta and solid-fuelled Scout rockets between 1962 and 1979. Ariels 1 and 2 were built in the US and fitted with British experiments, while Ariels 3 and 4 were entirely British-built. Each studied galactic radio signals and the properties of the Earth's ionosphere. The later Ariels 5 and 6, in contrast, were two of the earliest X-ray astronomy satellites.

HISTORY FOCUS

BRITAIN'S SPACE SWANSONG

The first Black Arrow test flight in June 1969, codenamed R0, veered off course due to a steering problem, but a successful second test eight months later (right) cleared the way for a satellite launch attempt in September 1970. Unfortunately this R2 launch failed to reach orbit when the second stage shut down early. Despite the project's cancellation in July 1971, permission was given to launch the R3 vehicle. With just one chance to get it right, the launch slipped to late 1971, but this time everything went perfectly, and the Prospero satellite entered orbit on 28 October 1971. Despite this, there was to be no reprieve, and Britain gained the distinction of being the first country to abandon its satellite launch capability.

The birth of Vostok

With their lead in the Space Race now established, the Soviet engineers turned their attention to the next great challenge – developing a vehicle suitable for manned spaceflight.

Although Mikhail Tikhonravov had first sketched out plans for manned spacecraft in the late 1940s, it was not until the mid-1950s that the Council of Chief Designers began to discuss the idea seriously. By 1955, some five different spacecraft designs were under consideration, and proposals for a series of suborbital spaceflights launched with the R-5 rocket got as far as recruiting volunteer cosmonauts, before being shelved as work on the R-7 ICBM took priority.

Off the drawing board

In early 1957, with plans for the first satellite launch well under way, Korolev created a new planning group at OKB-1, where talented young engineers would work on designs for a manned spacecraft that could be launched with the R-7. The kindergarten group, as it was known, developed a proposal for a two-element spacecraft, with a spherical Descent Module attached to the front of a conical Instrument Module, similar in design to Tikhonravov's Object D satellite (Sputnik 3). The Instrument Module would be abandoned in orbit, and the cosmonaut would stay seated in the Descent Module throughout the flight, ejecting and parachuting to the ground at the last moment.

Originally known as Object OD-2, the spacecraft was soon given the more evocative name of Vostok (East). By the time Korolev gave it his seal of approval in June 1958, the design had evolved, with plans for both manned and unmanned versions (in the unmanned version, later known as Zenit, the Descent Module would carry reconnaissance equipment). In May 1959, after some fierce political wrangles over which element of the programme took priority, the State Commission finally authorized a production schedule that aimed to achieve the first manned spaceflight by late 1960.

Fate had other ideas, however – test launches got going in May 1960, but after a series of setbacks to other aspects of the space programme in late 1960 (see pp.54, 64–65) it was to be the spring of 1961 before Korolev was allowed to risk a manned launch.

LIFTING FOR LAUNCH
Like all rockets launched at Tyuratam, the R-7s were transported horizontally along a purpose-built railway system and raised to point skywards at the launch pad.

IN PRODUCTION
The production-line system established for the Vostok capsules has remained largely unchanged to this day. Here engineers work on fitting a Vostok spacecraft into its launch shroud.

TECHNOLOGY

BAIKONUR COSMODROME

The main Soviet launch centre, the Baikonur Cosmodrome, was actually located near the village of Tyuratam, 370km (230 miles) from the town of Baikonur – the facility's location was deliberately misrepresented in an effort to confuse Western intelligence. Construction got under way in mid-1955 at a frenetic pace – a small army of construction workers and engineers was transported to a remote site in the deserts of Kazakhstan (then a Soviet Republic), where they developed facilities for the preparation and launch of rockets. Meanwhile, the railway from Tyuratam itself was improved and extended to carry rockets and other equipment transported from the factories and design bureaux around Moscow.

19 May 1960
The first unmanned test flight of the Vostok capsule, disguised as Sputnik 4, becomes stranded in orbit.

19 August 1960
A second Vostok test, Sputnik 5, carries dogs Strelka and Belka into space and successfully returns them to Earth after a day in orbit.

1 December 1960
Sputnik 6 is launched into orbit but burns up on re-entry the next day.

9 March 1961
Sputnik 9, carrying a dummy cosmonaut and a dog named Chernushka, makes a successful test flight.

25 March 1961
Another successful test flight, this time under the name Sputnik 10, clears the way for a manned launch by the Soviets.

STAR OF THE EAST
Sited in central Kazakhstan, well away from foreign borders, the Tyuratam launch site covers an area of 6,721 square km (2,593 square miles). A further 104,279 square km (40,262 square miles) of land downrange of the launch facilities was cleared in case of any rocket failures.

1956 1957 1958 1959 1960 1961 1962 1963 1964 1965 1966 1967 1968 1969 1970 1971 1972 1973 1974 1975 1976 1977 1978 1979 1980 1981 1982 1983 1984 1985 1986 1987 1988 1989 1990 1991 1992 1993 1994 1995 1996 1997 1998 1999 2000 2001 2002 2003 2004 2005 2006 2007 2008 2009 2010 2011 2012 2013 2014 2015 2016 2017 2018 2019 2020

August 1959
Cosmonaut selection teams visit air bases throughout the Soviet Union, searching for candidate astronauts among their top pilots.

March 1960
Construction work begins on the training complex at Star City.

11 June 1960
The cosmonaut training centre at Star City is established. Korolev has already chosen his shortlist of candidate Vostok pilots.

18 June 1960
The 20 successful cosmonaut candidates visit Kaliningrad to see the Vostok capsule under development.

19 September 1960
Korolev submits a formal proposal for a manned spaceflight for official approval by the Communist Party's Central Committee.

6 January 1961
A shortlist of six candidates for the first manned Vostok flight is drawn up.

March 1961
Gherman Titov and Yuri Gagarin are selected as the two potential pilots for Vostok 1.

Cosmonaut training

With the development of a manned Soviet space capsule finally approved in May 1959, the search for suitable cosmonauts could finally get underway.

Although the formal search for cosmonaut candidates did not begin until late 1959, there had already been some discussion of the qualities needed in a good cosmonaut, and, just as in the United States (see p.70), the selectors soon realized that jet pilots were most likely to meet the basic requirements.

In August 1959, a group of experts in aeronautical medicine began visiting air bases across the Soviet Union and interviewing candidates to fly what they mysteriously described as a "completely new type of aircraft". It was established whether the pilots had any interest in space travel, though other important selection criteria were as mundane as height and weight – many were rejected simply because they were too tall or too heavy.

Out of about 3,000 interviewees, the selection panel drew up a list of 102 potential cosmonauts, who were then sent for intensive and sometimes harrowing medical testing. Aside from numerous X-rays and physiological tests, the experts assessed their psychological well-being and ability to cope with stress and isolation. The most difficult challenge was the isolation chamber, where candidates would live and work for several days at a time, subjected to a cycle of day and night determined at the whim of the operators, with long periods of silence punctuated by occasional deafening noise. By the end of this process, the shortlist of candidates had dwindled to 40, but as the deadline approached this was further reduced to an initial wave of 20.

CRUDE BUT EFFECTIVE
Gherman Titov spins on apparatus used to familiarize cosmonauts with rapid acceleration. The Soviet cosmonaut programme also used rapidly spinning centrifuges for training.

Into training

When the candidate cosmonauts arrived in Moscow for training, they found themselves in the custody of an intimidating figure – the newly appointed Head of Cosmonaut Training Nikolai Kamanin (see panel, opposite). Kamanin's regime was a mixture of tough physical exercise, academic lectures, and practical training. At first, the lectures focused almost entirely on the biomedical aspects of flight, but this had little appeal for a group that mostly had engineering backgrounds, and so Kamanin and Korolev brought in engineers from OKB-1 to talk about spacecraft and launch-vehicle design, orbital mechanics, and astronomy. Training exercises included parabolic flights to simulate weightlessness, ejector seat tests, countless parachute jumps, and long periods inside the isolation chamber. In addition, the cosmonauts had to get used to the newly designed spacesuits,

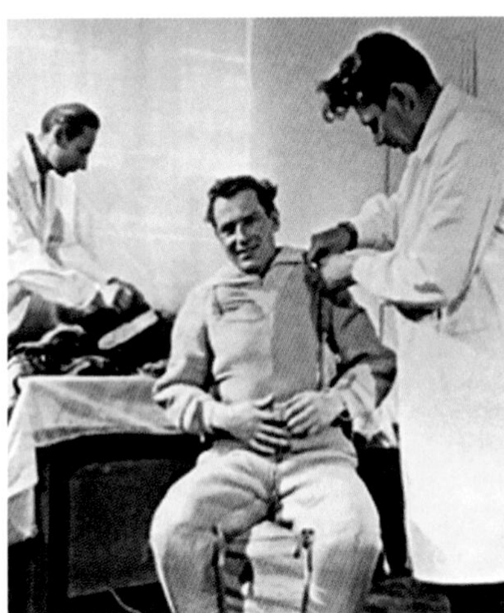

SUITING UP
A trainee is fitted with the underlayers of his spacesuit. Soviet suits were developed from those used in high-altitude balloon flights.

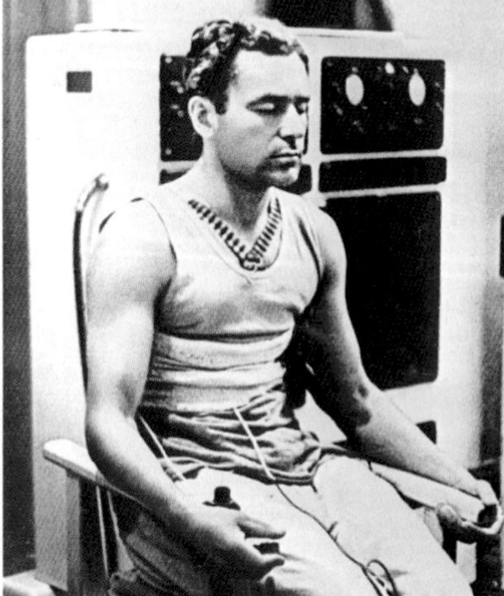

WIRED FOR TESTS
Gherman Titov concentrates during medical tests – trainees were often subjected to violent vibrations or extremes of temperature.

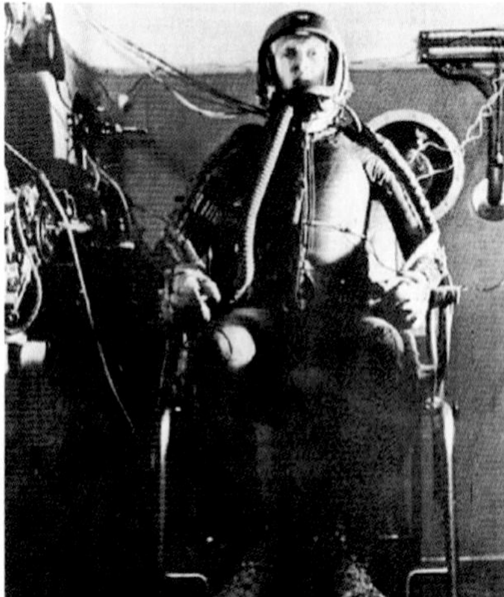

UNDER PRESSURE
In the altitude pressure chamber, cosmonauts were subjected to very low air pressures to test how they fared at high altitude.

FIRST COSMONAUTS
A rare group photograph captures the first nine Soviet spacefarers at Star City in 1964. At rear, left to right, are the Sputnik cosmonauts Bykovsky, Titov, Gagarin, Nikolayev, and Popovich. In front are Boris Yegorov, Konstantin Feoktistov, and Vladimir Komarov (the crew of Voskhod 1), along with Valentina Tereshkova.

modified from high-altitude pressure suits, which they would have to wear throughout the flight. At first the trainees were based at a Moscow airfield, but by June new accommodation was ready at the specially built town of Zvezdny Gorodok (Star City), just outside Moscow. The new facilities included a flight simulator for the Vostok spacecraft itself, but with equipment limited it was decided to focus on an even smaller number of cosmonauts.

The lucky members of this "group for immediate preparedness", named on 30 May 1960, were Yuri Gagarin, Anatoly Kartashov, Andrian Nikolayev, Pavel Popovich, Gherman Titov, and Valentin Varlamov. Kartashov and Varlamov were invalided out after accidents, to be replaced by Valery Bykovsky and Grigori Nelyubov. Gagarin and Titov soon emerged as Korolev's favourites – though he took to the entire group and called them "my little swallows".

Delays and setbacks

Even as the cosmonauts were training and the Vostok capsules rolling off the production line, rows continued about whether and when a manned launch should go ahead. The military continued to lobby for concentration on the unmanned Vostok-based spy satellites at the expense of what they saw as the publicity stunt of human spaceflight.

Meanwhile, two early attempts to launch a Mars probe failed (see pp.54–55) and a third ended in the disaster known as the Nedelin catastrophe (see p.66). There was also concern about the biological effects of spaceflight after Belka, one of two dogs carried aboard Sputnik 5, was sick in orbit. Contrary to rumours spread in the West about Soviet recklessness, Khrushchev and his engineers were very unwilling to risk a cosmonaut until they could expect to get him back in one piece.

With a manned launch planned for the spring of 1961, there was one further tragic setback to come. On 23 March, the youngest of the trainees, 24-year-old Valentin Bondarenko, died when a fire broke out in the oxygen-rich atmosphere of the isolation chamber. It was just days before the expected launch window opened, and the leading cosmonauts were not immediately told what had happened. A week later, on 30 March, final authorization for a launch was agreed.

BIOGRAPHY

NIKOLAI KAMANIN

General Nikolai Kamanin (1908–82) was a highly respected and formidable figure when made Head of Cosmonaut Training in 1960 (a role he held until 1971). After training as a pilot and an early career as a polar explorer, he became famous in 1934 for his role in the rescue of passengers and crew from an ice-bound steamship trapped in the Arctic Ocean – an event which saw him named a Hero of the Soviet Union. At Star City, he was disliked by many of the cosmonauts for his harsh regime and manner, but he had some progressive ideas – he was keen to train a group of women for space, and supported calls for civilian cosmonauts.

Vostok spacecraft

Designed by Mikhail Tikhonravov's "kindergarten" team, the first manned spacecraft combined a spherical manned Descent Module with an unmanned Instrument Module and retrorocket unit. Unmanned versions were flown under cover of Sputnik launches 4 onwards (sometimes known as korabl sputniks, from the Russian for "ship"), and the manned spacecraft was launched half-a-dozen times. Although plans for later manned flights were scrapped, unmanned Vostok variants carrying reconnaissance cameras and other experiments continued to fly for three decades.

THE VOSTOK ROCKET
The lower stages of the Vostok rocket were identical to those of the reliable R-7 Semyorka rocket, with four engine blocks powered by an RD-107 engine, surrounding a central core powered by the RD-108. Here, one of the side blocks is being attached to the core during assembly. Both engines were powered by kerosene and liquid oxygen.

RD-107 ENGINE
The RD-107 engine, developed by Valentin Glushko, used a turbopump to feed combustion chambers similar to those on the V-2. Four such chambers fed the four main rocket nozzles, while two gimballed vernier engines on the side helped steer the rocket.

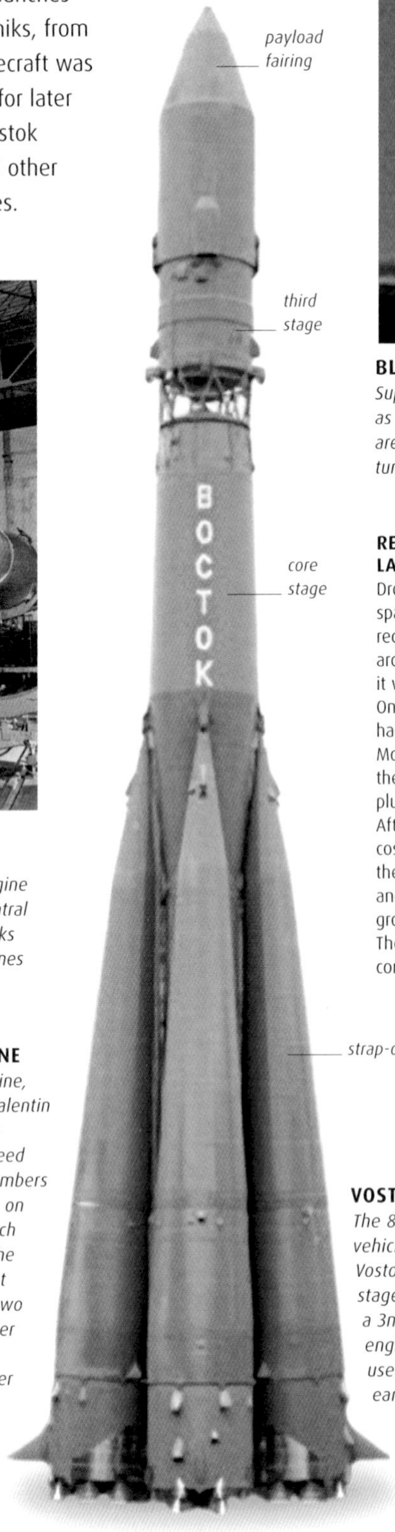

payload fairing

third stage

ВОСТОК

core stage

strap-on booster

VOSTOK ROCKET
The 8K72K launch vehicle developed for Vostok used the lower stages of an R-7 with a 3m (10ft) third-stage engine similar to that used to boost the early Luna probes.

BLAST-OFF FROM BAIKONUR
Support gantries fall away from each side of a Vostok rocket as it blasts its way into the sky. Flames from the exhaust are directed into the pit below, where they escape along tunnels so they cannot threaten the rocket.

RE-ENTRY AND LANDING SEQUENCE
Dropping the Vostok spacecraft out of orbit required turning it around in space so it was facing backwards. Once the retrorockets had fired, the Instrument Module separated and the Descent Module plunged towards Earth. After re-entry, the cosmonaut ejected from the Descent Module and parachuted to the ground separately. The Instrument Module continued in orbit.

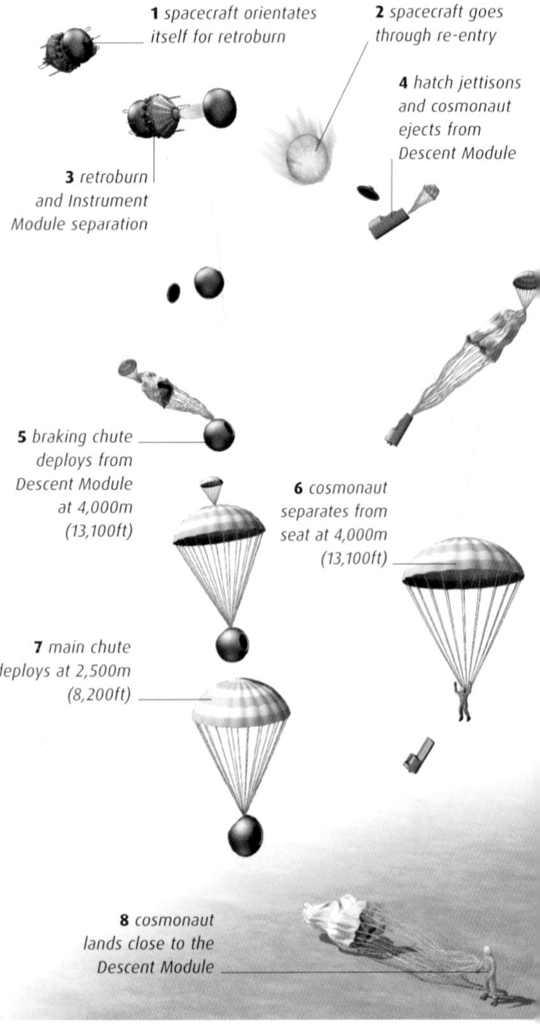

1 *spacecraft orientates itself for retroburn*

2 *spacecraft goes through re-entry*

3 *retroburn and Instrument Module separation*

4 *hatch jettisons and cosmonaut ejects from Descent Module*

5 *braking chute deploys from Descent Module at 4,000m (13,100ft)*

6 *cosmonaut separates from seat at 4,000m (13,100ft)*

7 *main chute deploys at 2,500m (8,200ft)*

8 *cosmonaut lands close to the Descent Module*

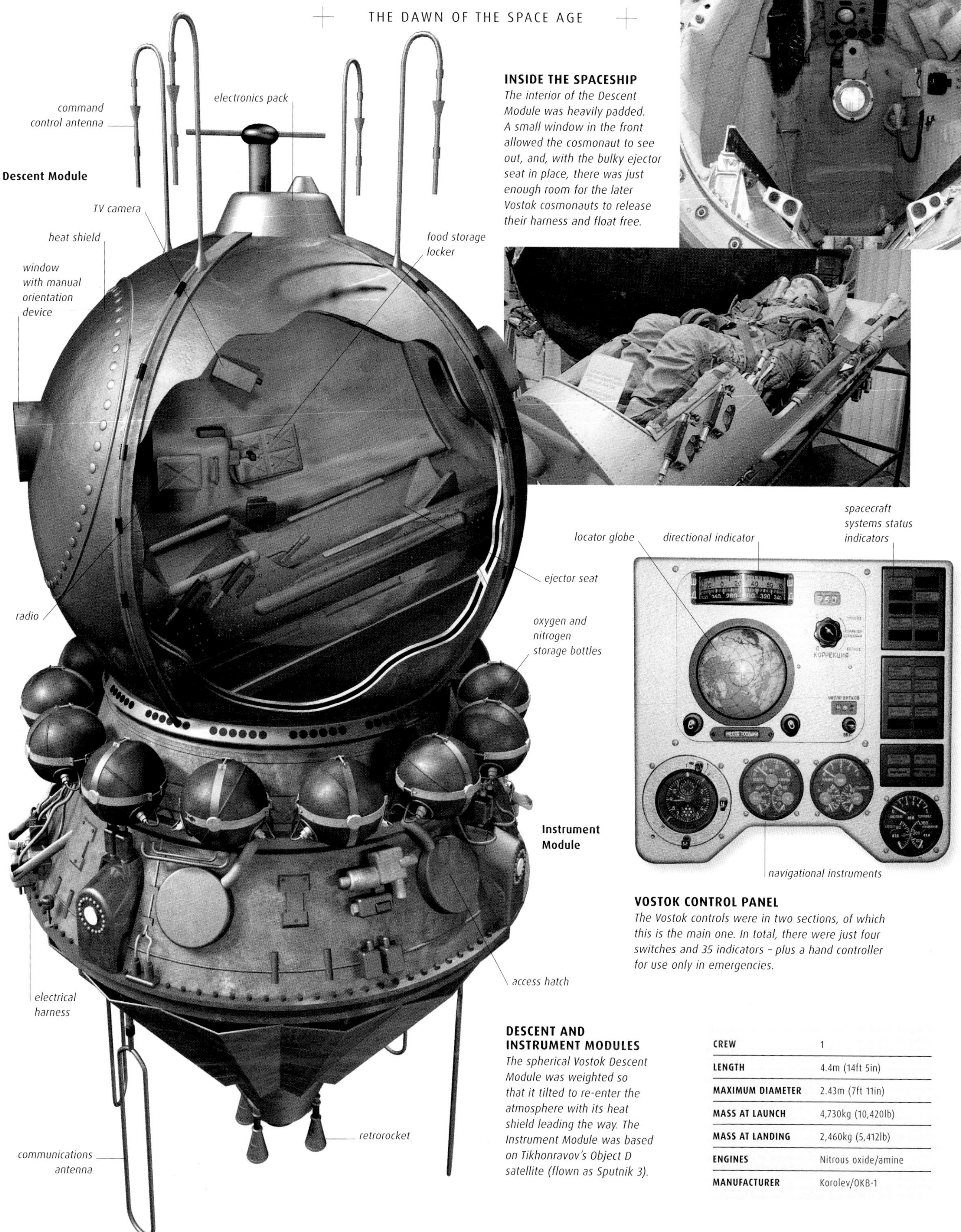

INSIDE THE SPACESHIP
The interior of the Descent Module was heavily padded. A small window in the front allowed the cosmonaut to see out, and, with the bulky ejector seat in place, there was just enough room for the later Vostok cosmonauts to release their harness and float free.

VOSTOK CONTROL PANEL
The Vostok controls were in two sections, of which this is the main one. In total, there were just four switches and 35 indicators – plus a hand controller for use only in emergencies.

DESCENT AND INSTRUMENT MODULES
The spherical Vostok Descent Module was weighted so that it tilted to re-enter the atmosphere with its heat shield leading the way. The Instrument Module was based on Tikhonravov's Object D satellite (flown as Sputnik 3).

CREW	1
LENGTH	4.4m (14ft 5in)
MAXIMUM DIAMETER	2.43m (7ft 11in)
MASS AT LAUNCH	4,730kg (10,420lb)
MASS AT LANDING	2,460kg (5,412lb)
ENGINES	Nitrous oxide/amine
MANUFACTURER	Korolev/OKB-1

EXPERIENCE

DISASTER ON THE LAUNCH PAD

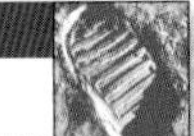

The Nedelin catastrophe

The explosion that rocked Baikonur Cosmodrome on 24 October 1960 was the greatest disaster in the history of rocketry, taking the lives of 126 Soviet space and missile personnel.

MITROFAN NEDELIN
After a proud military career that saw him fight against the fascists in the Spanish Civil War and command artillery in the Ukraine, all that remained to identify Marshal Nedelin was his gold star as a Hero of the Soviet Union.

It took many years for the truth about what became known as the Nedelin catastrophe to emerge – the Soviet state was habitually secretive, especially when it came to such sensitive matters. US spy satellites revealed that something had happened at the cosmodrome, but there was no way of knowing quite what – most analysts believed that, coming shortly after the failed launches of Mars 1 and Mars 2 and towards the end of the 1960 launch window for Mars, the explosion had been a disastrous third attempt to launch a Mars probe.

The reality was that the rocket at the heart of the inferno was a prototype Soviet ballistic missile, devised by Sergei Korolev's former deputy and now rival Mikhail Yangel. The disaster had been caused by the impatience of one man and a callous disregard for safety procedures. Once again, the Soviet Union had planned a spectacular, but this time things had gone very badly wrong.

TRIGGER SEQUENCE
Cameras designed to trigger when the missile's engines fired captured the explosion and its horrific aftermath frame by frame. By the end, only charred wreckage of the R-16 remained.

THE FIREBALL STRIKES
The explosion laid waste an area 120m (400ft) in diameter. Many died instantly, but others were trapped by the fence around the pad. The fire raged for two hours before it could be brought under control.

"Above the pad erupted a **column of fire**. In a daze we watched the flames burst forth **again and again until all was silent** … the bodies were in unique poses – all were without clothes or hair. It was **impossible to recognize anybody**."

Unnamed worker at a nearby observation point

"... cameras ... recorded the scene ... The men on the scaffolding dashed about in the fire and smoke; many jumped off and vanished into the flames."

Soviet nuclear physicist and dissident Andrei Sakharov, Memoirs, 1990

The rocket on the pad was the R-16 ICBM. While Korolev's R-7 had proved itself a formidable launch vehicle, it had been dismissed as a weapon of war, because it could not be stored with its fuel onboard. Yangel's R-16 was supposed to get around that, and it had been given top priority. Marshall Mitrofan Nedelin, head of Soviet Strategic Rocket Forces, took personal charge and was eager to launch the rocket before the anniversary of the Bolshevik Revolution, on 7 November. On 21 October, therefore, despite the protestations of many engineers who thought it was not ready for launch, the R-16 was moved to the launch pad at Baikonur's Site 41. By 24 October, all was not going well. Nedelin had already refused a request the previous day that the fuel should be drained and the rocket removed from the pad. When a further delay was announced, he insisted on going to the pad in person to see what was happening. With countless checks and procedures to run through in so short a space of time, something was more than likely to go wrong – and it did. At 18:45, the routine resetting of a timer in the command bunker caused the rocket's second stage to ignite, firing directly into the fuel-laden first stage below and creating a devastating fireball.

Mikhail Yangel survived only because he was having a cigarette break with some colleagues at the time of the explosion. The incident robbed him of his ambitions to play a key part in the space programme, but the R-16 did make it into space, just over three months later in February 1961.

"... **people ran** to the side of the other pad, towards the bunker ... but on this route was a strip of new-laid tar, which **immediately melted.** Many got stuck in the hot sticky mass and became victims of the fire."

Andrei Sakharov, 1990

"... a fire took place which caused the destruction of the tanks with components of the propellant. The **casualties numbered up to 100 or more people**, including fatalities – several dozen. Chief Marshal Nedelin was present ... now, the search for him is going on."

Mikhail Yangel notifies the Kremlin

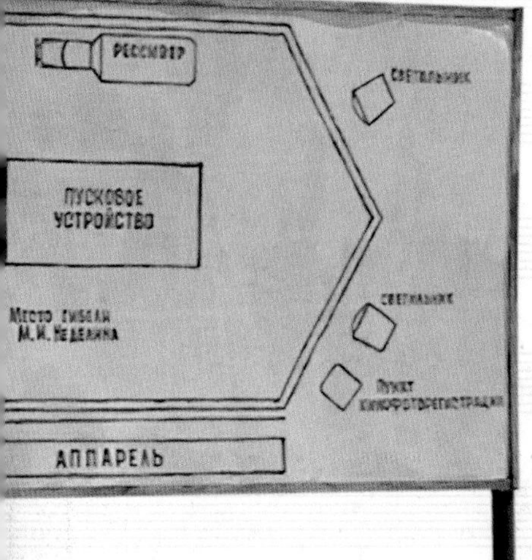

LAUNCH PAD LAYOUT
A present-day map shows how Site 41 looked at the time of the disaster. The R-16 stood at the centre of the hexagon, and the circle shows where Marshal Nedelin was sitting at the time of the explosion.

MEMORIAL TO THE DEAD
This memorial was only erected many years after the disaster but is traditionally visited by officials before every launch.

1955 1956 1957 **1958** 1959 1960 1961 1962 1963 1964 1965 1966 1967 1968 1969 1970 1971 1972 1973 1974 1975 1976 1977 1978 1979 1980 1981 1982 1983 1984 1985 1986 1987 1988 1989 1990 1991 1992 1993 1994 1995 1996 1997 1998 1999 2000 2001 2002 2003 2004 2005 2006 2007 2008 2009 2010 2011 2012 2013 2014 2015 2016 2017 2018 2019 2020

18 March 1958
Maxime Faget presents a paper on his ballistic launch profile for manned spaceflight to a conference at NACA's Ames laboratories.

29 July 1958
Congress approves the creation of a National Aeronautics and Space Administration.

1 October 1958
NASA officially comes into existence.

7 October 1958
T. Keith Glennan publicly announces NASA's manned spaceflight programme, and the formation of the Space Task Group.

17 December 1958
The manned spaceflight programme is named Project Mercury, from a suggestion by Abe Silverstein.

12 January 1959
McDonnell Corporation is awarded the contract to manufacture the Mercury capsules.

9 April 1959
The seven astronauts selected to take part in Project Mercury are presented to the world.

Mercury takes wing

In late 1958, the newly formed NASA announced its manned space programme to the world. But while Soviet engineers could rely on massive rockets to go straight to orbit, NASA would have to proceed more slowly.

Of course, some US engineers had been weighing the options for manned spaceflight since before the launch of Sputnik. Much of the early design work was done at NACA's Pilotless Aircraft Research Division, where forward-thinkers such as Maxime Faget and Robert Gilruth worked out a mission profile that involved lobbing a conical, wingless spacecraft into orbit, its underside fitted with a protective casing, called ablative shielding, that would burn away as it re-entered the atmosphere. Once it reached terminal velocity in the lower atmosphere, parachutes would bring it to a gentle landing.

When NASA opened for business in October 1958, one of Administrator T. Keith Glennan's first decisions was to establish the Space Task Group, a panel of talented scientists and engineers inherited from the various bodies absorbed by NASA, whose role would be to make manned spaceflight a reality. Faget was among those selected in the group, and Gilruth was appointed as its chairman.

A pivotal meeting came on 15 December 1958, when Glennan met with ABMA engineers, including Wernher von Braun, at Redstone Arsenal. Although the Huntsville unit would not be formally absorbed into NASA until 1960 (when it became the Marshall Space Flight Center), they were determined to play a pivotal role in the US manned space programme, offering not just help in the design of launch vehicles to put men in Earth orbit, but a complete step-by-step plan to put Americans on the Moon by 1967.

Steps to orbit

On 17 December, Glennan announced Project Mercury to the world. The plan was to develop the programme incrementally, as different launch vehicles came on stream. A relatively small rocket called Little Joe would carry dummy capsules to high altitude. Redstones and Jupiters would perform ballistic "hops" into space, and finally the US Air Force's giant Atlas ICBM, still in development, would take the capsule into orbit.

But as the programme got started, one question remained – who would be the astronauts? Debate raged within NASA about selection criteria, and it was Eisenhower who ultimately decreed that the astronauts should be test pilots, although some had more radical ideas (see panel, below left). On 9 April 1959, the United States was presented with seven new heroes – the Mercury astronauts who were intended to be the first humans to travel into space.

IN A SPIN
The Multiple Axis Test Inertia Facility (MASTIF) was a Mercury simulator mounted on gimbals that allowed trainees to practise controlling its motion in three different axes.

HISTORY FOCUS

THE MERCURY WOMEN

Early in Project Mercury, it looked as though the USA might pick a woman as its first astronaut. The idea was proposed by Dr. Randolph Lovelace in early 1959, while he was assessing medical criteria for spaceflight. Lovelace argued that women were smaller and lighter than men, required less oxygen and coped better with stress. A group of talented female pilots including Jerrie Cobb (left) travelled to the Lovelace Clinic in secret and underwent rigorous testing in which they proved themselves just as good as male candidates. A couple went on to undertake further training. However, the social attitudes of the time ultimately ensured that none of them would ever make it to space.

THE LUCKY SEVEN
The test pilots selected to be Mercury's astronauts inspect a model of the Mercury-Redstone vehicle that would carry manned US missions into space.

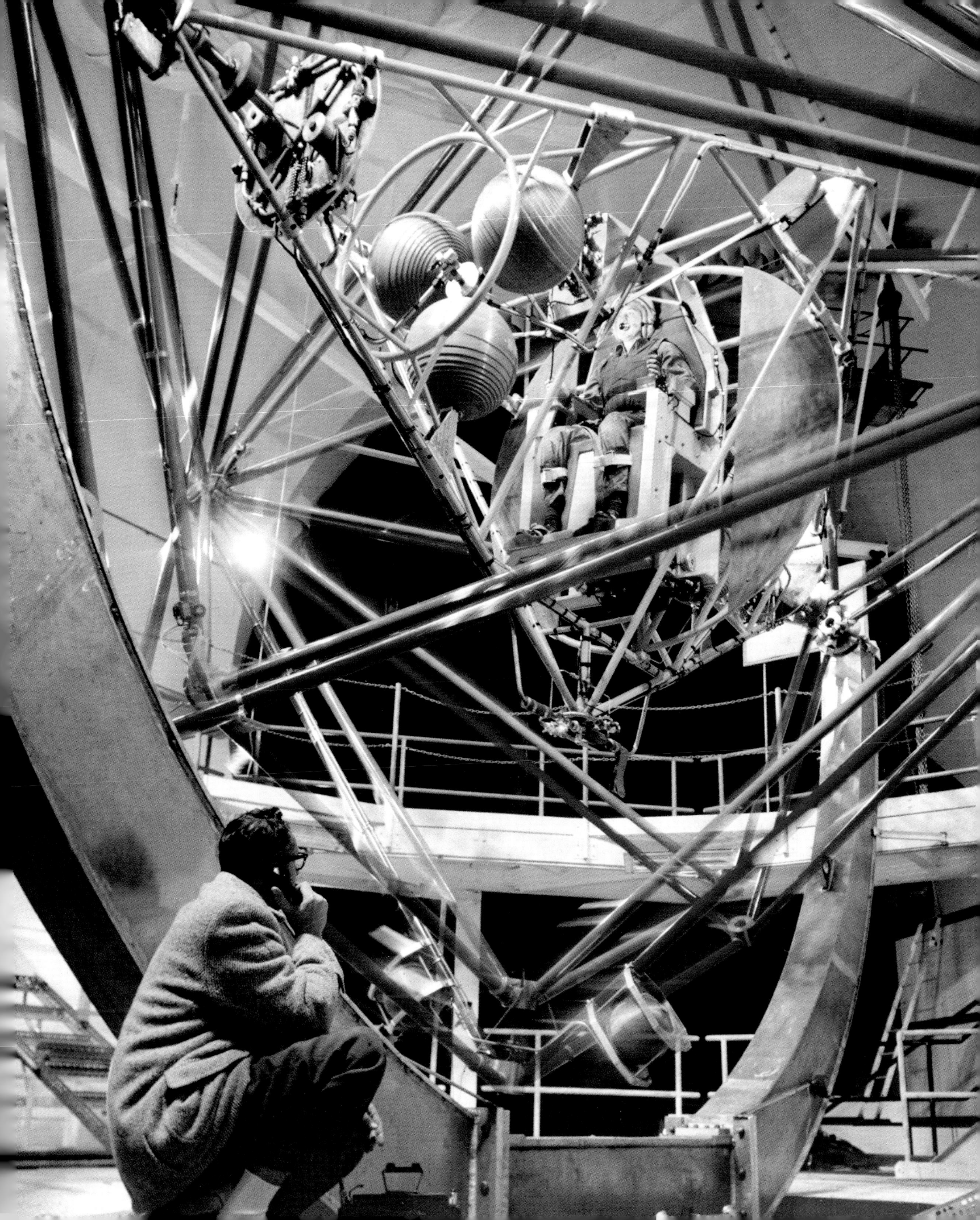

The Mercury Seven

The group of astronauts NASA selected to take America into space would become national heroes – but they would also be subjected to a tough training regime to prepare them for orbit.

SUITED UP
The silver spacesuits designed for NASA by the B.F. Goodrich Company, and modelled here by John Glenn, became an iconic image of the Mercury programme.

President Eisenhower made the decision over Christmas 1958 that only military test pilots should be considered for the first phase of the astronaut programme. Despite his military background, he was keen on maintaining a civilian aspect to the space programme (hence the formation of NASA), but in this case he saw that test pilots would have many of the characteristics needed.

So in early January 1959, NASA's selection committee began sifting through the records of 508 potential candidates. Some 110 of these were called for interviews and written tests, and by February the number had been reduced to just 32. All these candidates were then subjected to a battery of medical tests, and 14 were dismissed due to various concerns, leaving 18 that were considered potential astronauts. Politics came into play with the final choice – seven launches were planned for the first phase of the programme, and when Gilruth sat down with the selection committee on 1 April 1959, they were careful to choose three astronauts each from the Navy and the Air Force. The seventh was to come from the Marines.

Seven face the press

NASA Administrator T. Keith Glennan introduced his astronauts to the assembled media a week later. They were Scott Carpenter, L. Gordon Cooper, John Glenn, Virgil "Gus" Grissom, Walter Schirra, Alan Shepard, and Donald K. "Deke" Slayton.

From the start, NASA was careful to court the press and public with carefully arranged photo opportunities and announcements designed to keep interest high. The seven were photographed in their futuristic silver spacesuits, on survival exercises in the desert, and inspecting prototypes under construction at the McDonnell factory. While most of the candidates were taciturn military men, Glenn stood out in the media as the most charismatic and self-deprecating of the group. However, competition among the seven was intense – when

BIOGRAPHY

CHRIS KRAFT

Virginia-born Chris Kraft (b.1924) was an aeronautical engineer at NACA before its absorption into NASA. In 1958 he was put in charge of developing flight-control systems for Project Mercury. Early satellite launches could be controlled from the blockhouses at Cape Canaveral, but longer-duration manned flights would require constant monitoring from Earth by a large team working at a dedicated Mission Control. Kraft developed technology and procedures used for NASA flight control to this day and became NASA's first Flight Director during Project Mercury. He stepped back from this role in the 1960s but remained in overall charge of flight control.

ROCKET PLANE
While the Mercury Seven were still in training, the hypersonic X-15 rocket plane was setting a series of aviation records. Several of its pilots would earn USAF astronaut wings, and there was a friendly rivalry between test pilots and astronauts.

1 April 1959
The Space Task Group's selection panel chooses seven potential astronauts for the Mercury programme.

9 April 1959
The Mercury Seven are introduced to the world at a press conference.

27 April 1959
The Mercury astronauts report for duty.

July 1959
Training begins using the MASTIF gimbal rig.

December 1959
The astronauts begin weightless flight training.

February 1960
The Mercury Seven study celestial navigation at Morehead Planetarium, North Carolina.

February 1960
The seven begin "water egress training".

1 April 1960
The astronauts complete centrifuge training.

July 1960
The seven undergo survival training in the Nevada Desert.

asked at that first press conference who believed they should be the first into space, each raised his hand, except for Glenn and Schirra, who raised both.

The right stuff

In between press calls, the astronauts underwent intensive training. As well as the survival exercises in water and on land (just in case they made landfall over inhospitable territory), there were lessons in astronomy, space science, and engineering, alongside parachute jumps, parabolic flights to train for weightlessness, and endless simulations. And on top of all this, they had to maintain their flying skills. One thing above all disappointed the astronauts – in the initial plans, the Mercury capsule was so completely automated that it left them with little to do. Despite all their training, it seemed they would be little more than passengers, reduced to hoping that the engineers on the ground had got it right.

Teasing from test-pilot colleagues, and NASA's plans to use chimps in place of astronauts on some test flights, reinforced the feeling, and Deke Slayton was only half joking when he talked about engineers who believed they would have a far simpler job if they "didn't have to worry about the bloody astronaut".

HIGH FLIERS, TOP GUNS
The Mercury Seven assemble for a portrait in front of a Convair F-106 B, one of several high-performance aircraft purchased by NASA for use by the astronauts in maintaining their finely honed flying skills.

FLYING THE "VOMIT COMET"
A pair of Mercury astronauts enjoy a few seconds of weightlessness in the cabin of a converted C-131 aircraft. The so-called Vomit Comet flew to high altitudes, then dropped on a parabolic flightpath so that its occupants went into freefall.

Despite the media push for manned spaceflight, there were still many who felt there was no point in sending a man to do a machine's job, including scientists such as James Van Allen and some military strategists. Fortunately for the astronauts, there were also influential believers, such as von Braun and some politicians, who saw the value of spaceflight in shaping their new vision of America.

And so many of Mercury's systems were redesigned to give the pilot more to do. While the entire system was designed to operate automatically or under control from the ground, the astronauts would be able to take control of thrusters to adjust spacecraft altitude and to manually trigger the retrorocket burn for re-entry.

BOILERPLATE CAPSULE
The first dummy capsules were built in-house at NASA's Langley division for test launches with the Little Joe rocket. They were designed to mimic the weight and aerodynamic characteristics of the completed Mercury.

CAPSULE TESTING
Model capsules were tested in wind tunnels with scaled-down parachutes (above). The inflatable ring that would cushion the splashdown and keep the spacecraft afloat was also tested (right). Meanwhile, engineers set to work on increasingly complex boilerplate capsules (above right).

TECHNOLOGY

NASA'S FIRST MANNED SPACECRAFT

Mercury capsule

NASA's first manned spacecraft was just one-third the weight of the Soviet Vostok – it had to be in order to be launched by the weaker US rockets of the time. Yet both vehicles had to address the same problems of life support in orbit, re-orientation in space, and re-entry into the atmosphere. A variety of designs were pitched by potential contractors, but NASA's Space Task Group already knew they wanted a conical, wingless capsule that would re-enter the atmosphere on a ballistic trajectory. In January 1959, McDonnell Aircraft Corporation were awarded the prime contract.

FLIGHT CONTROLS
Compared to the minimalism of Vostok, the Mercury capsule was lined with banks of switches and controls. This was largely thanks to the persistent lobbying of the Mercury Seven astronauts, who argued that there was little point in sending highly trained pilots into space as mere passengers (or "spam in a can", as a sceptical Chuck Yeager memorably put it).

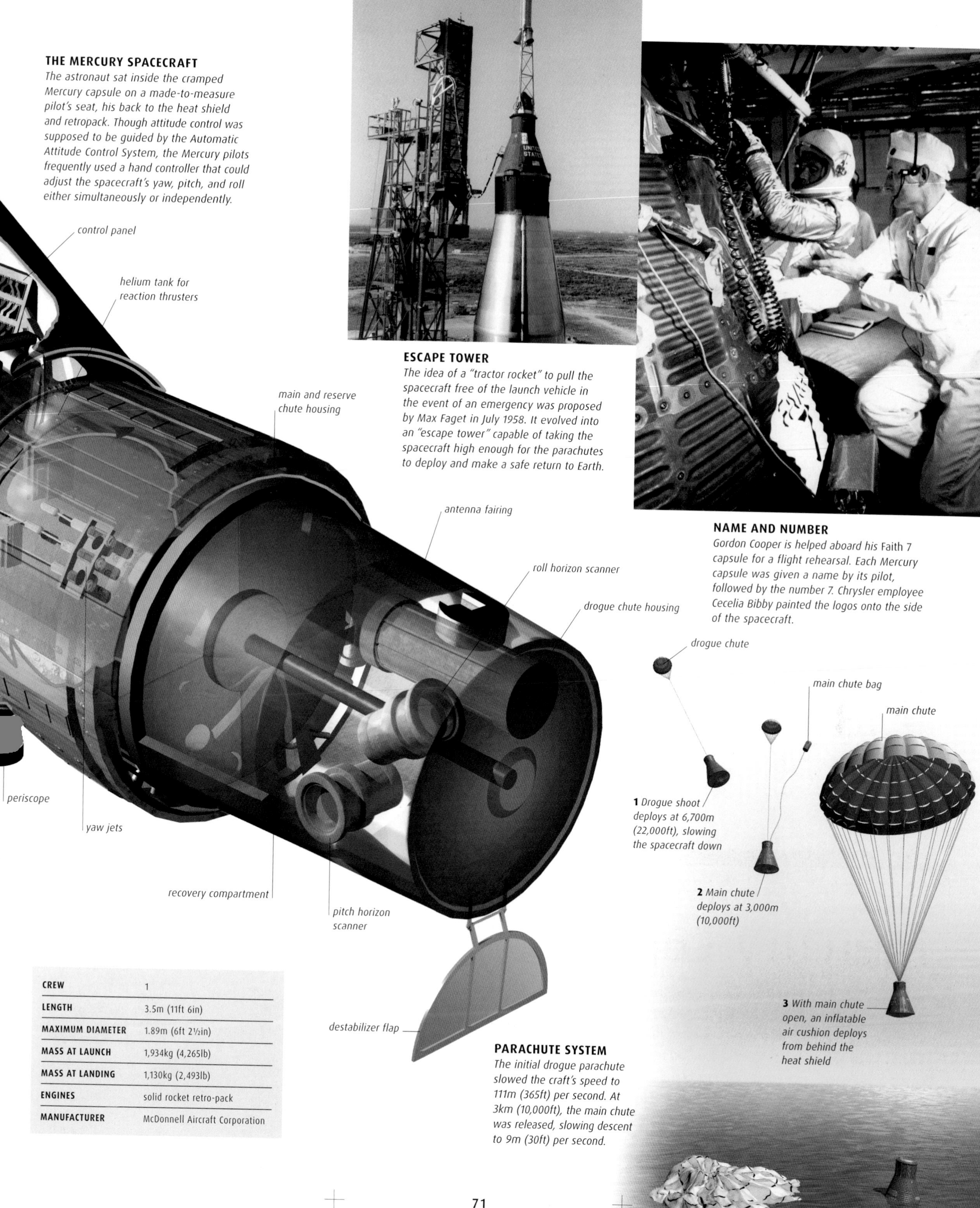

THE MERCURY SPACECRAFT

The astronaut sat inside the cramped Mercury capsule on a made-to-measure pilot's seat, his back to the heat shield and retropack. Though attitude control was supposed to be guided by the Automatic Attitude Control System, the Mercury pilots frequently used a hand controller that could adjust the spacecraft's yaw, pitch, and roll either simultaneously or independently.

ESCAPE TOWER

The idea of a "tractor rocket" to pull the spacecraft free of the launch vehicle in the event of an emergency was proposed by Max Faget in July 1958. It evolved into an "escape tower" capable of taking the spacecraft high enough for the parachutes to deploy and make a safe return to Earth.

NAME AND NUMBER

Gordon Cooper is helped aboard his Faith 7 *capsule for a flight rehearsal. Each Mercury capsule was given a name by its pilot, followed by the number 7. Chrysler employee Cecelia Bibby painted the logos onto the side of the spacecraft.*

PARACHUTE SYSTEM

The initial drogue parachute slowed the craft's speed to 111m (365ft) per second. At 3km (10,000ft), the main chute was released, slowing descent to 9m (30ft) per second.

CREW	1
LENGTH	3.5m (11ft 6in)
MAXIMUM DIAMETER	1.89m (6ft 2½in)
MASS AT LAUNCH	1,934kg (4,265lb)
MASS AT LANDING	1,130kg (2,493lb)
ENGINES	solid rocket retro-pack
MANUFACTURER	McDonnell Aircraft Corporation

BIG JOE RIDES OUT

An Atlas launch vehicle with a Mercury boilerplate capsule on top, Big Joe was a test of Mercury's ablative heatshield. On 9 September 1959, the rocket launched its payload to a height of 145km (90 miles) above the Atlantic. When the capsule was recovered, the shield proved to have coped with its ordeal well.

1956 1957 1958 1959 **1960** 1961 1962 1963 1964 1965 1966 1967 1968 1969 1970 1971 1972 1973 1974 1975 1976 1977 1978 1979 1980 1981 1982 1983 1984 1985 1986 1987 1988 1989 1990 1991 1992 1993 1994 1995 1996 1997 1998 1999 2000 2001 2002 2003 2004 2005 2006 2007 2008 2009 2010 2011 2012 2013 2014 2015 2016 2017 2018 2019 2020

Tests and space chimps

Before NASA would commit to send an astronaut into space, they tested the Mercury technology thoroughly with a series of unmanned launches and a number of flights with primate passengers.

9 May 1960
First in a series of trials, known as "beach aborts", at Wallops Island test the Mercury capsule's escape systems.

29 July 1960
The launch of a boilerplate Mercury capsule aboard an Atlas rocket ends in a crash after 59 seconds.

8 November 1960
The first of the Little Joe launches takes off from Wallops Island to test the spacecraft's structural integrity. However, a rocket fault destroys the spacecraft 15 seconds after launch.

19 December 1960
The Mercury-Redstone 1A mission launches an unmanned capsule on a suborbital flight.

31 January 1961
Mercury-Redstone 2 takes off, carrying Ham the chimp on a suborbital flight.

24 March 1961
The successful test flight of Mercury-Redstone mission MR-BD qualifies the rocket for manned flight.

While the Mercury Seven spent their days in training or parading before the press, the engineers at McDonnell and the Space Task Group's offices at Langley Field, Virginia (the former NACA laboratories), were labouring to complete the vehicle that would eventually take them into space.

Little Joe launches

To test the basic principle of the conical Mercury capsule, a number of bare-bones boilerplate models were produced. These could be launched on top of the relatively cheap Little Joe booster rocket – a two-stage launch vehicle in which each stage was itself a cluster of four solid-fuelled rockets.

The boilerplate capsules and their rockets were fitted with instruments to record the stresses and temperatures encountered in each flight. Low-altitude flights tested the escape system, while higher trajectories allowed the engineers to see how Mercury behaved as it re-entered the atmosphere and to test the performance of prototype heat shields. Two of the later Little Joe missions carried passengers – rhesus monkeys called Sam and Miss Sam – in order to test their ability to survive the forces experienced in a real Mercury mission. Both animals survived with no obvious ill effects.

While Soviet engineers typically sent dogs on their early test flights, NASA's medical experts felt that primates would provide the best data about the stresses of space travel – if a monkey could survive a Mercury launch and re-entry in good condition, then it seemed likely that a human could do. And there was another advantage of using primates – some were smart enough to be trained, allowing doctors to assess how they fared mentally in orbit.

FIT FOR AN APE
Ham and the later space chimp Enos (shown here) both wore custom-made spacesuits for their flights. In Ham's case, the suit probably saved his life when a loose valve resulted in a sudden drop in cabin pressure.

The first ape in space

A group of six chimps were trained for test flights aboard the Mercury capsule, and a male called Ham (see panel, right) was selected for a suborbital flight aboard a Redstone-launched Mercury on 31 January 1961. Other tests with the Redstone had gone well, but the Atlas vehicle needed to put the capsule into

SIMULATOR TRAINING
While Ham took to the skies, the Mercury Seven were still stuck on the ground – here John Glenn is undergoing a simulated mission on the Mercury Procedures Trainer at Langley Field, Virginia.

orbit was still hitting problems, and it was clear that NASA's first human spaceshot would have to be on a suborbital trajectory. Ham's flight was a dress rehearsal for the human flight. It hit a number of snags, but the stresses the chimp overcame convinced the experts that a Mercury mission was survivable even if things did go wrong. However, von Braun insisted on another unmanned launch of the Redstone, infuriating Alan Shepard, who was slated to pilot the first suborbital flight (several of the seven already felt slighted by the fact that an ape was taking the lead in the space programme instead of them). A final test of the Mercury-Redstone configuration on 24 March went perfectly, but as it turned out problems with Shepard's *Freedom 7* capsule would delay the launch still further. In the meantime, the Soviets were about to seize the initiative, and the headlines, once again.

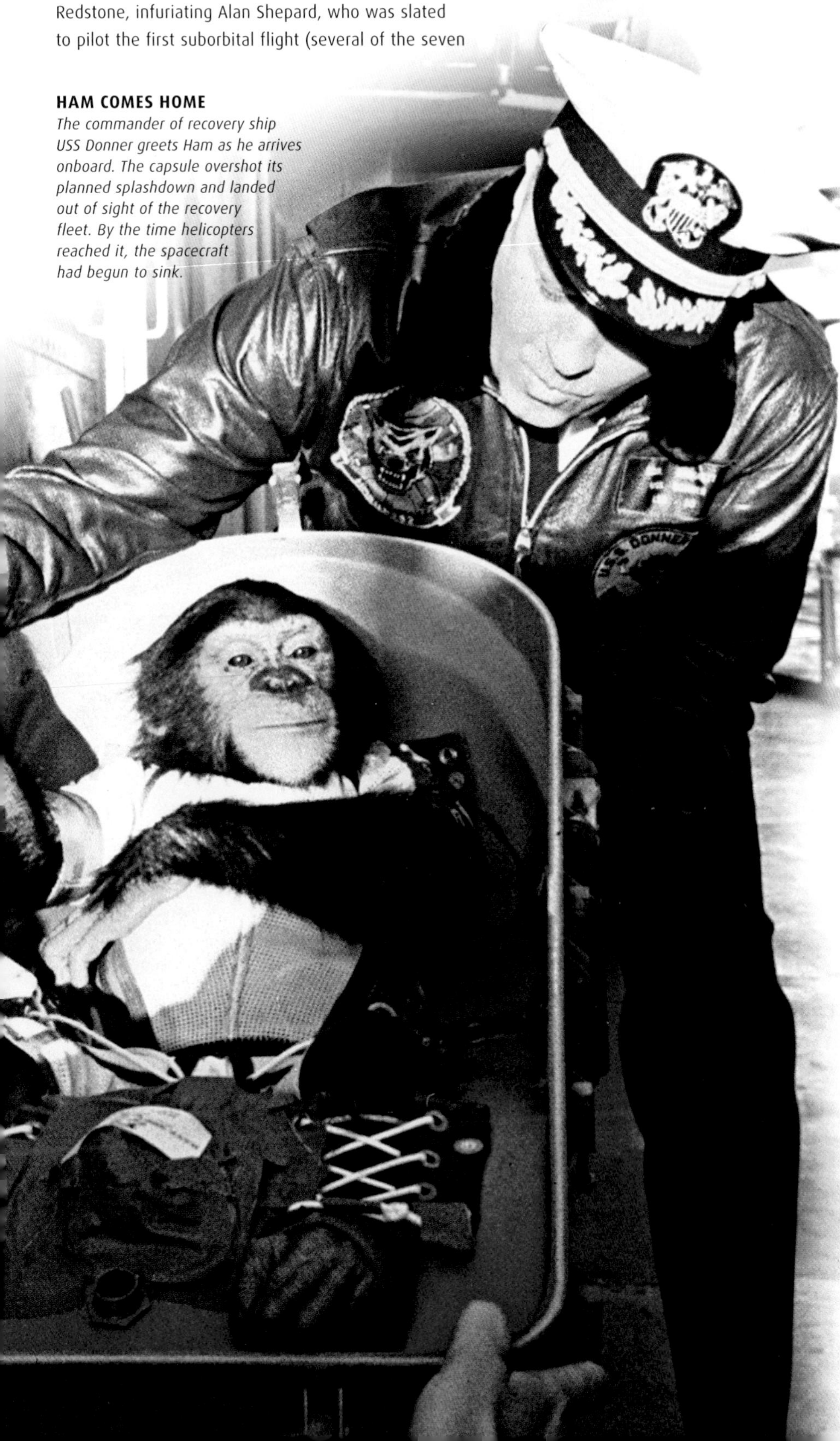

HAM COMES HOME
The commander of recovery ship USS Donner greets Ham as he arrives onboard. The capsule overshot its planned splashdown and landed out of sight of the recovery fleet. By the time helicopters reached it, the spacecraft had begun to sink.

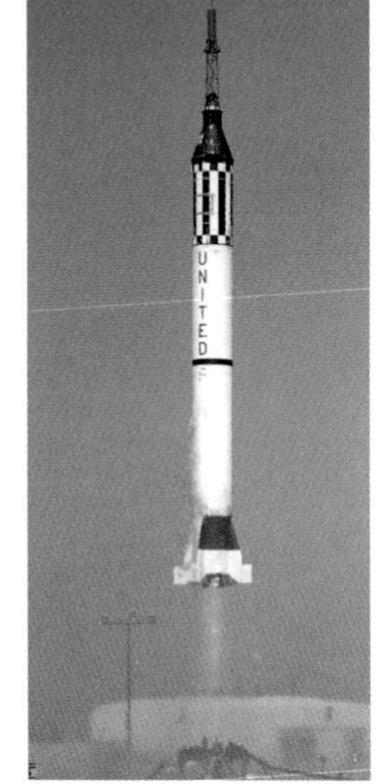

CHIMP ASCENDING
During the launch of Mercury-Redstone 2, the main engine burned through its fuel supply faster than expected, and Ham had to endure far higher g-forces than intended, peaking at 15 g. He survived unscathed, but the fault also meant that Ham was weightless for more than six-and-a-half minutes – two minutes longer than planned.

HAM GRITS HIS TEETH
Cameras monitored Ham's reactions throughout the flight – during ascent and re-entry, he experienced extreme acceleration, but he quickly recovered and performed his tasks well.

BIOGRAPHY

HAM THE SPACE CHIMP

Captured as an infant in the wild, Ham (1956–83) became part of a colony of chimpanzees established in the late 1950s at Holloman Air Force base in New Mexico. Six of these chimps were recruited by NASA. They were trained to pull a series of levers in the correct order to receive a reward – if they got it wrong, they suffered a mild electric shock. After his flight, Ham retired to the National Zoo at Washington, D.C. and then to North Carolina, where he died in 1983. Most of the colony were not so lucky, but the last were retired in 1997 and settled at a sanctuary in Florida.

The first man in space

The launch of the first cosmonaut into space on 12 April 1961 shook the world almost as much as the launch of Sputnik 1 three-and-a-half years earlier. Yet Yuri Gagarin's short flight in Vostok 1 almost ended in disaster.

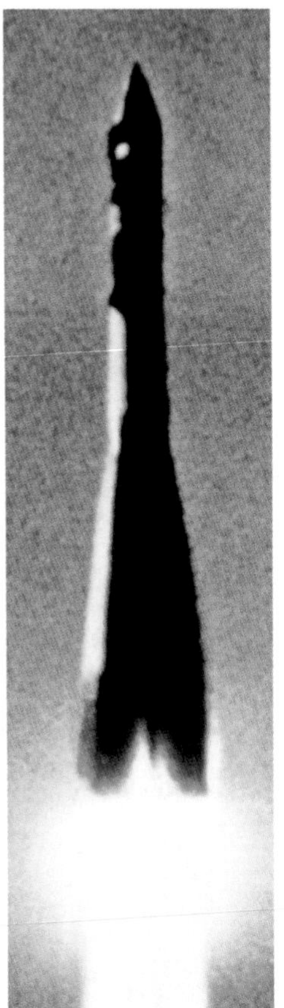

VOSTOK LIFTS OFF
The R-7 launch vehicle carrying Vostok 1 blasts free of the launch pad at Tyuratam in Kazakhstan. Onboard, Gagarin (left) had little freedom to move and still less to influence his spacecraft.

By the end of March 1961, the first Soviet manned spaceflight had been authorized, but who would be on board? The choice was down to two men – the evenly matched friends Gagarin and Titov.

When the State Commission met on 7 April, they agreed that Titov was probably the fitter of the two trainees. However, Gagarin had done a better job of treading the fine line between unquestioning obedience and independent thought, and it was this that won him the coveted prize. The fact that he had a peasant background similar to Khrushchev's own (see panel, right) may also have given him an edge over the middle-class Titov. The choice was made in closed session, but the commission sat again the next day, with Gagarin, Titov, and the cameras present, to repeat its choice in public.

At 5:00am, Moscow time, on 11 April, the R-7 launcher with Vostok 1 attached rolled out along the track leading from the assembly complex to the launch pad at Tyuratam. A day of rehearsal and exhaustive testing followed, and that night the medical team monitored the sleep patterns of Gagarin and his reserve through sensors inside their mattresses (a counterproductive measure, since the two men convinced the doctors they were having a restful night by lying rigid in their beds and barely sleeping at all).

Into orbit

On the morning of 12 April, Korolev personally woke his cosmonauts at 5:30am. After a final series of medical checks, both astronauts were suited up, and Titov had to suffer the agony of riding with Gagarin in the bus to the launch site, then waiting on stand-by as his comrade was secured inside the Vostok capsule. Only then was Titov taken to an observation bunker to remove his spacesuit.

Meanwhile, engineers worked to secure Gagarin, plugging him into a variety of monitors and life-support systems. The controls of Vostok 1 were locked on autopilot and could be freed only by entering a three-digit code, intended to be sent from ground control in an emergency. However, no fewer than four people revealed the code to Gagarin before he was sealed into the spacecraft. After a last-minute glitch with a circuit monitoring the hatch closure, countdown began in earnest. At 9:06am the R-7's main engines fired and slowly began to lift Gagarin towards orbit, on a 108-minute flight that would make history (see over).

Vostok's orbit initially carried it northeast across Siberia. Here, an accurate flightpath was calculated by controllers from telemetry signals received at a series of remote listening posts. Then the spacecraft swept southeast across the Pacific Ocean and onto the night side of the Earth.

Even as Gagarin passed out of radio contact into the western hemisphere, Radio Moscow was already announcing the latest Soviet success. The announcement was rather premature, for the most dangerous part of the flight was yet to come.

As Vostok 1 began re-entry above Africa, explosive charges blew the main links between the Instrument and Descent Modules – but a thick bundle of wires did not detach as planned, and Vostok 1 began to spin wildly as it plunged back into the atmosphere. Fortunately, hot gas building up around the spacecraft eventually burned through the cable, and Gagarin's capsule came free. Seven kilometres (4½ miles) above the ground, automatic pressure sensors blew the hatch and launched Gagarin into the air in his ejector seat. He parachuted to the ground as planned in Saratov province, southern Russia.

BIOGRAPHY

YURI GAGARIN

Yuri Gagarin (1934–68) was the son of peasant farmers, born on a collective farm near Smolensk. After showing early academic promise, he studied engineering at college, before enrolling as a fighter pilot in 1957. Selected for cosmonaut training in 1960, he soon became Korolev's favourite among the "Top Six" trainees (the two are pictured together here), which eased his selection as the first cosmonaut. After his historic flight, however, Gagarin had trouble adjusting to his new-found celebrity. Paraded as a Hero of the Soviet Union, he fell into depression, and died in a jet crash while training for a return to space on Soyuz 3.

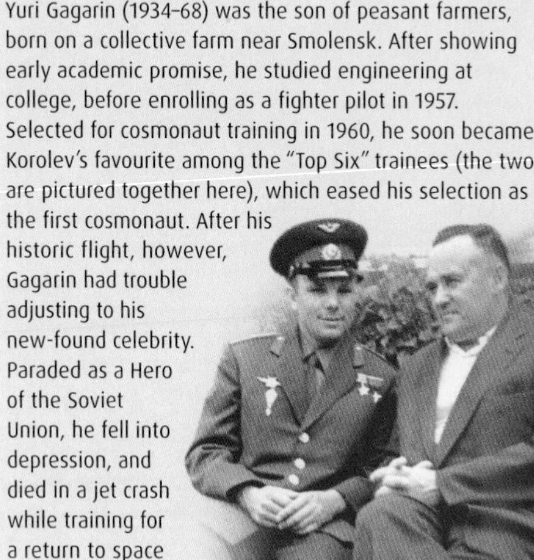

9 March 1961
A Vostok test capsule is successfully launched.

25 March 1961
A second Vostok test capsule is successfully launched. The same day, Gagarin and Titov practise boarding the Vostok 1 capsule on the launch pad.

27 March 1961
Following the successful test flights, a manned launch is approved.

3 April 1961
Gagarin and Titov rehearse donning their spacesuits and are filmed boarding the Vostok capsule.

7 April 1961
The cosmonauts rehearse emergency procedures in case of a launch failure. Meanwhile, a closed session of the State Commission selects Gagarin to be the pilot for Vostok 1, with Titov as backup.

8 April 1961
A filmed meeting of the State Commission repeats its decision for the cameras, with Gagarin and Titov present.

11 April 1961
Gagarin and Titov undergo final pre-launch medical tests. Gagarin is given the all-clear for his historic flight.

12 April 1961
Yuri Gagarin, aboard Vostok 1, becomes the first man to travel in space.

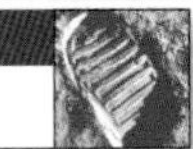

EXPERIENCE
MAN ENTERS SPACE

108 minutes

FAMILY MAN
Gagarin met Valentina Goryacheva while in training at Orenburg Pilot's School. They married in 1957 and had two daughters, Elena and Galya.

Yuri Gagarin's ascent into history began at 9:06am on 12 April 1961, Moscow time, as the R-7 rocket carrying Vostok 1 fired its engines and slowly lumbered off the pad at Baikonur. Before leaving Moscow, the first cosmonaut had recorded a message that might, in other circumstances, have been his epitaph:

"**Dear friends, known and unknown to me**, my dear compatriots and all people of the world! Within minutes from now, a mighty Soviet rocket will boost my ship into **the vastness of outer space** ... My whole life is now before me as a single breathtaking moment."

Two minutes into the flight, Gagarin felt a jolt of acceleration as the R-7's booster rockets separated and the core continued alone. After five minutes, that too was exhausted. By the time it too had separated, the payload shroud surrounding Vostok 1 had already fallen away. As the final rocket stage pushed him towards orbit, Gagarin reported back:

"I can see the Earth. The **visibility is good** ... I almost see everything. There's a certain amount of space under cumulus cloud cover. **I'm continuing the flight – everything is good**."

Four minutes later, the engines cut out and the upper stage fell away, leaving Gagarin in an elliptical orbit with a period of just over 89 minutes. As Gagarin lost radio contact with Baikonur, he briefly made contact with the Kolpashevo tracking station, then was alone for three minutes as his spacecraft swept above northern Siberia. Last contact with his homeland came as he flew into the Pacific night, maintaining contact with a station at Khabarovsk on the east coast.

Even as Gagarin finally lost contact, Radio Moscow and the Soviet news agency TASS were reporting the successful launch:

"The Soviet Union has successfully launched a manned spaceship-satellite into an orbit around the Earth. Present aboard the spaceship is the **pilot cosmonaut, Yuri Alekseyevitch Gagarin**, an Air Force pilot, 27 years of age. The spaceship was launched about 9am, Moscow time, April 12, 1961. **The spaceship is named Vostok** and weighs 4,725 kilograms, including the pilot but excluding the last stage of the carrier rocket. The hermetically sealed cabin of the spaceship is equipped with a two-way radio, TV, and a telephone-type communication system."

FINAL WAVE
Yuri Gagarin is photographed for the last time before being sealed inside Vostok 1 by chief constructor Oleg Ivanovsky. The fully automated spacecraft had a keypad to release the manual controls, with a code that would be sent from the ground in an emergency. But Nikolai Kamanin, Sergei Korolev and Ivanovsky had all informed Gagarin of the code before launch.

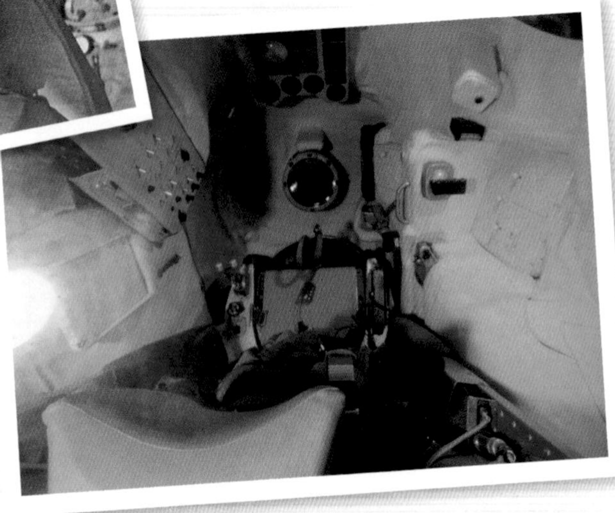

FROM BUS TO CAPSULE
Gagarin and his backup Gherman Titov rode together in the bus to the launch pad. By 7:10, Gagarin was seated in the cramped Vostok cabin, and an hour later, he was sealed in to await the launch itself.

HAPPY LANDINGS
As the recovery team moved in to retrieve the charred Vostok 1 Descent Module, Gagarin was already being greeted as a hero by well-wishers who had heard reports on the radio. Later he had time for some brief reflection before being thrust into the limelight once again.

"To be the first to enter the cosmos, to engage, single-handed, in an unprecedented duel with nature – **could one dream of anything more?"**

Yuri Gagarin, recorded in Moscow before the launch

Gagarin flew back into daylight over the Atlantic at 10:10, and Vostok 1 automatically aligned itself for its re-entry burn. As he flew on towards Africa, the lone cosmonaut continued transmitting his regular status messages, though there was no one within range to receive them. Fifteen minutes into the new day, and still 8,000km (5,000 miles) from landing, Vostok fired its engines for a 42-second braking manoeuvre. As the Instrument and Descent modules separated, the main electrical cable linking them stuck in place, and Gagarin found himself shaken and spun around as the linked modules plunged into the atmosphere.

Finally, after ten minutes, heat from re-entry burned through the cable and the Instrument Module fell away at around 10:35. The Descent Module now righted itself, but there were still 20 minutes of ballistic descent to endure, during which Gagarin almost blacked out under forces of up to 8*g*. Passing over the Black Sea, Vostok 1 re-entered Soviet airspace, slowing all the while. Finally reaching the required altitude of 7km (4½ miles) at 10:55, the spacecraft automatically ejected the escape hatch. Seconds later, Gagarin was blasted free on his ejector seat as the craft deployed its descent parachute.

"When they saw me in my spacesuit and the parachute dragging alongside as I walked, they started to **back away in fear.** I told them, don't be afraid. I am a Soviet like you, who has **descended from space,** and I must find a telephone to call Moscow!"

Gagarin describes his meeting with a farmer and her daughter

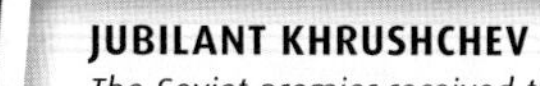

JUBILANT KHRUSHCHEV
The Soviet premier received the news of Gagarin's safe return with delight, and shortly afterwards Radio Moscow reported: "At 10:55 Cosmonaut Gagarin safely returned to the soil of our motherland." Two days after the flight, Gagarin flew to Moscow to meet the Soviet leadership and was greeted as a national hero by the largest crowds the city had seen since the end of the Second World War.

Vostok sets the pace

The later Vostok missions achieved a series of space firsts designed to keep the Americans on the back foot in the Space Race. Vostok 2, for example, saw cosmonaut Gherman Titov spend an entire day in space.

1961 1962 1963 1964 1965 1966 1967 1968 1969 1970 1971 1972 1973 1974 1975 1976 1977 1978 1979 1980 1981 1982 1983 1984 1985 1986 1987 1988 1989 1990 1991 1992 1993 1994 1995 1996 1997 1998 1999 2000 2001 2002 2003 2004 2005 2006 2007 2008 2009 2010 2011 2012 2013 2014 2015

6 August 1961
Gherman Titov, aboard Vostok 2, becomes the first person to spend a day in orbit.

11 August 1962
Andrian Nikolayev is launched into orbit in Vostok 3.

12 August 1962
Pavel Popovich, aboard Vostok 4, enters orbit close to Nikolayev – for the first time there is more than one person in space at the same time.

15 August 1962
Vostoks 3 and 4 re-enter the atmosphere and land within minutes of each other.

14 June 1963
Vostok 5 carries Valery Bykovsky into orbit.

16 June 1963
Vostok 6 joins Bykovsky's capsule in orbit. Valentina Tereshkova becomes the first woman to travel in space.

19 June 1963
Vostoks 5 and 6 successfully return to Earth.

The decision to aim for a day-long flight was partly driven by necessity – Vostok's inclined orbit and the Earth's slow rotation meant that, within a few hours of launch, the capsule would no longer be over Soviet territory. Mission planners had to choose between a three-orbit mission, which would be over in five hours, or an entire day in space. The publicity offered by a day-long flight doubtless swung the decision.

However, there was also a scientific motivation as no one knew what the effects of extended weightlessness might be on the human body. A day-long flight would allow time for the cosmonaut to eat and sleep in orbit, as well as test how he coped psychologically. One problem became clear a few hours after the launch of Vostok 2 on 6 August 1961.

Group flights

As Titov attempted to sleep, he became nauseous. However, he had no problem with eating and drinking, briefly took control of the spacecraft, and used a film camera to record the view from his cockpit window. Re-entering Earth's atmosphere, Vostok 2 experienced similar separation problems to those on Vostok 1, but Titov made a safe return to Earth by parachute after 17 orbits. Although later missions would continue to extend flight times, this was unlikely to provide the stream of propaganda demanded by the Kremlin, and longer missions alone could provide little in the way of new scientific data or engineering results. With the limitations of the Vostok spacecraft itself, there was only one way to proceed – a so-called "group flight".

After a series of delays (largely caused by problems in development of Zenit, a spy satellite that took priority for R-7 launches in late 1961), specialist training for a group of cosmonauts got under way in early 1962. Nevertheless, when Vostok 3 blasted off, piloted by Andrian Nikolayev, on 11 August 1962, the world had little idea of what to expect. Precisely 23 hours 32 minutes later, as Vostok 3's orbital path took it back across Tyuratam, Pavel Popovich's Vostok 4 rose to join it, arriving in orbit just 6.5km (4 miles) away. The cosmonauts were able to establish

THE FIRST WOMAN IN SPACE
Despite rumours that Tereshkova suffered from severe space sickness in orbit, and was uncommunicative with ground controllers, Vostok 6 still fulfilled all of its goals.

VOSTOK 1: GAGARIN THE HERO
After his return to Earth, Gagarin was fêted as a national hero and sent on a world tour that saw him treated like a movie star. Here, he studies newspaper coverage of his flight with his wife, Valentina, and Nikita Khrushchev.

VOSTOK 2: GHERMAN TITOV
Titov, seen here in the bus with his backup Nikolayev (right), was the fittest of the first trainees and the one who had coped best in the isolation tests. This made him the obvious choice for the first longer duration spaceflight.

VOSTOK 3: ANDRIAN NIKOLAYEV
Nikolayev's tolerance of isolation training earned him the nickname Iron Man and the chance to set a new endurance record. He later married fellow cosmonaut Valentina Tereshkova, though the marriage eventually collapsed.

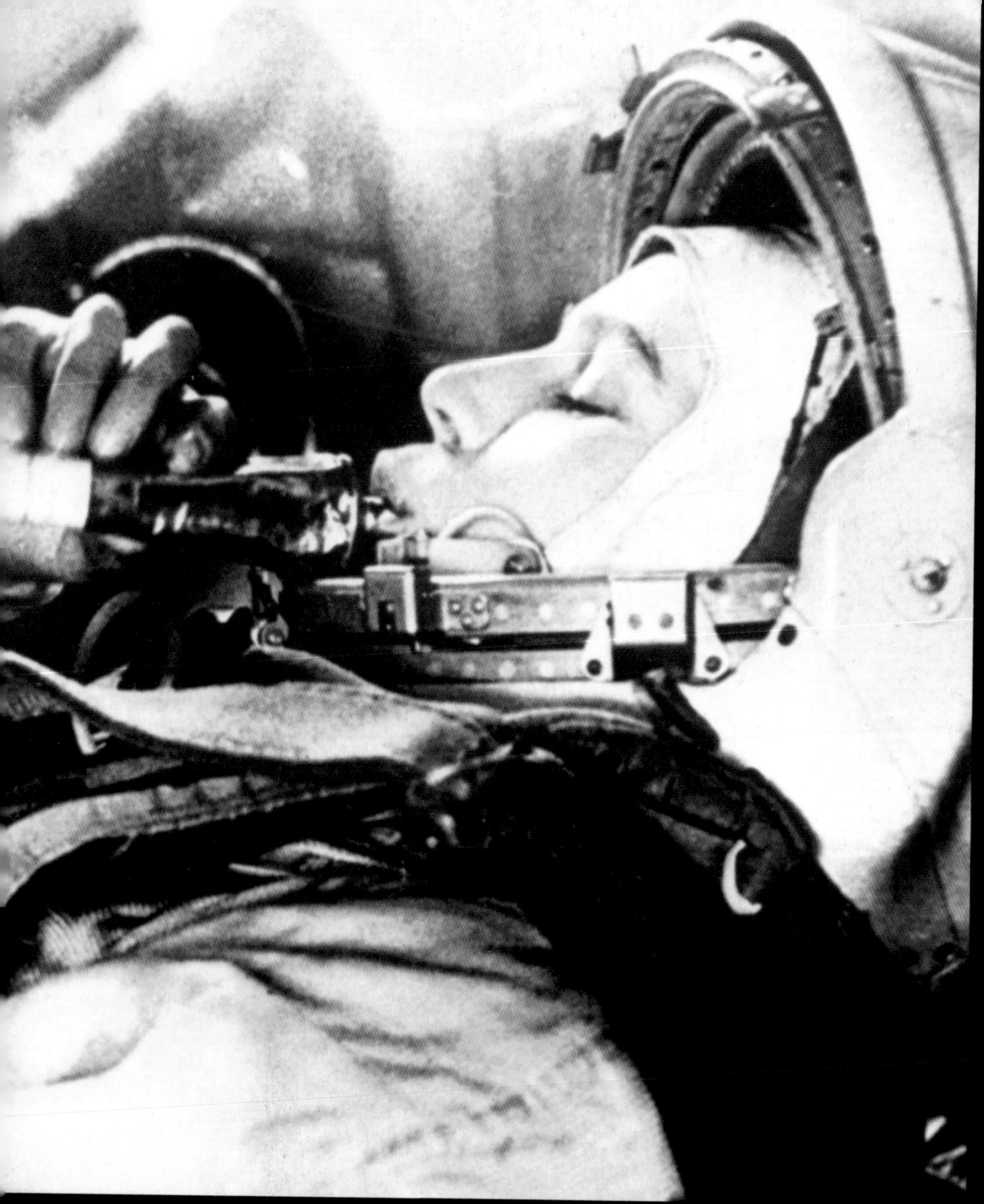

BIOGRAPHY

GHERMAN TITOV

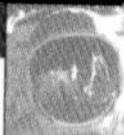

Just a month short of his 26th birthday when he took to the skies aboard Vostok 2, Gherman Stepanovich Titov (1935–2000) is still the youngest person to have travelled in space. Born in a small village in the Altai region, he trained as a pilot at Stalingrad, before his recruitment to the Soviet cosmonaut programme. His day-long flight in 1961 made him an international celebrity, and the Kremlin propagandists sent him on tours around the world. From 1962, he retrained as a test pilot alongside some of the other cosmonauts, and led the team training to fly the planned Spiral spaceplane. However, the spacesickness he suffered aboard Vostok 2, coupled with a tendency to clash with his superiors, meant that he never flew in space again.

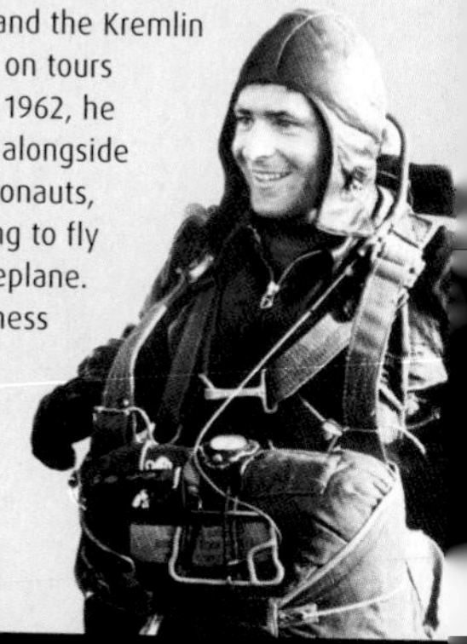

direct radio contact between the spacecraft before they gradually drifted apart. When they returned to Earth within minutes of each other on 15 August, Nikolayev had been in space for four days.

Although Korolev wanted to continue the Vostok programme, he was overruled – work on the Voskhod modification was to take priority. But the last paired Vostok mission saw another propaganda coup, as Valery Bykovsky's Vostok 5, launched on 14 June 1963, was joined in orbit two days later by Valentina Tereshkova, the first female cosmonaut, in Vostok 6. Tereshkova was an expert parachutist and one of a group of women trained for flight by Kamanin and Korolev. The two cosmonauts returned to Earth on 19 June, marking the end of the Vostok programme.

VOSTOKS 3 AND 4: THE WORLD LISTENS IN
Muscovites crowd around a car radio reporting the formation flight of Vostoks 3 and 4. Although the Vostoks were unable to manoeuvre in orbit, it would be three years before the US could improve on this early space rendezvous.

VOSTOK 4: PAVEL POPOVICH
Popovich's flight was intended to continue after Vostok 3's return to Earth, but confusion broke out when he reported storms (groza) below. This was an agreed codeword to indicate nausea, and so an early landing was ordered.

VOSTOKS 5 AND 6: TERESHKOVA/BYKOVSKY
Nikita Khrushchev parades Valentina Tereshkova and Valery Bykovsky in Moscow's Red Square following the successful mission of Vostoks 5 and 6. Bykovsky's five days in orbit set a record for lone spaceflight that stands to this day.

Mercury rising

In the aftermath of Gagarin's Vostok flight, the US was once again looking outpaced in the Space Race. An early and successful start to the Mercury programme was vital – and, fortunately, NASA delivered.

2 May 1961
A planned launch of the *Freedom 7* capsule is scrubbed due to cloud cover. The identity of the astronaut onboard is revealed to the press.

4 May 1961
The launch of *Freedom 7* is delayed for a second time due to bad weather.

5 May 1961
Freedom 7 finally launches at 09:34, on a 15-minute suborbital flight. Shepard is the first American in space.

25 May 1961
President John F. Kennedy makes his famous speech to Congress, vowing that an American will walk on the Moon before the decade is out.

18 July 1961
Weather delays the planned launch of *Liberty Bell 7*.

21 July 1961
After another delay, Gus Grissom's *Liberty Bell 7* capsule is finally launched successfully into space. However, a hatch malfunction after splashdown floods the capsule and Grissom is lucky not to drown.

The Soviet announcement of Vostok 1's successful mission had not surprised the rest of the world as much as the sudden appearance of Sputnik 1, but it was still galling for the Americans to be beaten by a matter of weeks. Alan Shepard in particular was furious at what he saw as delays caused by overcautious management. In the White House, meanwhile, recently inaugurated President John F. Kennedy found himself on the receiving end of the same accusations of complacency that he had levelled at the Eisenhower administration.

The first manned Mercury spacecraft was finally ready for launch on 2 May 1961. The name of *Freedom 7*'s astronaut had been a closely guarded secret, and it was only after a cancellation due to poor weather that Alan Shepard's identity was revealed to the media. Further delays followed, until conditions were right for launch on 5 May. Even then, there were repeated delays in the countdown. Eventually Shepard's patience ran out – "Why don't you light the damned candle, 'cause I'm ready to go!", he snapped at mission control.

And so, at 9:34am, Mercury-Redstone 3 lifted off the pad carrying its pilot into history. Compared to Vostok 1's trip, *Freedom 7*'s flight into space was just a short hop, lasting only 15 minutes 22 seconds, but it punched above its weight thanks to NASA's decision to broadcast the entire event. Forty-five million Americans watched live on their televisions, and the images helped convince the world that the US was keeping up with the Soviets. In one way, they were actually ahead – Shepard was the first astronaut to return from space aboard his capsule, though this was not known at the time because of the secrecy surrounding Vostok 1's return.

WAITING TO GO
Grissom looks cheerful moments before climbing aboard Liberty Bell 7. *The window in the enlarged escape hatch can be seen behind him.*

BIOGRAPHY
ALAN SHEPARD

New Hampshire-born Alan Bartlett Shepard (1923–98) saw active service during the Second World War aboard the destroyer *USS Cogswell*. He then trained as a naval pilot, gaining his wings in 1947, before qualifying as a test pilot in 1950. He was selected for the Mercury programme in 1959. After becoming the first American in space, he just missed the chance of a second Mercury flight – NASA had briefly considered a three-day mission before deciding to concentrate on the Gemini programme. He became Chief of NASA's Astronaut Office in 1963 and lost the opportunity to command the first Gemini mission due to a problem with his inner ear. After surgery in 1969, he joined the Apollo programme and in 1971 commanded Apollo 14. He retired from NASA in 1974 and went into business, serving on the board of several corporations.

Liberty Bell cracks

On 8 May, the Mercury astronauts enjoyed a celebratory dinner at the White House, while back at NASA thoughts were already turning to a second suborbital flight. Almost everything on *Freedom 7* had worked perfectly, and if all went well with the next mission, the third flight could be sent into orbit.

Gus Grissom named his capsule *Liberty Bell 7*, and in tribute to the original Liberty Bell (rung in 1776 in Philadelphia before the reading of the Declaration of Independence), a crack was painted down one side.

This was to prove prophetic – *Liberty Bell*'s flight on 21 July went perfectly, but shortly after splashdown the explosive bolts holding a new and larger escape hatch in place triggered accidentally, flooding the capsule and sending it sinking to the bottom of the Atlantic. Grissom himself, his spacesuit filling with water, was lucky to escape with his life.

GRISSOM STEPS OUT
Gus Grissom strides purposefully towards the Mercury-Redstone 4 rocket, with technicians and support personnel gathered at its base, on the morning of 21 July 1961.

EXPERIENCE
ALAN SHEPARD'S SUBORBITAL HOP

Almost in orbit

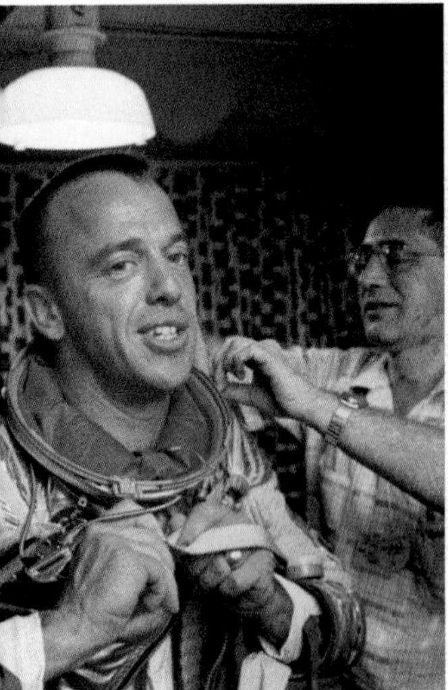

ONE OF A KIND
Alan Shepard's crooked grin appears during rehearsals for fitting his pressure suit. Shepard was a man of contradictions with a changeable personality – perhaps fittingly described as mercurial.

Unlike the flight of Vostok 1 – which was prepared for and carried out in almost total secrecy until it was actually under way – the launch of Mercury-Redstone 3 from Cape Canaveral carrying Alan Shepard's *Freedom 7* capsule happened amid a blaze of publicity on 5 May 1961.

Shepard was woken in the early hours of the morning. After breakfasting with his backup, John Glenn, and other members of the team, he was given a final medical examination and pronounced fit to fly. Before he suited up, biosensors were placed on his skin in a variety of places. By 3:55am EST, he was boarding the transit bus for the journey out to the launch pad. As he later recalled:

"The excitement really didn't start to build until the trailer – which was carrying me, with **a spacesuit with ventilation and all that sort of stuff** – pulled up to the launch pad."

Shepard entered the spacecraft at 5:20am – inside he found a note that had been left by Glenn, which read "No handball playing here!". The launch was set for 7:25 – barring delays, he had 125 minutes to wait. But at T-15 the first in a series of holds was called. After an hour's wait, and with another hour at least to go, Shepard had a problem, which he communicated to Gordon Cooper in launch control. As the communications transcript records, the capsule's electrical supply had to be turned off while Shepard relieved himself:

Alan Shepard: Gordo!
Gordon Cooper: Go, Alan.
AS: Man, I got to pee.
GC: You what?
AS: You heard me. I've got to pee. I've been up here forever ... tell them to turn the power off!
GC: Okay, Alan. Power is off. Go to it.

MAN IN A CAN
Shepard later commented: "It's a very sobering feeling to realize that one's safety factor was determined by the lowest bidder on a government contract."

FOUR STEPS TO SPACE
Shepard and Glenn suited up together with the assistance of suit technician Joe Schmitt, but only Shepard got to make the fateful walk to the gantry elevator, carrying his portable air unit. Once there, he was helped into the tiny capsule by the ground crew. Schmitt then shook his hand while the ground crew wished the astronaut "Happy landings!"

"I think **all of us certainly believed the statistics** which said ... probably 88% chance of mission success and maybe **96% chance of survival**. And we were willing to take those odds."

Alan Shepard, February 1991

Four hours and 14 minutes after Shepard boarded *Freedom 7*, launch control "lit the candle" and Shepard soared skywards. For 45 seconds the ascent was smooth, but then vibrations began to build up as the rocket approached the sound barrier. Two minutes into the flight, Shepard was experiencing maximum acceleration of around 6*g*. Another 20 seconds, and the Redstone engine beneath him shut down. Still soaring skywards, *Freedom 7* jettisoned its launcher and escape tower as its external temperatures rose to 104°C (220°F). Shepard described what he could see:

"On the periscope ... **What a beautiful view.** Cloud cover over Florida – three to four tenths near the eastern coast. Obscured up to Hatteras ... I can see [Lake] Okeechobee. Identify Andros Island. Identify the reefs."

The capsule had automatically turned itself around by the time it reached its peak altitude of 187km (116 miles). Shepard now assumed manual control to fine-tune the capsule's attitude and fire the retrorockets. As he plummeted back down the re-entry curve, he jettisoned the retropack strapped across the capsule's heat shield. Plunging back into the atmosphere, Shepard felt the strain of up to 11.6*g* before the drogue parachute was deployed at 6,400m (21,000ft). At 3,000m (10,000ft) the main parachute slowed the capsule further, dropping it back to a splashdown at a relatively sedate 10.5m (35ft) per second.

"The rocket had **worked perfectly**, and all I had to do was survive the re-entry forces. **You do it all**, in a flight like that, in a rather short period of time, **just 16 minutes** as a matter of fact."

Alan Shepard, February 1991

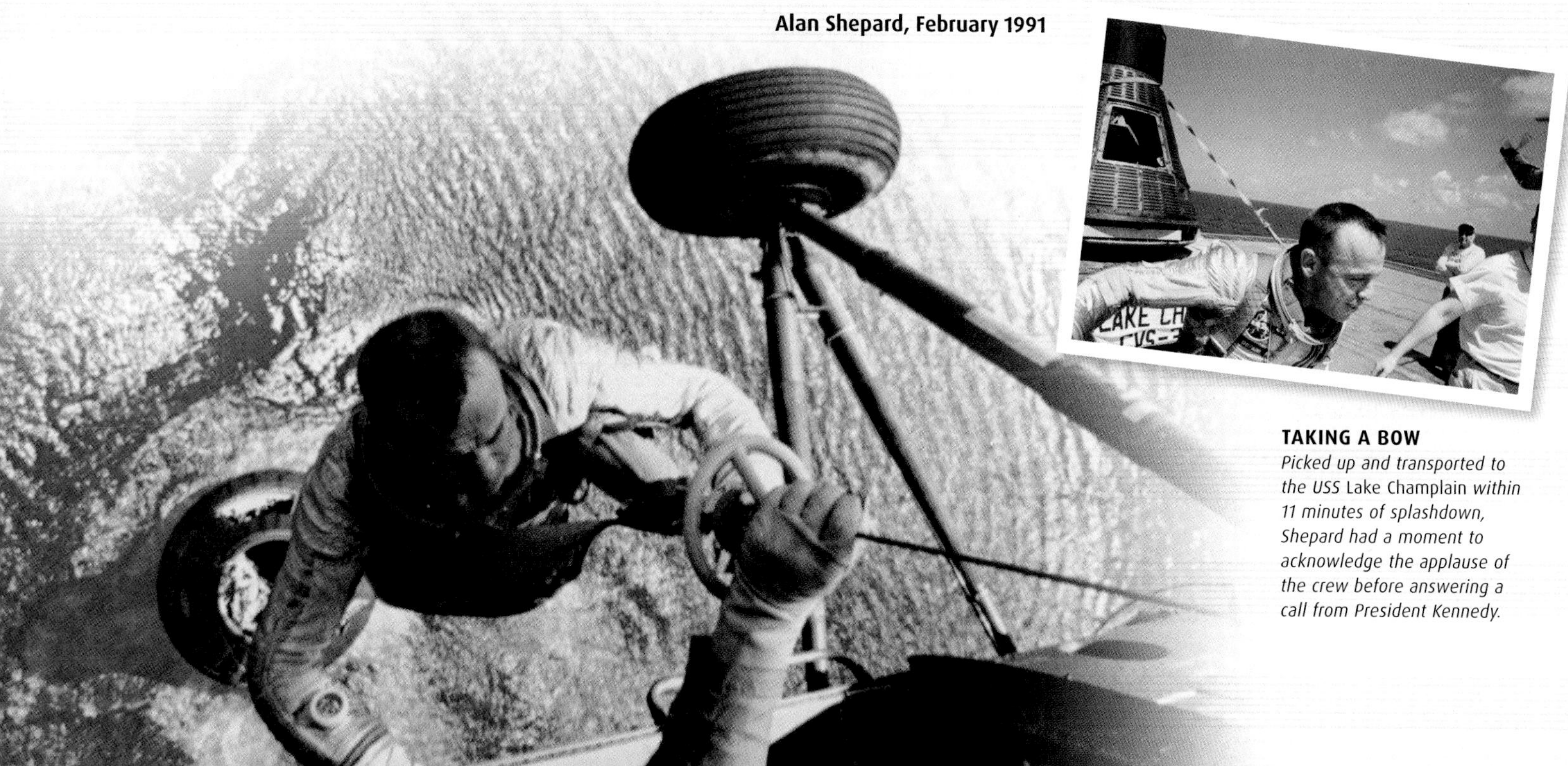

TAKING A BOW
Picked up and transported to the USS Lake Champlain *within 11 minutes of splashdown, Shepard had a moment to acknowledge the applause of the crew before answering a call from President Kennedy.*

1955 1956 1957 1958 1959 1960 1961 1962 1963 1964 1965 1966 1967 1968 1969 1970 1971 1972 1973 1974 1975 1976 1977 1978 1979 1980 1981 1982 1983 1984 1985 1986 1987 1988 1989 1990 1991 1992 1993 1994 1995 1996 1997 1998 1999 2000 2001 2002 2003 2004 2005 2006 2007 2008 2009 2010 2011 2012 2013 2014 2015 2016 2017 2018 2019 2020

29 November 1961
John Glenn is selected as pilot for the first orbital Mercury flight, with Scott Carpenter as his backup.

23 January 1962
The first in a series of postponements hits the scheduled launch of *Friendship 7*. A series of hitches, caused by bad weather but also by a fuel leak, eventually delays the launch for almost a month.

20 February 1962
An Atlas rocket finally launches Glenn and *Friendship 7* into orbit. The flight lasts a little less than five hours and mostly goes smoothly, although re-entry is a more traumatic experience than Glenn had expected.

1 March 1962
Four million people line the streets of New York for a tickertape parade to honour Glenn.

Mercury in orbit

John Glenn's three historic orbits around the Earth in February 1962 briefly put NASA back on a more even footing with its Soviet rivals. However, the first orbital Mercury was not entirely flawless.

Despite the near-disastrous ending of Grissom's *Liberty Bell 7* mission in July 1961, the Mercury capsules had proved themselves reliable in flight. Meanwhile, the Atlas ICBM also seemed to have overcome its early glitches – now it was time to combine the two and put Mercury into orbit.

However, any hopes that NASA might have had of levelling the score with its Soviet rivals were dashed on 6 August by Gherman Titov's successful day-long flight aboard Vostok 2 (see p.80). Despite pressure from some quarters to rush to a manned launch, Gilruth's Space Task Group, now in the process of relocation from Langley to the new Manned Spaceflight Center at Houston, Texas, continued to insist on a steady process of qualification. The first unmanned launch of a Mercury capsule on an Atlas rocket took place on 13 September and went flawlessly, but before NASA would trust a man to orbit, it insisted on a full dress rehearsal with another space chimp.

As a result, Enos, a male chimp like Ham, was launched into space on 29 November aboard Mercury-Atlas 5. The flight went well, despite a few problems with control of the spacecraft's attitude in orbit, and Enos coped magnificently with 181 minutes of weightlessness and higher G-forces than even Ham had tolerated. The Mercury-Atlas combination was now ready to take its first human passenger.

BIOGRAPHY

JOHN GLENN

Ohio-born John H. Glenn (1921–2016) was militarily the senior member of the Mercury seven, a highly decorated captain in the Marine Corps with experience in the Second World War and the Korean War. His charismatic personality made him a particular media favourite among the seven, and after his return to Earth he retired from NASA to follow a career in business and politics, eventually as Democratic Senator for Ohio (1975–99). In 1998 he finally returned to orbit at the age of 77, becoming the oldest person to travel in space during a nine-day mission aboard the Space Shuttle *Discovery* (see p.207).

A saga of delays

NASA was understandably eager to launch Glenn into orbit before the end of 1961, but a series of snags conspired against them. Problems with the hardware during testing at Cape Canaveral pushed the launch back, and a provisional date of 16 January was finally set. Problems with the Atlas fuel tanks held that up until 23 January, and poor weather then caused a further week of delays. While fuelling the rocket

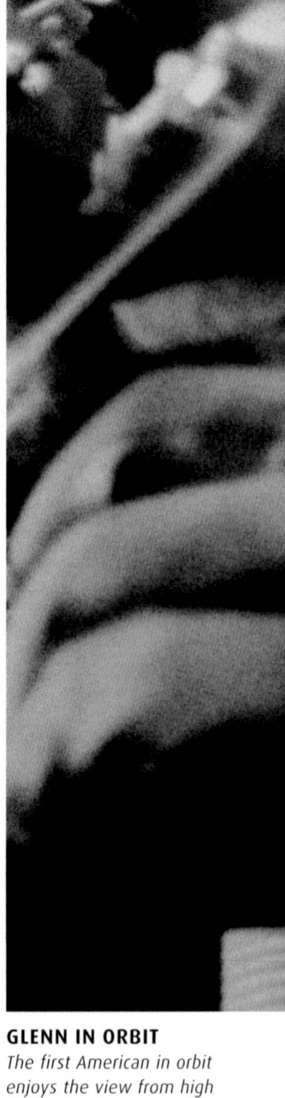

GLENN IN ORBIT
The first American in orbit enjoys the view from high above the Earth (note the reflection in his visor). Throughout the flight, an automatic camera recorded his every action.

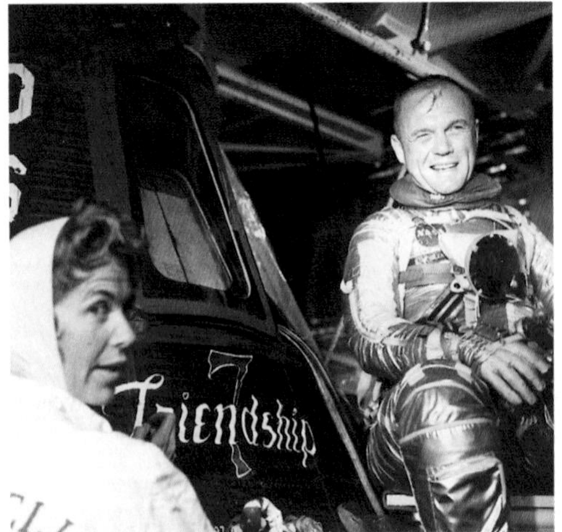

FRIENDSHIP 7
Each astronaut got to choose the name of his own craft. Here, Glenn poses with Chrysler employee Cecelia Bibby, the painter responsible for each of the emblems.

GLENN GIVES THE SALUTE
An ebullient Glenn finally boarded his spacecraft at 6:03am. He had already been awake for four hours, and had to wait almost four more before launch.

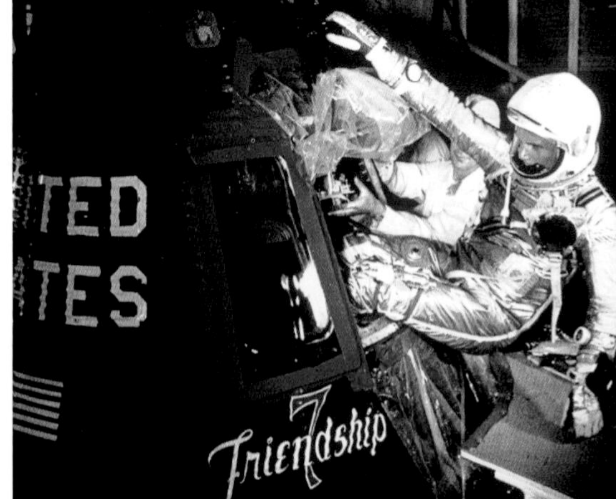

CLIMBING ABOARD
Once Glenn had clambered into position, 70 bolts secured the hatch in place. Halfway through the process, a broken bolt was found, so the whole process had to be restarted.

SUNSET FROM SPACE
As Glenn flew over the Indian Ocean, he described the "beautiful" sunset. It lasted longer than he had expected, and even over the night side of the Earth an arc of blue daylight lingered on the horizon.

in preparation for a 1 February launch, engineers discovered a serious fuel leak, and repairs took a further fortnight, by which point the weather had closed in again. It was not until 19 February that it began to clear, and preparations could be made for a launch the next day.

The flight of *Friendship 7*

At 09:47am local time on 20 February, Glenn's spacecraft finally soared into the clear blue skies above Cape Canaveral. The launch went perfectly, and *Friendship 7* was soon in an orbit between 159 and 265km (99 and 165 miles) above the Earth.

The capsule's flight path took it east across the Atlantic, over tracking stations in the Canary Isles and Nigeria, then out above the Indian Ocean and across the night side of Earth. As it left Australia behind and flew into sunrise over the Pacific, Glenn reported seeing tiny glowing specks dancing outside the capsule. The mystery of these "fireflies" would eventually be solved by Scott Carpenter (see over).

Glenn had successfully used the capsule's thrusters to turn his spacecraft around over the Atlantic so he was facing forwards, but a problem began to develop with the automatic control of the yaw thrusters as he approached the west coast of the USA, and he now had to maintain the capsule's attitude manually. During the second 89-minute orbit, a number of further problems developed. The need for manual control had drained the capsule's fuel supply quite rapidly, and Glenn was told to let it drift to preserve fuel for later. More seriously, during the first pass over Cape Canaveral, a capsule sensor indicated that the heat shield and landing cushion were no longer securely in place. In fact, it seemed that they were held on only by the retrorocket pack strapped across the shield. After some analysis, Flight Director Chris Kraft decided that the pack should be kept in place, rather than jettisoned as planned, during re-entry.

As *Friendship 7* flew back across the Pacific on its final orbit, Glenn adjusted its attitude before firing the retrorockets to drop out of orbit over California. Re-entry with the rocket pack still in place required more manual control than usual, but Glenn proved himself up to the task and was treated to a spectacular light show as fiery fragments of the pack trailed behind the spacecraft. Splashdown in the Atlantic fell 60km (40 miles) short of predictions, but Glenn's capsule was soon being hoisted aboard the recovery ship *USS Noa*, safe after 295 minutes in orbit. Later tests showed that the sensor, not the shield itself, had been faulty.

MARINE AWARD
Glenn was awarded a special medal by the US Marine Corps to mark his successful orbital flight.

HISTORY FOCUS

NASA'S FIRST COMPUTERS

From 1943, NACA employed a large staff of (mostly female) mathematicians to work on aeronautical calculations at its Langley Research Center in Virginia. Segregation meant that these "human computers" were separated into white and African-American groups. With the establishment of NASA in 1958, segregation ended and the computers supplied vital calculations for the development of Project Mercury, as dramatized in the 2016 movie *Hidden Figures*. Katherine Johnson (b.1918), seconded to the Flight Research Division, made the trajectory calculations for all the early Mercury flights. Famously, John Glenn refused to launch until she had verified the electronically calculated profile of his *Friendship 7* mission.

MISSILE ROW
This 1964 aerial shot of Cape Canaveral shows the view north along the coast and up Missile Row, a range of launch pads used for testing the Redstones and early ICBMs, from which many of the Mercury missions blasted off. NASA's larger Apollo-era pads are under construction in the distance.

Later Mercury missions

Having finally reached orbit almost a year behind the Soviet Union, NASA used the remaining Mercury missions to extend American experience in space and investigate the possibilities of science in orbit.

SPECTACULAR AURORA
Mercury-Atlas 7 blasts into the early morning sky in May 1962. Scott Carpenter's spacecraft, Aurora 7, *got its name from the street where Carpenter lived as a child.*

When John Glenn was allocated the first Mercury orbital flight in late 1961, Deke Slayton was told he would be the second American into orbit. But by the time Glenn had made his historic flight, fate had intervened – doctors discovered that Slayton had a slightly erratic heartbeat. The unlucky astronaut was grounded, and so it was Scott Carpenter whose *Aurora 7* capsule entered orbit on 24 May 1962. This three-orbit flight was essentially a repeat of Glenn's, but this time the astronaut could concentrate on science rather than the condition of his spacecraft. Carpenter made a brief study of how fluids behaved in the weightless conditions of orbit, ate a meal, and photographed the Earth from above. He also accidentally solved the mystery of Glenn's orbital "fireflies" – approaching his third dawn, Carpenter accidentally knocked his head against the cabin wall, and dislodged a shower of sparkling ice crystals from the exterior. Snags with guidance and alignment systems caused problems during re-entry, and Carpenter splashed down more than 400km (250 miles) off target. By the time the recovery crews reached him, he had escaped through the top of the capsule and was floating on a life raft.

LOCATION MAP
Cape Canaveral lies along the Florida coast, at the southern end of the US eastern seaboard. This location ensures relatively reliable weather, and the islands to the east are ideal for tracking launches.

TECHNOLOGY
CAPE CANAVERAL

NASA's major launch site grew up at Cape Canaveral in Florida for a variety of reasons. The site, to the north of an established US Air Force base, was first earmarked for ballistic missile tests in 1949, as the range of missiles being tested began to outgrow landlocked ranges such as White Sands. Its location later made it an obvious place to attempt satellite launches – as the Earth rotates, areas close to the equator move more quickly than those near the poles, and a vehicle launched eastwards at relatively low latitudes receives a substantial speed boost to help it into orbit. The launch complex, today known as Kennedy Space Center, is actually two separate establishments – NASA's civilian and commercial launch pads are in the northern half, on Merrit Island, while the Air Force operates a military launch facility to the south. In reality, though, civilian launches often take place from the military pads, and vice versa.

TECHNOLOGY
THE MERCURY-ATLAS LAUNCHER

The Atlas D launcher used in the later stages of the Mercury programme was a modified version of the US Air Force's Atlas ICBM, with a distinctive configuration sometimes referred to as "1.5 stage". Three engines at the base of the rocket all drew fuel and oxidant from the same tanks and fired in parallel at launch, before the two flanking engines were discarded. Modified Atlas rockets still form the backbone of today's US space programme – their unusual single-skin design, in which the outer hull acts as the fuel-tank wall, reduces the rocket's weight and increases its range.

Sigma and *Faith*

Carpenter's re-entry problems turned the emphasis of Walter Schirra's *Sigma 7* mission back to engineering. During six orbits of the Earth on 3 October, much of the time was spent testing the automatic control systems. Schirra also tested elastic devices for exercise in space, attempted to steer his capsule by the stars, and made the first live TV broadcast from space. Re-entry this time was perfect, and *Sigma 7* splashed down in the Pacific Ocean.

The last Mercury mission, piloted by Gordon Cooper, was also by far the longest – 22 orbits over a period of 34 hours. The capsule, *Faith 7*, needed modification to support an astronaut for this long, and so launch did not take place until 15 May 1963. During his flight, Cooper was occupied with a range of experiments, including the release and tracking of a strobing microsatellite from the capsule. He also studied the Earth from orbit – his reports of seeing individual roads and houses on the ground below made some on the ground think he was suffering from hallucinations, but they ultimately paved the way for the modern science of remote sensing.

Following Cooper's successful Pacific splashdown, some at NASA pushed for a three-day Mercury mission. But it was time to move on. In May 1961, the President had given NASA a new and exciting goal – they were going to the Moon.

15 March 1962
Scott Carpenter is moved up the Mercury flight roster to replace Deke Slayton on the second orbital Mercury mission, after the discovery that Slayton has a minor heart defect.

24 May 1962
Carpenter and *Aurora 7* launch on time at 07:45. During the mission, the spacecraft uses more manoeuvring fuel than expected, and as a result splashes down 400km (250 miles) off target. Carpenter has to board his life raft and wait several hours for rescue.

3 October 1962
Walter Schirra's six-orbit flight aboard *Sigma 7* is completed without a hitch.

14 May 1963
The scheduled launch of Gordon Cooper's *Faith 7* is postponed due to a problem at a radar tracking station.

15 May 1963
Faith 7 launches successfully. Cooper completes 22 orbits of the Earth and successfully deploys a tethered balloon for studying conditions around the Earth.

21 May 1963
The end of the Mercury programme is marked by a White House reception at which the astronauts and other key project personnel are honoured by President Kennedy.

NOVEMBER 1967: TARGET MOON
The first of NASA's giant Saturn V rockets stands ready to launch the Apollo 4 test mission. Within two years similar rockets would launch men all the way to the Moon.

THE RACE TO THE MOON

THROUGHOUT THE 1960S, the Space Race became an outlet for all the Cold War rivalries that threatened to spill over into open conflict on Earth. Fully aware that the perception of Soviet superiority in space would have political repercussions around the globe, America's new, young President galvanized his nation with what seemed an insanely ambitious goal – to put American astronauts on the Moon by the end of the decade.

The challenge of a lunar landing was the endgame of the entire Space Race – the target lay at the very limits of 1960s technology, stretching the ingenuity and engineering skill of each side, while remaining tantalizingly achievable. Through the early stages, the rival powers continued an open race to each new spectacular, but cracks were starting to appear in the Soviet programme, and their lunar effort eventually fell apart in almost total secrecy. John F. Kennedy would not survive to see it, but America would live up to his challenge and eventually emerge as the ultimate victor in the Space Race.

Kennedy's challenge

With America soundly beaten in the early stages of the Space Race, in 1961 President Kennedy announced an ambitious programme to overtake the Soviet Union and land the first man on the Moon.

20 January 1961
John F. Kennedy is inaugurated as President of the United States.

14 February 1961
James E. Webb is appointed as new NASA Administrator.

12 April 1961
Cosmonaut Yuri Gagarin makes his historic first flight into orbit.

14 April 1961
A meeting of top officials tells Kennedy that NASA's best chance of catching the Soviets is in the race to the Moon.

8 May 1961
Vice President Lyndon Johnson formally delivers to the President a set of recommendations on America's future in space, based on his discussions with Webb and others.

10 May 1961
Kennedy and his senior advisers ratify Johnson's recommendations.

25 May 1961
President Kennedy announces America's lunar ambitions in a speech to Congress.

John F. Kennedy owed his presidency in part to the Space Race – he had turned early Soviet triumphs to his advantage, accusing Eisenhower's administration of complacency. His inauguration coincided with the arrival of a new administrator at NASA. In February 1961, James E. Webb took charge of both the agency and the Mercury programme. Hugh Dryden, former head of NACA, remained as his deputy. In addition, one of Kennedy's first acts as President was to establish a National Space Council, headed by Vice President Lyndon Baines Johnson.

Racing for the Moon

Despite initial popularity, Kennedy's honeymoon with the American electorate ended within just a few months, amid a number of political setbacks including Yuri Gagarin's historic flight. Although Alan Shepard's forthcoming suborbital hop would help to restore American pride, it was clearly a poor retaliation to the Soviet achievement. Something had to be done that would focus the country's gaze on a more distant goal, allowing them to see past the immediate impression of Soviet space superiority. Just two days after Gagarin's flight, on 14 April, Kennedy summoned senior members of his administration and NASA to a policy meeting. At what point, Kennedy asked, might America finally manage to overtake its rivals? The power of Russian launchers meant that the Soviets would almost certainly be the first to put a multi-cosmonaut spacecraft into orbit. The same might go for any plans to launch a large, semi-permanent space station. When it came to the Moon, though, the competition seemed more balanced. While it was likely that a Soviet crew would be first to circle the Moon, the task of landing astronauts and returning them safely to Earth required so much new technology that the US would have a chance to catch up. If an all-out effort was made, the chances of an American being the first to set foot on the Moon were probably about fifty-fifty.

This was good enough for Kennedy, though details of the mission itself still had to be worked out. In early May, Johnson, Webb, and others met to draft a political justification of why America should race the Soviet Union to the Moon. This formed the basis for the President's historic announcement to Congress on 25 May, when he proclaimed: "I believe that this nation should commit itself to achieving the goal, before this decade is out, of landing a man on the Moon and returning him safely to the Earth. No single space project in this period will be more impressive to mankind or more important for the long-range exploration of space; and none will be so difficult or expensive to accomplish."

BIOGRAPHY

DEKE SLAYTON

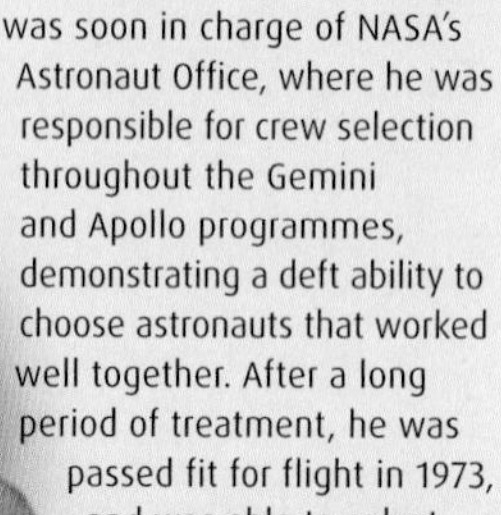

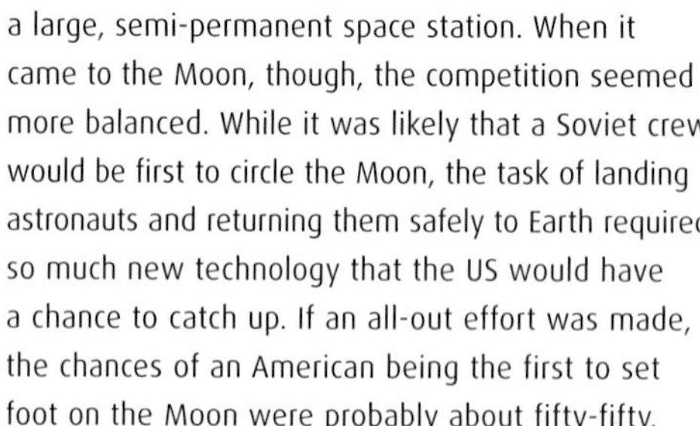

Donald K. (Deke) Slayton (1924–93) was the only one of NASA's original seven astronauts not to fly on a Mercury mission. Invalided out of the programme (and the Air Force) after the discovery of a heart irregularity, Slayton was soon in charge of NASA's Astronaut Office, where he was responsible for crew selection throughout the Gemini and Apollo programmes, demonstrating a deft ability to choose astronauts that worked well together. After a long period of treatment, he was passed fit for flight in 1973, and was able to select himself for the Apollo-Soyuz mission of 1975.

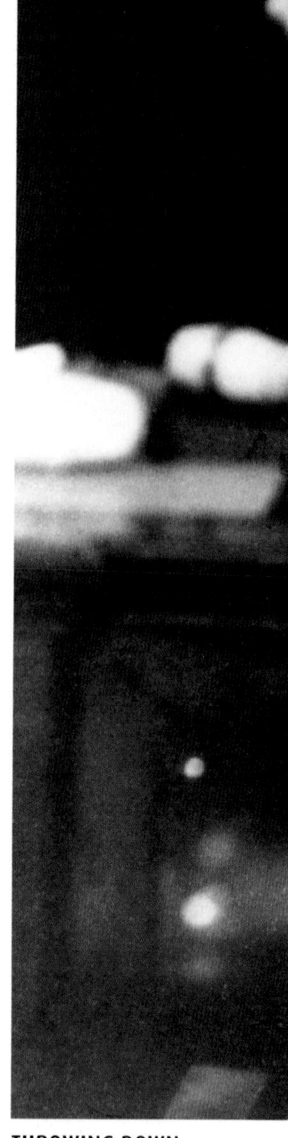

THROWING DOWN THE GAUNTLET
Kennedy's commitment to space travel was more political than personal – he understood what it could mean to the American people in the midst of the Cold War. After his historic address to Congress, he soon found that the politicians on Capitol Hill agreed with him.

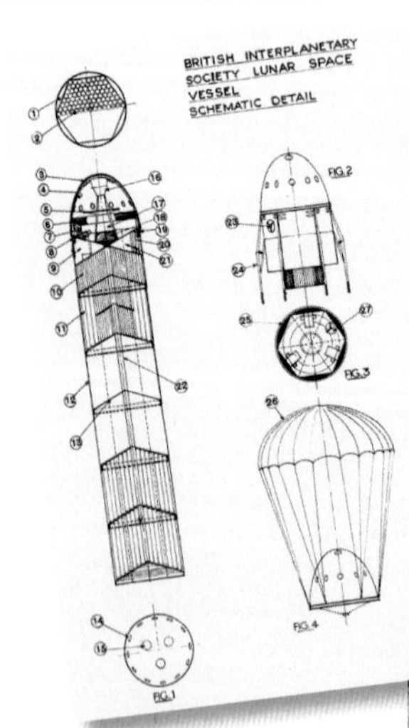

BRITISH VISION
In 1937, members of the British Interplanetary Society made a detailed study of how a manned lunar mission might be carried out. Their spacecraft designs bore a striking resemblance to Apollo.

"We choose to go **to the Moon in this decade** and do the other things, not because they are easy, but **because they are hard ...**"

US President John F. Kennedy, Houston, Texas, 12 September 1962

With such an ambitious goal ahead, NASA had to change priorities. Until now, the unspoken assumption had been that the exploration of space would roughly follow the template laid out by von Braun's *Colliers* articles of the mid-1950s, with colonization of Earth orbit as a prelude to the lunar voyage. Now there would be an all-out race for the Moon, and that would require new spacecraft and new skills. With the Mercury capsule relatively limited, an intermediate trainer spacecraft would be necessary – a vehicle that could be used for practising orbital manoeuvres, rendezvous and docking in space, and other techniques. This spacecraft would be called Gemini.

EYES ON THE SKY
Wernher von Braun's rocket team would play a key role in the US Moon programme. Here von Braun explains his Saturn launch system to Kennedy in November 1963.

TECHNOLOGY

THE FIRST FLYING SPACESHIP

Gemini spacecraft

Gemini has been called the first true spaceship, because its revolutionary design allowed it to change orbits and actually "fly" in space, rather than just following the trajectory into which it was initially launched. It was also the first spacecraft with a docking capability. Conceived after Apollo, Gemini's design was in many ways more advanced than the spacecraft that succeeded it. Even after its last flight in 1966 Gemini had a long afterlife, with proposals for new projects based on the spacecraft continuing into the 1970s.

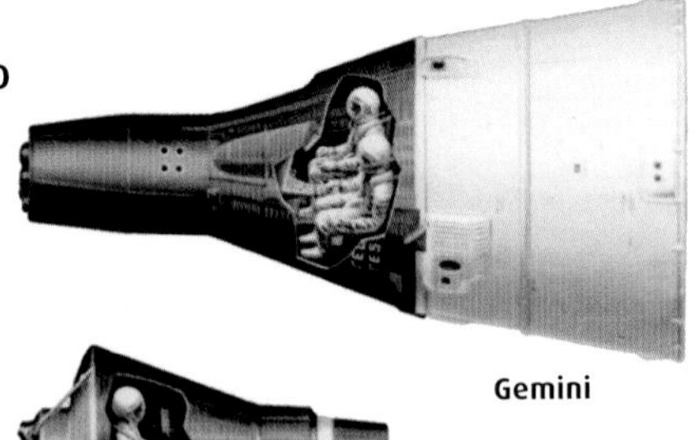

CAPSULES COMPARED
Although the Gemini re-entry module was the appropriate size for a two-man version of Mercury, it was only one element of a larger spacecraft, with a retrograde section and equipment module attached behind it.

THE GEMINI SPACECRAFT
Gemini had three major sections: the re-entry module, the retrograde section, and the equipment module. While the crew were confined to the re-entry module, vital supplies of power and oxygen came from the equipment module. The retrograde module contained thrusters for changing orbit and retrorockets that were used to trigger re-entry.

CREW	2
LENGTH	5.6m (18ft 4in)
MAXIMUM DIAMETER IN ORBIT	3.05m (10ft)
MASS AT LAUNCH	3,763kg (8,297lb)
MASS AT LANDING	1,983kg (4,371lb)
ENGINES	4 x solid fuel retrorockets
MANUFACTURER	McDonnell Aircraft Corporation

EASY MAINTENANCE
Hatches dotted all around the Gemini equipment module allowed components and consumables to be removed and replaced with ease.

propellant tanks
drinking water tank
retrograde rocket
orbit attitude control thruster
rear shielding and insulation
launch-vehicle mating cable
orbit attitude-control thruster
fuel cells
battery pack
communications equipment
manoeuvre thruster
Equipment module
Retrograde module

PARAGLIDER TEST
One early concept for Gemini would have seen it fly back to a ground-based landing beneath a glider called a Rogallo Wing. However, tests showed that the wing would not always deploy reliably, and so the concept was abandoned in favour of a splashdown.

GEMINI FUEL CELL

Gemini was the first spacecraft to use fuel-cell technology, which generated electricity by chemically combining hydrogen and oxygen to form water. This allowed it to operate for much longer than its predecessor.

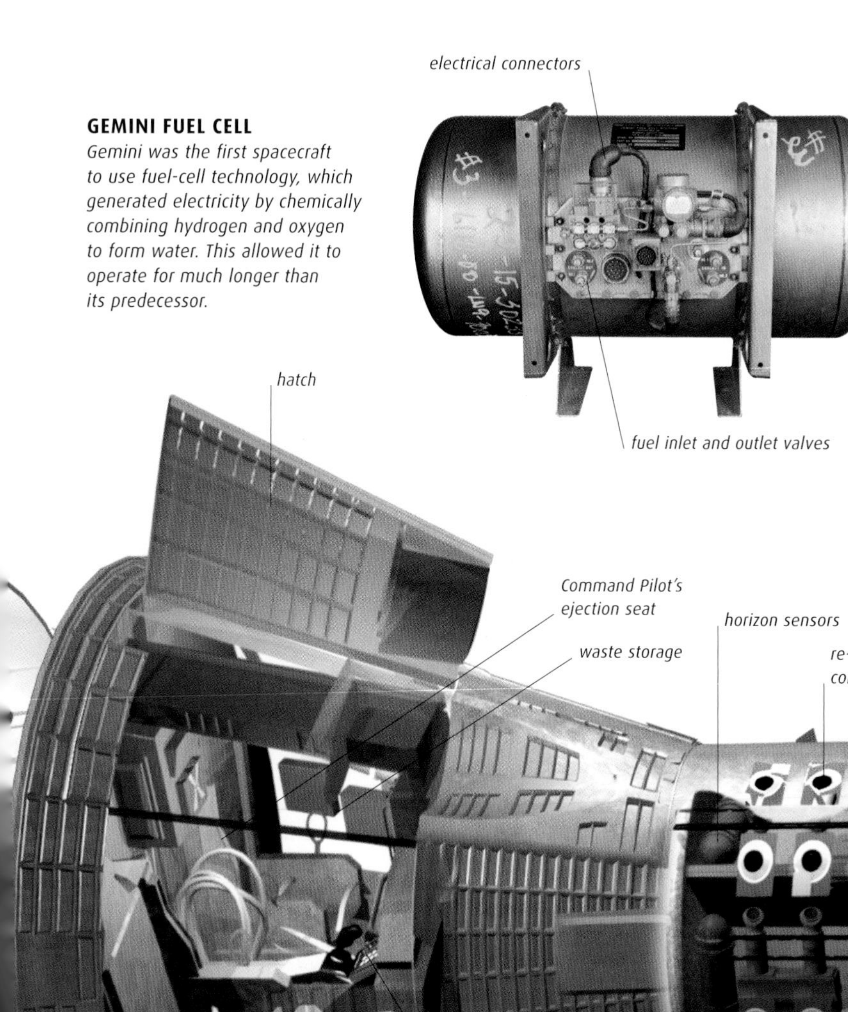

Re-entry module

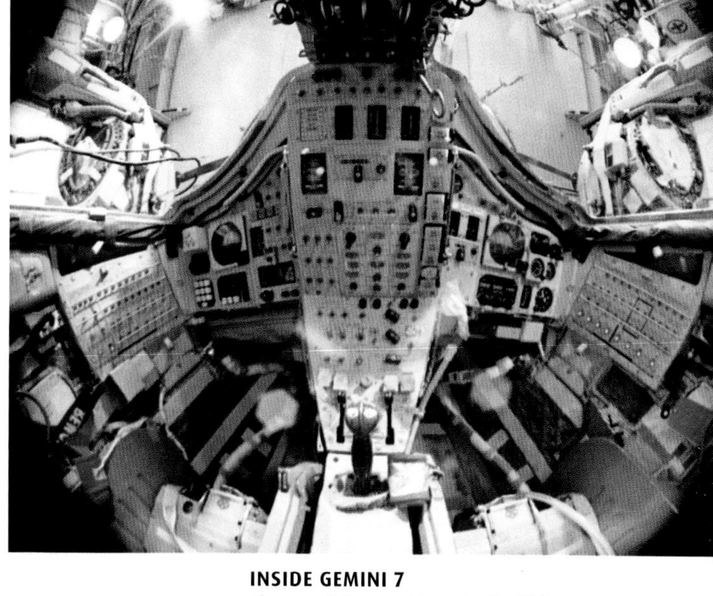

INSIDE GEMINI 7

The capsule's resemblance to the flight deck of a jet was partly due to Jim Chamberlin, a Canadian engineer who joined NASA after working on the Avro Arrow fighter-interceptor aircraft project.

STACKING GEMINI 4

After the Gemini spacecraft arrived at Kennedy Space Center (above), it was stacked onto a Titan launch vehicle and raised using an erector tower. There was no escape tower on top of the capsule – unlike in Mercury and Apollo, Gemini astronauts would have used their ejector seats for emergency escapes.

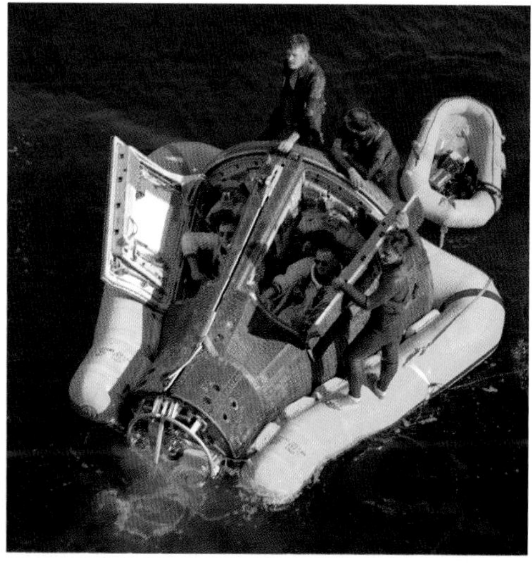

RECOVERY AT SEA

Gemini was suspended from its parachute at two points, allowing it to splash down horizontally. The weighting of the spacecraft kept it upright in the water until the recovery crew arrived and attached a flotation collar.

Voskhod

After the announcement of the US Gemini project, Sergei Korolev was determined to maintain a Soviet lead in the Space Race. The result was Voskhod, a hurriedly modified and risky three-man spacecraft.

5 February 1964
During a tour of OKB-1 to review progress on the Soyuz spacecraft, Korolev surprises both cosmonauts and colleagues with his plans for a three-man version of Vostok.

13 March 1964
Development of the Voskhod project is approved by the Soviet government's Military-Industrial Commission.

15 August 1964
Following testing of the landing systems, the Council of Chief Designers meets to approve Voskhod 1 as ready for spaceflight.

8 September 1964
A test drop of the Voskhod re-entry module from 10km (33,000ft) ends in a spectacular crash. Nevertheless, Korolev overrides the concerns of others about the retrorocket system.

6 October 1964
A Voskhod test launch goes almost flawlessly. The capsule returns safely to Earth a day later.

12 October 1964
Voskhod 1 is launched from Baikonour on a day-long flight.

When the Gemini programme was announced in December 1961, it created a dilemma for Soviet politicians and engineers alike. The true successor to the Vostok capsules was Korolev's ambitious three-man Soyuz complex (see p.128), but this was still in the early stages of development, and the signs were that Gemini would certainly be ready long before it took flight. Faced with the prospect of losing the lead in the Space Race long before the final chase to the Moon, Korolev took a desperate gamble – one that would put the lives of his cosmonauts at more risk than ever before, but which would ultimately fool the watching world and maintain Soviet prestige. The Chief Designer apparently took the decision to develop a makeshift three-man capsule without consulting his superiors.

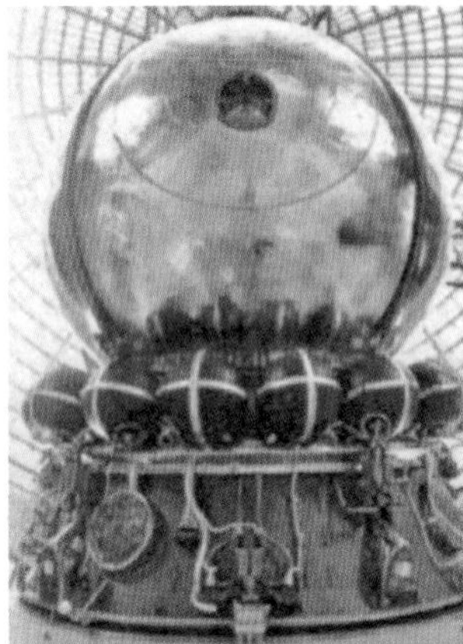

MINOR MODIFICATION
Externally, Vostok and Voskhod looked to be near-twins. Even the new retrorockets were attached to the descent parachute rather than to the capsule.

Korolev's gambit

During a visit to inspect work on the Soyuz capsule in February 1964, Korolev announced to the assembled cosmonauts that there would be no more lone Vostok flights – instead the capsules already under construction would be converted into new configurations. One variant would squeeze three people into the cramped space, while the other would carry two cosmonauts in spacesuits and incorporate an airlock system allowing them to leave the spacecraft and float free in space.

The trade-offs needed to meet these requirements would make the flights much more risky for the cosmonauts – the ejector seat would be replaced by couches, and a new retrorocket would have to slow the re-entry module's descent as it neared the ground, enabling the cosmonauts to land safely inside. Most dangerous of all, cosmonauts in the three-man capsule would not have room to wear spacesuits. Dressed instead in jumpsuits, they would have no protection from the vacuum of space if the capsule lost pressure. Several of Korolev's colleagues voiced doubts about the plan, including Kamanin. However, within a month the project had been given the go-ahead. Khrushchev was never told of the safety fears, but even if he had been, it is unlikely that a man with such an eye for the spectacular would have rejected the proposal.

Choosing the crew

Crew selection was a long process. At first Yuri Gagarin himself was to have commanded, but Kamanin was unwilling to risk a national hero on such a dangerous mission and Vostok 4 backup pilot Vladimir Komarov was finally selected. For the first time, Voskhod would allow people other than trained pilots into space. All agreed that sending a doctor into orbit would have benefits, while as an incentive to his engineers, Korolev decreed that one of them would have the chance to fly in the completed capsule. In the end, the successful candidates were medical specialist Boris Yegorov and engineer Konstantin Feoktistov.

Despite an overambitious initial target of a flight by August, Voskhod launched on 12 October 1964. The mission, which lasted for just over 24 hours, went relatively smoothly and provided the Soviets

INSIDE VOSKHOD
The Vostok capsule, on which the Voskhod design was based, was very cramped with three people aboard. It was also difficult to read the instrument panel in the new capsule.

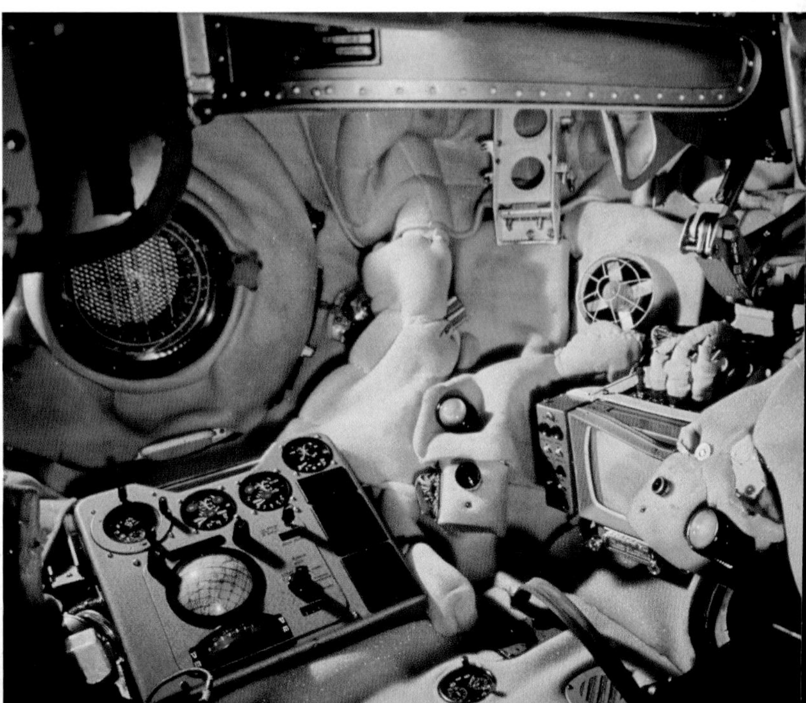

with yet another propaganda coup. Carefully worded statements gave the impression that Voskhod was a major advance in spacecraft design, rather than the rush job that it was in reality, but the mission still provided useful insights into how a crew could work together in space. By the time America was ready to launch its first Gemini, a second Voskhod spectacular – the first spacewalk – was almost ready. Ironically, though, Korolev's great sponsor, Khrushchev, was in no position to crow about these new triumphs – even as Voskhod 1 circled the Earth, he was deposed in a coup, to be replaced by Leonid Brezhnev.

WELL-PADDED CELL
One concession to the risks of landing was the padding attached to every surface inside the capsule. However, in the event of a parachute or retrorocket failure, it would have made little difference.

THE VOSKHOD THREE
The first Voskhod crew of test pilot Vladimir Komarov (left) and civilians Boris Yegorov (centre) and Konstantin Feoktistov (right) were able to send back some dramatic views of the Earth as seen from space (above).

1964

8 April 1964
A test launch puts an unmanned Gemini capsule into orbit, still attached to its upper rocket stage.

19 January 1965
An unmanned suborbital hop tests the capsule's performance during atmospheric re-entry.

23 March 1965
Gemini 3 carries Gus Grissom and John W. Young on three orbits around Earth.

3 June 1965
Gemini 4 launches, carrying James McDivitt and Ed White on a 62-orbit mission that includes the first American spacewalk.

7 June 1965
Gemini 4 returns safely to Earth.

21 August 1965
Gemini 5, with Gordon Cooper and Charles Conrad aboard, launches.

29 August 1965
Gemini 5's safe return from a record-breaking eight-day mission establishes the practicality of a mission to the Moon and back.

Gemini takes flight

The early Gemini missions gave NASA its first experience of long-duration spaceflight. They also achieved a number of other American space firsts, paving the way for the more advanced later missions.

Gemini's journey from concept to completion took place at breakneck speed – the first test launch came on 8 April 1964, less than 30 months after the start of the programme. Gemini had to meet this timescale in order to fulfil its role as a bridge between the Mercury flights and the Apollo missions due to start in 1967.

Although the Gemini capsule resembled a scaled-up Mercury (the programme was originally named Mercury Mark 2), it marked a major leap forwards – it was perhaps the first true spaceship thanks to its ability to change orbits and manoeuvre in space.

First flights

The first test launch in 1964 was designed to check the spacecraft's function in orbit. In place of a crew, it carried instruments that sent back data on conditions during launch and in orbit. Gemini 1 was not designed for recovery, and so a second capsule was sent on a brief suborbital hop in January 1965 to assess re-entry conditions. By 23 March 1965, Gemini 3 was ready for the first manned launch.

Deke Slayton, now in charge of NASA's Astronaut Office, wanted to mix experienced astronauts with newer recruits. The first Gemini crew, therefore, were Gus Grissom (of *Liberty Bell 7* fame) and John W. Young. In a reference to his earlier misadventures, Grissom named the capsule *Molly Brown*, after a survivor from the "unsinkable" *Titanic* disaster. This was to be the last spacecraft named by its pilot.

Molly Brown's voyage proved less eventful than her namesake's – it lasted just three orbits, but during that time the astronauts were able to test their new engines and change their orbit in space for the first time. They also enjoyed a corned-beef sandwich that Young had smuggled aboard – much to the annoyance of mission control.

TECHNOLOGY

GEMINI'S TITAN LAUNCHER

All the Gemini missions were launched using Titan II rockets, derived from the US Air Force's Titan ICBM. This two-stage rocket, 33.2m (101ft) tall with the Gemini capsule in place, consisted of a first stage with dual rocket motors that burned a mix of unsymmetrical dimethylhydrazine (UDMH) and nitrogen tetroxide. The motors were relatively simple because this combination of fuel and oxidant is hypergolic – it combusts on contact, with no need for an ignition system. The second stage had a single engine burning the same fuel combination, and all three engines had gimbal-mounted exhaust nozzles that could be tilted to change direction.

BIOGRAPHY

GENE KRANZ

Flight director Gene Kranz (b.1933) was in charge at NASA's new Houston-based Mission Control during many historic missions. Born in Ohio, Kranz trained as a fighter pilot in the USAF reserve after graduation, before joining the McDonnell Aircraft Corporation to work on surface-to-air missile development. He then joined NASA's Space Task Group and became procedures officer for the early Mercury flights, charged with ensuring the smooth transfer of power from Cape Canaveral's launch control to the Mercury control room. By the time of Gemini 4 he had been promoted to flight director (one of several who worked with a team of controllers on shifts throughout each mission). He became the best known of NASA's control staff thanks to his role in the Apollo 13 mission.

Longer and further

The second manned flight was far more ambitious. James McDivitt and Ed White stayed in orbit aboard Gemini 4 for more than four days, and White became the first American to walk in space (see over). The astronauts also carried out a variety of other experiments and attempted unsuccessfully to rendezvous with the spent upper stage of their Titan rocket.

Gemini 5, launched in August 1965, pushed the limits still further. New fuel cells allowed Gordon Cooper and Pete Conrad to remain in orbit for eight days, conducting various experiments. This flight also saw the debut of the crew-designed mission patch (see top left), though officials insisted the words "Eight days or bust" were removed in case the spacecraft had to be brought home early.

ENDURANCE FLIGHT
Command Pilot Pete Conrad is photographed by crewmate Gordon Cooper shortly after launch on their record-breaking eight-day mission aboard Gemini 5 in August 1965.

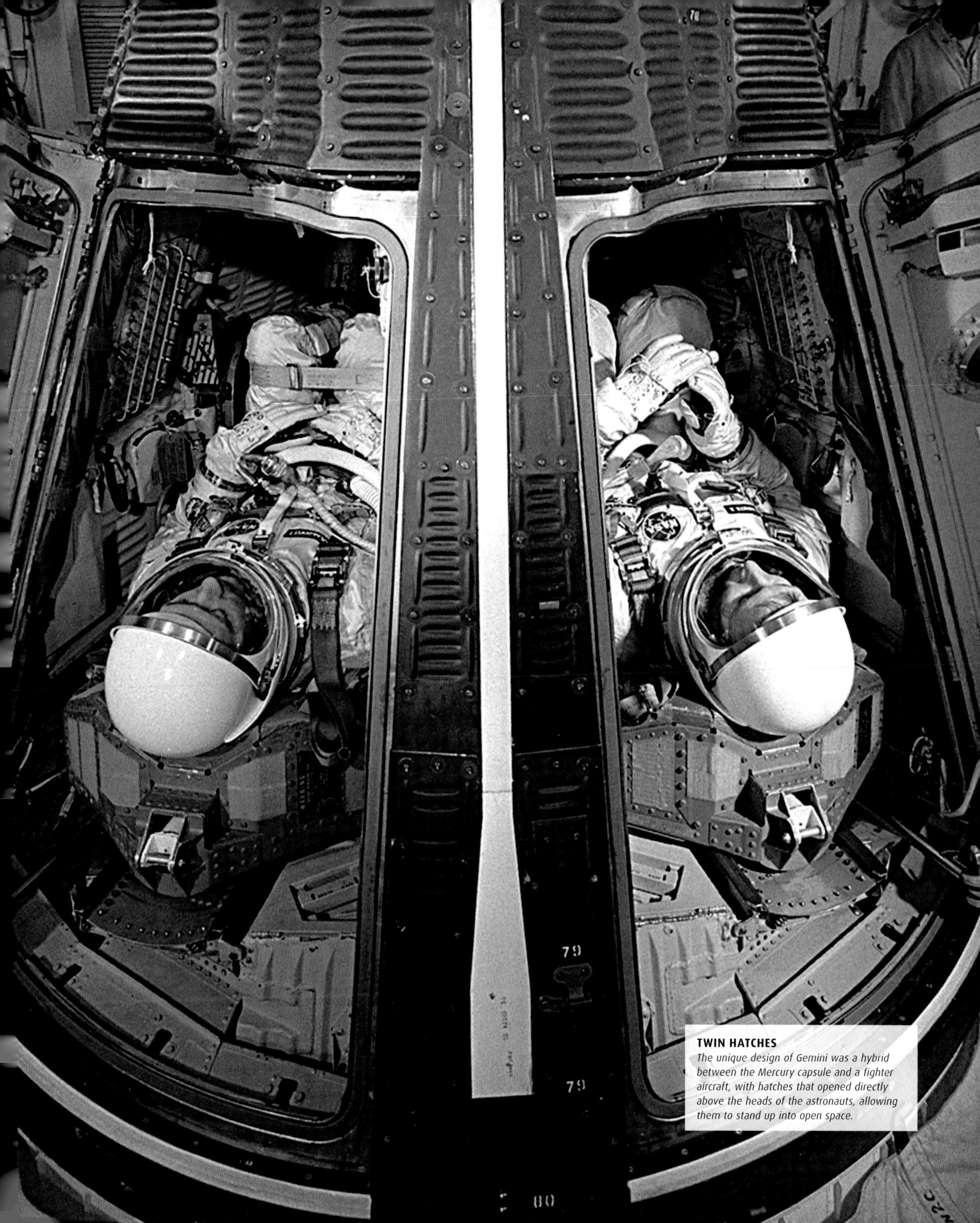

TWIN HATCHES
The unique design of Gemini was a hybrid between the Mercury capsule and a fighter aircraft, with hatches that opened directly above the heads of the astronauts, allowing them to stand up into open space.

SPLASHDOWN PRACTICE
John Young straddles a Gemini capsule while Gus Grissom looks on from a life raft during water egress training at Ellington Air Force Base, Texas. Grissom had experience of watery escapes from his Mercury days and at first wanted to name his Gemini 3 capsule Titanic *– an idea that was firmly vetoed by NASA officials.*

FLOATING FREE
Ed White's 15-minute spacewalk went more smoothly than Alexei Leonov's and was captured in stunning photographs by James McDivitt. However, the fuel supply in his manoeuvring gun soon ran out and he, too, had difficulties re-entering his spacecraft.

The first spacewalks

With future missions expected to become more complex, the time soon came to test the ability of astronauts and cosmonauts to operate in open space, beyond the confines of their spacecraft.

Sergei Korolev devised the idea of a two-man Vostok variant with an inflatable, detachable airlock at the same time as the three-man version used on Voskhod 1. Initially known as Vykhod (Exit), the mission's name was changed to Voskhod 2 at a late stage – the authorities felt the original name gave away the mission's purpose and could lead to embarrassment if the spacewalk did not go ahead.

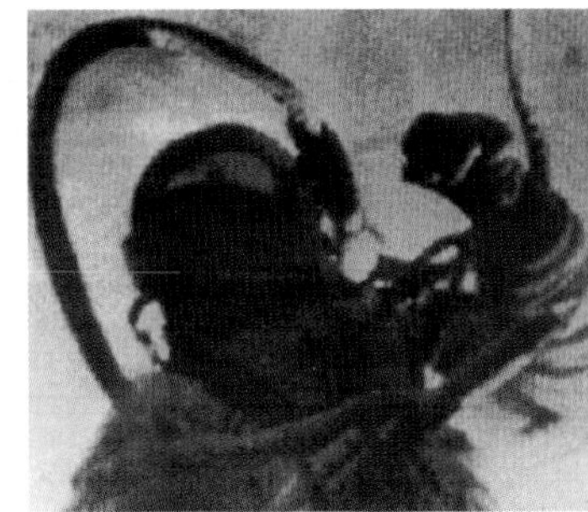

SOVIET SPACEWALKER
The cramped design of Voskhod 2 meant that Pavel Belyayev's photographs of his crewmate Leonov were far less spectacular than the American images.

Four cosmonauts trained for the mission, but Alexei Leonov was always frontrunner for the first spacewalk. Pavel Belyayev was selected as the commander, remaining inside the craft throughout the mission. After numerous tests, including a fully automated launch and spacewalk by a suited dummy, Voskhod 2 launched from Baikonur on 18 March 1965. During the second orbit, the airlock inflated and Leonov made his way outside for a historic but ultimately nerve-shredding 12 minutes floating above the Earth (see over). The return to Earth did not go smoothly either – a failure of the main retrorocket meant that Belyayev had to manually fire the backup on the following orbit, but the awkward cabin layout contributed to a 46-second delay and an overshoot of the original landing zone. To compound this, there was a repeat of the separation problems that plagued Vostoks 1 and 2. The capsule finally made a bumpy landing in the snowy forests of the Perm region, some 368km (229 miles) off target, and the cosmonauts had to spend a freezing night inside the spacecraft, surrounded by curious wolves, before rescuers on skis arrived the following day.

White walks in orbit

NASA's plans for Extra-Vehicular Activity, or EVA, were brought forward after the latest Soviet spectacular – astronaut Ed White would now leave the spacecraft during Gemini 4, in June 1965. Fortunately the Gemini design, with twin hatches above each of the astronaut couches, needed no modifications to allow for easy exit – the astronauts simply depressurized the cabin and opened White's hatch, allowing him to step out into space with none of the complexities of the Voskhod spacewalk. Although White, like Leonov, remained attached to the spacecraft by a tether, he took with him a hand-held "jet gun" that squirted pressurized gas from a nozzle, allowing him to push himself around until the fuel supply was exhausted.

White's EVA produced far more spectacular images than those from Leonov's spacewalk and helped NASA overcome the impression of trailing the Soviets yet again. Although no one knew it at the time, Leonov's spacewalk was to be the last Soviet spectacular – with Khrushchev ousted, the Voskhod project was cancelled, freeing the designers of OKB-1 to concentrate fully on the development of Soyuz.

BIOGRAPHY

ED WHITE

Born in Texas, Edward Higgins White (1930–67) was one of NASA's second astronaut group. After studying aeronautical engineering, he became a USAF pilot and later test pilot, before joining NASA. He was a star among his group and, having flown on Gemini 4, was scheduled to fly again on Gemini 10 but instead took a promotion to the Apollo 1 prime crew in 1967. He died with his crewmates in the fire that engulfed the Apollo 1 capsule during training (see pp.118–19).

"I'm **coming back** in ... and it's **the saddest moment** of my life."

Ed White, on being told to re-enter the Gemini capsule, 3 June 1965

13 April 1964
The go-ahead is given to develop a two-man Vostok variant with an inflatable airlock, known initially as Vykhod.

24 September 1964
Khrushchev visits the Baikonur Cosmodrome to see a demonstration of the Vykhod EVA technique.

9 February 1965
Pavel Belyayev and Alexei Leonov are selected as prime crew for the renamed Voskhod 2.

22 February 1965
An unmanned test mission ends early after two ground stations send conflicting commands to the Cosmos 57 spacecraft.

18 March 1965
Voskhod 2 launches. On its second orbit, Alexei Leonov makes history's first spacewalk.

29 March 1965
A meeting of NASA officials chaired by Robert Gilruth decides that Ed White's planned stand-up EVA on Gemini 4 should be upgraded to a full, tethered spacewalk.

3 June 1965
Gemini 4 carries James McDivitt and Ed White into orbit, where White performs a spacewalk.

EXPERIENCE
THE FIRST SPACEWALK

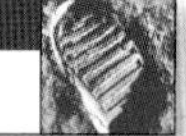

ALEXEI LEONOV
Leonov's relaxed personality and sense of humour made him popular among the early cosmonaut trainees. During the Vostok 1 flight, Gagarin found time to send "regards to Blondin" – a reference to Leonov's fair hair.

Alone in the darkness

The flight of Voskhod 2 saw the first attempt by a cosmonaut to leave his ship and walk in orbit, protected only by a spacesuit. Although ultimately a triumph, Alexei Leonov's ten-minute foray into open space almost ended in disaster.

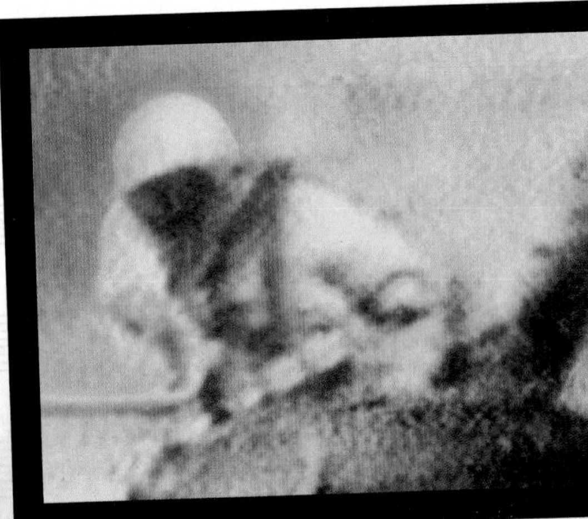

During Voskhod 2's second orbit around the Earth, Commander Pavel Belyayev began to inflate the Volga airlock. Meanwhile, Leonov had donned the backpack that would supply his suit with oxygen during the spacewalk. The suit itself was a modified version of the standard Vostok pressure suit, called Berkut (meaning "golden eagle"). The backpack blew oxygen into the suit, while a relief valve allowed air to vent into space, carrying carbon dioxide, heat, and moisture with it – a design feature that soon would prove vital. Leonov now climbed into the airlock while Belyayev sealed the hatch behind him and drained away the air, allowing his comrade to float out into open space, to the very limit of the 5m (16ft) cord that attached him to the spacecraft. As he later recalled:

"I was surprised that the Earth looked **very much like a globe or a map**, and that the Black Sea was really black – the darkest sea on Earth … I wondered who called the Black Sea 'black' and how did he know it? **I saw it from outer space!"**

While Leonov was delighted by the experience of floating free in orbit, he was already experiencing problems. The heat was intolerable, and the suit had inflated in the vacuum – Leonov was unable to reach the switch on his trouser leg that would have activated a high-quality Swiss camera mounted on his chest. The most difficult part of the spacewalk was just beginning.

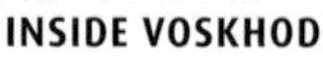

INSIDE VOSKHOD
The Descent Module of the Voskhod was cramped with two astronauts in full spacesuits. Belyayev also wore a pressure suit in case something went wrong with the Volga airlock system.

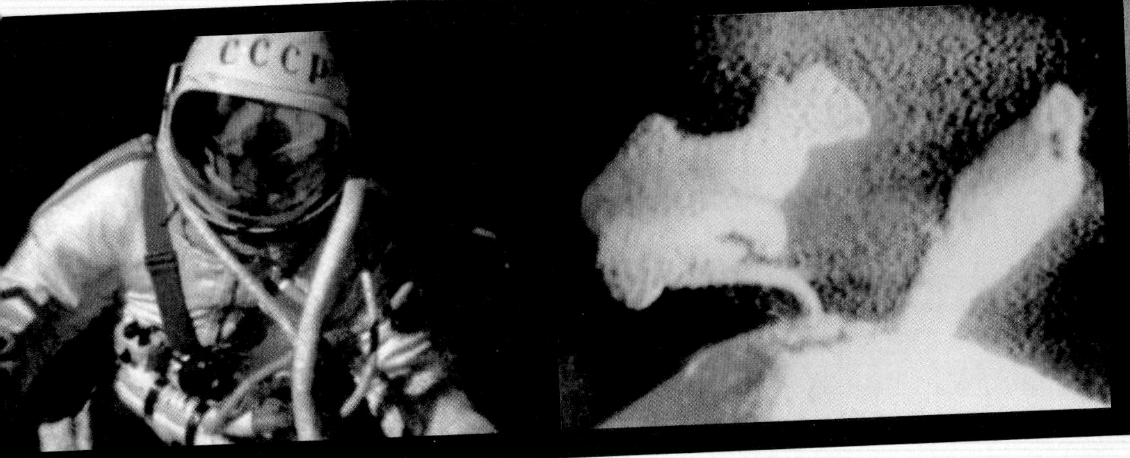

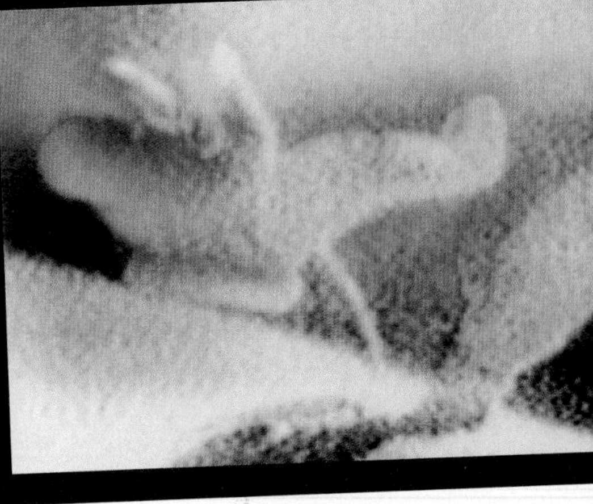

FLOATING IN SPACE
Most of the pictures from Leonov's spacewalk (above and left) were grainy images transmitted from the television cameras – Leonov did not have a chance to retrieve film from the higher-quality camera mounted on the airlock. Leonov's daughter was frightened by the sight of her father alone in space, while his ageing grandfather criticized him for fooling around!

"The Earth was **absolutely round** ... I never knew what the word round meant until I saw Earth from space."

Alexei Leonov, 1980

After ten minutes in space, Leonov attempted to re-enter the airlock. The set procedure was to grip the airlock collar and push in feet first – but in the overinflated spacesuit, he found the gloves had ballooned away from his fingers and his feet had slipped out of his boots. Leonov tried to clamber in head first, but he could not fit into the airlock. The only solution was to open up the pressure relief valve and drain the suit of air.

"... I had to take a decision to lower the pressure **inside the space suit**, but by how much? Too much would have led to a boiling of blood in the body, which would have finished me off. But I had to do it. I didn't report this **down to Earth**."

COMMEMORATION
Leonov's walk in space was heralded as a Soviet triumph on a par with the flight of Gagarin and was depicted in coins, medals, stamps and badges.

Once inside, he had to twist around in the narrow space to close the external hatch behind him so that the airlock could be repressurized. An exhausted Leonov re-entered the Voskhod capsule 20 minutes after leaving it.

"Building manned orbital stations and exploring the Universe **are inseparably linked** with man's activity in open space."

Alexei Leonov, 1980

CELEBRATED EXPLOIT
The USA was gripped by Leonov's exploits – more so as his personality emerged at post-mission press conferences. NASA hurriedly added a spacewalk to its Gemini 4 mission.

18 МАРТА 1965 ГОДА ЧЕЛОВЕК ВЫШЕЛ В ПРОСТОРЫ ВСЕЛЕННОЙ.

SOVIET HERO
A commemorative postcard shows Leonov floating like a superhero above the Earth – the truth of his struggle to re-enter the airlock did not emerge for several decades. Leonov has recently revealed he was also given a suicide pill in case he came adrift in space.

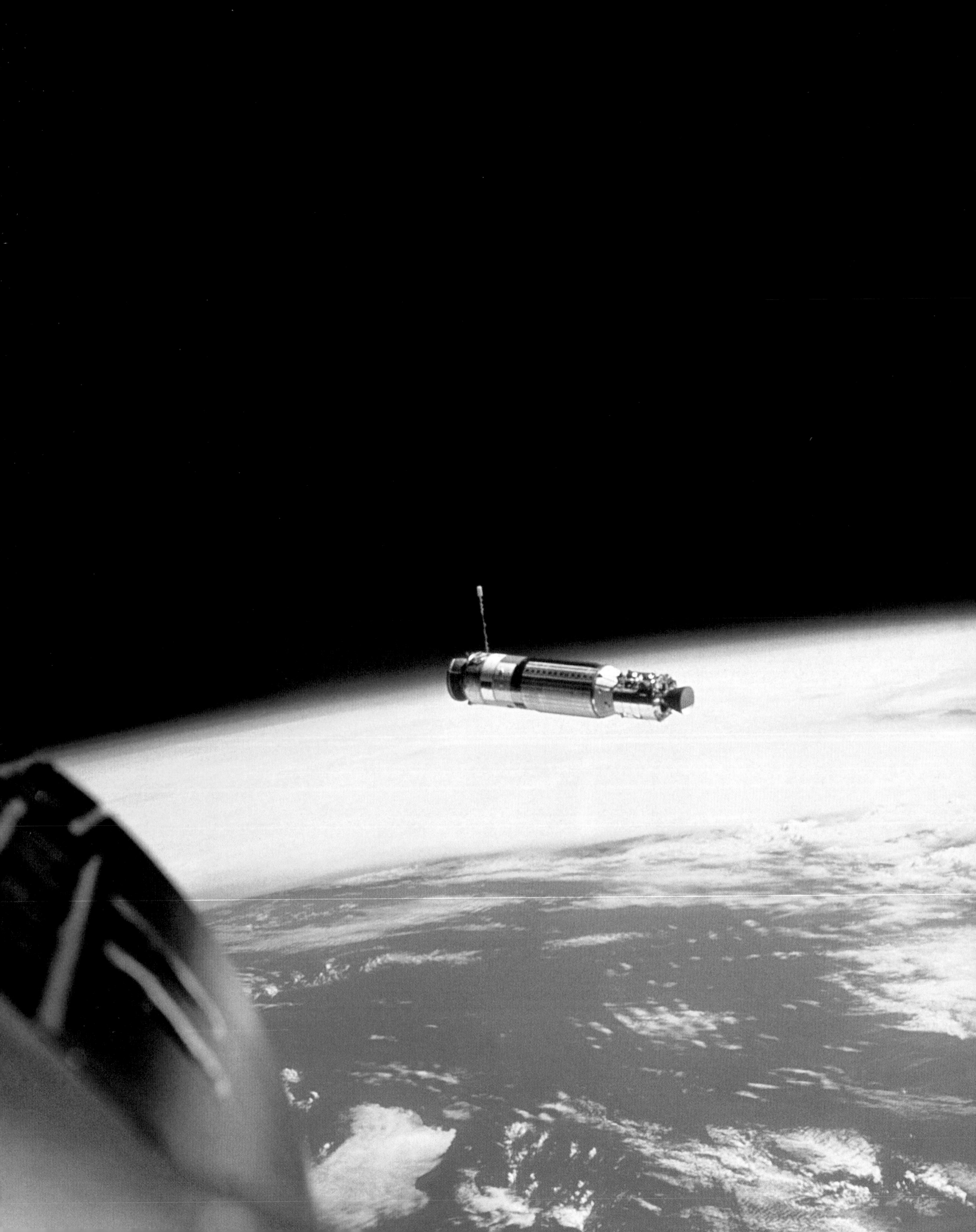

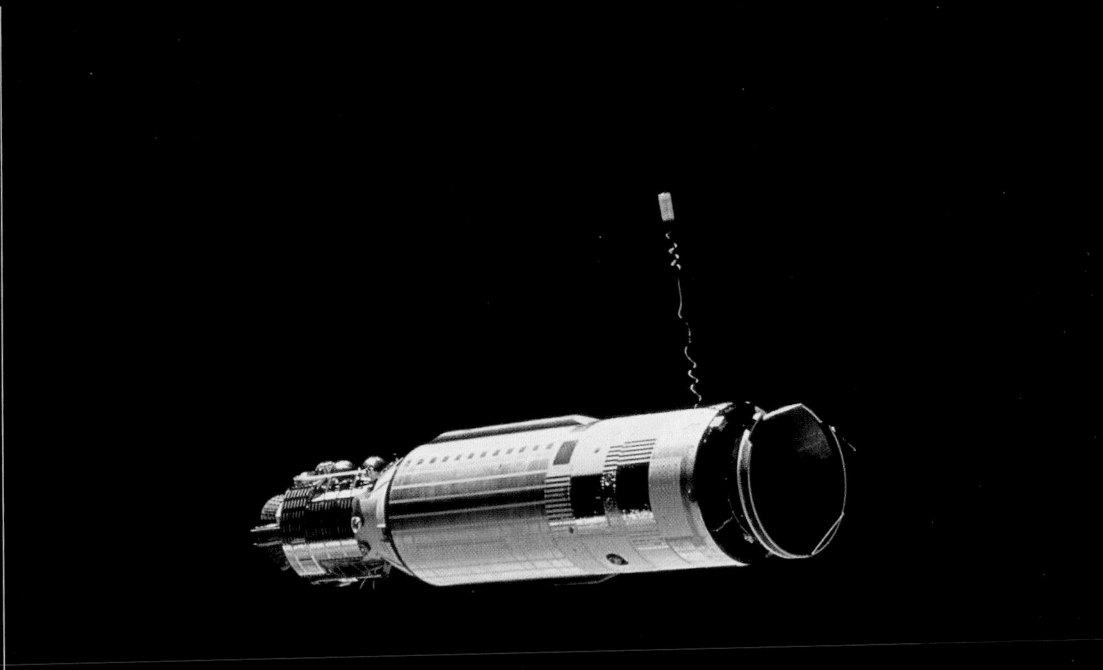

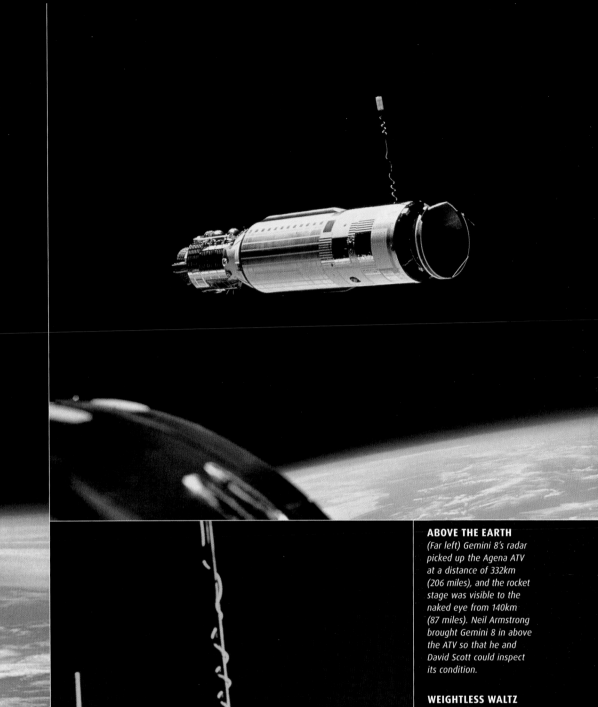

ABOVE THE EARTH
(Far left) Gemini 8's radar picked up the Agena ATV at a distance of 332km (206 miles), and the rocket stage was visible to the naked eye from 140km (87 miles). Neil Armstrong brought Gemini 8 in above the ATV so that he and David Scott could inspect its condition.

WEIGHTLESS WALTZ
(Above) After inspecting the ATV from a distance of 46m (150ft), Armstrong used Gemini 8's manoeuvring system to line up with the target vehicle, then edged towards it at a speed of 8cm (3in) per second.

READY TO DOCK
(Left) Finally, Gemini 8's nose cone edged into the ATV's docking adapter. The mechanism engaged first time – the first docking in space. But within minutes, Armstrong and Scott would be fighting for their lives as their spacecraft began to spin out of control.

Orbital ballet

Any practical plan to reach the Moon would involve rendezvous and docking in space. Rehearsing these manoeuvres was to be a key part of the later Gemini missions.

23 August 1965
Gemini 5 practises a phantom rendezvous in orbit, after fuel-cell problems prevent rendezvous with its intended target.

25 October 1965
The planned launch of Gemini 6 is scrubbed after its Agena docking target fails to reach orbit.

4 December 1965
Gemini 7 successfully launches on a 14-day endurance mission for Jim Lovell and Frank Borman.

12 December 1965
The planned launch of Gemini 6 aborts due to an engine fault on ignition. Since the crew did not eject from the spacecraft, the launch can be rescheduled.

15 December 1965
Gemini 6A (previously Gemini 6) launches and makes a rendezvous with Gemini 7.

16 March 1966
Gemini 8 achieves the first successful space docking shortly after launch, but a malfunction brings the mission to a premature end.

Simple fuel economics meant that the most direct route to the Moon – a launch directly from Earth of a spacecraft that could simply touch down in one piece on the Moon and carry enough fuel to blast off and return to Earth – was out of the question. NASA's mission planners came up with two practical alternatives (see p.117), but both would involve precision flying to bring together the parts of a lunar spacecraft in orbit.

Early attempts

Early tests of Gemini's flight controls showed how much there was to learn. Gemini 3 proved that the capsule could change its orbit, but Gemini 4's attempts to catch up with its own discarded upper stage met only with frustration – as the capsule fired its thrusters, instead of catching up with the target, White and McDivitt found that they drifted into a higher orbit (see panel, right). Gemini 5 released a small pod into a different orbit to use for target practice, but a fuel cell problem meant that the crew could not risk wasting energy to catch it. Fortunately Buzz Aldrin, a trainee astronaut and orbital mechanics expert, came up with a new test – a phantom rendezvous in which Gemini 5 flew to a precise point without burning excessive energy.

TECHNOLOGY

SWITCHING ORBITS

Changing orbits in space is simply a matter of firing engines to either slow down or speed up the spacecraft. In the example shown here, two spacecraft start out side by side in a circular orbit around the Earth. The gold craft briefly fires its thrusters, but instead of moving faster in its existing orbit, it is pushed into an elliptical orbit, with a higher apogee (the point in the orbit furthest from Earth). The further a spacecraft is from the object it is orbiting, the slower it moves, and this, combined with the greater distance it travels, causes the gold craft to lag further and further behind, even though both craft still have a common perigee (the point closest to Earth).

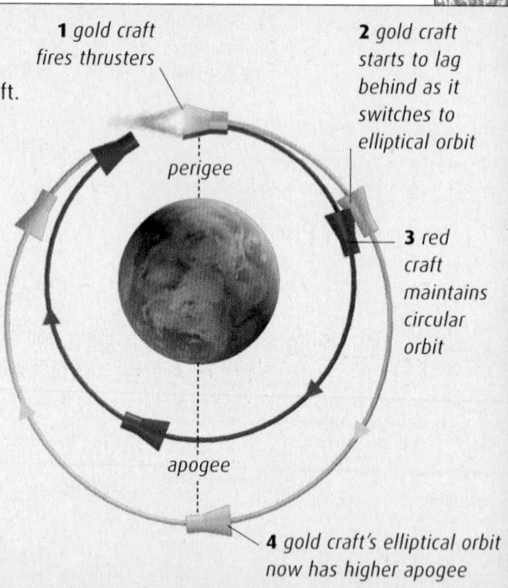
1 gold craft fires thrusters
2 gold craft starts to lag behind as it switches to elliptical orbit
3 red craft maintains circular orbit
4 gold craft's elliptical orbit now has higher apogee
perigee
apogee

Dual flights

The next plan was to rendezvous with an unmanned Agena Target Vehicle or ATV (an adapted rocket stage) that had already been launched into orbit. This was to have been the mission of Gemini 6, but even as Walter Schirra and Thomas Stafford sat in their Titan rocket on 25 October 1965, ready to chase their target into orbit, the ATV launcher exploded and the mission had to be scrubbed.

HISTORY FOCUS

IN A SPIN

Shortly after Gemini 8 docked with its ATV (see p.106) the crew of David Scott (left) and Neil Armstrong faced a serious problem, as a jammed thruster set the joined spacecraft spinning rapidly. Separating from the ATV only made the problem worse, and Armstrong had no choice but to shut down the thruster system and fire the re-entry engines to stabilize the spacecraft. The plan worked, but it brought the spacecraft back to an emergency splashdown just 10 hours after launch.

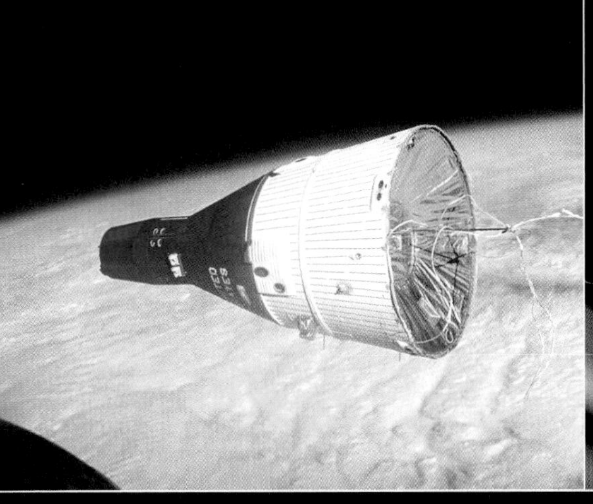

MEETING IN SPACE
Schirra and Stafford took this photograph of Gemini 7 as their own Gemini 6A capsule flew towards the first fully controlled orbital rendezvous. Gemini 7 had already been in orbit for 11 days.

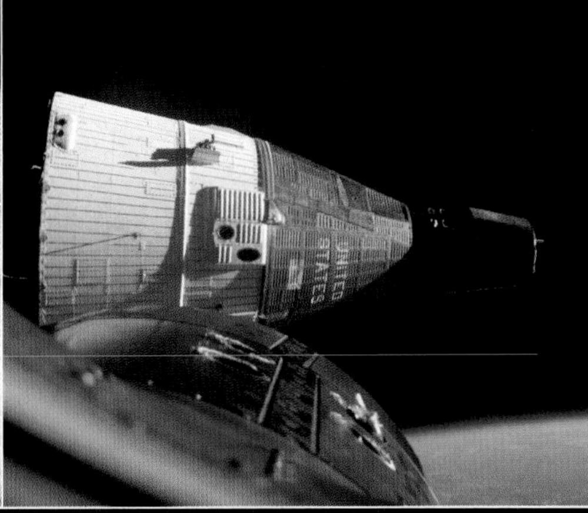

CLOSING IN
At their closest approach, the Gemini spacecraft were so close that the astronauts could communicate by holding handwritten signs up to the windows of their capsules, 260km (160 miles) above the Earth.

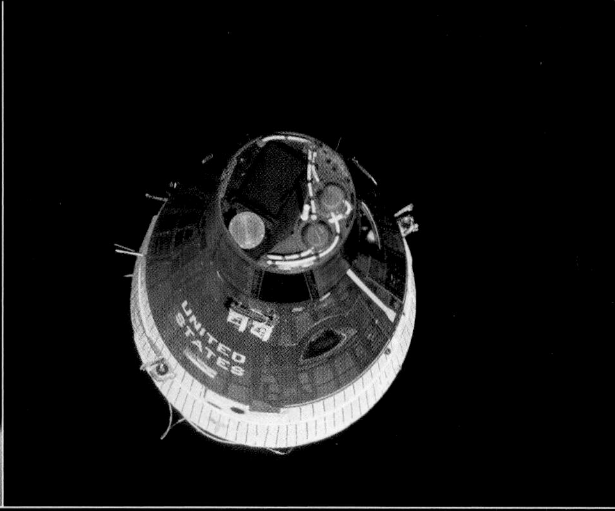

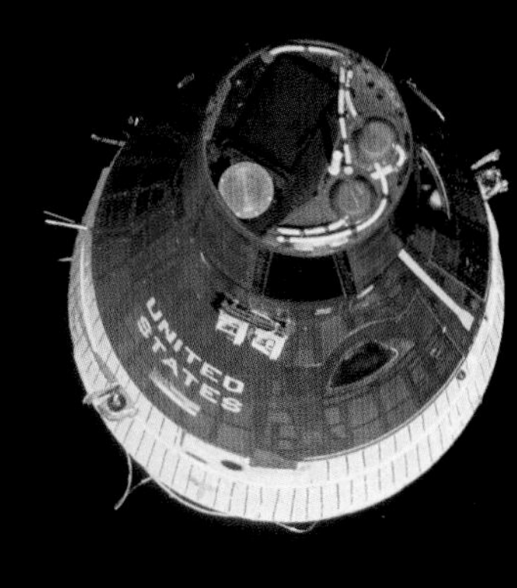

SEEING IS BELIEVING
During their closest approach, Jim Lovell, aboard Gemini 7, asked, "How's the visibility?". "Pretty bad," replied Walter Schirra on Gemini 6A, "I can see through the window and see you fellows inside!"

Frank Borman and Jim Lovell, whose Gemini 7 mission was intended largely as a two-week space-endurance test, now made an audacious suggestion – why not have Schirra and Stafford rendezvous with them instead? NASA officials took some persuading, but eventually Gemini 7 launched on 4 December, with a renamed Gemini 6A following 11 days later. Schirra skilfully steered his spacecraft to within 30cm (1ft) of the waiting target, and the astronauts waved to each other, but there was no way for them to dock.

The normal launch schedule resumed in March 1966, when Neil Armstrong and David Scott's Gemini 8 performed the first ever space docking with the ATV. However, the mission was short-lived, as the docked vessels developed a dangerous spin that brought the flight to an early end (see panel, opposite).

GEMINI 6A IN ORBIT
A head-on view of Gemini 6A from Gemini 7 reveals the "business end" of the spacecraft, the docking adapter that allowed a Gemini capsule to mate with an Agena ATV that could boost it into a higher orbit.

1955 1956 1957 1958 1959 1960 1961 1962 1963 1964 1965 1966 1967 1968 1969 1970 1971 1972 1973 1974 1975 1976 1977 1978 1979 1980 1981 1982 1983 1984 1985 1986 1987 1988 1989 1990 1991 1992 1993 1994 1995 1996 1997 1998 1999 2016 2017 2018 2019 2020

Learning to fly

The final phase of the Gemini programme saw a series of increasingly ambitious flights, dockings, and EVAs as America's astronauts geared up for the coming Apollo missions.

28 February 1966
The planned Gemini 9 crew, Elliott See and Charles Bassett, are killed when their T-38 jet crashes.

17 May 1966
Gemini 9's Agena ATV fails to reach orbit.

1 June 1966
The Augmented Target Docking Adapter (ATDA), a new target for Gemini 9, is launched.

3 June 1966
Gemini 9A launches to a rendezvous with the ATDA, but docking is not possible.

18 July 1966
Gemini 10 is launched and docks successfully with its ATV.

20 July 1966
Gemini 10 makes a second rendezvous, with the abandoned Gemini 8 ATV.

12 September 1966
Gemini 11 launches and docks successfully with its ATV, setting a new world altitude record.

15 November 1966
Gemini 12 successfully splashes down.

Even before launch, Gemini 9 was cursed by bad luck – the original crew of Elliott See and Charles Bassett died in February 1966 when their plane crashed in fog while attempting to land at the McDonnell plant in St. Louis, Missouri, where their spacecraft was under construction. As a result, the backup crew of Thomas Stafford and Eugene Cernan took their place.

The mission plan involved a rendezvous with an Agena ATV, but as had happened before, the target vehicle failed to reach orbit. A replacement was built and launched in just two weeks, and the Gemini mission (renamed 9A to indicate the use of a backup crew) flew to meet it two days later, on 3 June. The rendezvous went perfectly, but the astronauts discovered that the shroud around the target vehicle had not separated properly, and so docking would be impossible. Nevertheless the mission continued, with the crew practising a variety of manoeuvres in orbit.

The capsule carried with it a prototype Astronaut Manoeuvring Unit (AMU), or rocket pack, to aid astronauts in orbit. The intention was for Cernan to test the device, but it had been mounted on the hull near the back of the spacecraft, and Cernan's efforts to reach it were hampered by a lack of handholds. When he did get there, he discovered that donning the AMU would mean severing his main tether to the cabin, and he would probably be unable to re-attach it. In the end, an exhausted Cernan, his visor steamed up by his exertions, decided that it was too risky. He returned to the capsule after a 128-minute spacewalk, the mission incomplete but with many lessons learned.

Geminis 10 and 11

Fortunately, Cernan's misadventure was the last major setback for Gemini. Little more than a month later, Gemini 10, crewed by John Young and Michael Collins, achieved a very successful double rendezvous – docking first with their own ATV and then using its engines to boost their orbit for a close rendezvous with the ATV abandoned during Gemini 8. Collins was even able to spacewalk across to the dormant vehicle.

Gemini 11, launched on 12 September 1966, was also a success. Pete Conrad and Richard Gordon were able to dock with their booster just 85 minutes after launch, and Gordon made a spacewalk to attach a tether from the Gemini capsule to the Agena, before they propelled themselves to a new record altitude of 1,374km (854 miles). As they returned to a lower orbit, they undocked from the ATV so that the two spacecraft began to spin around their common centre of mass. The result was a weak form of artificial gravity for the two astronauts.

ADAPTER IN ORBIT
The Augmented Target Docking Adapter (ATDA) designed for Gemini 9 floats high above the Earth, its shrouding still half-attached. This unusual view of the device led to Thomas Stafford's memorable comment, likening it to an "angry alligator".

Last hurrah

The final Gemini mission was the most ambitious of all, and fortunately a great success. After a series of delays caused by spacecraft glitches, Buzz Aldrin and Jim Lovell ascended to orbit on 11 November 1966. Their four-day mission was a dry run for many of the techniques needed for Apollo, and they practised docking and undocking the ATV, and manoeuvring with it attached. Plans to use it to boost them into a higher orbit had to be abandoned, however, because of concerns over the ATV's condition after launch. Perhaps the most important achievements of the mission were Aldrin's EVAs. Extra grips had been fitted to the capsule to help with weightless manoeuvres, and Aldrin was able to spend over two hours on the end of an umbilical tether, carrying out various tests and finally proving that an astronaut could perform useful work outside a spacecraft.

THWARTED PLANS
As Gemini 9A came in sight of the hastily launched ATDA docking target, Thomas Stafford exclaimed "Look at that moose!" The protective shroud was frustratingly close to coming free, but edging the Gemini capsule's nose into the open end was too risky.

RIDING HIGH
Astronaut Dick Gordon of Gemini 11 straddles the Agena ATV, while attaching the tether that would later be used to generate the first artificial gravity in space.

"It looks like an **angry alligator** out here rotating around."

Gemini 9A astronaut Thomas Stafford describes the defective target vehicle, 3 June 1966

STAND-UP EXPERIMENTS
Gemini 12's first EVA lasted for just under two-and-a-half hours and was a stand-up – Buzz Aldrin stood up in the capsule's hatch and performed various experiments, as well as setting up ultraviolet and movie cameras.

PARTICLE COLLECTOR
Here Aldrin is retrieving a micrometeorite collector. The device was used to collect small particles drifting through Earth orbit. These were later analyzed for signs of any organisms capable of living in space.

FLOATING FREE
During his second EVA, Aldrin floated free, tethered to his spacecraft by an umbilical line. Here, he is conducting experiments at a workstation attached to the ATV and taking photographs of star fields.

Surveying the Moon

To pave the way for Apollo, NASA embarked on an ambitious programme of lunar discovery. This involved three different series of spaceprobes – Rangers, Surveyors, and Lunar Orbiters.

23 August 1961
Ranger 1, an ill-fated engineering test for NASA's series of lunar crash-landers, fails to reach its intended orbit around the Earth.

26 January 1962
Ranger 3, the first of the Ranger probes intended for the Moon, is launched but misses its target completely.

31 July 1964
Ranger 7 becomes the first successful probe in the series, sending back pictures right up to its impact with the Moon.

2 June 1966
Surveyor 1 touches down in the Oceanus Procellarum – the first successful soft-landing on the Moon.

10 August 1966
Lunar Orbiter 1 is launched into orbit around the Moon, the first of five mostly successful orbiter probes that capture details on the lunar surface as small as 2m (6ft) across.

31 January 1968
Lunar Orbiter 5 crashes onto the Moon after a successful mission.

When President Kennedy announced America's lunar ambitions in May 1961, the closest thing to a successful moonshot achieved by NASA had been Pioneer 4's relatively near miss in March 1959 (see p.52). The Moon still held many mysteries – and while some were scientific puzzles of interest mainly to astronomers, others might have a direct bearing on any expedition attempting to land there.

For example, were the Moon's plentiful craters volcanic or formed by impacts from space? If they were volcanic, would there still be seismic activity on the Moon? If they were caused by impacts, then would the surface be stable, or so badly pulverized that it could not support the weight of a spacecraft?

Automatic probes

In order to answer these questions, NASA developed three series of automatic probes, each of which would add more to our understanding of the Moon and help to answer some of the questions that still hung over the Apollo programme. The first of these were the relatively unambitious Rangers. These were succeeded by the more accomplished Surveyors and Lunar Orbiters.

The aim of the Ranger programme, which began in January 1961, was to crash-land probes on the Moon, sending back photographs up to the moment of impact. However, problems plagued the early launches. Rangers 1 and 2, intended for testing in Earth orbit, never even got that far after their launch vehicles failed. Ranger 3 missed the Moon entirely, while Ranger 4 hit its target, but was crippled at launch and did not return any data. Ranger 5 was both disabled and missed its target, while Ranger 6 had a near-perfect flight except for a failure of its cameras.

Things came right in 1964 when Ranger 7 returned more than 4,300 pictures before impact just south of Copernicus crater. Rangers 8 and 9 were also successes, crashing in the Sea of Tranquillity and the Alphonsus crater respectively. They gave a good look at the lunar surface and solved a mystery – some of the Moon's craters were so small that they could only be the result of asteroids collisions.

Soft-landers and orbiters

The far better success rate of the more ambitious Surveyors and Lunar Orbiters shows how much NASA learned during the early 1960s. The Surveyors were

omni-directional antenna
aperture for six cameras
instrument housing
solar batteries
solar panel

RANGER CRASH-LANDER
Ranger was the first NASA spaceprobe to be stabilized on all three axes, instead of spinning to remain stable. This was achieved by imparting small amounts of thrust from nitrogen gas jets. This allowed the use of large flat solar panels tilted toward the Sun, instead of the earlier spinning drum design, and produced a large increase in electrical power to the probe.

Ranger 7 – 31 July 1964

When Ranger 7's cameras were activated during its final approach to the Moon, they returned this image – the first picture of the Moon from a US spacecraft.

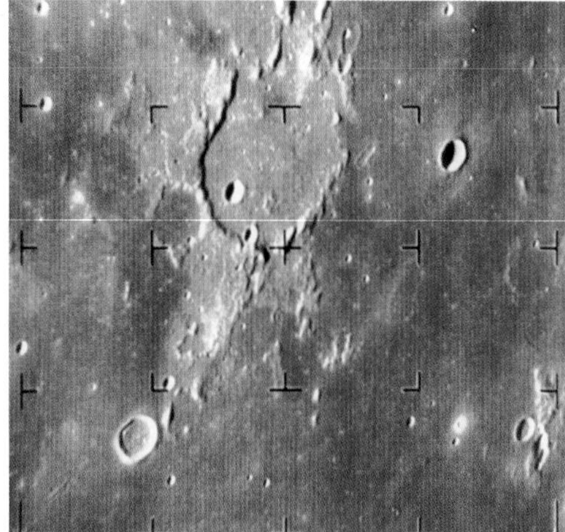

Ranger 7 – 31 July 1964

From an altitude of 1,335km (829 miles), just over eight minutes from impact, Ranger 7 returned this image of Guericke, a battered impact crater 63km (39 miles) across.

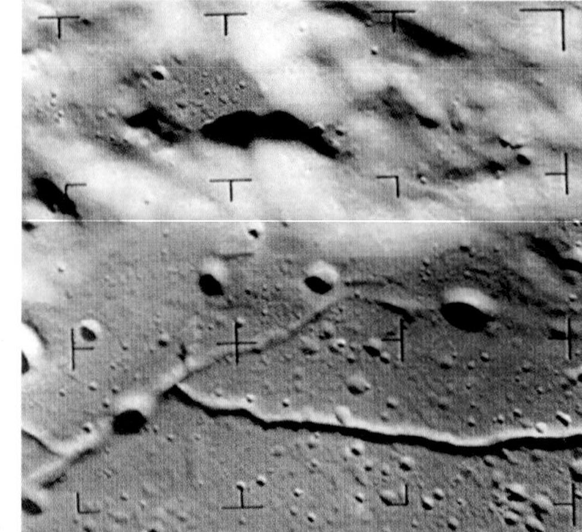

Ranger 9 – 24 March 1965

Television audiences received live pictures as Ranger 9 plunged into the crater Alphonsus. From 650km (400 miles) up, lava channels were clearly visible on the crater floor.

EARTH FROM THE MOON
On 23 August 1966, Lunar Orbiter 1 sent back this "first take" of an iconic image – Earthrise over the Moon. In fact, this is Earthset – the spacecraft, on its 16th orbit, was just about to pass behind the lunar far side, out of sight of the Earth.

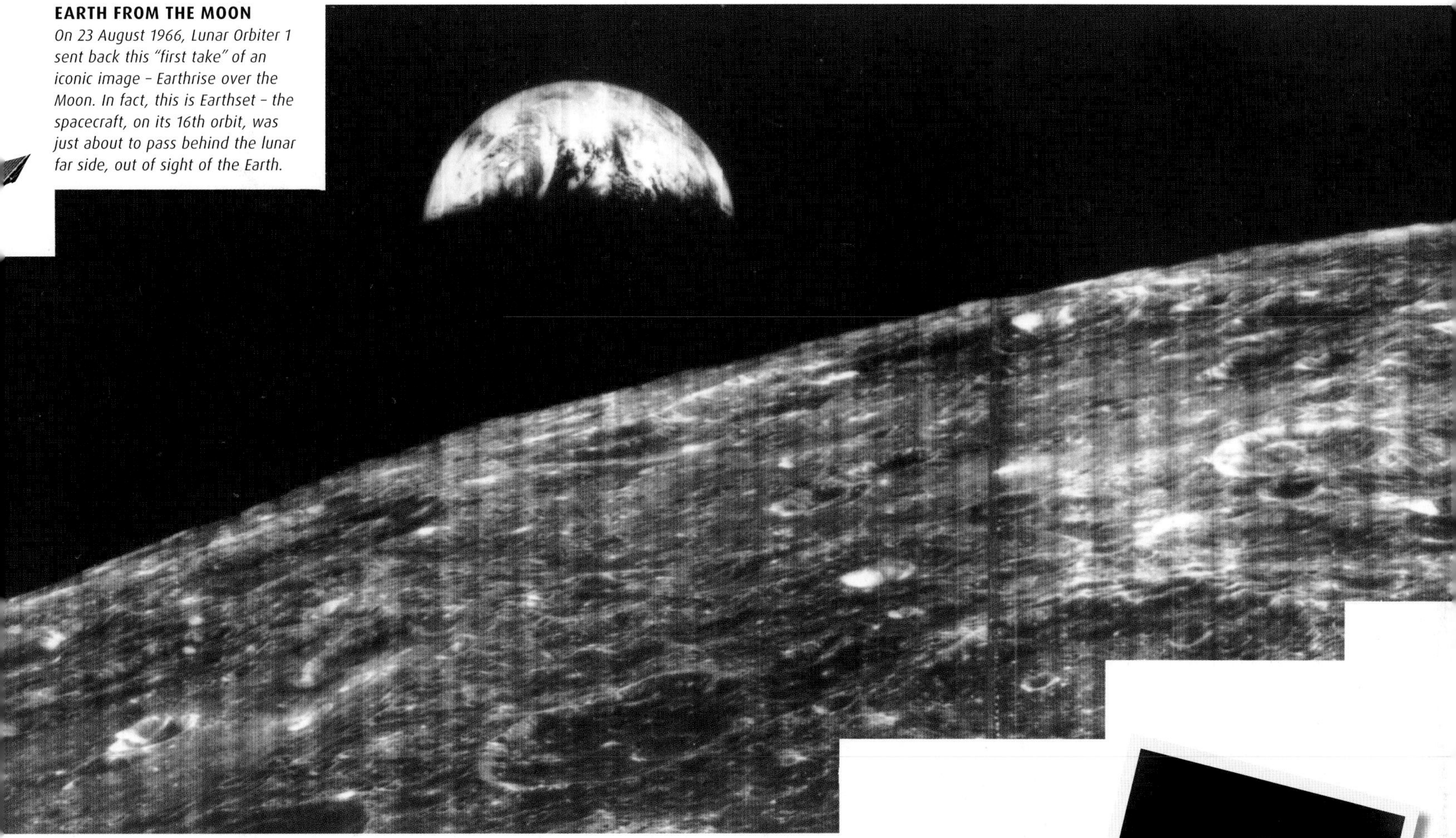

designed to make soft landings in targeted areas of the Moon and send back data about lunar conditions. Surveyor 1 touched down in the Sea of Storms on 2 June 1966. To everyone's relief, it did not sink without trace, but instead sent back images and data about the Moon's surface chemistry. Six more Surveyors followed over the next 20 months, with only two failures – Surveyors 2 and 4, both of which crash-landed.

In parallel with the Surveyors, NASA launched a series of five Lunar Orbiters, spacecraft that would become satellites of Earth's own satellite. Placed in different orbits around the Moon, the first three concentrated on imaging possible landing sites for the Apollo missions, while the last two completed a broader scientific survey that saw 99 per cent of the Moon's surface, on both near and far sides, mapped at relatively high resolution.

TECHNOLOGY

RETROROCKETS

A retrorocket is an engine attached to a spacecraft that provides a retrograde (decelerating) force when fired. The most familiar use of retrorockets is during descent to the surface of a planet or a moon. Parachutes of a reasonable size can only slow an object's descent by a certain amount, and require a substantial atmosphere to generate drag, so retrorockets are a necessity for soft landing on an airless body such as the Moon. The other main use of retrorockets (as on the Lunar Orbiter, right) is to modify a spacecraft's trajectory – perhaps slowing it down so that, instead of flying past a body at high speed, it is caught up by its gravity and pulled into orbit, or slowing it further, so that it drops out of orbit towards a landing site.

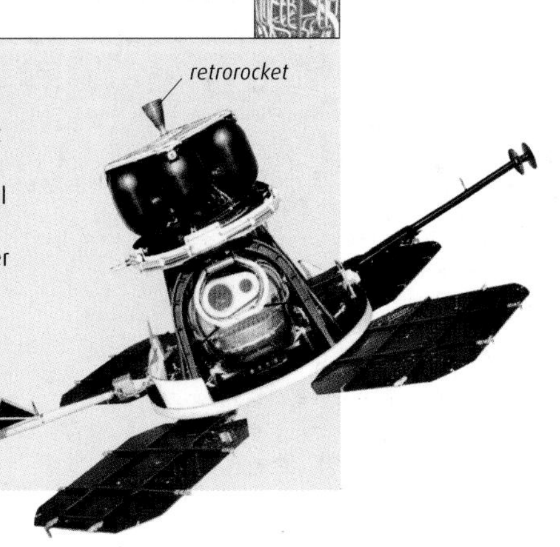

VISITORS FROM EARTH
In November 1969, the astronauts of Apollo 12 landed within walking distance of Surveyor 3 in the Sea of Storms and inspected it to see how it had fared during 30 months on the Moon.

Gemini 6

Gemini 5

Gemini 8

Gemini 9

Gemini 10

Apollo 7

Apollo 8

Apollo 9

Apollo 10

Apollo 11

Apollo 12

Apollo 13

Apollo 14

Apollo 15

Apollo 16

Apollo 17

NASA mission patches

The familiar patches used to identify NASA missions did not exist in the Mercury era, but the patches from Gemini and Apollo onwards provide a colourful chronicle of the American adventure in space.

The first mission patch came about at the direct request of the Gemini 5 crew. After the fuss about Gus Grissom wanting to name Gemini 3 *Molly Brown* (see p.82), NASA decided it would no longer allow astronauts to name their spacecraft. Gordon Cooper felt strongly that the astronauts should put some kind of personal stamp on each mission and submitted an aviation-inspired handmade "mission patch" to NASA Administrator James Webb. The idea was approved, but on the condition that the words "8 days or bust" were covered up – NASA did not want to tempt fate or snide remarks from the media if the flight failed to last for its intended duration.

Throughout the Gemini and Apollo missions, astronauts tackled the prospect of designing their patches with enthusiasm – sometimes recruiting family members to help, sometimes working with official NASA artists. Among the most striking is the Apollo 8 patch, sketched up by Jim Lovell during a flight with Frank Borman shortly after they had learned their mission would be the first around the Moon. From Apollo 9 onwards, the need to distinguish between the LM and CSM meant that astronauts were once again allowed to name their spacecraft. The later Apollo patches often reference the spacecraft names – most notably on the famous Apollo 11 Eagle patch. The elegant Apollo 15 patch was designed by Italian dress designer (and former aeronautical engineer) Emilio Pucci, at the special request of the crew.

The post-Apollo patches tended to be based on ideas that came from NASA centres, which were then selected and tweaked at the request of the crew. Curiously, the Skylab patches are wrongly numbered – they ignore the fact that Skylab 1 was actually the station's unmanned launch.

Apollo–Soyuz

Skylab 1 (2)

Skylab 2 (3)

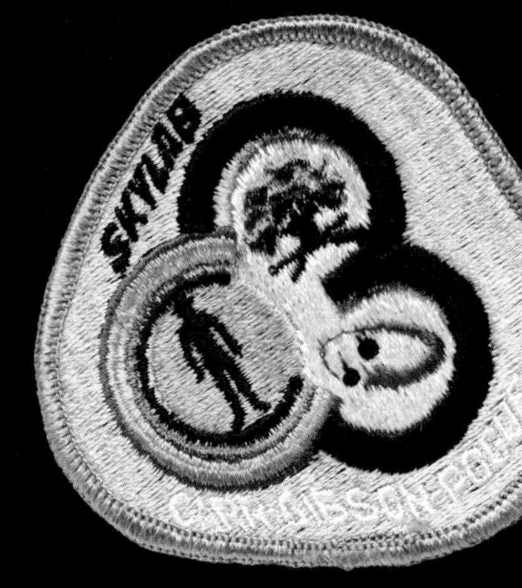

Skylab 3 (4)

1960

Planning Apollo

In the wake of President Kennedy's monumental announcement, NASA's experts turned their attention to working out how they would turn Project Apollo into a reality.

September 1960
John C. Houboult begins lobbying within NASA to promote the advantages of lunar orbit rendezvous for a manned Moon mission.

27 October 1961
A test of the Saturn I launch vehicle marks the first launch of the Apollo programme.

21 December 1961
NASA selects the Saturn C-5 as its Moon rocket. In the preceding days, contracts have been awarded for construction of various stages.

6 February 1962
Robert Gilruth and the Space Task Group conclude that an LOR mission is the best way to reach the Moon.

7 June 1962
Wernher von Braun backs an LOR mission.

11 July 1962
James Webb announces NASA's decision to base Apollo around an LOR mission profile.

7 November 1963
Pad Abort Test 1 sees the first test of a boilerplate Apollo CSM.

The challenge of putting a spacecraft on the Moon and returning it safely was not a trivial one, and the first decision that had to be made was precisely how NASA would get there. Three clear options soon emerged, each with its own advantages and disadvantages. These were direct ascent (DA), Earth orbit rendezvous (EOR), and lunar orbit rendezvous (LOR) (see illustration, opposite).

While direct ascent was the simplest approach, it required far more fuel than the other options and would involve the construction of truly monstrous rockets, so large that their construction and testing would almost certainly push the project past its 1969 deadline. The choice therefore narrowed down to the two options that involved in-flight docking and separation. EOR kept these delicate exercises in the relative safety of Earth orbit, but it also involved taking a large fuel-laden spacecraft down to the lunar surface. LOR shifted the in-flight manoeuvres away from the Earth, and risked stranding the astronauts, but it considerably reduced the size of the lunar lander required. In the end, the LOR advocates, led by John C. Houboult of Langley Research Laboratory, won the vital backing of Wernher von Braun for their approach, and in July 1962 the decision to go with an LOR mission was formally approved.

Rockets for the Moon

More debate surrounded the choice of rocket that would take Apollo to the Moon. NASA was already developing its own massive launcher, called Nova, when von Braun's ABMA team formally joined the organization. The Huntsville group brought with them plans for a heavy-lift launcher called Saturn.

The vehicle's first stage was already being tested, and plans for the upper stages were well developed, so a NASA committee was established to evaluate them, under experienced engineer Abe Silverstein. They recommended a variety of configurations, including a giant called the Saturn C-5. NASA ultimately decided that this rocket (later known as the Saturn V) would be easier to get into production by the deadline than their own Nova. Since the Moon rocket itself would not be ready for several years, they would also develop the Saturn C-1 (later Saturn I), a simpler variant more reliant on existing technology, for testing lunar spacecraft hardware in Earth orbit. Von Braun was delighted – his team at Huntsville's newly established Marshall Space Flight Center would be at the heart of the US lunar effort.

UNDER CONSTRUCTION
Technicians at North American Aviation's factory in Downey, California, prepare to fit the heat shield to the Command Module of Apollo CSM 012, the module that should have hosted Apollo 1.

SATURN I CLUSTER
The lower stage of a Saturn I, seen here being unloaded from its transport barge, reveals its secret – the basis of von Braun's first Saturn rocket was in fact a cluster of eight Redstone rockets surrounding a central Jupiter, each with improved engines.

FLYING THE BEDSTEAD
A Moon landing would require precise use of the LM's retrorockets. To practise descent, astronauts used the ungainly Lunar Landing Research Vehicle, soon nicknamed the Flying Bedstead.

A lunar spaceship

An LOR mission called for a spacecraft with at least two separate elements – a mothership that would remain in orbit around the Moon and a lander that would go down to the surface. To simplify re-entry, NASA ultimately opted to use three distinct elements. The mothership would consist of two sections: a conical Command Module, in which the astronauts would spend most of the journey to and from the Moon; and a cylindrical Service Module that would contain equipment such as the rocket motors needed for manoeuvring in lunar orbit. For almost all of the mission, the two would be united and referred to as the Command and Service Module or CSM. The third element, the Lunar Excursion Module, usually known simply by the initials LM, was a separate spacecraft that would only ever have to fly in a vacuum and so rapidly evolved an ungainly, spider-like shape in which function overruled form.

As usual, NASA operated a contracting process that invited interested companies to bid for the manufacturing work. The contract for the CSM ultimately went to North American Aviation, while the LM was to be built by Grumman Aerospace. By 1966, the initial "Block 1" batch of CSMs was being readied for mounting on rockets and testing on the pad – but the Apollo programme was about to suffer a tragic setback.

THREE ROUTES TO THE MOON
The three practical methods considered by NASA for reaching the Moon each required major technological advances.

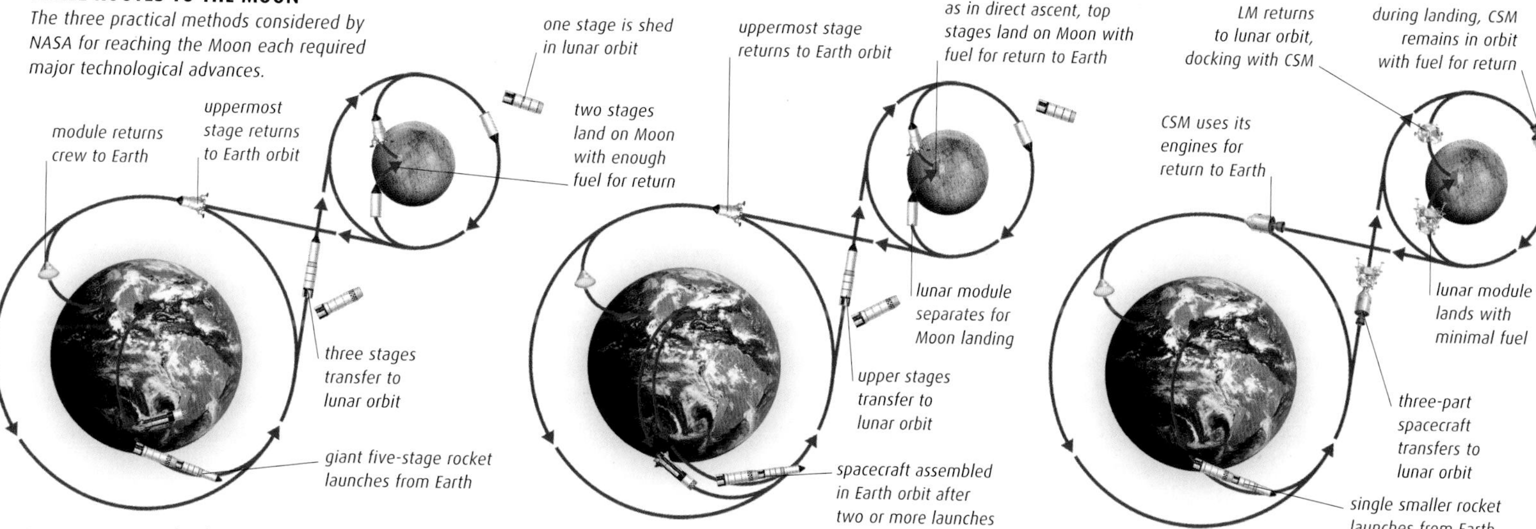
module returns crew to Earth
uppermost stage returns to Earth orbit
one stage is shed in lunar orbit
two stages land on Moon with enough fuel for return
three stages transfer to lunar orbit
giant five-stage rocket launches from Earth
uppermost stage returns to Earth orbit
as in direct ascent, top stages land on Moon with fuel for return to Earth
lunar module separates for Moon landing
upper stages transfer to lunar orbit
spacecraft assembled in Earth orbit after two or more launches
LM returns to lunar orbit, docking with CSM
during landing, CSM remains in orbit with fuel for return
CSM uses its engines for return to Earth
lunar module lands with minimal fuel
three-part spacecraft transfers to lunar orbit
single smaller rocket launches from Earth

DIRECT ASCENT (DA)
An enormous multi-stage rocket is launched directly towards the Moon. The upper two stages land upright on the lunar surface, still with enough fuel to transport the crew back to Earth.

EARTH ORBIT RENDEZVOUS (EOR)
A lunar spacecraft docks with a huge propellant tank already launched into orbit. The tank has fuel for the journey to the Moon, and the spacecraft touches down with enough fuel for its return.

LUNAR ORBIT RENDEZVOUS (LOR)
A three-stage rocket puts the spacecraft on course for the Moon. Once in lunar orbit, a lunar module descends to the surface while during landing, the CSM remains in orbit with fuel for return.

Apollos 1 to 6

Meeting the 1969 deadline for a Moon landing called for a breakneck development and testing programme, but Apollo was almost derailed by tragedy at its birth.

Even though the Saturn launcher had been in development for some time, launching a lunar mission by 1969 would be an enormous challenge. One thing was clear to Dr. George Mueller, NASA's Associate Administrator for Manned Space Flight – the old philosophy of step-by-step rocket testing, altering just one component at a time between tests, would slow things down immeasurably.

Instead, Mueller opted for the bold approach of all-up testing – launching a complete spacecraft and studying how an entire system worked together. Even if a few repeats of certain missions proved necessary, this would drastically reduce the number of launches required during the testing process. Under the traditional system, it could have taken 20 launches of different Saturn V and Apollo elements to reach a manned lunar landing. Following Mueller's proposals, it might take as few as six Saturn Vs, preceded by a number of Saturn I and IB launches to test boilerplate capsules in Earth orbit.

The first elements of Apollo hardware to be ready for testing were the CSM and the Saturn IB launch vehicle. A manned launch of these components was planned for February 1967, and so it was that on

LOST CREW
In October 1966, the crew of the fated AS-204 (later known as Apollo 1) rehearsed procedures for their splashdown in the Gulf of Mexico. Pictured from left to right on the deck of the NASA's Motor Vessel Retriever *are Ed White, Gus Grissom, and Roger Chaffee.*

FIRST LAUNCH
A huge S-I rocket stage is lifted to the launch pad during preparations for the first launch of the Apollo programme in October 1961. Designated SA-1, this was one of a series of early tests of the clustered rocket.

BIOGRAPHY

GUS GRISSOM

Virgil "Gus" Grissom (1926–67) was among the brightest and the best of NASA's original Mercury Seven astronauts. Born in Mitchell, Indiana, he studied engineering before joining the US Air Force. After seeing action in Korea, he was a test pilot prior to his selection for astronaut training. The sinking of the *Liberty Bell 7* Mercury capsule hung like a cloud over Grissom – several people argued that he had panicked and blown the capsule hatch himself. Despite the criticism, Grissom went on to command Gemini 3, and NASA continued to show its confidence in him with his selection for the first Apollo mission.

27 January Gus Grissom, Ed White, and Roger Chaffee were sealed into the CSM on Launch Complex 34 at Kennedy Space Center for a mission simulation. (Shortly after John F. Kennedy's assassination in November 1963, NASA's launch site had been named in his honour and Cape Canaveral itself also bore the Kennedy name until 1973.)

Tragedy strikes

Almost six hours into the test, an electrical spark started a catastrophic fire in the capsule (see over). The flames spread rapidly in the CSM's pure oxygen atmosphere, and difficulties opening the hatch sealed the fate of the crew, who died from smoke inhalation as they struggled to escape.

NASA immediately established a review board that looked into every aspect of the spacecraft design and reached some damning conclusions. Although built by North American Aviation, Apollo's design was ultimately determined by NASA, and several suggestions made by the contractor to improve safety had been overruled. The board recommended a huge number of changes, including a less volatile atmosphere, removal of flammable materials, and more than 1,400 wiring improvements.

Despite the shock of losing three astronauts, there was little talk of abandoning Apollo, and NASA had soon put together a recovery plan. Grissom, White, and Chaffee's AS-204 test mission was retrospectively renamed Apollo 1, and a series of unmanned tests would follow, with Apollo 7 as the next manned launch if all went well.

Unmanned launches

The peculiarities of NASA's mission numbering system meant that the next official Apollo launch was Apollo 4, the first all-up test of a Saturn V launcher. The sight of the mighty rocket thundering into the Florida skies in November 1967 did a great deal to restore American belief in the Apollo programme in the wake of the fire, and the mission went flawlessly, with an empty CSM splashing down just a few kilometres off target after running through a series of orbital manoeuvres.

In January 1968, Apollo 5 saw a Saturn IB launch an unmanned lunar module for testing in Earth orbit, and in April, a second unmanned Saturn V launch, despite some glitches, was considered enough of a success to qualify the rocket for manned flights.

VIEW FROM THE TOP
This stunning view from the top of the Vertical Assembly Building shows the Saturn V rocket stacked in preparation for the Apollo 4 launch. Despite the size of the VAB, the fully stacked Saturn V, on top of its crawler transporter, was too tall to fit in, and the lightning conductor at the rocket's tip had to be erected after it left the building.

27 January 1967
The crew of AS-204 are killed by a fire in their CSM during training.

5 April 1967
The Apollo 204 Review Board delivers its report to NASA Administrator James Webb.

4 May 1967
NASA renumbers the Apollo missions, designating AS-204 as Apollo 1 at the request of Gus Grissom's widow.

9 November 1967
Apollo 4 sees the successful launch of the first Saturn V rocket.

22 January 1968
The Apollo 5 mission conducts tests of the LM engines in Earth orbit.

4 April 1968
Apollo 6, the second test launch of Saturn V, develops "pogoing" oscillations during launch due to uneven engine burn, then suffers a failure of two second-stage engines. However, with some modifications the Saturn V is eventually approved for crewed spaceflight.

9 October 1968
Modifications to the Apollo spacecraft following the Apollo 1 fire are formally completed.

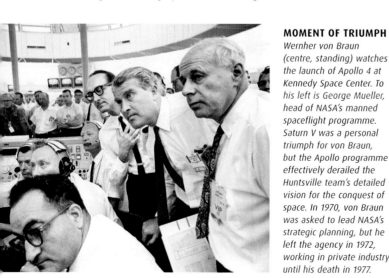

MOMENT OF TRIUMPH
Wernher von Braun (centre, standing) watches the launch of Apollo 4 at Kennedy Space Center. To his left is George Mueller, head of NASA's manned spaceflight programme. Saturn V was a personal triumph for von Braun, but the Apollo programme effectively derailed the Huntsville team's detailed vision for the conquest of space. In 1970, von Braun was asked to lead NASA's strategic planning, but he left the agency in 1972, working in private industry until his death in 1977.

TECHNOLOGY

THE VEHICLE ASSEMBLY BUILDING

Building the world's largest rocket called for a truly vast hangar, one of the world's largest buildings. Construction work on the Vertical (later Vehicle) Assembly Building or VAB got underway in 1962. The towering building is still one of the world's biggest enclosed spaces – 160m (525ft) high, with four times the volume of the Empire State Building. It contained construction equipment and cranes to assemble the Apollo components, which arrived by barge and aircraft from the various contractors, on top of a huge crawler transporter that then carried the rocket to its pad. Used throughout the Space Shuttle programme, it will soon house NASA's new Space Launch System.

EXPERIENCE

FIRE ON THE LAUNCH PAD

Apollo 1 ... a tragic start

The fire that devastated the Apollo capsule during the AS-204 ground test claimed the lives of three of America's best and brightest astronauts. And the investigations that followed it laid NASA open to criticism of its management and safety procedures for the first time.

DOOMED CREW
Ed White, Gus Grissom, and Roger Chaffee pose for a photograph with a model of the Command Module in which they would die.

On 27 January 1967, the first Apollo crew, assigned at the time to an orbital test flight designated Apollo 204, boarded their Command and Service Module for a routine test that would, if successful, have paved the way for the first manned Apollo launch a few weeks later. Gus Grissom, Ed White, and Roger Chaffee were to test the CSM's operation in "plugs out" mode – with all external power supplies withdrawn to simulate conditions in space. The test would include a full launch rehearsal.

The spacesuited crew boarded the Command Module at 13:00 local time, but a series of problems with the oxygen supply and faulty communications equipment caused repeated holds in the simulated countdown. By 18:30, the "launch" had reached T-10 minutes, but was again on hold. Then at 18:31 there came a chilling cry over the radio, probably from Roger Chaffee:

"We've got **fire in the cockpit!"**

Seconds later came another cry: "We've got a bad fire – let's get out!". On the television monitors, Ed White was briefly seen struggling to open the hatch, but its design was complex – even in practice, no astronaut had succeeded in getting it open in the suggested 90-second timeframe. Outside, the ground crew also struggled to open the hatch, but they were beaten back by fire as the capsule ruptured. By the time they got the hatch open, the astronauts were dead.

"The catastrophe having occurred at Cape Kennedy on 27 January 1967 is a tragedy not only for the United States of America. The sorrow of American people is shared **by peoples of all countries**. In reality, cosmonauts are somehow representatives of the whole Earth, of the entire mankind **in the boundless Cosmos**, no matter what country has dispatched them."

Soviet Embassy press release, 1 February 1967

ROAD TO DISASTER
After months of ground rehearsals of every aspect of the mission, the Saturn I launch vehicle was stacked on the pad at Launch Complex 34, then the CSM, in its fairing, was winched into place. A launch date was set for 21 February 1967. On 27 January, the astronauts boarded the spacecraft as if preparing for a real launch.

"If we die, do not mourn for us. This is a risky business we're in and **we accept those risks.** The space programme is too valuable to this country to be halted for too long if a disaster should ever happen."

Gus Grissom, interviewed three weeks before the fire

INTERNAL DAMAGE
Highly flammable materials used inside the Command Module fed the fire and produced the toxic fumes that ultimately suffocated the crew.

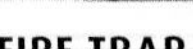

FIRE TRAP
The CSM had a two-part hatch with an inner section that opened inwards – unlike earlier and later spacecraft, it did not have explosive bolts for an emergency escape. Long before the 90-second nominal escape time, the capsule itself was a blazing inferno.

The investigation that followed the fire was exhaustive and its conclusions about NASA's design specification for the Apollo CSM damning. Reconstruction of events in the capsule suggested that the fire had started in exposed wiring beneath Grissom's couch, spreading rapidly in the pure oxygen atmosphere. The nation went into mourning for their lost astronauts, while the CSM was grounded for extensive redesign. Most significantly, the atmospheric mix onboard was changed, the miles of wires and cables were given improved insulation, and the hatch was redesigned to permit opening in just 10 seconds.

"You know, I suppose you're much more likely to accept the loss of a friend in flight, but **it really hurt to lose them in a ground test.** That was an indictment of ourselves. I mean because we didn't do the right thing somehow."

Neil Armstrong, 2001

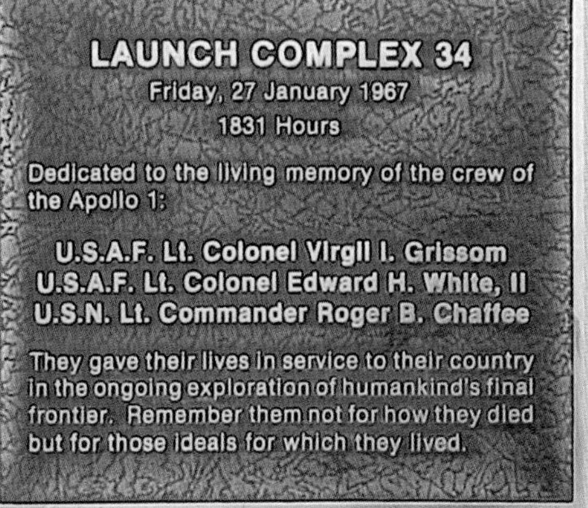

LAUNCH COMPLEX 34
Friday, 27 January 1967
1831 Hours
Dedicated to the living memory of the crew of the Apollo 1:
U.S.A.F. Lt. Colonel Virgil I. Grissom
U.S.A.F. Lt. Colonel Edward H. White, II
U.S.N. Lt. Commander Roger B. Chaffee
They gave their lives in service to their country in the ongoing exploration of humankind's final frontier. Remember them not for how they died but for those ideals for which they lived.

MEMORIAL PLAQUE
Launch Complex 34 is now abandoned and dismantled save for the concrete platform where this plaque commemorates the three astronauts who perished.

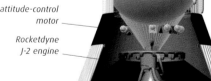

STACKING APOLLO 11
The CSM and LM were stacked at the top of the rocket, with the LM protected by an aerodynamic fairing and stowed beneath the CSM. An escape tower supported a rocket to pull the CM free of the vehicle in an emergency. Once en route to the Moon, the CSM separated from the S-IVB and turned around to dock with the LM and pull it free.

S-IVB THIRD STAGE
The upper stage of the Saturn V performed two main roles – it ignited directly after second-stage separation to reach a low Earth orbit and then, after several orbits, it reignited to put Apollo on course for the Moon.

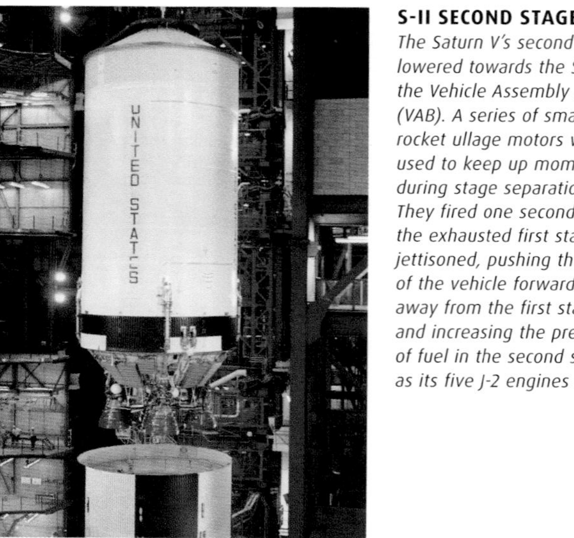

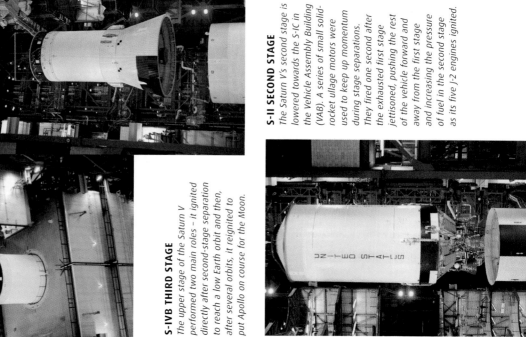

S-II SECOND STAGE
The Saturn V's second stage is lowered towards the S-IC in the Vehicle Assembly Building (VAB). A series of small solid-rocket ullage motors were used to keep up momentum during stage separations. They fired one second after the exhausted first stage jettisoned, pushing the rest of the vehicle forward and away from the first stage and increasing the pressure of fuel in the second stage as its five J-2 engines ignited.

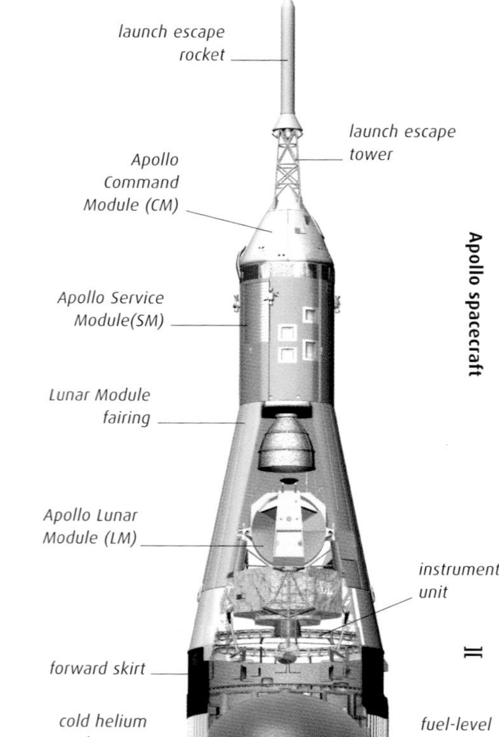

SPECTACULAR DEBUT
The first launch of a Saturn V was on the unmanned Apollo 4 test flight. For an all-up test, it went remarkably well – in 13 launches, the Saturn V never suffered a truly disastrous failure.

MOUNTING THE SPACECRAFT
The elements of the Apollo spacecraft were combined at ground level and enclosed in a protective shroud before being hoisted to the top of the VAB and set on top of their launch vehicle.

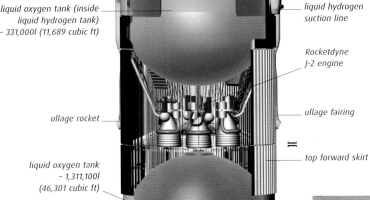

S-1C ASSEMBLY

The first stage, seen here being lifted into place in the VAB prior to the Apollo 8 launch, was powered by five Rocketdyne F-1 engines burning liquid oxygen and kerosene from a pair of huge tanks.

WERNHER'S BABY

Wernher von Braun stands by the massive F-1 engines at the back of a Saturn V launcher (in fact a test vehicle kept at Huntsville). The challenge of reaching the Moon saw him finally abandon the clustered rocket concept of the Saturn I in favour of much larger propellant tanks supplying multiple engines.

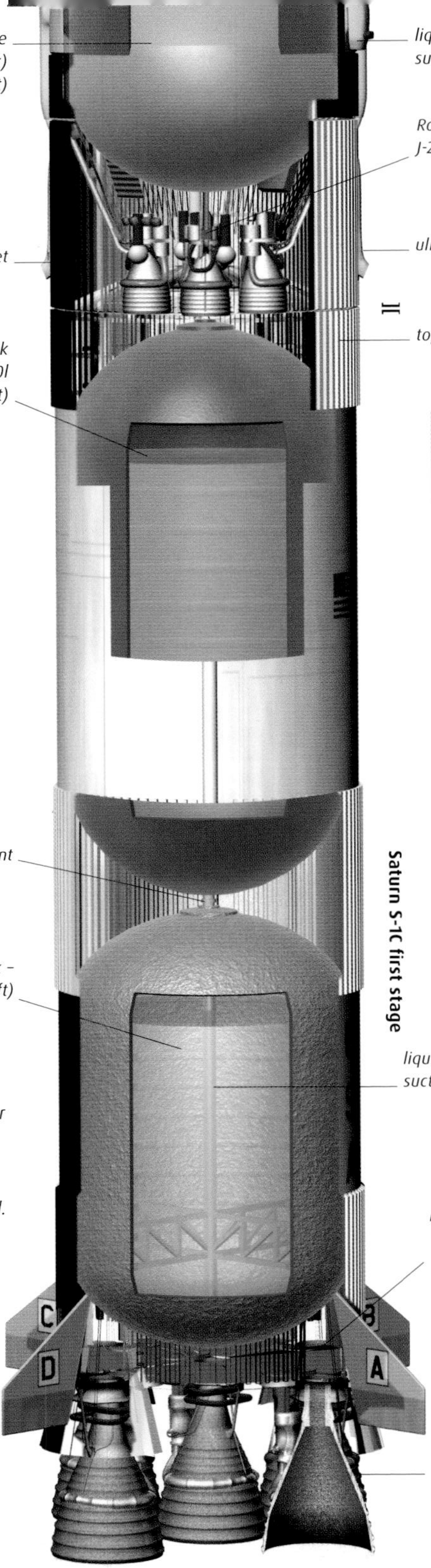

TECHNOLOGY

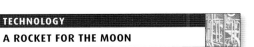

A ROCKET FOR THE MOON

Saturn V

Although a lunar mission using Saturn I rockets might have been feasible, the lunar-orbit rendezvous mission that NASA eventually settled on required a much larger and more powerful launch vehicle. This rocket was the Saturn V, the largest rocket ever to fly successfully. Masterminded again by the rocket team at Huntsville's Marshall Space Flight Center, it built on, but did not replicate, the technology of the Saturn I. Fifteen of these giants were ultimately built.

liquid oxygen tank (inside liquid hydrogen tank) – 331,000l (11,689 cubic ft)

ullage rocket

liquid oxygen tank – 1,311,100l (46,301 cubic ft)

fuel vent

RP-1 (kerosene) tank – 810,000l (28,605 cubic ft)

liquid hydrogen suction line

Rocketdyne J-2 engine

ullage fairing

top forward skirt

Saturn S-1C first stage

liquid oxygen suction line

thrust structure prevents engines from pushing into tank above

outer Rocketdyne F-1 engines pivot to steer first stage

SATURN V LAUNCHER

The complete Saturn V stood the height of a building of over 30 storeys. Transported to the pad empty, the weight of its propellants caused it to shrink by 20cm (8in) as it was fuelled.

HEIGHT	110.6m (363ft)
MAXIMUM DIAMETER	10.1m (33ft)
WEIGHT AT LAUNCH	3,038,500kg (6,699,000lb)
UNFUELLED WEIGHT	183,395kg (404,317lb)
ENGINES	5 x Rocketdyne F-1 5 + 1 x Rocketdyne J-1
THRUST AT LAUNCH	3,440,344kgf (7,584,582lbf)
MANUFACTURERS	Boeing, North American, Douglas

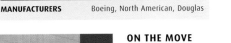

ON THE MOVE

A prototype S-IC stage is moved into the Propulsion and Vehicle Engineering Laboratory at Marshall Space Flight Center, for testing under stresses and loads similar to those encountered during launch.

DESIGN MOCK-UPS

One of the first tasks at Huntsville was to design engineering models for testing weight characteristics, dimensions, and aerodynamics. This is a full-size mock-up of the S-IC first stage.

1955 1956 1957 1958 1959 1960 1961 1962 1963 1964 1965 1966 1967 1968 1969 1970 1971 1972 1973 1974 1975 1976 1977 1978 1979 1980 1981 1982 1983 1984 1985 1986 1987 1988 1989 1990 1991 1992 1993 1994 1995 1996 1997 1998 1999 2000 2001 2002 2003 2004 2005 2006 2007 2008 2009 2010 2011 2012 2013 2014 2015 2016 2017 2018 2019 2020

10 February 1965
Sergei Korolev's N1 Moon rocket is approved for development.

14 January 1966
Sergei Korolev dies during routine surgery.

11 May 1966
Vasili Mishin is confirmed as the new head of OKB-1.

18 September 1968
The Zond 5 probe loops behind the Moon and returns to Earth safely, landing in the Indian Ocean, in an unmanned test of a lunar orbiter.

30 December 1968
After the successful manned mission of Apollo 8, the Soviet Union abandons the race to put a man on the Moon.

21 February 1969
The first of four test flights of the N1 rocket ends when a fire causes its engines to shut down shortly after launch.

23 November 1972
The last N1 launch ends in an explosion.

17 February 1976
The last elements of the Soviet manned lunar programme are abandoned.

The Soviet challenge

Throughout the 1960s, the Soviet designers suffered a series of setbacks that ultimately brought an end to their hopes of beating the Americans in the race to the Moon.

Despite years of denials and cover-ups, the Soviet Union was racing to keep up with Apollo until almost the last moment. Even before the first cosmonaut had flown, work had begun on the huge N1 rocket, a successor to the R-7 with enough power to match even the mighty Saturn V.

Like the Apollo planners, Soviet designers had to choose between three possible mission profiles (see pp.116–17). At first, Sergei Korolev chose Earth orbit rendezvous – in 1963, he set out a plan that would use three N1 launches and a single launch of the new Soyuz-A vehicle (see over) to build a 200-tonne spacecraft in Earth orbit.

But the Soviet system generated fierce competition between designers, and military rocket designer Vladimir Chelomei, in particular, was becoming increasingly prominent. Chelomei wanted to develop his own superbooster, the UR-700, outclassing even his own UR-500 Proton (see p.210). He believed this would allow an alternative Moon mission – the fuel-squandering direct ascent approach. With Korolev and Valentin Glushko also arguing over engine design and choice of propellant, the entire Soviet Moon programme was soon mired in bureaucracy and politics. It took two vital years for Korolev's lunar project to get the go-ahead, and by then engineering realities had stripped it down to a far less ambitious lunar-orbit rendezvous mission, using a single N1 launch and a spacecraft called the L3, combining a variant of the Soyuz spacecraft, with a new one-man lander called the LK.

Development problems

The N1 rocket project lagged behind schedule from the very start, largely thanks to wrangles between the designers and interference from bureaucrats. All this was made inevitable by the sheer scale of the project – it was simply too huge for OKB-1 to handle alone. And then, in January 1966, came the sudden death of Korolev during routine surgery (see panel, below). While

LUNAR GIANT
The enormous N1 rocket, 105m (344ft) tall and 17m (56ft) across, stands ready for a launch on its specially built pad at Tyuratam.

HISTORY FOCUS

THE DEATH OF THE CHIEF DESIGNER

Sergei Korolev died on 14 January 1966, after complications that developed during routine colon surgery. It was only after his death that the Soviet authorities finally allowed his identity to become widely known. Korolev's funeral left the Soviet space programme rudderless. His larger-than-life, driven personality, combined with a pragmatic recognition of the need to play the Soviet political game, meant that he was recognized as leader, sometimes reluctantly, even among the other chief designers. His pivotal role in the Soviet conquest of space has only become public since the 1970s, and he has since become acclaimed as a Russian national hero – the city formerly known as Kaliningrad, headquarters of OKB-1, is now named Korolev in his honour.

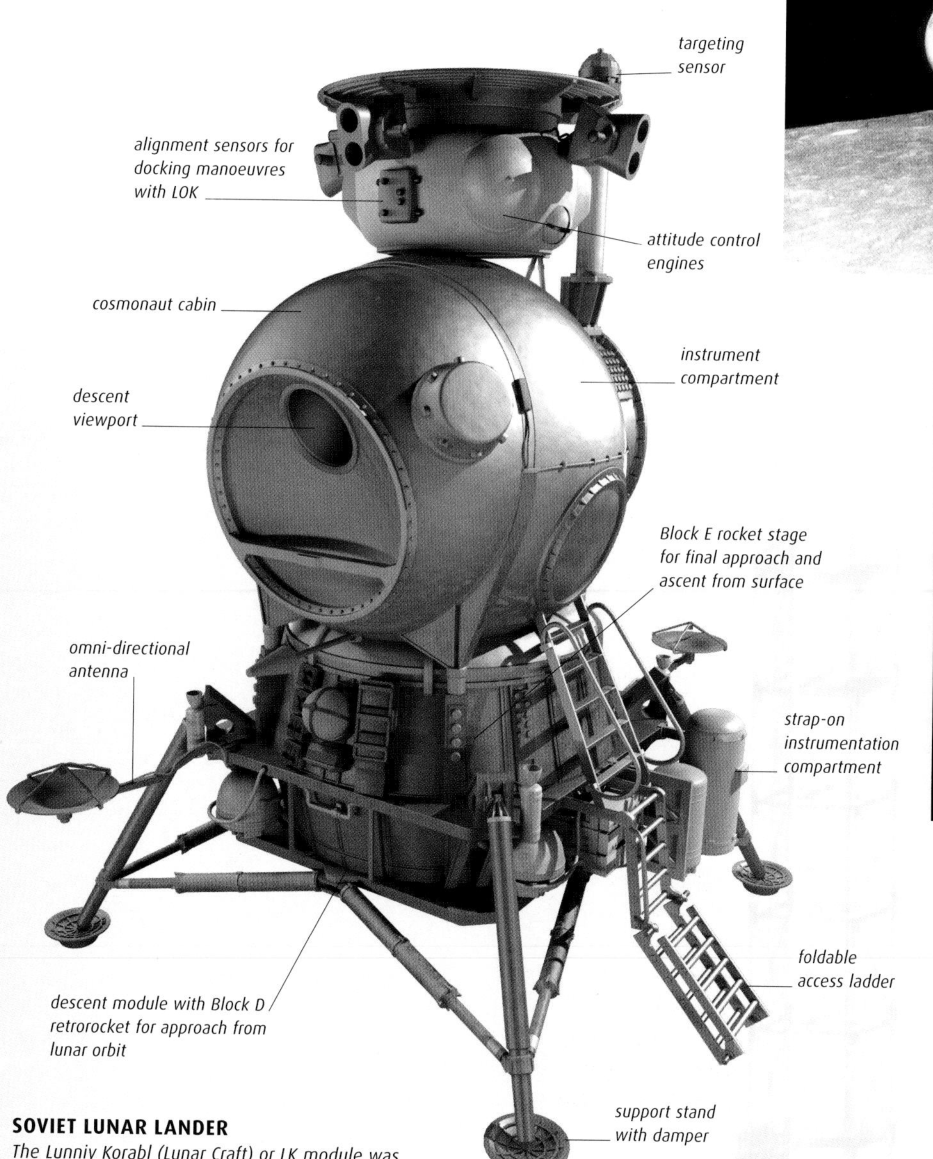

SOVIET LUNAR LANDER
The Lunniy Korabl (Lunar Craft) or LK module was a one-man lander that would have been docked with a Soyuz-derived LOK spacecraft for the trip to the Moon. Three successful unmanned tests in Earth orbit were listed as Cosmos satellites.

UNMANNED MOON SHOTS
The Luna programme continued throughout the 1960s. Key achievements included the first craft in lunar orbit and the sending back of the first pictures from the surface, in February 1966.

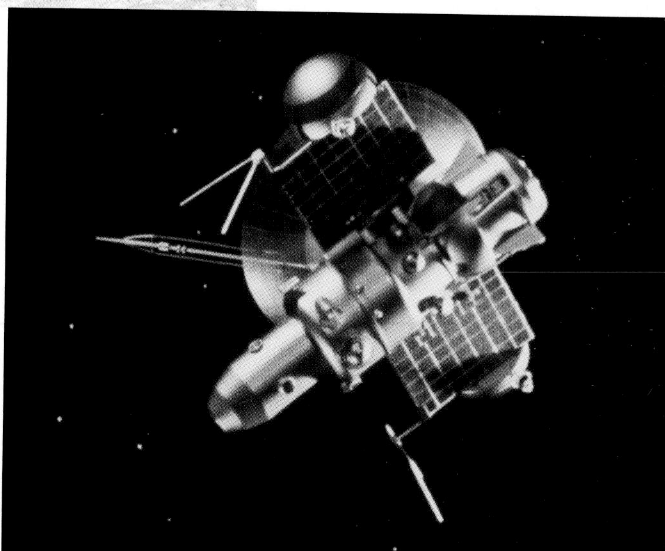

ZOND MISSIONS
One minor success of the Soviet lunar programme was the launch of the Zond probes – unmanned Soyuz spacecraft without an orbital module, designed to loop around the Moon and return to Earth in rehearsals for a similar manned flight.

the Soviets had many other capable designers, the loss of their leading light left the programme temporarily rudderless. In the aftermath, OKB-1 was restructured and there was a four-month delay before Vasili Mishin, Korolev's deputy, was appointed as his successor. All this time, the Soviets were lagging further behind Apollo, but the situation was so confused that, as late as February 1967, the Soviet government could talk of a landing in late 1968.

By the time that never-realistic deadline rolled around, the truth was unavoidable – the N1 and its LK lunar lander still languished in development, while Saturn V rockets shook the ground at Cape Canaveral and a manned American spacecraft orbited the Moon. Nikolai Kamanin, in particular, blamed the whole fiasco on infighting and a mistaken design philosophy that treated the cosmonauts as passengers and therefore developed overcomplex, fully automated spacecraft before worrying about clearing them for manned flight.

The big lie

On 30 December 1968, Soviet space leaders met to discuss their response to Apollo 8 and their impending defeat. Mstislav Keldysh proposed a novel solution – using Chelomei's Proton rocket to launch a robotic mission that would collect and return samples of lunar rock, hopefully ahead of a manned American landing. State-controlled media, meanwhile, would make it clear that the manned lunar programme had never existed – the Soviet Union would not dream of risking the lives of its heroic cosmonauts for a political stunt. Fortunately, the growing success of Soyuz provided them with an ideal cover story.

And so, in July 1969, the Luna 15 probe raced towards the Moon ahead of Apollo 11 – only to add to Soviet embarrassment by crashing during its final descent onto the Sea of Crises. In the following years, the Soviets would have greater success with unmanned probes (see p.258), but the truth would not emerge until after the country's collapse in 1991.

And even after Apollo 11, Soviet lunar plans did not die swiftly. The L3 spacecraft was abandoned in favour of development of the larger L3M, which might have landed on the Moon in the mid-1970s and paved the way for a Soviet Moonbase. The N1 project spluttered on until 1974, when it was finally cancelled in the aftermath of four launch failures.

Soyuz takes shape

While Soviet plans for the Moon stalled in the mid-1960s, work continued on a sophisticated new spacecraft. The Soyuz design ultimately proved so successful that it is still used, in modified form, to this day.

The Soyuz complex was the centrepiece of Sergei Korolev's original plan for an Earth-orbit rendezvous mission to the Moon. The Chief Designer envisaged a series of spacecraft with different roles that could function independently or together. Soyuz A was a manned vehicle, Soyuz B an unmanned craft with a powerful rocket motor, and Soyuz V a fuel tanker. Soviet bureaucracy and the confusion over rival lunar missions ultimately forced the unmanned engine block and tankers to be abandoned, and Soyuz A was redefined as a smaller, less ambitious modular spacecraft (though still on the scale of the Apollo CSM and capable of going to the Moon if it had been given the chance).

The Soyuz craft that began to roll off Soviet production lines in 1966 had three distinct parts. A spherical forward section, the orbital module, sat at the front of the capsule and would provide working space and house equipment needed in orbit. This connected via a hatch to a bell-shaped descent module that contained couches to accommodate three cosmonauts during launch and return to Earth. Behind the descent module's heat-shielded broad end was a cylindrical service module, containing the engines and other spacecraft systems. Solar panels that extended to either side of the service module could be used to recharge the spacecraft's batteries while in orbit.

A bad start

Unmanned Soyuz test launches were mostly disguised by listing them under the catch-all Cosmos programme of satellites. Although cosmonaut chief Nikolai Kamanin criticized the time wasted in developing fully automated spacecraft with systems that would become redundant once they had a crew onboard, most of the engineers felt this allowed for extensive safety checks before trusting a cosmonaut's life to the vehicle. In this case, it also added versatility – an unmanned Soyuz controlled from the ground might one day be used to resupply cosmonauts aboard an orbiting space station (an idea that reached fruition with the Progress supply ferry – see p.210).

Cosmonaut Vladimir Komarov was chosen to fly the first manned mission, Soyuz 1. The launch, on 23 April 1967, went well but problems began to develop soon afterwards. The initial plan was for Soyuz 2 to launch the next day with a three-man crew. The vehicles would rendezvous in space, and two cosmonauts would spacewalk to Soyuz 1, returning to Earth with Komarov. But a problem unfolding the solar panels left Soyuz 1 critically short on power and incapable of playing its part in

UNDER CONSTRUCTION
Engineers at OKB-1 inspect Soyuz 9 during assembly. The orbital module is at the top, then the descent module, with the service module at the base.

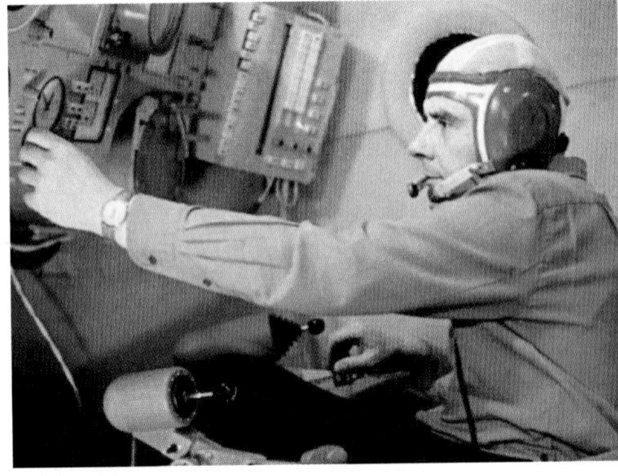

KOMAROV IN TRAINING
Cosmonaut Vladimir Komarov, veteran of the Voskhod 1 mission, works at a simulator console during preparations for his ill-fated flight aboard Soyuz 1.

BIOGRAPHY

VASILI MISHIN

Vasili Pavlovich Mishin (1917–2001) was among the Soviet scientists sent to Germany in 1945, where he first met Sergei Korolev. On his return to the Soviet Union, Mishin became Korolev's deputy at OKB-1 and helped the bureau and the Soviet Union to take the lead in the Space Race. Staff lobbied for his promotion to Chief Designer after Korolev's death, and Mishin oversaw the early Soyuz and Salyut programmes. However, he was ousted from the post in favour of Valentin Glushko after the failure of the N1 Moon rocket.

ZARYA CALLING
Controllers and visiting family members at Baikonur's Spaceflight Control Centre (often known by its callsign Zarya, meaning "Sunrise") take advantage of a brief communication window during the Soyuz 9 mission.

the manoeuvre. The second launch was scrubbed as mission controllers concentrated on getting Komarov down safely. Soyuz 1 was proving difficult to stabilize in orbit, and two re-entry attempts were aborted before the descent module finally plunged into the atmosphere during its 18th orbit. The return to Earth appeared to be going well, but disaster struck when the module's braking parachute failed to deploy properly. Komarov died instantly as his capsule hit the ground near Orsk in the Ural mountains.

The road to recovery

The programme was immediately suspended for a thorough safety overhaul, adding to the woes of the Soviet lunar effort. It was to be 18 months before a Soyuz took to the skies again, in October 1968. This was Soyuz 2, an unmanned vehicle that would act as a rendezvous target for Georgi Beregovoi, launched aboard Soyuz 3 the next day. After achieving the mission's limited aim of testing the spacecraft's manoeuvring and rendezvous systems, Beregovoi returned to Earth safely four days after launch, and the Soyuz programme began to inspire some confidence.

The following January, a joint flight of Soyuz 4 and 5 finally achieved the aims of the original Soyuz 1 mission. The two spacecraft rendezvoused and docked in orbit, and Yevgeny Khrunov and Alexei Yeliseyev spacewalked from Soyuz 5 to Soyuz 4, joining their comrade Vladimir Shatalov for the return to Earth, while Boris Volynov came home alone aboard Soyuz 5. The transfer was another indication of the spacecraft's versatility – the orbital module could be sealed off from the descent module and drained of air before an external hatch was opened for spacewalk access – effectively, it was functioning as a large airlock.

The next Soyuz mission, in October 1969, was an ambitious group flight of three spacecraft. The intention was for Soyuz 7 and 8 to dock while the crew of Soyuz 6 filmed the manoeuvre. The rendezvous in orbit went well, but faults in the spacecraft electronics prevented the docking, and the three crews returned to Earth. Soyuz 9, launched in June 1970, set a new space endurance record, of 18 days, for its two-man crew. This was intended to be the last independent Soyuz mission – flights would now rendezvous with the new Salyut space stations.

DOCKING IN SPACE
Soyuz 4 Commander Vladimir Shatalov demonstrates how Soyuz 4 and Soyuz 5 met in Earth orbit in January 1969 – the first time that two piloted spacecraft docked in space.

EATING BREAKFAST
Andrian Nikolayev samples breakfast aboard Soyuz 9. When the cosmonauts landed after 18 days, they were weakened by their experience, though they made a good recovery. Later flights showed that longer stays would allow the body to adapt better and that if they exercised properly the cosmonauts would actually return in better shape.

18 October 1965
The Soyuz programme is reoriented to the development of a smaller Earth-orbiting spacecraft.

23 April 1968
Soyuz 1 runs into trouble after launch and plans to rendezvous with another Soyuz are cancelled.

23 April 1968
Vladimir Komarov is killed when Soyuz 1's parachute fails after re-entry.

26 October 1968
Georgi Beregovoi is launched aboard Soyuz 3 on a mission to rendezvous with the unmanned Soyuz 2.

16 January 1969
Soyuz 4 and 5 rendezvous and dock in orbit, transferring crew between spacecraft during a spacewalk.

14 October 1969
Soyuz 6, 7, and 8 conduct the first rendezvous between three spacecraft in orbit, but Soyuz 7 and 8 fail to dock.

1 June 1970
Soyuz 9 is launched on an 18-day endurance mission for its crew.

TECHNOLOGY

THE SOYUZ ROCKET

The Soyuz spacecraft was considerably heavier than previous Soviet manned spacecraft. In order to lift it to orbit, Korolev's bureau developed a modified version of the R-7, which by the mid-1960s was an established, trustworthy launch vehicle. Like the Vostok rocket, the Soyuz has two main stages with four boosters around the lower stage – the main difference is that in this case the upper stage was provided with a more powerful engine. Despite its age, the launcher's reliability has ensured that it remains in service today.

TECHNOLOGY

RUSSIA'S MULTI-PURPOSE SPACECRAFT

Soyuz spacecraft

Sergei Korolev's lasting legacy to Russian spaceflight, the Soyuz entered service in 1967 and is still operating, in upgraded form, 40 years later. The multipurpose vehicle was the first Soviet spacecraft capable of docking with other craft in orbit, but the first version, model 7K-OK, did not allow the crew to move between craft. This ability arrived with the 7K-OKS (Soyuz 11), but after the loss of that mission's crew, the spacecraft was given an extensive redesign, producing the two-man 7K-T, which operated through the 1970s.

DOCKING IN SPACE
In January 1969, Soyuz 5 loomed large in the porthole of Vladimir Shatalov's Soyuz 4 spacecraft, moments before the first docking of two manned vehicles in space. Shatalov picked up two passengers from Soyuz 5, but they had to spacewalk between the orbital modules in order to reach him.

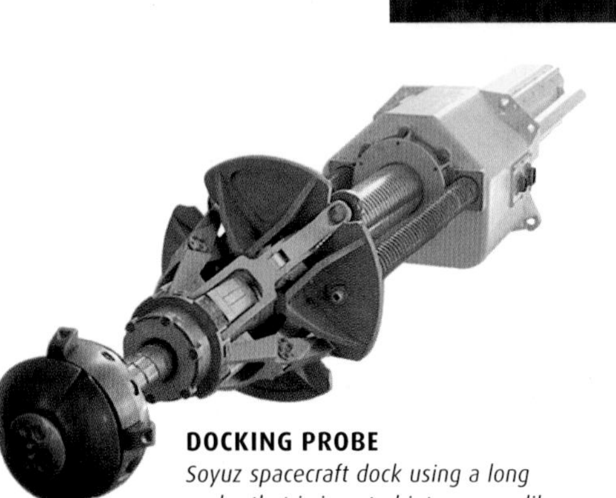

DOCKING PROBE
Soyuz spacecraft dock using a long probe that is inserted into a cone-like target mechanism on the other vehicle. The probe is then retracted, triggering latches that pull the spacecraft back in to form an airtight seal.

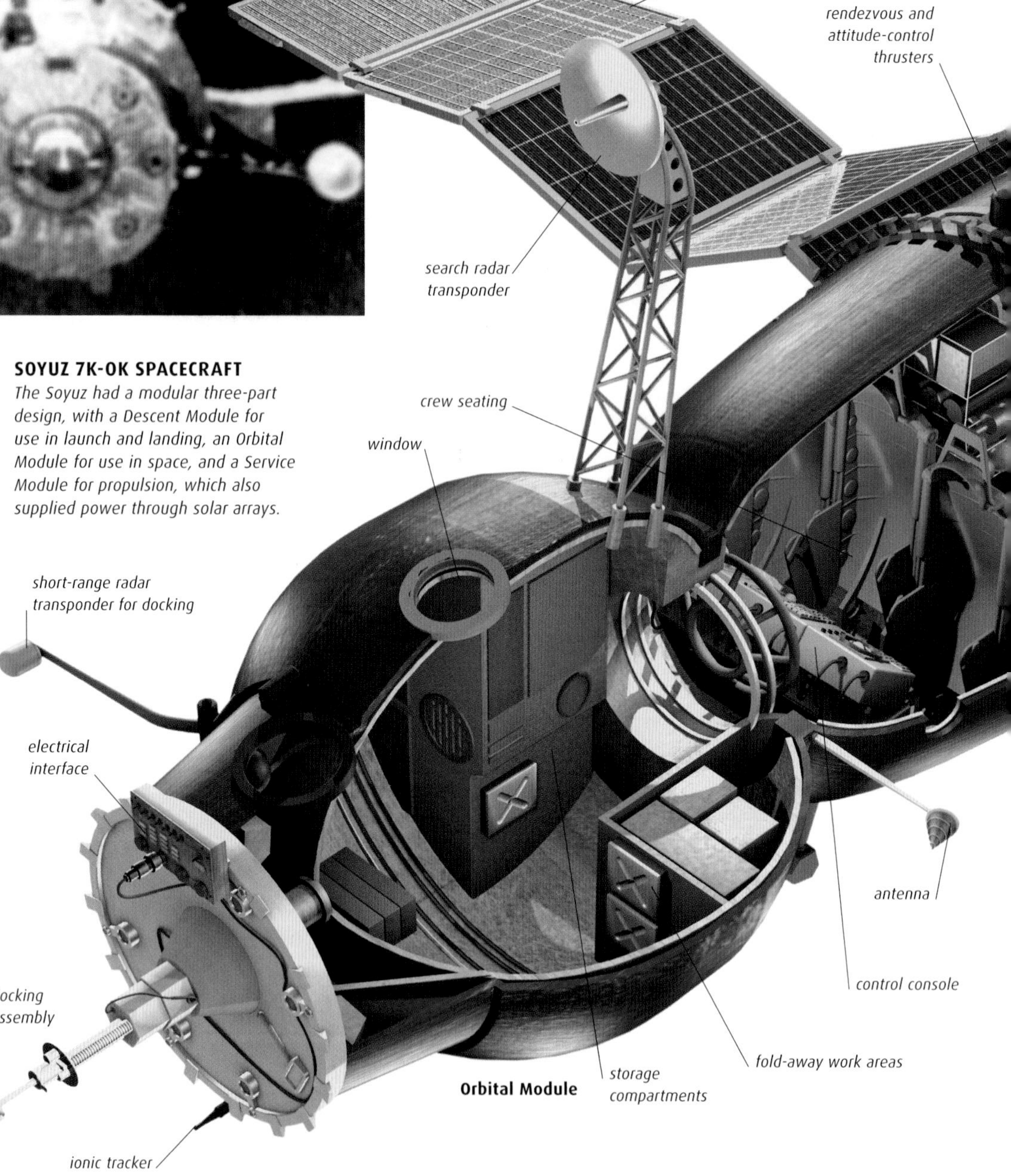

SOYUZ 7K-OK SPACECRAFT
The Soyuz had a modular three-part design, with a Descent Module for use in launch and landing, an Orbital Module for use in space, and a Service Module for propulsion, which also supplied power through solar arrays.

LENGTH	7.95m (26ft 1in)
MAXIMUM DIAMETER	2.72m (8ft 11in)
SOLAR ARRAY SPAN	9.9m (32ft 1in)
WEIGHT AT LAUNCH	6,560kg (14,460lb)
WEIGHT AT LANDING	2,810kg (6,190lb)
CREW	3
ENGINE	1 x main engine: nitric acid/ hydrazine; 36 x RCS thrusters: hydrogen peroxide
MANUFACTURER	Korolev (OKB-1)

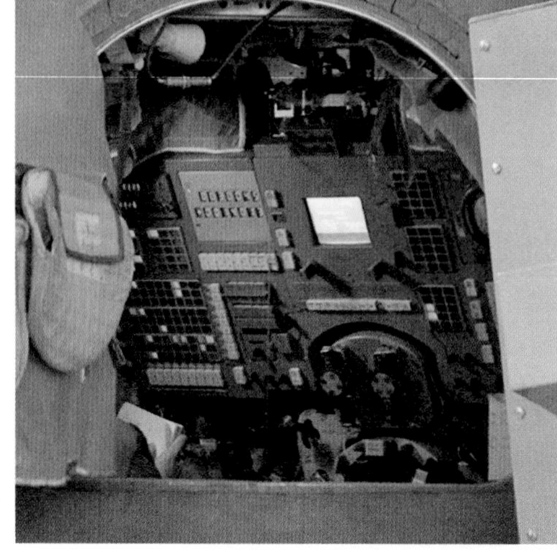

FLIGHT CONTROLS
Soyuz was the first Soviet spacecraft with its own propulsion, enabling it to change its orbit and make other manoeuvres in space. An increasingly sophisticated Igla rendezvous system used radio signals to guide the Soyuz towards a docking target. The more recent Kurs system allows the entire process to take place automatically and is used on Progress, the Soyuz-derived unmanned ferry *(see p.210).*

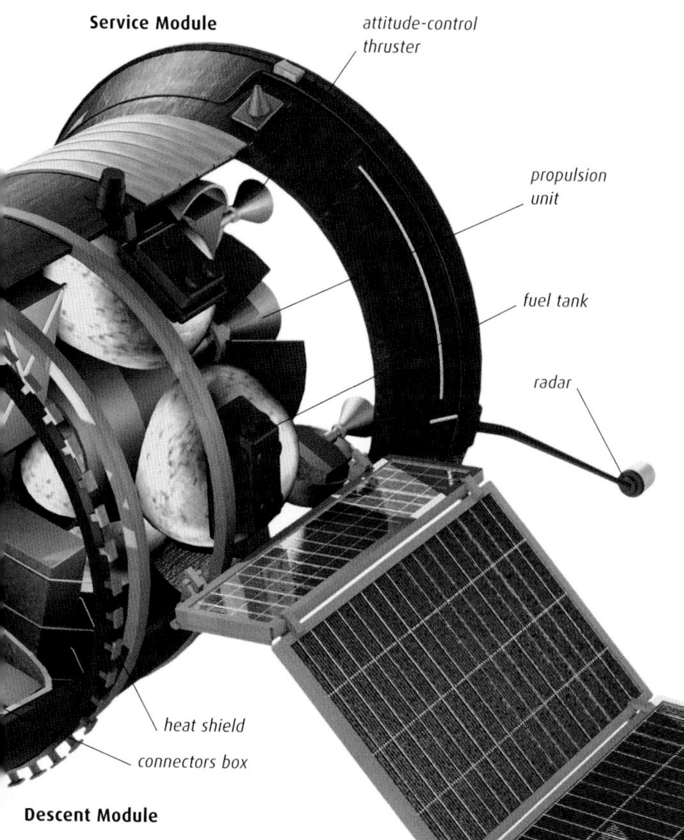

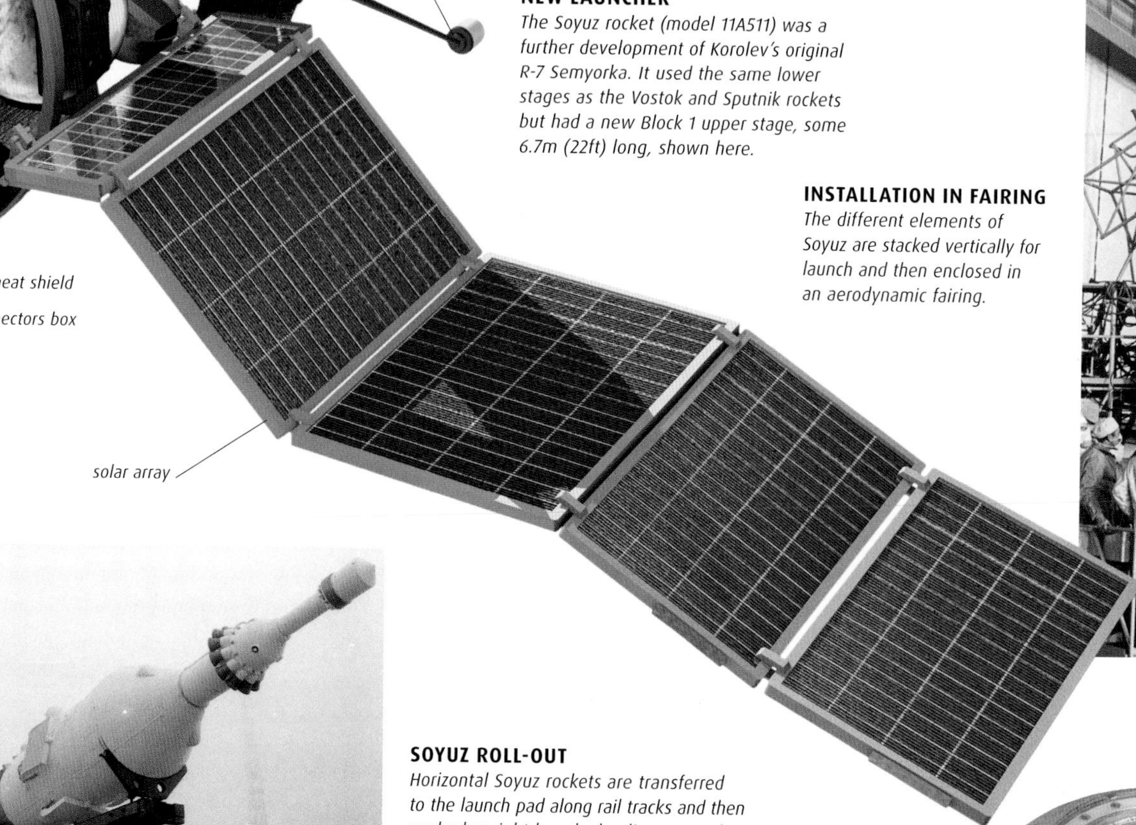

Service Module

attitude-control thruster

propulsion unit

fuel tank

radar

heat shield

connectors box

Descent Module

solar array

NEW LAUNCHER
The Soyuz rocket (model 11A511) was a further development of Korolev's original R-7 Semyorka. It used the same lower stages as the Vostok and Sputnik rockets but had a new Block 1 upper stage, some 6.7m (22ft) long, shown here.

INSTALLATION IN FAIRING
The different elements of Soyuz are stacked vertically for launch and then enclosed in an aerodynamic fairing.

SOYUZ ROLL-OUT
Horizontal Soyuz rockets are transferred to the launch pad along rail tracks and then pushed upright by a hydraulic ram on the transporter. Retractable gantries around the pad then swing across to lock the launch vehicle into place.

exit hatch

parachute cover

retrorocket housing

ablative shielding

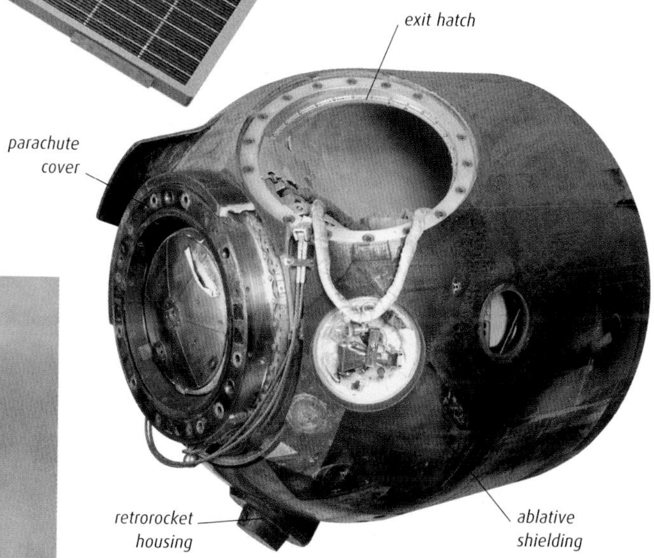

RETURN TO EARTH
The Descent Module is the only section that returns to Earth. As it approaches the ground on its parachute, retrorockets fire at the last moment to ensure a soft landing.

LAUNCH OF SOYUZ 3
The second manned flight of the Soyuz spacecraft blasts off from Baikonur Cosmodrome in October 1968, on its way to attempt the first Soviet rendezvous and docking in space.

Apollos 7 and 8

By late 1968, NASA officials were ready to let the spacecraft fly at last. The initial manned missions tested the Apollo Command Module in Earth orbit, and then took it on the journey to the Moon and back.

18 May 1967
The crew for Apollo 7 is officially named.

7 August 1968
George Low, Apollo Program Manager, suggests sending Apollo 8 around the Moon.

17 August 1968
NASA Administrator James Webb agrees to the new Apollo 8 mission, pending a successful flight for Apollo 7.

11 October 1968
Apollo 7 is launched into Earth orbit on an 11-day mission.

22 October 1968
After a fractious, though ultimately successful, mission, the Apollo 7 Command Module splashes down.

21 December 1968
The first manned Saturn V rocket launches Apollo 8 on its historic mission to the Moon.

24 December 1968
A precise engine burn puts Apollo 8 in lunar orbit.

27 December 1968
The Apollo 8 astronauts splash down safely.

In the aftermath of Apollo 1, Wally Schirra, Donn Eisele, and Walter Cunningham moved up to become the prime crew for the next manned mission. Despite the fact that the lunar module was not ready for flight, the crew included a Lunar Module Pilot (Cunningham), and so Apollo 7 became America's first three-man space mission. Its main aims were to test the Command and Service modules within a safe distance of Earth, over a period of 11 days – more than enough time to make it to the Moon and back.

War in space

Apollo 7 blasted into orbit on a Saturn IB launcher on 11 October 1968. It was to be a stressful mission for everyone involved – both in orbit and at mission control. Although the spacecraft itself worked without any major problems, the crew suffered during their orbital confinement. The stress of giving Apollo a thorough "shakedown" was bad enough, but much of the mission was to be broadcast on live television. Schirra came down with a cold soon after launch, and as it spread to his crewmates, all three became irritable. They insisted that they should be allowed to get on with the real work of the mission without worrying about TV cameras, so the first live broadcast from space was postponed until later in the flight.

This was only one of several occasions when tensions surfaced between the crew and Mission Control – Schirra called some of the tests he had to carry out "idiotic", while Chris Kraft labelled him "paranoid". Despite all this, the crew completed their major goals, and splashed down on time and on target. In the aftermath, Deke Slayton called the incident "the first space war", and it was no coincidence that the Apollo 7 crew were not selected for further missions.

From the Earth to the Moon

Despite the recriminations, Apollo 7 had been a resounding success. According to the original plan, the next step should have been a test of the Lunar Module in Earth orbit, but events had already conspired to give Apollo 8 a new objective. For one thing, the Lunar Module would not be ready for flight until early 1969. With a further test in lunar orbit

TESTING, TESTING
During their mission, the Apollo 7 astronauts practised traditional navigation by taking sightings from bright stars through the CSM's small but serviceable windows.

WRITING IN SPACE
Walter Cunningham takes notes aboard Apollo 7. Contrary to popular myth, NASA did not spend millions on a "space pen" – the pen was developed by a company that managed to get them adopted by NASA.

DOCKING PRACTICE
Once the CSM was free of its S-IVB upper stage (right), the crew practised the docking manoeuvres that would be needed to link up with the Apollo LM. The white disk just off centre (left) marked the docking target.

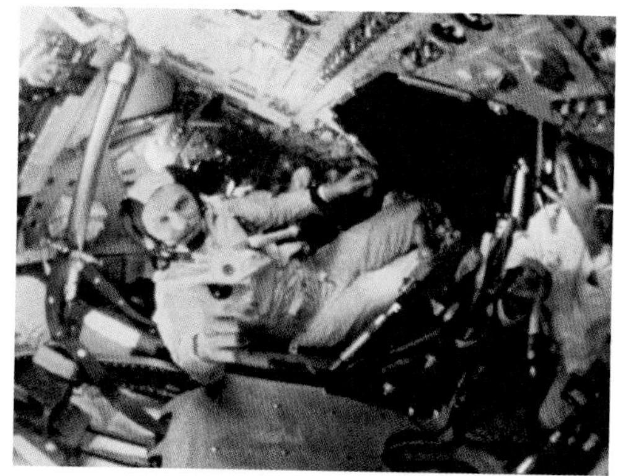

EN ROUTE TO THE MOON
After the cramped conditions of Mercury and Gemini, which saw the astronauts essentially confined to their seats, the Apollo CSM was relatively spacious. Here, Frank Borman enjoys floating free in zero gravity.

still needed ahead of a first landing attempt, and the ever-present risk of a problem that might require the insertion of a repeat flight into the schedule, Apollo would run very close to its deadline with three such crucial missions in one year.

Then there were the Soviets: no one knew the state of the rival lunar programme, but everyone assumed that it was more advanced than the reality. The unmanned Zond 5 had circled the Moon and returned to Earth weeks before Apollo 7's launch. Given that the CIA was also reporting a Soyuz spacecraft would soon link up with a group of fuel tankers in Earth orbit, it was starting to look like the Soviets were on the verge of a lunar mission involving an Earth-orbit rendezvous (EOR).

And so NASA decided that Apollo 8 would go around the Moon at Christmas 1968. It would beat the Soviets to another first and help to reveal any problems in the Command Module that might affect the later LM lunar-orbit rehearsal. When it blasted off on 21 December, with a crew of Frank Borman, Jim Lovell, and Bill Anders, Apollo 8 was also the first manned Saturn V launch. Fortunately for NASA, it was an almost flawless mission, its effect on the watching world almost as great as Armstrong and Aldrin's first steps on the Moon would be a few months later. With everything going well, the crew were given the go-ahead for the crucial engine burn that would put them into lunar orbit less than three days after leaving Earth. The burn had to take place over the Moon's far side, out of contact with Earth, and there was tension at Mission Control until the craft emerged, intact and in orbit, from its radio silence.

That Christmas, the astronauts became the first people to see Earthrise over the Moon, and to take in our fragile planet's isolation in space. The images they sent back were iconic, and matched with equally stirring words as they read the opening lines from the Bible's Book of Genesis – "In the beginning, God created the heavens and the Earth ..." – before wishing a Merry Christmas to the entire world.

BIOGRAPHY

FRANK BORMAN

Born in Indiana but raised in Arizona, Frank Borman (b.1928) began flying as a teenager. He graduated from the United States Military Academy in 1950 and served as a fighter and test pilot in the US Air Force before joining NASA in 1962. After flying on the Gemini 7 mission in 1965, he sat on the investigation board following the Apollo 1 fire, and then took command of Apollo 8. After leaving NASA and the USAF in 1970, he built a second career in the airline business, retiring in 1986 to enjoy his hobby of restoring and flying vintage aircraft.

AROUND THE FAR SIDE
The Apollo 8 crew (below) returned with the most detailed images yet from the lunar far side, including the spectacular Earthrise (above) and the dark-floored crater Tsiolkovskii (left).

SPIDER ABOVE THE EARTH

During day five of the Apollo 9 mission, McDivitt and Schweickart boarded the LM and practised separation and flying manoeuvres in Earth orbit. David Scott watched from the CSM and took this picture of the LM hanging above the horizon. It shows the legs in their unfolded position and the rarely seen surface probes extending out from each of the foot pads.

Apollos 9 and 10

By early 1969, the Apollo Lunar Module was finally ready for action, but there were still two crucial qualifying missions needed before the first attempt to put a man on the Moon.

DOCKED IN ORBIT
David Scott stands halfway out of the hatch of the Apollo 9 Command Module, 190km (118 miles) above the Earth. The body of the LM Spider dominates the foreground.

The history of spaceflight is full of "what ifs". If NASA had stuck with its original rosters, the crew of Apollo 9 should have flown on Apollo 8, circling the Moon at Christmas 1968. As it was, perhaps due to the fact that they had been training to fly the LM for more than two years, James McDivitt, David Scott, and Russell Schweickart were shifted back by one flight, lifting off with the first complete Apollo spacecraft on 3 March 1969, but destined to go no further than Earth orbit. With Apollo 8's lunar mission making plans for a further LM test in high Earth orbit redundant, this meant that there would be only one more mission - a dress rehearsal in lunar orbit - before the first landing attempt. It also put McDivitt's backup crew - led by Neil Armstrong - in line for a landing attempt aboard Apollo 11.

Gumdrop and *Spider*

Because the Apollo 9 spacecraft would be split in two for parts of the mission, NASA allowed the crew to choose individual callsigns for the CSM and LM - naming their spacecraft for the first time since Gemini 3. The LM was called *Spider* for obvious reasons, while the CSM was nicknamed *Gumdrop* after it arrived at Kennedy Space Center wrapped in blue cellophane. Throughout their ten-day mission, the Apollo 9 crew thoroughly tested both spacecraft, practising docking and undocking manoeuvres, flying the LM up to 179km (111 miles) from the CSM, and using both the ascent and descent engines in orbit. Schweickart also performed the first Apollo Spacewalk, testing the new spacesuit's integral life-support systems. Apart from a bout of space sickness for Schweickart, everything went well, and by the time the crew returned on 13 March, the Moon was almost within Apollo's grasp.

So far and so near

The final dress-rehearsal mission, Apollo 10, thundered into the skies above Cape Canaveral on 18 May, with the highly experienced crew of Thomas Stafford, John Young, and Eugene Cernan onboard. The flight to the Moon and injection into lunar orbit went without a hitch, and once in lunar orbit, Stafford and Cernan boarded the LM, named *Snoopy* after the dog in the popular *Peanuts* cartoon strip, and undocked from the CSM (named *Charlie Brown* after Snoopy's owner).

While Young remained in orbit high above the Moon, his colleagues flew the LM to within 16km (10 miles) of the lunar surface with no major problems, snapping high-resolution photographs of the Sea of Tranquillity, by now selected as the target for Apollo 11. After their close encounter with the Moon, they reunited with the CSM, discarded the LM, and prepared for the long voyage home.

By the time Apollo 10 returned to Earth four days later, even Soviet cosmonaut trainer Nikolai Kamanin was privately admitting that, barring disaster, the Americans would be on the Moon within weeks.

24 January 1969
The first of the Apollo Lunar Modules is finally approved for flight as part of the Apollo 9 mission.

3 March 1969
Apollo 9 launches on a ten-day mission in orbit above the Earth.

7 March 1969
The LM is flight-tested alone in Earth orbit.

13 March 1969
Apollo 9 makes a safe return to Earth.

18 May 1969
Apollo 10 launches on its dress-rehearsal mission to the Moon.

21 May 1969
Apollo 10 goes into lunar orbit.

22 May 1969
Aboard the LM *Snoopy*, astronauts Cernan and Stafford come within 16km (10 miles) of the lunar surface.

26 May 1969
The Apollo 10 Command Module splashes down in the Pacific Ocean.

9 June 1969
In the wake of Apollo 10's success, NASA confirms that Apollo 11 is on schedule for launch on 16 July.

LUNAR SEPARATION
The Apollo 10 LM Snoopy separates from the Command Module. Onboard, Cernan and Stafford begin tests of the LM's independent performance.

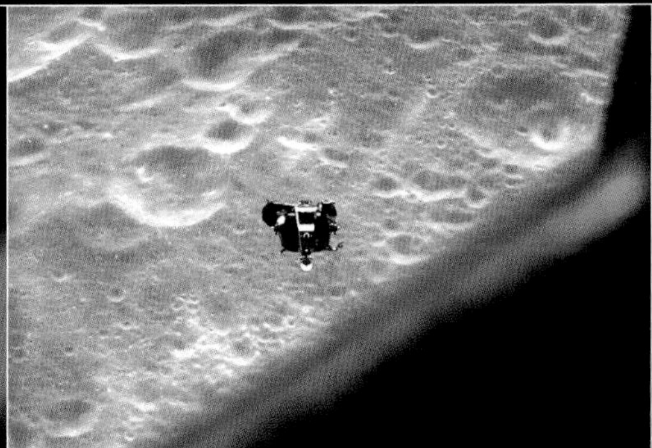

CLOSE TO THE SURFACE
After a 27-second engine burn, the LM has entered an elliptical orbit, with an altitude ranging from 113km (70 miles) to 15.6km ($9\frac{2}{3}$ miles) from the Moon's surface.

MISSION COMPLETED
Snoopy returns from its expedition, reuniting with Charlie Brown. So far as it went, the mission was a complete success, though the LM was not equipped for a landing.

Voyage to the Moon

The world watched in awe as Apollo 11 sped towards the Moon in mid-July 1969. Now the Space Race was no longer with the Soviet Union, but with Kennedy's self-imposed deadline – and NASA's own good fortune.

16 July 1969
Apollo 11 launches from Cape Canaveral on its way to the Moon. Within two hours, it is out of Earth orbit and on its translunar flightpath.

17 July 1969
Michael Collins takes star sightings to compare Apollo 11's theoretical flightpath to its actual one. A three-second mid-course engine burn corrects the trajectory.

18 July 1969
The crew transmit a 96-minute guided tour of their spacecraft back to Earth, where it is broadcast live on television.

19 July 1969
Following a successful retrorocket burn, Apollo 11 enters orbit around the Moon.

20 July 1969
The LM separates from the CSM and enters an orbit that spirals gradually closer to the Moon. Just over two hours later, *Eagle* touches down in the Sea of Tranquillity.

At 09:32 local time on 16 July 1969, the huge S-IC first stage of a Saturn V rocket thundered into life on Pad A of Kennedy Space Center's Launch Complex 39. Five F-1 engines gradually throttled up to full power, consuming 13,000 litres (3,500 gallons) of liquid hydrogen and liquid oxygen every second. Explosive bolts blew, separating the rocket from its support structure, and the Saturn V slowly lumbered into the sky. A million people lining the nearby highways and beaches cheered as the rocket soared higher, rapidly gaining speed. An estimated 600 million television viewers around the world were watching with them. Within 12 minutes, Apollo 11 was in orbit.

The spacecraft's crew – Commander Neil Armstrong, Lunar Module Pilot Buzz Aldrin, and Command Module Pilot Michael Collins – had fallen into the frame for the first lunar landing when the decision was made to send Apollo 8 around the Moon (see p.134). They had since been subjected to the most intensive training of their careers, and ferocious media scrutiny. By the time they walked to the pad that fateful morning, they were ready for almost anything. As a crew, they were not socially close – Collins was the most personable, Aldrin perhaps the most intense, while Armstrong had the remote brilliance of a top-gun test pilot. Nevertheless, all were consummate professionals, and as the training instructors conspired to throw countless disaster scenarios at them, they reached a point where each could trust the others with his life.

BIOGRAPHY

NEIL ARMSTRONG

Ohio native Neil Alden Armstrong (1930–2012) studied aerospace engineering on a naval scholarship from 1947, as part of which he began service as a US Navy pilot in 1949. After training, he saw action in the Korean War before returning to complete his degree, and upon graduation he applied to become a test pilot for NACA and later NASA. He applied for the second astronaut intake in 1962, and was on the backup crews for Geminis 5 and 11, as well as commanding Gemini 8 (see p.108). After backup duties on Apollo 9, Apollo 11 was his last spaceflight. He worked on the Apollo 13 inquiry before leaving NASA in 1971 to pursue interests in business and education. He also served on the commission investigating the loss of *Challenger* in 1986 (see pp.202–03).

En route to history

After just one-and-a-half orbits of the Earth, the S-IVB upper stage kicked the spacecraft onto its translunar trajectory. Once safely on its way, the CSM *Columbia* eased free of the rocket stage, turned through 180 degrees, and docked with the LM *Eagle*, which had nestled beneath it during the launch.

Safely freed from the rocket, Apollo 11 sped on towards the Moon. The journey took a little over three days – then came the vital retrorocket burn to slip into lunar orbit and the preparations for separation of the LM and CSM. Twenty-five hours after arrival, a 30-second burn on *Eagle*'s descent engine dropped it into an orbit that took it within 13km (8 miles) of the surface. Inside the LM, Armstrong and Aldrin stood side by side, held in place by elastic stays. Face-down to the surface, they watched the landscape roll past beneath them until Houston gave the final go-ahead to land. Now Armstrong used a fine-guidance controller to throttle the descent engines while Aldrin read the module's altitude and fuel. Both astronauts stared out of the windows, looking for a smooth area to land. Spotting a dusty plain, Armstrong eased the spacecraft down, making contact with the surface with barely 20 seconds of descent fuel remaining.

***EAGLE* IN ORBIT**
Michael Collins took this photograph of Apollo 11's Lunar Module as it began to draw away from his CSM on the long spiral down to the Moon. The LM executed a complete rotation outside the CSM's windows, while Collins looked for any signs of damage inflicted by the stresses of launch.

MAKING HISTORY
Work in Mission Control is all but forgotten as the controllers turn from their desks to watch pictures of the first Moon landing coming back from Apollo 11.

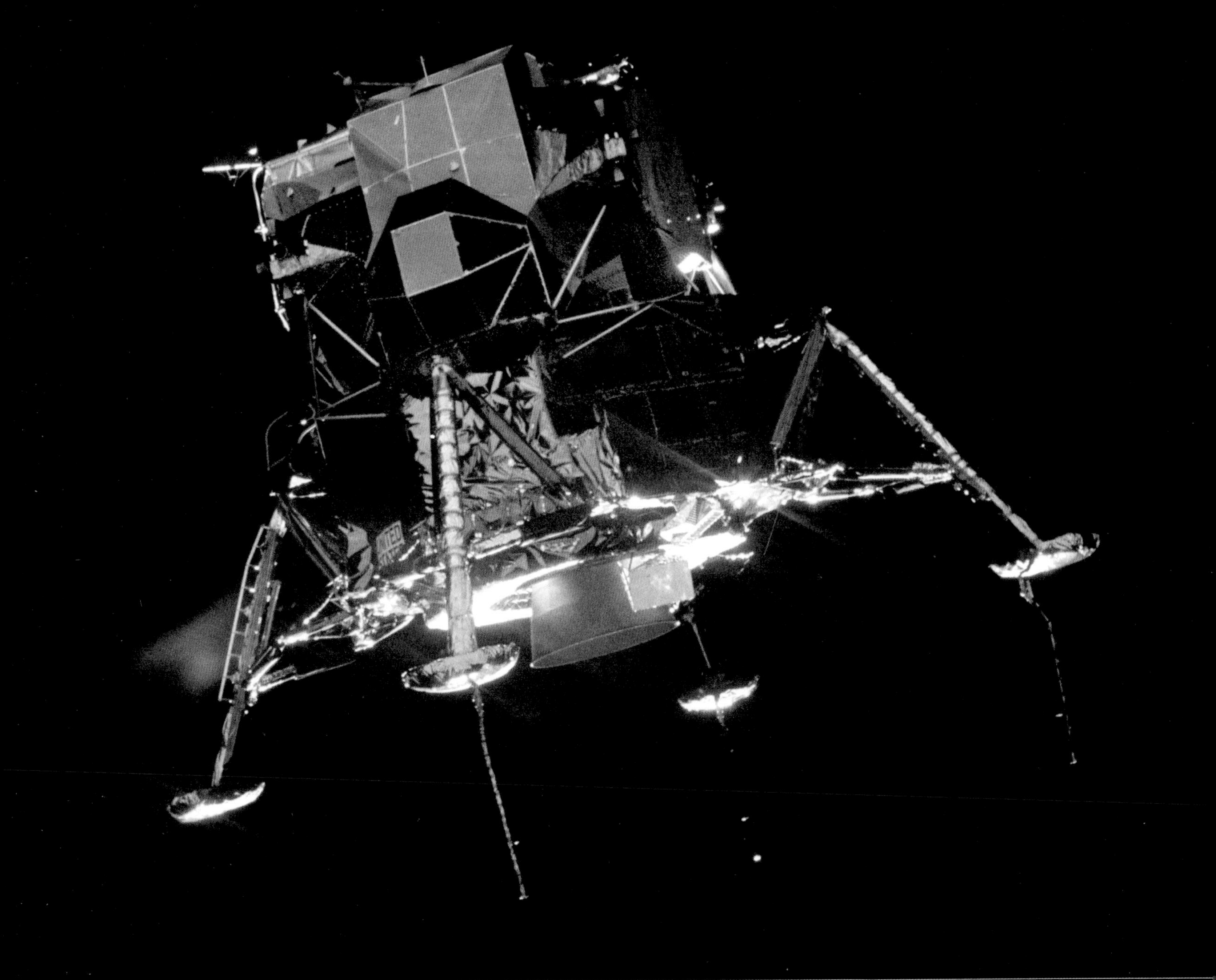

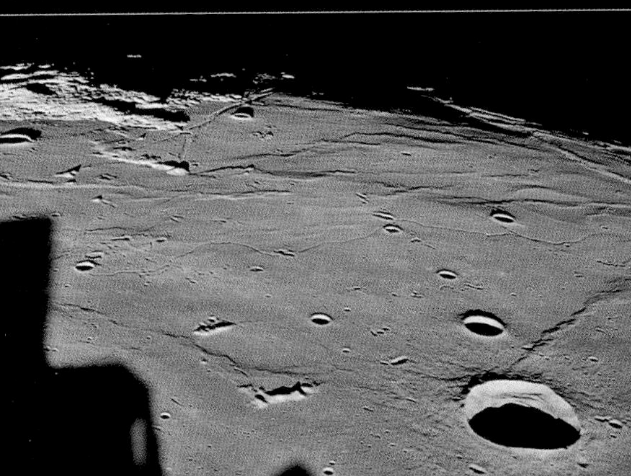

20 – 07– 69 17:44 GMT

Shortly after separation, Armstrong and Aldrin pass over their target – "Landing Site 2" in the Sea of Tranquillity. In this photograph, it lies just to the right of centre on the edge of the retreating night shadow.

20–07–69 19:08 GMT

Looking back at the CSM from their elliptical orbit, the LM astronauts can see it against the lightly cratered Sea of Fertility. At the lowest point in their orbit, Armstrong triggers the engines and the LM begins its final descent.

20–07–69 20:17 GMT

The astronauts soon realize they are landing several miles off target. As the automatic descent is heading for a field of large boulders, Armstrong takes manual control, bringing the LM down in a smoother area.

DESTINATION MOON
On the morning of 16 July 1969, Apollo 11's Saturn V launch vehicle lumbers clear of its own exhaust flames at the beginning of an epic 1.5-million-km (950,000-mile) voyage.

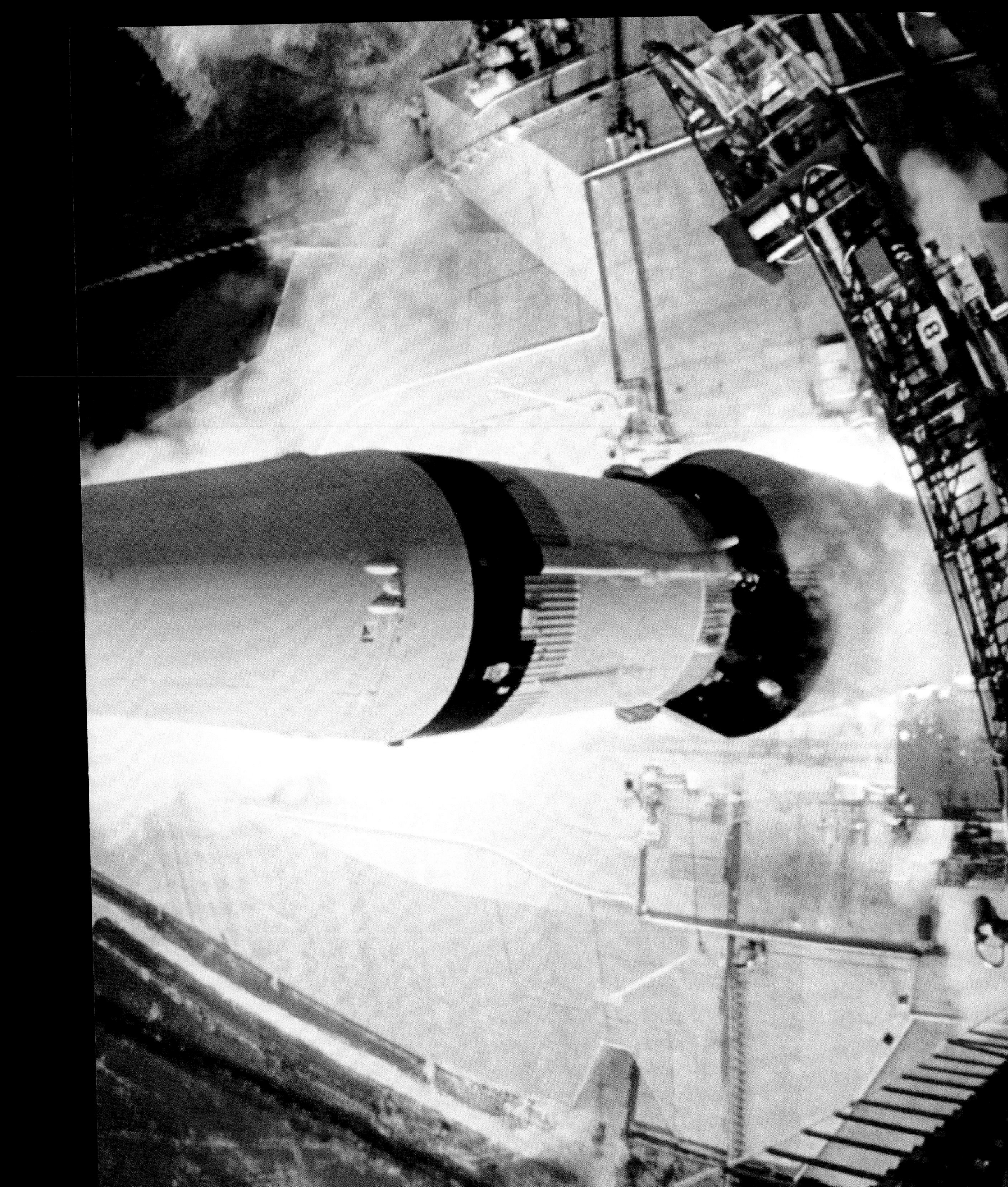

The Eagle has landed

Three days after their spectacular launch from Cape Canaveral, the Apollo 11 astronauts were ready to take the first human steps on another world.

21 July 1969
Neil Armstrong becomes the first person to step onto the surface of the Moon. He and Buzz Aldrin complete a spacewalk lasting just over two hours. The LM lifts off from the lunar surface 12 hours later, and docks safely with the CSM in lunar orbit.

22 July 1969
The LM and CSM separate in lunar orbit, and the CSM makes a successful transfer back into a trans-Earth orbit. Later in the day, a small mid-course correction burn is needed.

24 July 1969
The Command and Service modules separate at 16:21 GMT, and the Command Module re-enters the Earth's atmosphere, splashing down at 16:50. Little more than an hour later, the astronauts are aboard the recovery ship.

10 August 1969
The crew emerge from quarantine.

As *Eagle* settled onto the Sea of Tranquillity, the astronauts waited, tense, to see what effect the weight of the capsule would have on the rocks and dust below. Fortunately, the lander's pads settled barely an inch into the soil. Relaxing briefly, Armstrong and Aldrin stopped for a meal before it was time to don their spacesuits. Getting out of the LM while burdened by the suit and its backpack was difficult, even though the lower gravity reduced the weight of suit and pack from 86kg (190lb) to just 14kg (30lb). Armstrong stepped out first, shuffling backwards through the hatchway and climbing down the ladder, pausing to deploy the television camera that would send pictures back to the billion or more people watching on Earth. He finally stepped onto the Moon at 02:56 GMT on 21 July 1969.

Aldrin followed 19 minutes later, and for just under two hours, both astronauts collected rock samples, deployed science experiments, and tested conditions in the reduced gravity. They filmed their entire expedition (though only black-and-white television pictures could be relayed live back to Earth) and attempted to describe their surroundings in an effort to pin down their location – they had clearly landed several miles from their intended site. One curiosity they noted was the difficulty of estimating distance – the lunar horizon was much closer than Earth's, and without an atmosphere there was no haze to offer a visual clue to the distance and scale of the nearby hills.

BIOGRAPHY

BUZZ ALDRIN

Born and raised in New Jersey, Buzz (formerly Edwin) Aldrin (b.1930) served in the US Air Force during the Korean War before studying for a doctorate in astronautics, for which he wrote a thesis that earned him the nickname of Doctor Rendezvous among his fellows in NASA's third astronaut training group. His first mission was as backup on Gemini 9, and he made the first truly successful spacewalk on Gemini 12. After Apollo, he returned to the Air Force, but struggled with personal problems. Today, he is known as an author and advocate of manned spaceflight.

They also placed a flag, a plaque, and artefacts commemorating the lost astronauts of Apollo 1 and Soyuz 1, and they had a brief conversation with US President Nixon.

All too soon, it was time to return to the LM, first loading aboard a precious 22kg (48lb) of rock samples from the area around their landing site. Back aboard the LM, Armstrong and Aldrin shed their spacesuits, ate a meal, and attempted to sleep, stringing up hammocks across the interior of the LM. They were the first to discover that contamination with lunar material was unavoidable – the fine dust of the

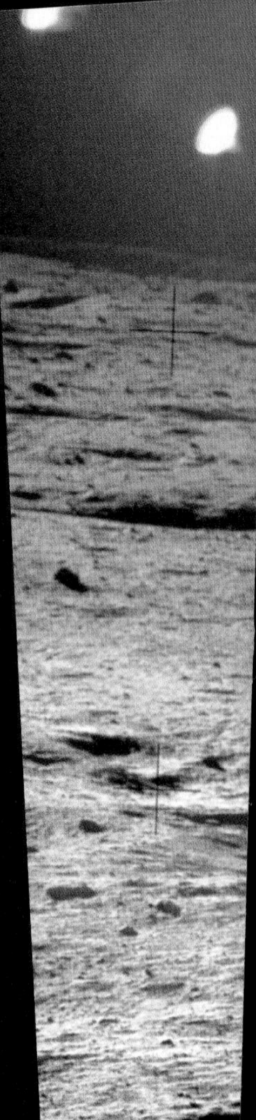

ALDRIN STEPS OUT
Buzz Aldrin poses in front of the LM during the Apollo 11 EVA. To his right is the flag-like collector of the Solar Wind Composition Experiment.

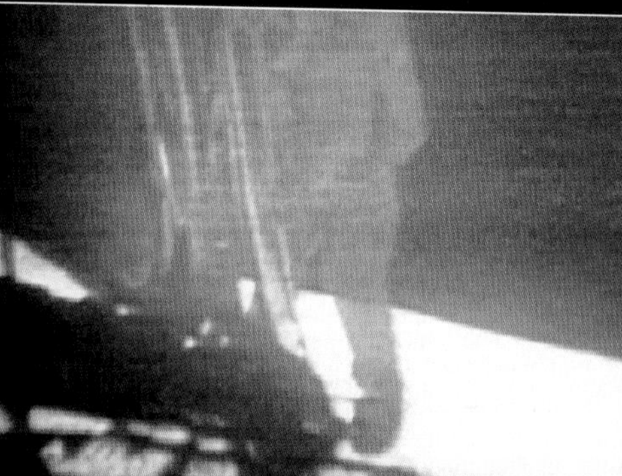

21 – 07– 69 02:56 GMT
Neil Armstrong steps from the LM ladder onto the lunar surface – an event recorded only by the monochrome television camera mounted on the side of the module.

21-07-69 03:15 GMT
Nineteen minutes later, Buzz Aldrin steps out to join Armstrong on the Moon. This time, Armstrong is on hand to take a high-quality 70mm photograph.

21-07-69 17:54 GMT
As the LM lifts off from the lunar surface, the American flag, planted a little too close to the lander, takes a buffeting from its exhaust.

“That´s one **small step** for (a) man; one **giant leap** for mankind.”

Neil Armstrong, as he steps onto the Moon, 21 July 1969

lunar surface stuck to everything, and tasted and smelled rather like gunpowder. Twelve hours later, they fired *Eagle*'s ascent engine, breaking free of the LM's lower stage and launching themselves on a trajectory to rendezvous with Collins and the CSM in lunar orbit. With the two spacecraft safely docked, the first task was to increase the atmospheric pressure in the CSM, and vent the stale, dusty air from the LM – that way, when the hatch between them was opened, air would flow into *Eagle*, minimizing the amount of dust drifting the other way. Then Armstrong and Aldrin passed their samples and other equipment through to Collins, before drifting through themselves, and swiftly changing their clothes and packing their dusty overalls into airtight containers.

With the astronauts back aboard *Columbia*, the faithful *Eagle* was cast adrift in an orbit that would eventually see it crash into the lunar surface. A brief engine burn then launched the CSM on a return flight to Earth that took two-and-a-half days. Re-entry was flawless, and Apollo 11 splashed down in the Pacific Ocean on 24 July, just 24km (15 miles) from the USS *Hornet*, where President Nixon was waiting.

The mission was not quite over, though – the astronauts still had to endure nearly three weeks in quarantine, first in a cramped trailer aboard the recovery ship and then in more comfortable conditions at Houston. The day of their release, 10 August 1969, was celebrated with parades across America – the beginning of a 25-nation, 45-day world tour in celebration of their “Giant Leap”.

TRIUMPHANT RETURN
Huge crowds greet the Apollo astronauts in New York on 13 August 1969. Later that day they would see similar celebrations in Chicago and Los Angeles.

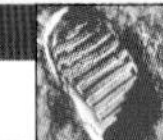

EXPERIENCE

THE FIRST MOON LANDING

APOLLO MANIA
Businesses capitalized on the US triumph by producing a host of Apollo memorabilia.

One small step ...

Six-and-a-half hours after the LM *Eagle* had touched down on the Moon, and watched by millions back on Earth, Neil Armstrong took his first historic steps onto the lunar soil. He was soon followed by Buzz Aldrin for a moonwalk lasting about two-and-a-half hours.

As they struggled into their bulky PLSS lift-support packs and readied themselves for the EVA, the astronauts were aware that they were running behind schedule. A series of strict checklists had been prepared during training on the ground, and although they followed these in minute detail, they found that there were many other issues to consider when doing it for real. Finally, at 02:30am, they opened the hatch. As Armstrong backed out onto the LM's "porch", and down the ladder, he paused to release the external television camera that would capture his first steps on the new world.

"I'm at the foot of the ladder. The LM footpads are only depressed in ... about one or two inches ... It's almost like a powder. Ground mass is very fine. I'm going to step off the LM now ... **That's one small step** for (a) man, one giant leap for **mankind** ... Yes, the surface is fine and powdery. I can kick it up loosely with my toe."

Neil Armstrong, 2:56am GMT, 21 July 1969

WALKING ON THE MOON
The astronauts spent some time in the LM studying their surroundings and deciding on the placement of experiments before they stepped outside. Aldrin passed a high-quality camera out to Armstrong, and the television camera was relocated from the LM to a tripod where it could capture most of their moonwalk in detail.

FOOTSTEPS IN THE DUST
Considering their weight on the Moon, the astronauts made quite deep prints in the lunar surface regolith. They are likely to remain for millions of years.

“Beautiful view!
Magnificent desolation.”

Buzz Aldrin steps onto the Moon, 21 July 1969

TEST LANDING
Since this was the first lunar landing, one important part of the moonwalk was for the astronauts to inspect the condition of the LM and report back to Houston – fortunately, it had not suffered any damage in the landing.

Fifteen minutes later, Aldrin was on the porch and ready to join Armstrong, who was by now equipped with a high-quality Hasselblad camera. Once on the surface, he spent a little time analyzing different methods of getting around in the one-sixth gravity, finally concluding that a loping stride was the best solution. After Armstrong had relocated the television camera to a tripod, the astronauts planted the US flag together, and received a special transmission via Houston:

AT WORK ON THE MOON
Towards the end of their moonwalk, Armstrong and Aldrin deployed the Early Apollo Scientific Experiments Package (EASEP), including a seismometer, an experiment to measure the build-up of lunar dust, and a solar-powered transmitter to send the results back to Earth.

President Nixon: Hello, Neil and Buzz. I'm talking to you by telephone from the Oval Room at the **White House** ... For every American, this has to be the proudest day of our lives ... And as you talk to us from the **Sea of Tranquillity**, it inspires us to redouble our efforts to bring peace and tranquillity to Earth. For **one priceless moment** in the whole history of man, all the people on this Earth are truly one; one in their pride in what you have done, and one in our prayers that you will return safely to Earth.
Neil Armstrong: Thank you, Mr. President. It's a great honour and privilege for us to be here representing not only the United States, but men of peace of all nations.

Resuming work, the astronauts set up a package of experiments on the surface. Aldrin then collected a pair of core samples from the lunar soil, while Armstrong loped to the rim of a nearby crater to photograph their surroundings. They spent the remainder of their allocated time collecting and cataloguing rock samples, before first Aldrin, and then Armstrong, climbed back aboard the LM.

“The blue colour of my boots has completely **disappeared** now into this ... **greyish-cocoa colour.”**

Buzz Aldrin at work on the Moon, 21 July 1969

A MESSAGE TO THE FUTURE
The plaque on the leg of the LM descent stage, left on the Moon for posterity, reads: “Here men from the planet Earth first set foot upon the Moon, July, 1969 AD. We came in peace for all mankind”.

14 November 1969
Apollo 12 is twice struck by lightning during its launch.

19 November 1969
The LM *Intrepid* touches down on the surface of the Moon in the Ocean of Storms. Conrad and Bean make their first moonwalk.

20 November 1969
The astronauts make a second moonwalk, visiting the landing site of Surveyor 3, before returning to orbit and redocking with the CSM *Yankee Clipper*.

21 November 1969
Yankee Clipper fires its engines to begin the return to Earth.

24 November 1969
The Apollo 12 Command Module splashes down in the South Pacific. During landing, a dislodged camera strikes Bean on the head and briefly knocks him out.

LIGHTNING STRIKES
Two bolts of lightning struck the Saturn V rocket during launch, scrambling the transmission of data from the spacecraft. Fortunately, quick-thinking flight controllers figured out a way to reboot the system and restore the flow of data.

Apollo 12

The second manned landing on the Moon offered an opportunity to refine the rough edges of the first mission, and to begin the scientific programme in earnest, with two much longer moonwalks.

It was only after the success of Apollo 11 that NASA drew up an itinerary of landing sites for later manned missions. Locations were chosen to provide rock samples and other data for a wide range of different lunar terrains. The target for Apollo 12, planned for launch later in the year, was to be the Ocean of Storms area.

There was another reason to visit this area – it had previously been targeted by the Surveyor 3 probe. Visiting an object that had been on the Moon for a known period of time (in this case some 30 months) could provide valuable information about present-day lunar conditions. The probe could also act as target for a pinpoint landing attempt – Apollo 11 had landed 6.5km (4 miles) from its planned target, and more precise landings would also be required for future operations.

BIOGRAPHY

ALAN BEAN

Texas-born Al Bean (b.1932) studied aeronautical engineering before joining the US Navy in 1955. He joined NASA in 1963, and was on the backup crew for Gemini 10, but was only added to the Apollo crew rosters after fellow astronaut Clifton Williams was killed in a jet crash during training. After training for the Apollo Applications Program, he commanded Skylab 3. Bean remained in NASA as an administrator until 1981, when he retired from the agency to become a full-time space artist.

Stormy voyage

Apollo 12 launched into thunderous skies above Cape Kennedy on 14 November 1969, with Pete Conrad, Richard Gordon, and Al Bean aboard. Aside from a scare during launch as lightning struck the spacecraft, the flight to the Moon went without a hitch, and on 19 November, Conrad and Bean boarded the LM *Intrepid*, leaving Gordon aboard the CSM *Yankee Clipper* in lunar orbit. Descent to the chosen landing site, nicknamed "Pete's Parking Lot", was initially automatic, but when the astronauts saw the area looked rougher than anticipated, Conrad took manual control and brought the LM down in a safe area nearby.

After five hours, Conrad stepped onto the surface. His first task was to collect and store a contingency soil sample in case some emergency forced a hasty departure. As Bean joined Conrad on the surface, both astronauts found themselves covered in a film of dust from the powdery rock.

IN THE COMMAND MODULE
Dick Gordon enjoys the relatively spacious confines of the Yankee Clipper *during Conrad and Bean's surface expedition. Gordon was able to spot both* Intrepid *and Surveyor 3 from orbit, confirming Conrad's pinpoint landing before the astronauts left the LM.*

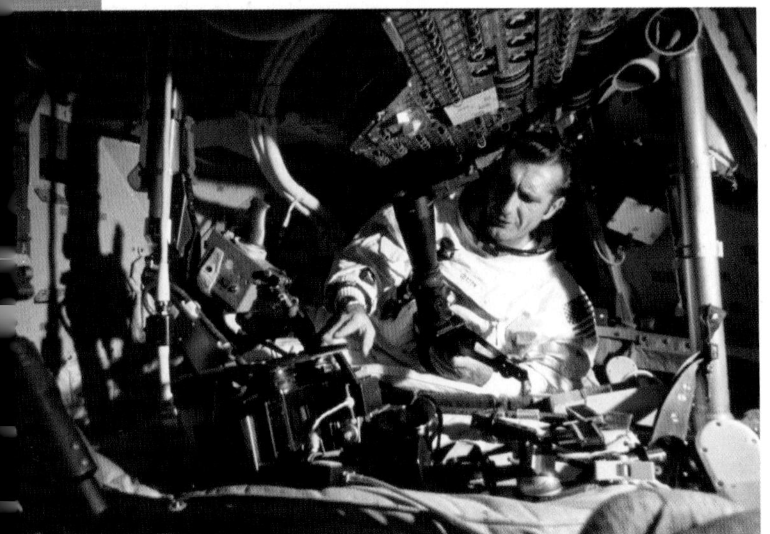

Working on the surface

Conrad and Bean made two separate moonwalks. The first was to set up scientific instruments around the LM – Apollo 12 and subsequent missions all carried an ALSEP (Apollo Lunar Surface Experiments Package) more comprehensive than the instruments deployed by Armstrong and Aldrin.

The next morning, the astronauts again stepped out on the surface, for a lunar hike of a little over 1km (2/3 mile) in total. They followed a route prepared by geologists back at Mission Control, collecting samples and answering questions from Earth as they went. Despite geological training, they admitted to finding it hard to fathom the history of the landscape around them. After two hours, they arrived at Surveyor 3, which they photographed before removing pieces for later analysis. Surveyor's television camera eventually proved to contain bacteria that had strayed into it back on Earth, and survived intact throughout their time on the Moon.

Intrepid and *Yankee Clipper* reunited in lunar orbit 37 hours after they had separated. Dick Gordon took one look at his crewmates and told them they were not coming aboard the CSM in their grimy state – so once the samples were transferred, Conrad and Bean had to strip naked and float through the hatch wearing only their headsets. *Intrepid* was then set free to plummet back onto the Moon, where the reverberations of its impact were picked up by the ALSEP seismometer for more than an hour. Eleven orbits later, *Yankee Clipper* fired its engines for the long, but uneventful, journey back to Earth.

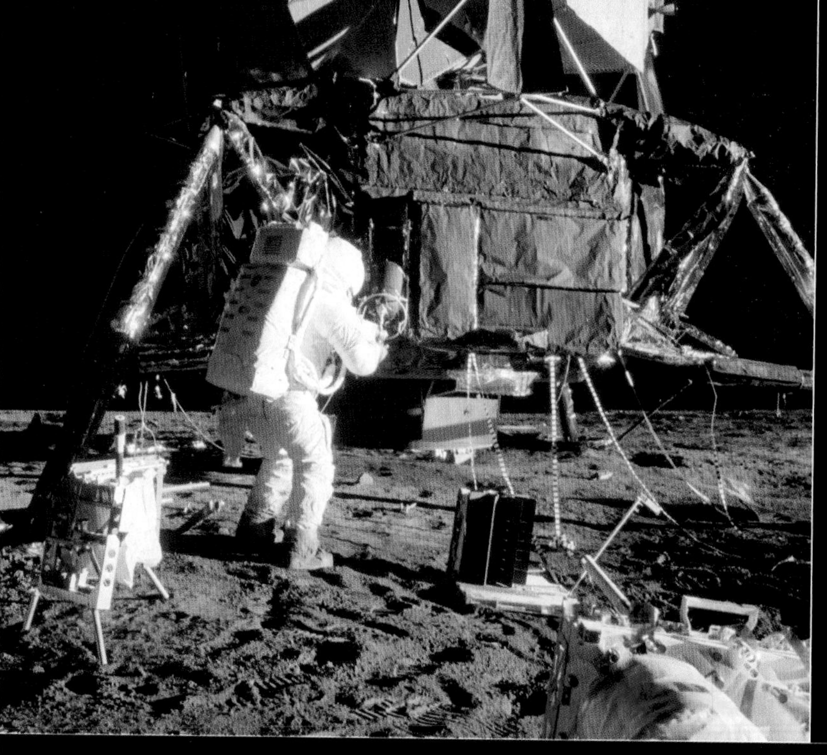

NUCLEAR ENERGY
Al Bean carefully extracts a Radioisotope Thermoelectric Generator (RTG) from its storage place in the Lunar Module. This was the first use on the Moon of an RTG – a simple electrical generator powered by radioactive decay.

"**Whoopee!** Man, that may have been a small one for Neil, but it's **a long one for me!**"

Pete Conrad, 11cm (4½in) shorter than Neil Armstrong, takes his first step onto the Moon

ON THE OCEAN OF STORMS
Al Bean poses for a photograph alongside a tool carrier during an Apollo 12 moonwalk. Pete Conrad can be clearly seen reflected in his visor (Bean's own camera is mounted on his chest control pack).

MINOR FAILURE
Plans to transmit colour TV pictures from the Moon were thwarted when the camera refused to work, burned out by accidental exposure to the Sun.

ECLIPSING THE SUN
On the way home, the astronauts captured this spectacular image as the Earth passed in front of the Sun.

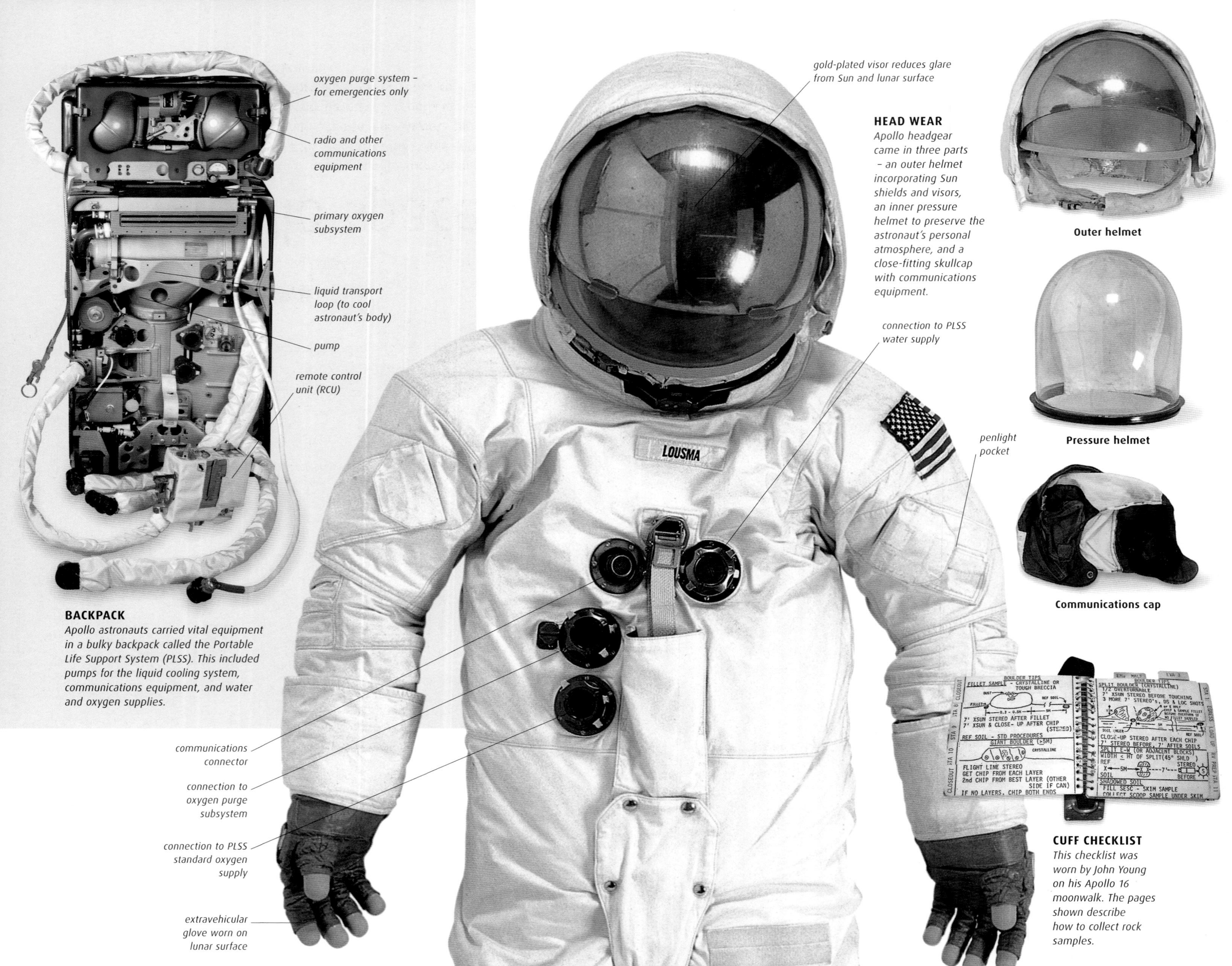

BACKPACK

Apollo astronauts carried vital equipment in a bulky backpack called the Portable Life Support System (PLSS). This included pumps for the liquid cooling system, communications equipment, and water and oxygen supplies.

HEAD WEAR

Apollo headgear came in three parts – an outer helmet incorporating Sun shields and visors, an inner pressure helmet to preserve the astronaut's personal atmosphere, and a close-fitting skullcap with communications equipment.

CUFF CHECKLIST

This checklist was worn by John Young on his Apollo 16 moonwalk. The pages shown describe how to collect rock samples.

TECHNOLOGY

A SURVIVAL SUIT FOR THE MOON

The Apollo spacesuit

Spacesuits for the Apollo missions had a number of differences from previous designs. They had to be more robust because of the extra risks that came with working on the surface of the Moon for extended periods, yet also more flexible and lighter because of the variety of tasks the astronauts might have to carry out and the need to operate in gravity rather than weightlessness. The solution was a basic spacesuit with optional extras that were worn during lunar excursions.

SAMPLE COLLECTION

Astronauts used adapted tongs and a brush to collect samples from the Moon. To avoid contamination, samples were sealed inside containers before they were taken onboard the LM. They were reopened in a laboratory back on Earth.

mechanism for sealing container

Sample-return container

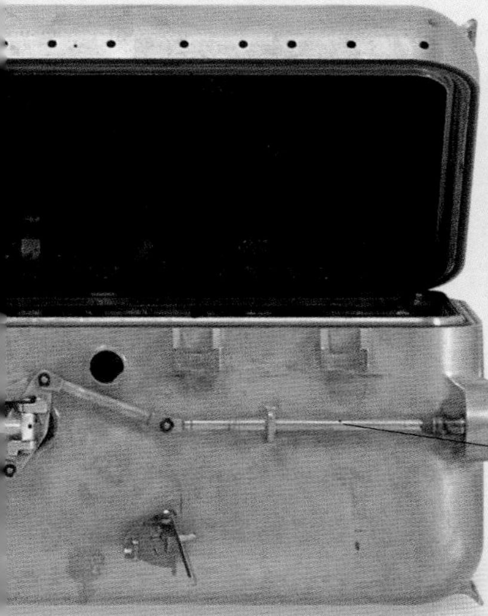

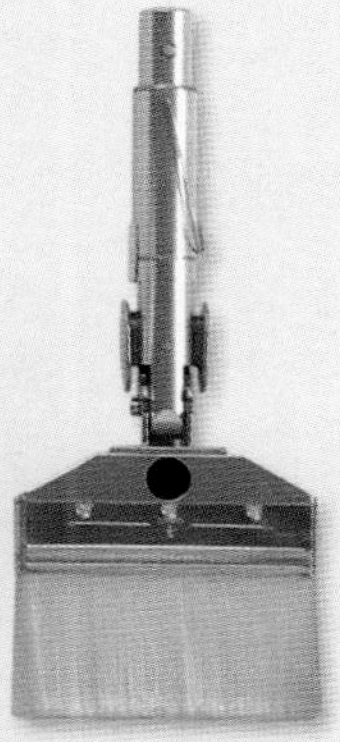

Brush

handle designed for operation wearing spacesuit gloves
rigid wire grippers

Tongs

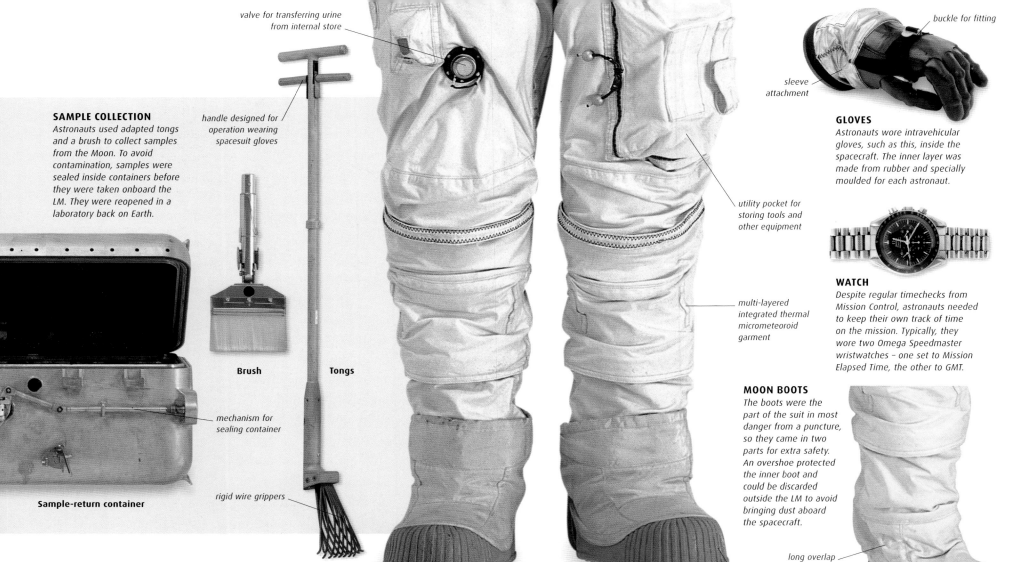

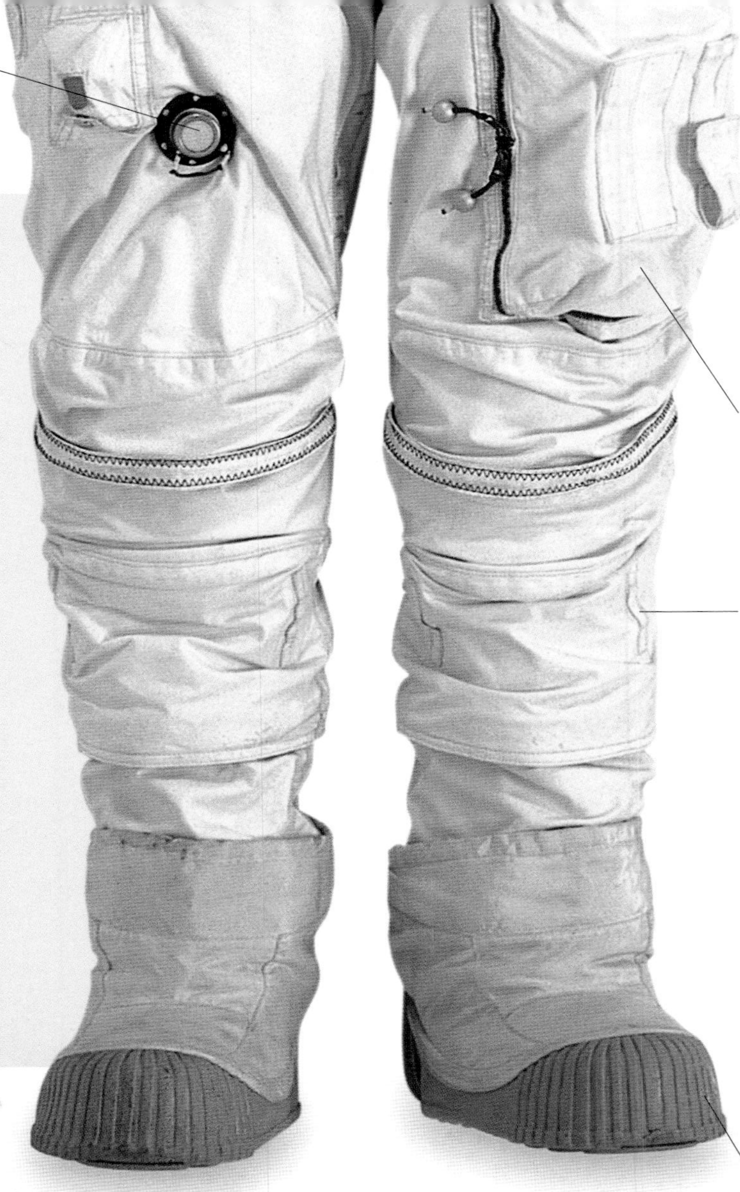

valve for transferring urine from internal store
utility pocket for storing tools and other equipment
multi-layered integrated thermal micrometeoroid garment
overshoe

EXTRAVEHICULAR MOBILITY UNIT

The Apollo spacesuit or EMU had several layers. Closest to the skin was a liquid-cooled undergarment. Next came a nylon pressure garment to seal the astronaut from the vacuum of space, then several aluminium-coated layers for further heat protection.

DATE	1967
ORIGIN	United States
WEIGHT ON EARTH	86kg (190lb)
WEIGHT ON MOON	14kg (30lb)

buckle for fitting
sleeve attachment

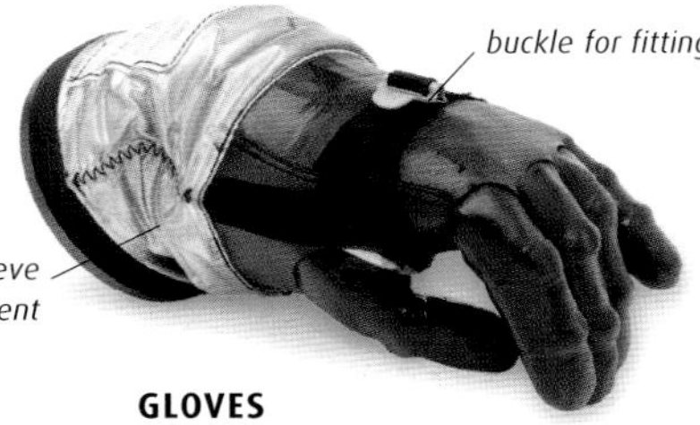

GLOVES

Astronauts wore intravehicular gloves, such as this, inside the spacecraft. The inner layer was made from rubber and specially moulded for each astronaut.

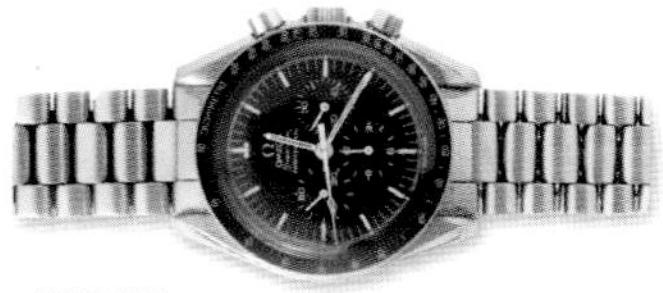

WATCH

Despite regular timechecks from Mission Control, astronauts needed to keep their own track of time on the mission. Typically, they wore two Omega Speedmaster wristwatches – one set to Mission Elapsed Time, the other to GMT.

MOON BOOTS

The boots were the part of the suit in most danger from a puncture, so they came in two parts for extra safety. An overshoe protected the inner boot and could be discarded outside the LM to avoid bringing dust aboard the spacecraft.

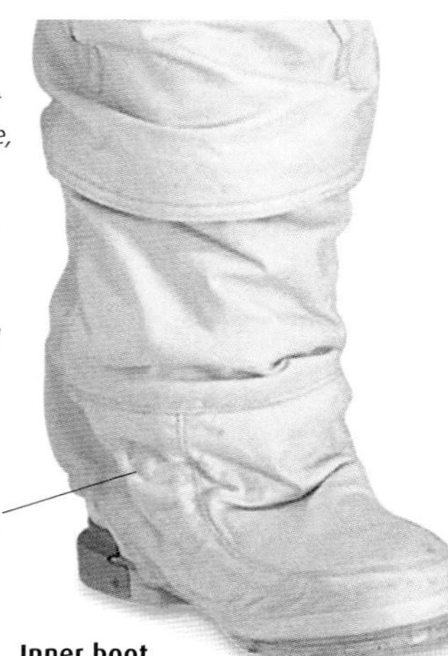

long overlap with suit leg

Inner boot

boot-strap

Overshoe

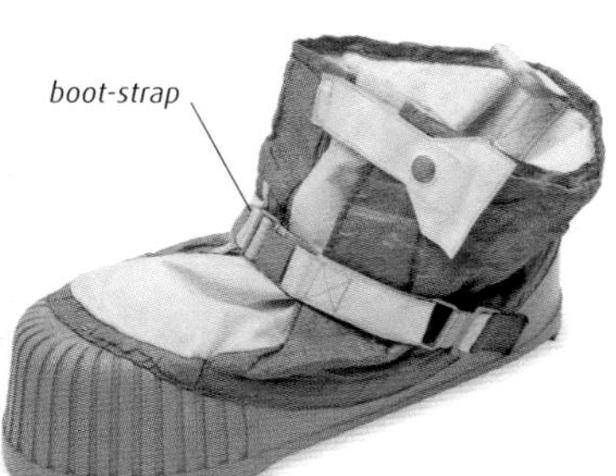

Apollo 13

After two successful missions, the American public was starting to take the Moon for granted. But when an explosion struck Apollo 13 in April 1970, the world was gripped by the drama.

11 April 1970
Apollo 13 blasts off from Cape Canaveral on its way to the Moon.

13 April 1970
An explosion cripples the CSM, which loses much of its air supply and the ability to generate power.

14 April 1970
Shortly after the explosion, the crew power up and board the LM, intending to use it as an escape lifeboat. Later in the day, they make a vital mid-course correction burn.

15 April 1970
Apollo 13 briefly loses contact with Earth as it passes over the far side of the Moon, some 254km (158 miles) above the lunar surface.

16 April 1970
A second mid-course engine burn accelerates Apollo 13's return.

17 April 1970
The Apollo 13 crew board the Command Module and make a successful re-entry and splashdown in the Pacific Ocean.

Shortly after launch, the crew of Apollo 13 settled in for a long translunar flight. For this mission, veteran astronaut Jim Lovell was joined by first-timers Fred Haise and Jack Swigert (a late substitute for Ken Mattingly, who had been exposed to German measles). Everything was going well until, some 56 hours into the flight and with the spacecraft already closer to the Moon than the Earth, the crew triggered a routine stir of the CSM's oxygen tanks. There was the sudden thump of an explosion in a vacuum, and the astronauts were alarmed to see pressure in one of the main oxygen tanks dropping and electrical power to the CSM rapidly failing (see over). With masterful restraint, Swigert famously reported "Okay, Houston, we've had a problem here."

In fact, the astronauts and their controllers knew very well that the "problem" was life-threatening. The Service Module was clearly crippled, unable to provide power or oxygen to the Command Module.

PROBLEMS FROM THE OUTSET
Pogo oscillations (see p.119) caused one of the second-stage engines to cut out during launch, forcing the others to burn longer.

All thought of landing on the Moon was abandoned, but the crew could not simply turn the spacecraft around – the CSM did not carry enough fuel for such a manoeuvre, even if it could still fire its engines. The way back to safety lay in flying onwards and looping around the Moon (see panel, opposite).

The immediate question was how to survive that long. The Command Module had limited batteries and oxygen supplies built into it, but only enough for a few hours of independent existence during and after re-entry – clearly these would have to be preserved.

Fortunately, NASA had a contingency plan to use the LM, attached but dormant at the front of the CSM, as a lifeboat in this kind of emergency. Racing against time, the crew powered up the LM *Aquarius*, moved supplies into it, and sealed themselves in, switching off nearly all systems on the CSM *Odyssey* to conserve what little power remained.

Getting back to Earth

While the astronauts suffered in the cramped conditions aboard an LM designed to accommodate just two men for two days, Houston ground controllers led by Gene Kranz (see p.100) feverishly calculated their path back to safety. Conditions on *Aquarius* soon became cold and damp, but more problematically the chemical filters designed to

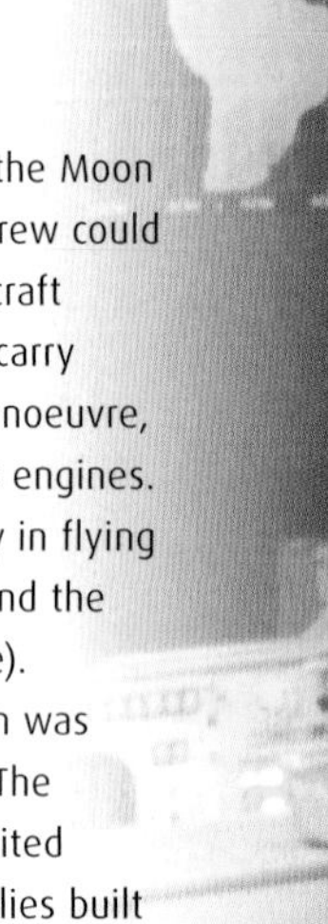

CRISIS TALKS
Apollo 13 Lunar Module Pilot Fred Haise talks with the Mission Operations Control Room at Houston during the crew's final television transmission, shortly before they boarded the LM. Gene Kranz is in the foreground with his back to the camera, wearing a distinctive white waistcoat.

WAVING GOODBYE
Jim Lovell leads Jack Swigert and Fred Haise to the van for transfer to the Apollo 13 launch pad, little knowing the perils that await in space.

"Flight control will **never lose an American in space**. You've got to believe ... that this crew is coming home."

Gene Kranz, briefing mission controllers at Houston

LIFEBOAT AQUARIUS
The Apollo 13 crew, now aboard the Command Module, took this snapshot of their trusty LM as it drifted away from them shortly after separation on the final approach to Earth.

remove toxic carbon dioxide from the air were near exhaustion. Using simulators at Houston, Mattingly (who had not, after all, fallen ill) and the others found a way to convert the incompatible filters from the CSM for use on the LM, using only materials the astronauts had to hand. With no access to the CSM engines, the crew had to improvise their course-correction burns, calculating their precise position and trajectory by the stars and using the descent engine on the LM as a retrorocket for the entire spacecraft. A passage behind the Moon put the spacecraft on a return course for Earth, and the crew were able to re-enter the Command Module on 17 April as they approached Earth.

Fortunately, the Command Module separated without problems and, as millions of television viewers around the world held their breath, the crew splashed down safely in the Pacific Ocean.

EXPLOSION DAMAGE
After abandoning the Service Module, the crew were able to see how the explosion had blown out one side of the module. The fault was eventually traced to poorly insulated wiring.

TECHNOLOGY

AROUND THE MOON TO SAFETY

The mission controllers at Houston soon worked out that the only way of bringing the Apollo 13 astronauts back to Earth was through a daring manual engine burn. This would put them into a free return trajectory – an orbit that would use the Moon's gravity like a slingshot, effectively catching the spacecraft, swinging it around the far side, and throwing it back towards the Earth without the need for a long engine burn. Nevertheless, some thrust was needed to adjust Apollo 13's approach to the Moon, and because there was no way of knowing how badly the CSM engine might be damaged, the burn used the LM's descent engine instead. Approaching Earth, the Service Module was, unusually, jettisoned with the Command Module still attached to the LM. This gave the astronauts a chance to photograph the damage.

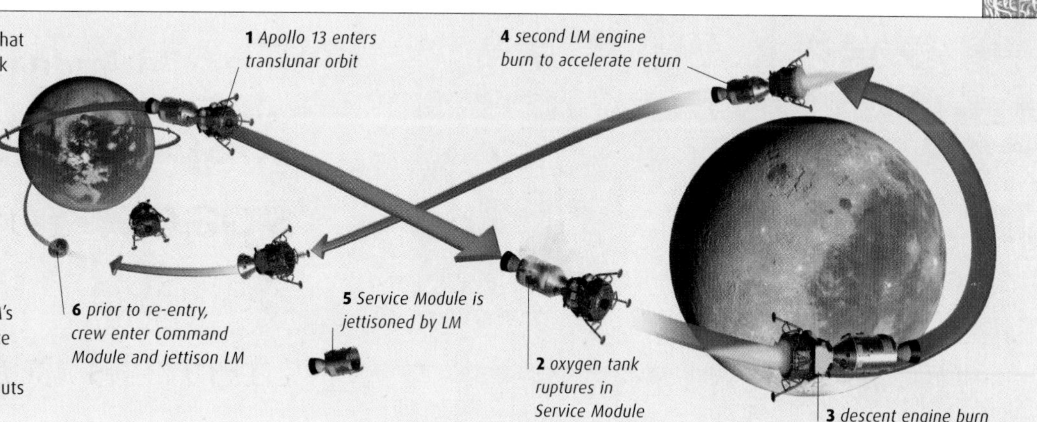

1 *Apollo 13 enters translunar orbit*
2 *oxygen tank ruptures in Service Module*
3 *descent engine burn*
4 *second LM engine burn to accelerate return*
5 *Service Module is jettisoned by LM*
6 *prior to re-entry, crew enter Command Module and jettison LM*

"Houston, we've had a problem ..."

Two days and eight hours into their mission, and more than 320,000km (199,000 miles) from Earth, the crew of Apollo 13 were thrown into crisis as an explosion left them with a crippled Service Module and a deteriorating oxygen and power supply. Their immediate response was crucial to their chances of survival.

On the evening of 13 April 1970, the crew of Apollo 13 broadcast a guided tour of their spacecraft back to Earth. Five minutes after transmission ended, they were still scattered through the spacecraft. Command Module Pilot Jack Swigert was in the CM, Lunar Module Pilot Fred Haise in the LM (specially opened for the tour), and Commander Jim Lovell in the tunnel between the two, helping Haise to close up the LM. Jack Lousma, on shift as Capcom at Houston, relayed a few routine requests – NASA wanted them to photograph Comet Bennett, and the engineers wanted some data on the high-gain antenna. Swigert reeled off the data from the instrument panels in front of him. The next request was to stir the tanks of low-temperature cryogenic oxygen – a routine procedure that prevented impurities in the liquid from settling to the bottom of the tank. As Swigert activated the stirring motor, a loud bang was heard throughout the spacecraft. Lovell looked accusingly at Haise – during the TV show he had startled them by triggering a repressurization valve that made a similar noise. But Haise's expression was deadly serious. The communications transcript records the moment:

Capcom: 13, we've got one more item for you, when you get a chance. We'd like you to stir up your cryo tanks. In addition, I have shaft and trunnion ... for looking at the Comet Bennett, if you need it.

JS: Okay. Stand by.

JS: Okay, Houston, we've had a problem here.

Capcom: This is Houston. Say again, please.

Jim Lovell: Houston, we've had a problem. We've had a main B bus undervolt.

Capcom: Roger. Main B undervolt.

Capcom: Okay, stand by, 13. We're looking at it.

Fred Haise: Okay. Right now, Houston, the voltage is – is looking good. And we had a pretty large bang associated with the Caution and Warning there. And if I recall, main B was the one that had had an amp spike on it once before.

"It looks to me, looking out the hatch, that we are **venting something.** We are venting something into the – **out into the space**"

Jim Lovell in the aftermath of the accident

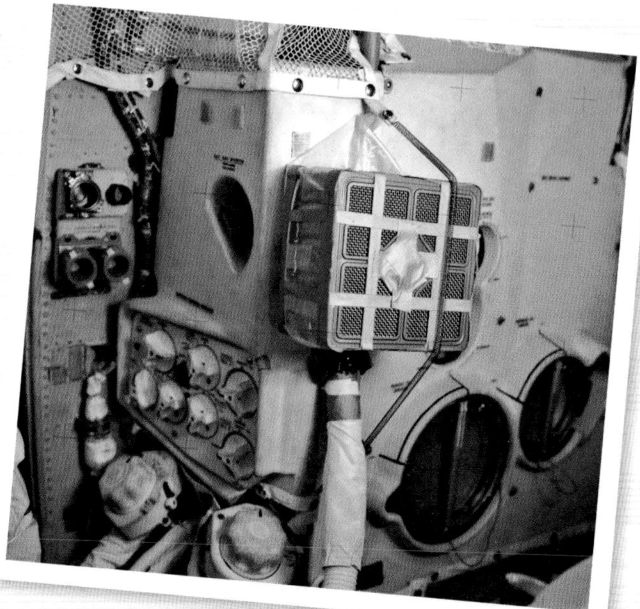

IMPROVISED LIFE-SAVER
As carbon dioxide built up in the LM, Mission Control told the astronauts how to build a "mailbox" device that would allow them to use chemical scrubbers from the Command Module in the incompatible LM system.

LUCKY ESCAPE
Alan Shepard, who might have commanded Apollo 13 but for his ear problem, listens in to the unfolding drama at Mission Control.

The "main bus undervolt" that flashed in the CM meant that one of the Service Module's power circuits was rapidly draining. As the crew struggled to reconcile the readings of their own instruments with the telemetry received on the ground, they realized that one of their two oxygen tanks was empty, and two of their three fuel cells were running flat. While staff at Houston frantically tried to work out what was happening, Lovell concentrated on stabilizing the spacecraft as it bucked from side to side. Peering out of the tiny CM windows, he spotted something that dashed his hopes of landing on the Moon – a jet of gas issuing from the side of the Service Module.

"Okay, Aquarius … down here we're **getting regrouped**, trying to work on your control modes and … taking a look at consumables as opposed to flight plan, and so forth, and as soon as we get all that information, we'll **pass it up to you**. We also have the 14 backup crew over in the simulators looking at dock burns and also trying to see what kind of alignment procedure they can come up with for **looking at stars** out the window. So if you ever are able to see any stars out there and think you can do an alignment … why [not] let us know?"

Jack Lousma, Apollo 13 Capcom

IMMORTALIZED ON CELLULOID
Ron Howard's acclaimed 1995 film, Apollo 13 *stuck closely to the actual sequence of events and helped to reignite popular interest in the Apollo missions.*

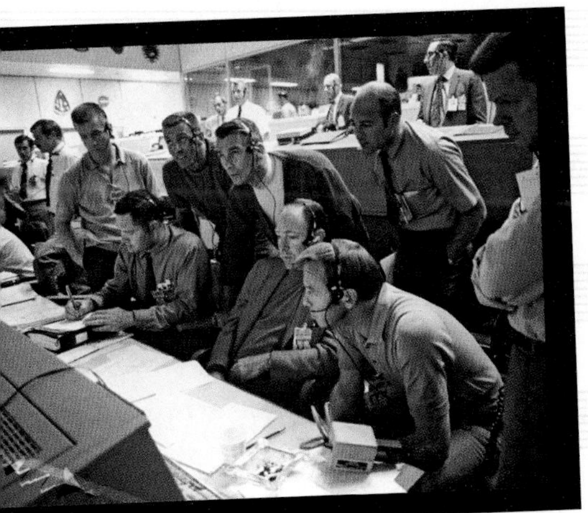

WORKING TOGETHER
Staff from all four Mission Control shifts collaborated to get the astronauts home. Pictures from the spacecraft were soon cut for power reasons, and the team had to rely on voice communications only. This made it even more difficult to relay instructions for building the "mailbox" filter that would keep the crew's air breathable.

Worse was to come, as the crew realized that pressure in the second oxygen tank was now also slowly dropping. The problem now became clear – the explosion had created a major rupture in the fuel cell system, into which they were still pumping precious oxygen. Not wanting to risk anything that might worsen the condition of the Service Module, Houston now sent Lovell and Haise back to the LM in an attempt to draw power from it. However, this idea was rapidly abandoned in favour of a lifeboat procedure – transferring the crew and their consumables to the LM *Aquarius* and powering down the CSM completely in order to swing around the Moon and back to Earth. Within three hours of the explosion, the crew of Apollo 13 were aboard *Aquarius* and ready for the long trip around the Moon and back to safety.

RETURN TO EARTH
(Left) Odyssey *splashed down in the Pacific Ocean at 12:07pm Houston time on 17 April. The sight of its return was greeted with jubilation in a Mission Control packed with NASA personnel including (in the front row, from left) Apollo 13 Flight Directors Gerald Griffin, Gene Kranz, and Glynn Lunney.*

SAFE AND SOUND
Navy divers were first to the scene, helping the crew aboard a life raft (top), while, back at Mission Control, Gene Kranz allowed himself a well-deserved cigar (middle). Within an hour of splashdown (from left) Haise, Swigert, and Lovell were being welcomed aboard the USS Iwo Jima *(bottom).*

Apollo 14

After the near-disaster of Apollo 13, NASA's next mission was vital to restoring its own confidence, and the prestige of the US space programme. Fortunately, it was a resounding success.

2 September 1970
NASA abandons plans for Apollos 15 and 19 in the face of budget cuts.

24 November 1970
Robert Gilruth reports that all recommendations of the Apollo 13 Review Board have been implemented.

31 January 1971
Apollo 14 blasts off. After some trouble docking with the LM, it makes a successful burn to put it on course for the Moon.

4 February 1971
Apollo 14 enters into lunar orbit.

5 February 1971
The LM *Antares* lands in the Fra Mauro area of the Moon. Alan Shepard and Ed Mitchell make their first moonwalk.

6 February 1971
The astronauts make a second moonwalk, and later takeoff from the lunar surface to reunite with the CSM for their return to Earth.

9 February 1971
The Apollo 14 Command Module splashes down in the Pacific Ocean.

TEEING OFF
As he returned from the second moonwalk, Shepard surprised everyone by producing a couple of smuggled golf balls and a club fashioned from various lunar tools. The first golf stroke on the Moon was a slice, but the second sent the ball for several hundred metres – the low gravity probably helped, since Shepard could use only one hand.

By the time Apollo 14 launched, on 31 January 1971, NASA had been away from the Moon for more than a year. During the hiatus, two more missions – the original Apollos 15 and 19 – had been lost to budget cuts, but Apollo 13's landing site at Fra Mauro, on the edge of the Oceanus Procellarum, was still a scientific priority, and so Apollo 14 was directed there instead. Leading the crew was a true veteran of the space programme. Alan Shepard, grounded since his historic *Freedom 7* flight, had undergone painful surgery to correct his ear problem, and Deke Slayton saw to it that his old Mercury Seven colleague would go to the Moon. Shepard's crewmates – Stuart Roosa and Edgar Mitchell – were both rookies.

Although launch and departure from Earth orbit went well, Apollo 14 hit a significant snag on its way to the Moon. As Shepard attempted to join the CSM *Kitty Hawk* to the Lunar Module *Antares*, the docking mechanism persistently failed to engage. Fortunately, it took on the sixth try. More problems dogged the LM's final descent, and Shepard's first words on the surface seemed doubly appropriate: "... it's been a long way, but we're here."

Shepard and Mitchell remained on the Moon for 33 hours, during which they made two long moonwalks. After collecting their contingency sample of rock, they set up the ALSEP and other experiments that would remain on the surface. The astronauts also investigated the lunar regolith using a device that fired "thumper" charges into the ground. The way these were detected by the ALSEP seismometers revealed properties of the ground in the area.

Lunar expedition

The second moonwalk was a "geology traverse" similar to that carried out on Apollo 12. The astronauts were asked to inspect the 300m- (1,000ft-) high rim of the nearby Cone Crater and found the climb up its outer flanks quite arduous even in one-sixth Earth gravity – especially as they were also towing a lunar cart called the Mobile Equipment Transporter (MET) with them. As they approached the rim, the terrain grew increasingly steep and hilly, and frustratingly they could not find the crater edge itself. After much debate, they turned back, not realizing until later that they had come within about 30m (100ft) of their goal.

On their return to Earth, the astronauts had to endure the same fortnight of quarantine suffered by their earlier colleagues. After they too failed to show any signs of unexpected illness, the practice was discontinued for later missions.

TENSE MOMENTS
Screens in the Mission Operations Control Room at Houston show the view from Kitty Hawk *during its repeated attempts to dock with* Antares *in Earth orbit. The top of the LM can be made out, still inside the S-IVB upper stage.*

ALSEP EXPERIMENTS
The astronauts laid out the various ALSEP experiments according to a map carried on their cuff checklists. The box in the centre is a detector for charged particles from space, while red flags mark geophones used to detect sound waves from the "thumper" experiment.

***ANTARES* AT REST**
Ed Mitchell named the LM Antares *after the bright star most likely to be visible during the descent from orbit. The module landed on the edge of a depression and came to rest tilted over at an angle of 8°.*

26 July 1971
Apollo 15, first of the J-class Apollo missions, launches from Cape Canaveral.

29 July 1971
The spacecraft enters lunar orbit.

30 July 1971
The LM *Falcon* touches down on target close to Hadley Rille and the Lunar Apennine mountains.

31 July 1971
David Scott and James Irwin make the first of three expeditions onto the lunar surface. The LRV is used on the Moon for the first time.

2 August 1971
After a third moonwalk, Scott and Irwin lift off and return to the CSM in lunar orbit.

4 August 1971
Apollo 15 leaves lunar orbit on the return journey to Earth.

5 August 1971
Al Worden performs the first spacewalk beyond immediate Earth orbit, retrieving film and experimental data from the CSM's instrument bay.

DRILLING THE MOON
David Scott picks up a cordless drill during his lengthy efforts to make holes for the probes of the Apollo 15 heat flow experiment. Although the drill had a design flaw that was modified for the later missions, it seems that Apollo 15 was also hampered by landing on far more solid ground.

"**Man must explore**, and this is exploration at its greatest."

Apollo 15 Commander David Scott, on Hadley Rille

DIGGING A TRENCH
James Irwin's work digging a trench in the topsoil was easier than the drilling, but he still had to stop at a depth of 30cm (1ft).

70MM HASSELBLAD CAMERA
Specially modified Hasselblad electric cameras were carried on the chest packs of all the Apollo astronauts. The familiar + marks allowed measurements of distance to be made from the photographs.

Apollo 15

Back on firm ground after the success of Apollo 14, the next Apollo mission was a major advance, as an improved CSM, a longer stay on the Moon, and the addition of a lunar rover yielded great results.

Apollo 15 was the first of what were known as the J missions – the ultimate development of Apollo as it was initially planned. For these missions, modifications to the CSM (in this case *Endeavour*) included the addition of a palette of scientific instruments mounted on the Moon-facing side of the hull. These would turn the orbiting module into a powerful science satellite in its own right and give the CSM pilot an expanded role in the mission. The LM, meanwhile, now carried the Lunar Roving Vehicle (LRV) – a Moon car developed by Boeing, which was folded and stored on the side of the module during descent. This massively increased the range over which the astronauts could travel, but improvements to life-support systems also had a role to play, as they allowed a much longer stay on the lunar surface.

Apollo 15's crew was led by Neil Armstrong's Gemini 8 partner David Scott, with James Irwin as LM Pilot and Alfred Worden in the CSM. Apollo 15 blasted off on 26 July 1971 and arrived in lunar orbit without any major problems.

Climbing lunar mountains

The Lunar Module *Falcon* touched down near a long, winding valley known as Hadley Rille on 30 July, and during almost three days on the Moon, Scott and Irwin undertook three separate moonwalks. The first of these tested the LRV on a trip to the base of Mons Hadley Delta, a mountain in the Lunar Apennine range some 4km (2½ miles) away. The rover performed well, although Scott had some difficulty getting used to its rear-wheel steering. Travelling at an average of 10kph (6mph), they needed to test the brakes on several occasions when they unexpectedly found themselves on the edge of a crater.

During this expedition, they made two long stops to explore geology around Elbow Crater and at the base of the mountain itself. In preparation for their mission, these astronauts had undergone far more extensive geology field training than their predecessors (including simulated expeditions in the New Mexico desert), and the extra work paid off as the later Apollos generated huge amounts of data that would ultimately allow geologists back on Earth to piece together the Moon's history.

On the second day, the astronauts returned to the same area, but this time they climbed the mountain itself, collecting samples including one that was later nicknamed "the genesis rock" after studies back on Earth revealed that it was more than four billion years old – almost as old as the Moon itself. Back at the LM, the astronauts had also been at work setting up the ALSEP package and conducting other experiments. Drilling into the lunar rocks to collect samples proved far more difficult than expected, and the final LRV expedition had to be shortened – though it still took the astronauts to the edge of Hadley Rille itself.

After their return to the CSM on 2 August, the astronauts spent a further two days in lunar orbit, during which time they continued scientific work and deployed a small satellite into lunar orbit.

BIOGRAPHY

DAVID SCOTT

Texan David Randolph Scott (b.1932) was educated at West Point before enlisting in the US Air Force. After a tour of duty in Europe, he returned to study in America before joining NASA in 1963. His first flight on Gemini 8 (see p.108), was followed by the role of CM Pilot on Apollo 9. After Apollo 15, he took part in preparations for the Apollo-Soyuz Test Project (see p.174). He left NASA in 1975.

FIRST WHEELS ON THE MOON
The LRV is seen here shortly after its first deployment onto the Moon. The tyres were actually made of woven wire mesh with chevron-shaped treads riveted on – making them far more effective shock absorbers.

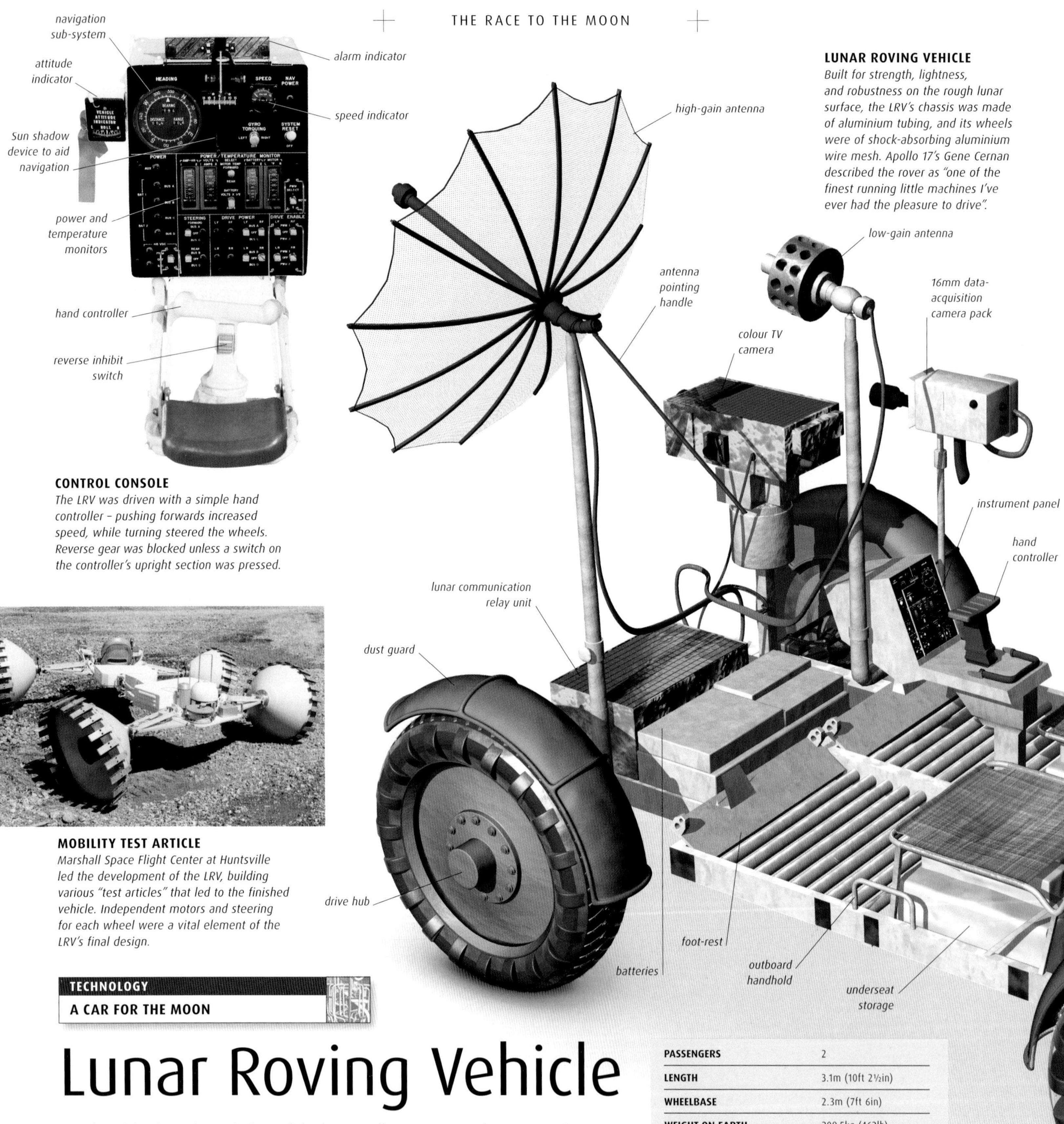

LUNAR ROVING VEHICLE
Built for strength, lightness, and robustness on the rough lunar surface, the LRV's chassis was made of aluminium tubing, and its wheels were of shock-absorbing aluminium wire mesh. Apollo 17's Gene Cernan described the rover as "one of the finest running little machines I've ever had the pleasure to drive".

CONTROL CONSOLE
The LRV was driven with a simple hand controller – pushing forwards increased speed, while turning steered the wheels. Reverse gear was blocked unless a switch on the controller's upright section was pressed.

MOBILITY TEST ARTICLE
Marshall Space Flight Center at Huntsville led the development of the LRV, building various "test articles" that led to the finished vehicle. Independent motors and steering for each wheel were a vital element of the LRV's final design.

TECHNOLOGY
A CAR FOR THE MOON

Lunar Roving Vehicle

Developed for the J-class missions of the later Apollo programme, the Lunar Roving Vehicle (or LRV) was built for NASA by Boeing. The design specification put a strict limit of 208kg (457.5lb) on the vehicle's weight, while at the same time requiring that it should support a heavy load of two astronauts, their equipment, and rock samples, and that it should be capable of many hours of operation at reasonably high speeds. Despite the demanding brief, Boeing delivered the first LRV in just 17 months.

PASSENGERS	2
LENGTH	3.1m (10ft 2½in)
WHEELBASE	2.3m (7ft 6in)
WEIGHT ON EARTH	209.5kg (462lb)
PAYLOAD ON LUNAR SURFACE	490kg (1,078lb)
PROPULSION	4 x battery-powered electric motors
MAXIMUM SPEED	18.6kph (11.5mph)
MANUFACTURER	Boeing

LINK TO EARTH

John Young, Commander of Apollo 16, adjusts the high-gain antenna on the LRV. The rover's communications system sent back a range of telemetry data while it was moving and could transmit live television pictures back to Earth when the vehicle was stationary.

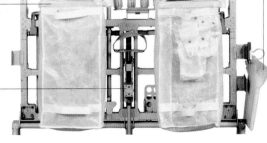

KIT CAR

Apollo 15 Commander David Scott admires the deployment mechanism that would lower the LRV from the side of his Falcon *Lunar Module for the first drive on the surface of the Moon.*

LUNAR GRAND PRIX

During the Apollo 16 mission, the astronauts put the LRV through a thorough performance test, pushing it to the limit and reaching a record speed of 18.6kph (11.5mph).

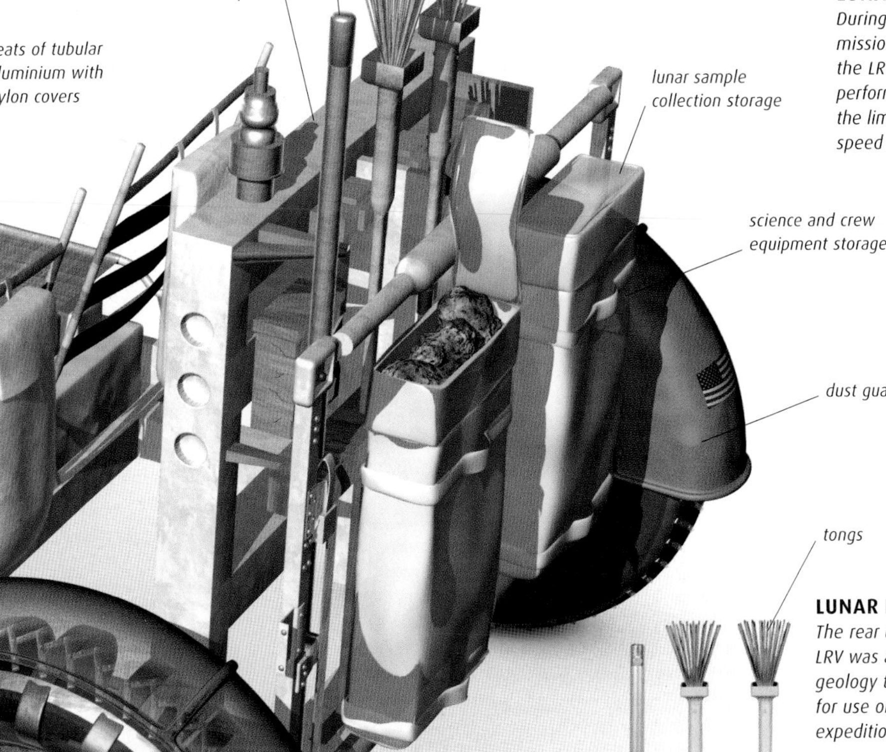

LUNAR HAND TOOL CARRIER

The rear upright section of the LRV was a rack holding various geology tools and sample bags for use on the rover's daily expeditions across the surface.

DEPLOYING THE ROVER

The LRV was stowed for flight on the side of "Quad 1" of the Lunar Module. It was designed to fold inwards for storage, leaving only the underside of the chassis exposed to damage during landing. It was lowered to the ground on a pulley system.

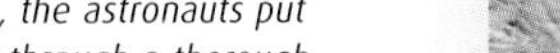

FLIP OPEN

From the porch, one astronaut released the LRV from its stowed position. His colleague then lowered it on pulleys.

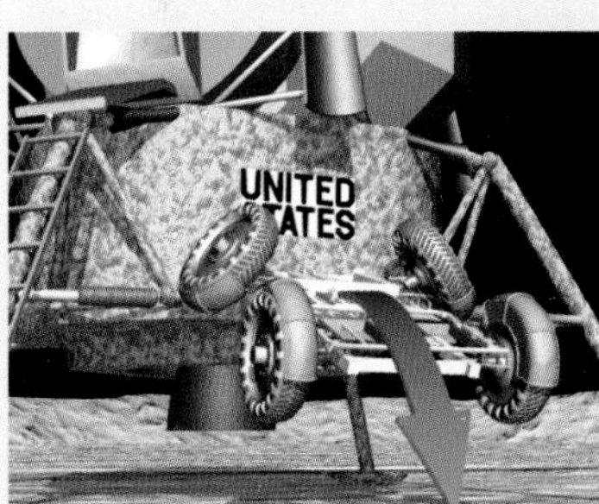

FOLD DOWN

As the rover opened, the rear wheels rotated and locked into place automatically.

ROLL OFF

With the rear wheels down, the front wheels were unfolded, and the front of the LRV was lowered to the ground.

IN THE SHADOW OF HADLEY
The LM pilot James Irwin tends to the sample holders on the LRV during the first Apollo 15 moonwalk. In the background, Mons Hadley Delta rises up, appearing deceptively close.

1961 1962 1963 1964 1965 1966 1967 1968 1969 1970 1971 1972 1973 1974 1975 1976 1977 1978 1979 1980 1981 1982 1983 1984 1985 1986 1987 1988 1989 1990 1991 1992 1993 1994 1995 1996 1997 1998 1999 2000 2001 2002 2003 2004 2005 2006 2007 2008 2009 2010 2011 2012 2013 2014 2015

Apollo 16

The second of the advanced Apollo-J missions targeted the Descartes region of the Moon and once again produced impressive scientific results.

16 April 1972
Apollo 16 blasts off from Kennedy Space Center.

19 April 1972
The spacecraft arrives in lunar orbit.

20 April 1972
A fault in the CSM almost forces the mission to abort during the Lunar Module's final descent to the Moon. However, the CM *Casper* is eventually allowed to continue, making the first landing in the lunar highlands.

21 April 1972
John Young and Charles Duke make the first of three highly successful moonwalks, retrieving the huge rock known as "Big Muley".

24 April 1972
After their third moonwalk, the astronauts return to the LM *Orion*. A few hours later, an engine burn puts the CSM on its return path to Earth – one day ahead of schedule for safety reasons after the problem with the CSM engines.

27 April 1972
Apollo 16 splashes down in the Pacific Ocean near Christmas Island.

The crew for the fifth lunar landing was led by experienced Apollo and Gemini astronaut John Young, with two unflown astronauts, Thomas K. (Ken) Mattingly and Charles Duke, as Command Module Pilot and Lunar Module Pilot respectively. Young, Duke, and Jack Swigert had originally been the backup crew for Apollo 13, with Mattingly scheduled as CMP for that mission. After Mattingly was exposed to German measles, Swigert and Mattingly had switched places in the crew roster.

After an uneventful trip to the Moon, the mission was almost aborted during the CM's final descent, on 20 April. A fault developed in one of the thrusters on the CM, *Casper*, hampering its manoeuvres in lunar orbit. Eventually Houston decided to press ahead with the landing the following day – but they cut short the stay in orbit after the surface mission, and Apollo 16 returned to Earth a day early.

Highland fling

The Moon's surface is divided into two distinct types of terrain: low-lying, dark, flat seas or *maria*; and bright, heavily cratered highlands. The previous Apollo landings had all been in or around the lunar seas and had confirmed that they were vast plains of solidified volcanic lava filling huge ancient impact basins. Mountain chains such as the Lunar Apennines, investigated by Apollo 15, appeared to have been thrown up around the edges of the largest impacts. But did this also explain the highlands? Apollo 16's LM, *Orion*, landed in the central highland region of

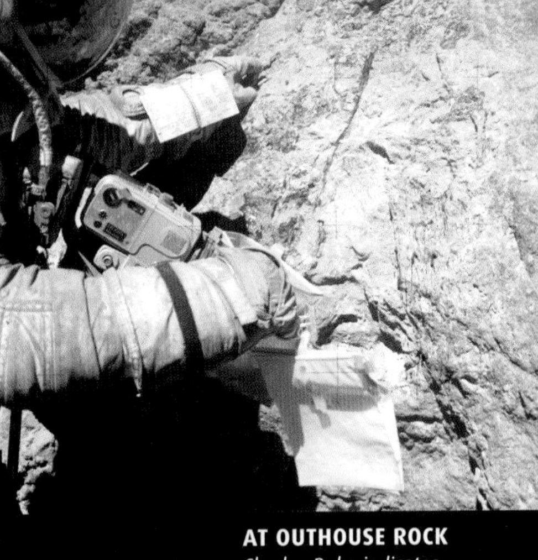

AT OUTHOUSE ROCK
Charles Duke indicates the location from which a sample was collected during the third moonwalk. Nearly all the rocks found in Descartes were breccias – rocks mixed up and re-formed by meteorite impacts – proving that the lunar highlands, too, were largely formed by impacts and not volcanism.

LUNAR MODULE ASCENT STAGE
After shedding its spidery legs on the surface of the Moon, the Apollo 16 LM Orion closes in on the CSM prior to redocking in lunar orbit.

LAST STOP
John Young stands in front of the LRV at the final stop on the astronauts' third moonwalk. The three-legged object in the foreground is part of a gnomon, a device for recording the size and colour of rock samples.

Descartes, far from any seas, where some geologists thought volcanic activity, not impacts, might have produced the mountainous plateau.

Out and about

The astronauts carried out moonwalks on three successive days. They spent more than 20 hours outside the LM and drove some 26.7km (16½ miles), taking the LRV to a top speed of 18kph (11mph). The terrain was heavily cratered and scattered with large boulders – among the samples the astronauts found was the largest lunar rock ever collected, an 11.7kg (25lb) monster nicknamed "Big Muley" after Bill Muehlberger, the mission's principal geologist.

Young and Duke also experienced some unusual problems. On the first EVA they drove directly away from the Sun and, with no shadows to help them, had trouble spotting obstacles against the glare of reflected sunlight. The orange-juice pouches that each astronaut carried inside his helmet to guard against dehydration were another source of irritation, as they kept leaking or clogging up.

Casper and *Orion* were reunited in the early hours of 24 April, and once astronauts and rock samples had transferred to the CSM, *Orion* was cast adrift, though a fault with the planned de-orbit burn meant that it did not impact immediately as intended but eventually crashed to the Moon about a year later. A small satellite was deployed before the CSM left orbit, carrying experiments to study any particles in lunar orbit, and the Moon's feeble magnetic field. During the return flight, Mattingly made a spacewalk, venturing outside the CSM to retrieve a film canister and experimental equipment.

SHADOW ROCK
During the course of their third moonwalk, Young and Duke came across this distinctive tilted rock. Samples taken from underneath its base produced soil that had not seen sunlight in perhaps a billion years.

BIOGRAPHY

CHARLES DUKE

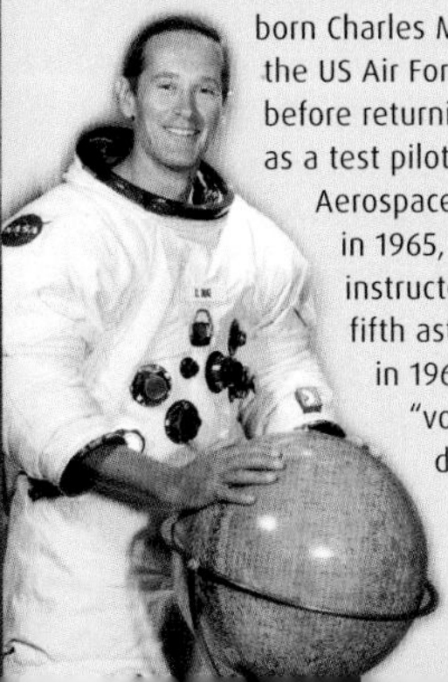

After studying at both the US Naval Academy and the Massachusetts Institute of Technology, North Carolina-born Charles M. Duke (b.1935) joined the US Air Force, serving in Germany before returning to America to train as a test pilot. Graduating from the Aerospace Research Pilot School in 1965, he stayed on as an instructor before joining NASA's fifth astronaut training group in 1966. He was Capcom, the "voice of Mission Control", during Apollo 11 and the backup LM pilot for both Apollos 13 and 17. He retired from NASA in 1975 to go into business.

Apollo 17

7 December 1972
Apollo 17 lifts off over two hours late after a countdown problem.

10 December 1972
The spacecraft arrives in lunar orbit, just ahead of its own S-IVB rocket stage, which is then deliberately crashed into the Moon. The impact is detected by seismometers set up by previous Apollo missions.

11 December 1972
The LM *Challenger* lands in the Taurus-Littrow valley. Later, Cernan and Schmitt make the first of three lengthy moonwalks.

14 December 1972
At the end of the third surface moonwalk, Cernan becomes the last man to step on the Moon in the 20th century.

16 December 1972
The CSM *America* heads for home.

19 December 1972
Apollo 17 splashes down in the Pacific Ocean.

21 December 1972
A reception at Houston marks the end of the Apollo programme.

The final Apollo mission was humankind's temporary farewell to the Moon. Even at the end of the programme, it was able to achieve another first – putting a qualified geologist on the lunar surface.

John F. Kennedy's ambition to reach the Moon – and just as importantly to beat the Soviet Union in the Space Race – had hung over his successors ever since the President's tragic assassination in November 1963. Apollo was a matter of national prestige, and the public would not forgive any politician who allowed it to slip from his priorities.

But with the race won, the question was – what next? By now, the United States was bogged down in the divisive and expensive Vietnam War, and the Apollo-era NASA budget was unsustainable in the long term. In fact, it was under pressure within months of *Eagle*'s landing, as first one and then two more of the original missions were axed, even though most of the hardware was ready for them.

While Apollo's sprint for the Moon ran roughshod over von Braun's 1950s plans for a stately march across the Solar System, its technology could still have given rise to a permanent lunar outpost, had the money and the will been there. But a more pragmatic administration felt that America's future in space lay closer to home. Apollo 17 would mark the end of lunar exploration for the foreseeable future.

Scientists in the valley

With fewer political constraints and no need to lay a basis for future colonization, the Apollo J missions could be governed almost entirely by scientific objectives, and so geologist Harrison "Jack" Schmitt took his place as Lunar Module Pilot on Apollo 17. Alongside him was Mission Commander Gene Cernan, and Command Module Pilot Ronald Evans.

After a spectacular night launch on 7 December 1972, the LM *Challenger* and the CSM *America* separated in orbit on 11 December, and Cernan and Schmitt touched down on the Moon shortly afterwards. This time the target was the Taurus-Littrow area, a flat-bottomed valley ringed by mountains, on the edge of the Sea of Serenity. The area seems to be one where lavas from the maria partially flooded a pre-existing valley.

Following the pattern of previous missions, there were three major moonwalks. Once again the focus was on collecting rock samples with the LRV, but the astronauts also deployed an unrivalled number of complex experiments as part of their ALSEP package. As on Apollo 15, several of these experiments involved drilling probes into the rock, and there were also core samples to be collected, but this time drilling went to plan and did not cause any delays. Meanwhile Evans conducted his own valuable experiments and observations from the orbiting CSM.

The valley was littered with boulders that seemed to have rolled down from the mountains around it, and these offered a valuable chance to study lunar bedrock that had not been altered by meteorite impacts. During their sampling expeditions, Cernan and Schmitt came across a bizarre patch of distinctly orange soil, which provoked a heated debate about its origins. It turned out to be a patch of naturally occurring glass (produced by a volcano) that had been uncovered during the formation of the nearby Shorty Crater. The astronauts also conducted a number of studies of the region's gravitational field – earlier probes and manned missions had already discovered regions of noticeably more powerful gravity, called mascons, apparently corresponding to denser areas of the lunar crust.

Last men on the Moon

The Apollo astronauts' final moments on the Moon included a small ceremony. They unveiled a plaque on the LM's landing strut to commemorate their visit and then packed away their samples and prepared for the return to Earth. Cernan was last aboard the LM, saluting the Stars and Stripes before declaring "... as we leave the Moon at Taurus-Littrow, we leave as we came and, God willing, as we shall return, with peace and hope for all mankind."

BIOGRAPHY

HARRISON SCHMITT

New Mexico-born Harrison Hagan "Jack" Schmitt (b.1935) gained a doctorate in geology from Harvard before joining the US Geological Survey at Flagstaff, Arizona. After his selection to NASA's first group of scientist-astronauts in 1965, he was closely involved in training the Apollo astronauts for geological fieldwork. Assigned to the backup crew of Apollo 15 alongside Richard Gordon and Vance Brand, he would have flown as LM pilot aboard Apollo 18, but the cancellation of that mission led to his replacing Joe Engle on Apollo 17. Schmitt retired from NASA in 1975 to enter politics and later business. He chaired the NASA Advisory Council, a group that advises the NASA Administrator, from 2005 to 2008.

"... America's challenge of today has forged **man's destiny of tomorrow**."

Gene Cernan, last Apollo astronaut to step on the Moon

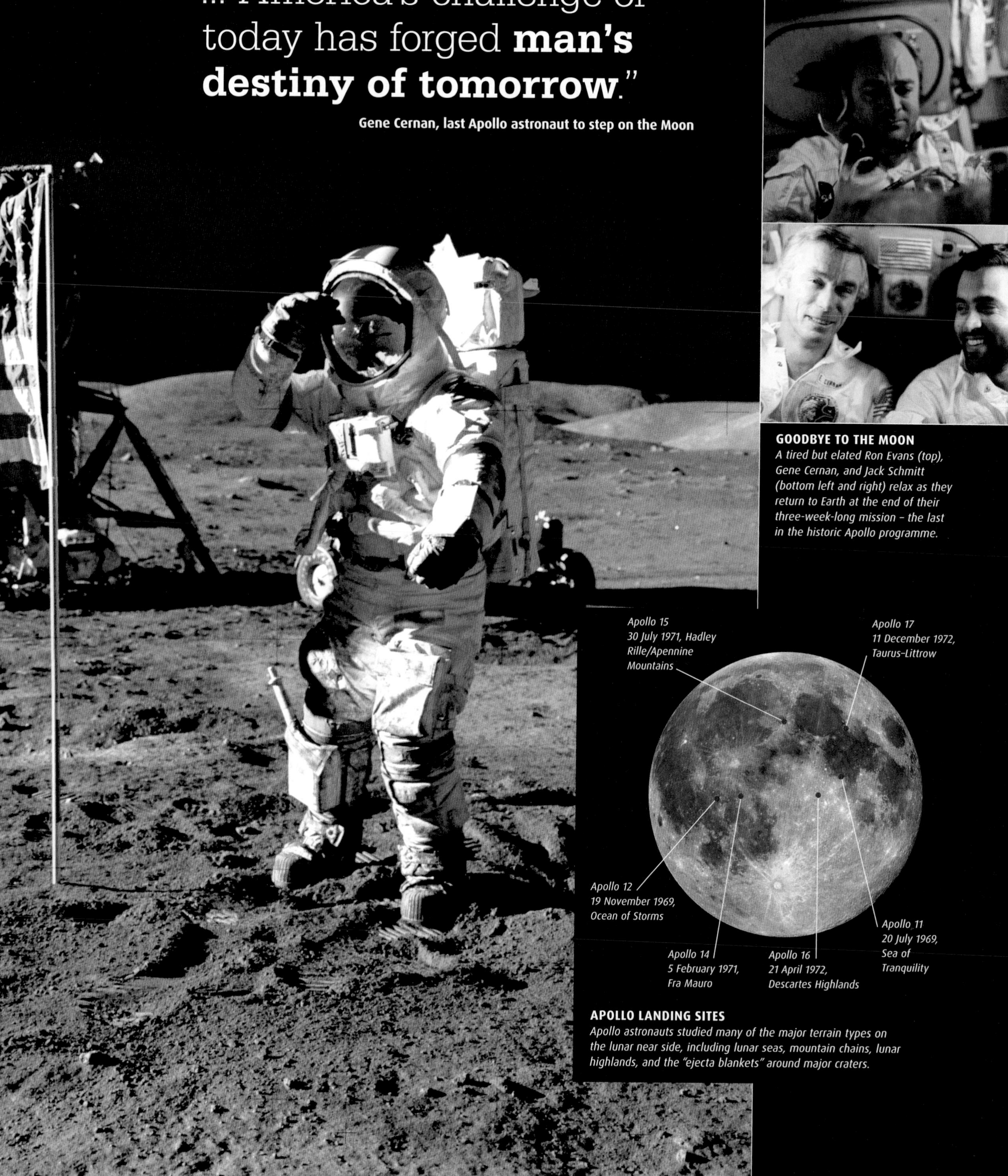

GOODBYE TO THE MOON
A tired but elated Ron Evans (top), Gene Cernan, and Jack Schmitt (bottom left and right) relax as they return to Earth at the end of their three-week-long mission – the last in the historic Apollo programme.

APOLLO LANDING SITES
Apollo astronauts studied many of the major terrain types on the lunar near side, including lunar seas, mountain chains, lunar highlands, and the "ejecta blankets" around major craters.

Apollo crews 1967–1972

After the tragedy of Apollo 1, Saturn rockets took men into space 11 times during the US lunar programme. The publicity photos reproduced here chronicle an era that cemented the image of the astronaut as hero.

APOLLO 1
Commander (left) *Virgil "Gus" Grissom (1926–67)*
CM Pilot (centre) *Ed White (1930–67)*
LM Pilot (right) *Roger Chaffee (1935–67)*
Mission *Never flew after catastrophic launch-pad fire*

APOLLO 7
LM Pilot *Walter Cunningham (b.1932)*
Commander *Wally Schirra (1923–2007)*
CM Pilot *Donn Eisele (1930–87)*
Mission *11–22 October 1968*

APOLLO 8
CM Pilot *Jim Lovell (b.1928)*
LM Pilot *William Anders (b.1933)*
Commander *Frank Borman (b.1928)*
Mission *21–27 December 1968*

APOLLO 12
Commander *Charles "Pete" Conrad (1930–99)*
CM Pilot *Richard Gordon (1929–2017)*
LM Pilot *Al Bean (b.1932)*
Mission *14–24 November 1969*

APOLLO 13
CM Pilot *John L. "Jack" Swigert (1931–82)*
Commander *Jim Lovell (b.1928)*
LM Pilot *Fred Haise (b.1933)*
Mission *11–17 April 1970*

APOLLO 14
CM Pilot *Stuart Roosa (1933–94)*
Commander *Alan Shepard (1923–98)*
LM Pilot *Edgar Mitchell (1930–2016)*
Mission *31 January – 9 February 1971*

APOLLO 9
Commander *James McDivitt (b.1929)*
CM Pilot *David Scott (b.1932)*
LM Pilot *Russell Schweickart (b.1935)*
Mission *3–13 March 1969*

APOLLO 10
LM Pilot *Gene Cernan (1934–2017)*
CM Pilot *John Young (1930–2018)*
Commander *Thomas Stafford (b.1930)*
Mission *18–26 May 1969*

APOLLO 11
Commander *Neil Armstrong (1930–2012)*
CM Pilot *Michael Collins (b.1930)*
LM Pilot *Buzz Aldrin (b.1930)*
Mission *16–24 July 1969*

APOLLO 15
Commander *David Scott (b.1932)*
CM Pilot *Alfred Worden (b.1932)*
LM Pilot *James Irwin (1930–91)*
Mission *26 July – 7 August 1971*

APOLLO 16
CM Pilot *Thomas K. "Ken" Mattingly (b.1936)*
Commander *John Young (1930–2018)*
LM Pilot *Charles Duke (b.1935)*
Mission *16–27 April 1972*

APOLLO 17
LM Pilot *Harrison "Jack" Schmitt (b.1935)*
Commander *Gene Cernan (1934–2017)*
CM Pilot *Ronald Evans (1933–90)*
Mission *7–19 December 1972*

JUNE 1973: SKYLAB OVER EARTH
NASA's space station shows the scars of its traumatic launch and hasty repairs – a lopsided arrangement of solar panels and an improvised foil sunshade deployed by the Skylab 2 mission.

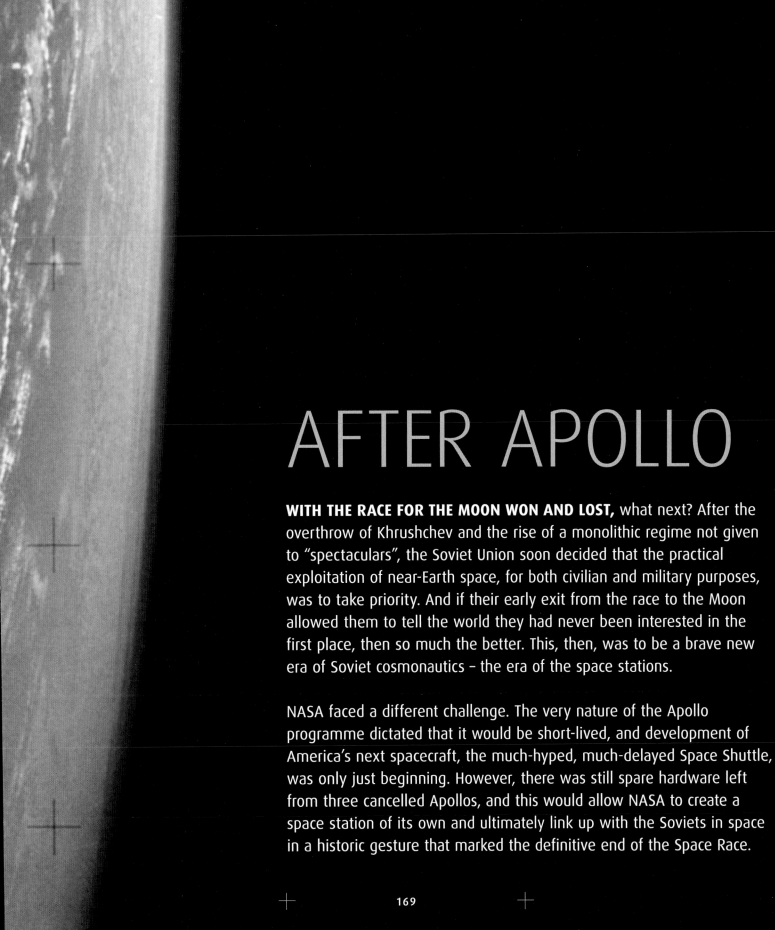

AFTER APOLLO

WITH THE RACE FOR THE MOON WON AND LOST, what next? After the overthrow of Khrushchev and the rise of a monolithic regime not given to "spectaculars", the Soviet Union soon decided that the practical exploitation of near-Earth space, for both civilian and military purposes, was to take priority. And if their early exit from the race to the Moon allowed them to tell the world they had never been interested in the first place, then so much the better. This, then, was to be a brave new era of Soviet cosmonautics – the era of the space stations.

NASA faced a different challenge. The very nature of the Apollo programme dictated that it would be short-lived, and development of America's next spacecraft, the much-hyped, much-delayed Space Shuttle, was only just beginning. However, there was still spare hardware left from three cancelled Apollos, and this would allow NASA to create a space station of its own and ultimately link up with the Soviets in space in a historic gesture that marked the definitive end of the Space Race.

14 May 1973
The launch of the unmanned Skylab space station atop a Saturn V rocket goes badly wrong and leaves the station crippled in orbit.

25 May 1973
The Skylab 2 launch carries the first crew to the station, with a complex repair mission to carry out before their scientific work can begin.

22 June 1973
Skylab 2 splashes down after 28 days in orbit – a new space endurance record for the crew.

28 July 1973
Skylab 3 begins a 59-day mission that sets another endurance record.

16 November 1973
Skylab 4 is launched on the third and final mission to the station.

8 February 1974
The Skylab 4 crew return to Earth after a record 84 days in orbit.

11 July 1979
The Skylab space station is destroyed re-entering the Earth's atmosphere. Parts land in Australia.

Laboratory in the sky

America's first space station, Skylab, was lofted into orbit in May 1973 in the final spectacular launch of a Saturn V rocket. Over the following year, three crews completed increasingly lengthy missions onboard.

Skylab 1, the space station's unmanned launch mission, almost ended in disaster. A shield designed to protect the station's walls from meteoroid impacts and the direct heat of the Sun deployed too early, and was ripped away by the supersonic air, carrying one of the two main solar panels with it. A loose metal strip then snagged across the remaining panel, preventing it from unfolding at all. By the time Skylab had reached orbit, it was already crippled – almost powerless and overheating badly. Only the smaller solar panels atop the Apollo Telescope Mount unfolded correctly.

Skylab on-line

The launch of Skylab 2, carrying the first crew of Pete Conrad, Joe Kerwin, and Paul Weitz, should have taken place the following day but was postponed while the engineers came up with a repair plan. Finally the trio rocketed skyward atop a Saturn IB 10 days late. After an unsuccessful attempt at untangling the remaining solar panel from onboard the Apollo CSM, they docked successfully and boarded the overheated station, pushing a hurriedly designed reflective parasol through a hatch and opening it up to provide some protection from the Sun. A complex spacewalk then finally released the snagged panel, providing the station with much-needed power and finally bringing it fully on-line.

With the station up and running, the astronauts were able to stay aboard for a total of 28 days, a new record for the longest stay in space. They continued to make repairs throughout the flight but also had time to carry out many experiments. These included observations and photography of the Earth and Sun, as well as medical experiments in which the astronauts themselves were guinea pigs, and five experiments suggested by high-school students back on Earth.

TECHNOLOGY

FOOD ADVANCES

The arrival of space stations increased the duration of missions from days to weeks and forced mission planners to reconsider the menus they were providing to spacefarers. Although much of the food carried aboard Skylab was still freeze-dried, the pastes and cubes supplied on early space missions were supplemented with frozen food that could be cooked in the station's galley area. Meals were prepared on metal trays that also acted as hotplates to warm the food. A magnetized upper surface held down the metal bowls and cutlery.

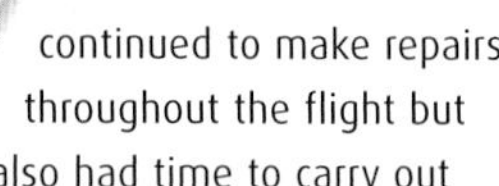

Science in orbit

Skylab 3, crewed by Al Bean, Jack Lousma, and Owen Garriott, returned to a station that had been empty for just over a month, in late July 1973. Aside from installing an improved sunshield and troubleshooting a problem with their own spacecraft's manoeuvring engines, the astronauts of this mission were able to concentrate on science. As well as studying their own condition during 59 days of weightlessness, they also recorded how a variety of smaller passengers coped, including mice, fruit flies, and spiders, as well as conducting a range of student experiments.

The last and longest of the Skylab missions saw Gerald Carr, Bill Pogue, and Edward Gibson spend 84 days in orbit. Although Skylab 4 involved a huge variety of experiments and observations – including studies of the giant Comet Kohoutek using the station's solar telescope – the crew spent much of the time in conflict with Ground Control, complaining that they were being worked too hard, while some mission controllers felt they weren't doing enough. Despite a highly successful mission, none of them was selected to fly into space again.

BIOGRAPHY

PETE CONRAD

Charles "Pete" Conrad (1930–99) was the third man to walk on the Moon, as commander of Apollo 12. Born in Pennsylvania, he studied aeronautics at Princeton before joining the US Navy and becoming a test pilot. He was selected for astronaut training in 1962 and first flew aboard Gemini 5 in 1965. A year later, he commanded Gemini 11. His first Apollo assignment was backup commander for the mission that became Apollo 9 – had it not been for the rearrangement of Apollos 8 and 9, he might well have been in line for the first manned landing on the Moon. After Skylab 2, Conrad quit both NASA and the Navy for a new career in business. He died following a motorcycle accident a month after his 69th birthday.

JETPACK TESTING
The large volume of Skylab allowed astronauts to test a rocket pack called the Automatically Stabilized Maneuvering Unit (ASMU). This was a prototype of the Manned Maneuvering Unit (MMU), later used on Space Shuttle missions *(see p.194)**.*

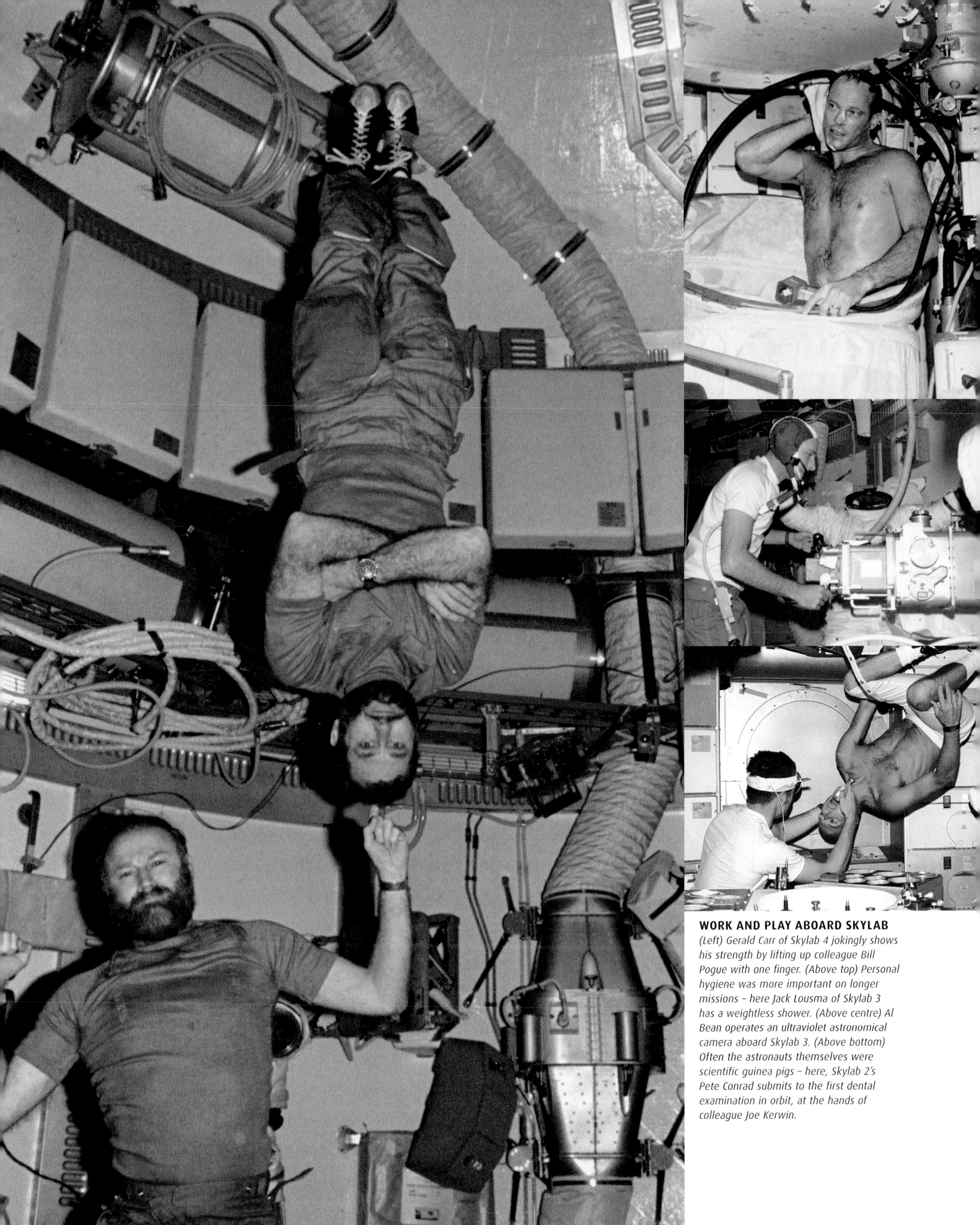

WORK AND PLAY ABOARD SKYLAB

(Left) Gerald Carr of Skylab 4 jokingly shows his strength by lifting up colleague Bill Pogue with one finger. (Above top) Personal hygiene was more important on longer missions – here Jack Lousma of Skylab 3 has a weightless shower. (Above centre) Al Bean operates an ultraviolet astronomical camera aboard Skylab 3. (Above bottom) Often the astronauts themselves were scientific guinea pigs – here, Skylab 2's Pete Conrad submits to the first dental examination in orbit, at the hands of colleague Joe Kerwin.

APOLLO TELESCOPE MOUNT

The Apollo Telescope Mount (ATM) was a large observatory originally planned for independent launch before being combined with Skylab. It is shown here during testing at Marshall Spaceflight Center.

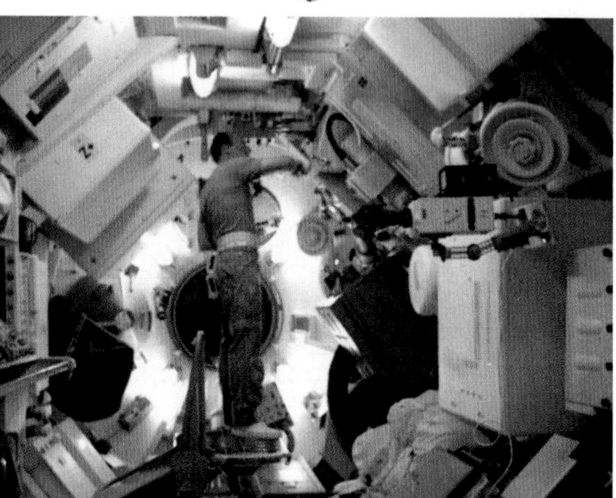

MULTIPLE DOCKING ADAPTER

Located at one end of Skylab, the Multiple Docking Adaptor (MDA) incorporated two docking points for Apollo CSMs: one on the main axis of the station for normal docking operations; and the other to one side for emergency use. Here, Pete Conrad is rehearsing procedures in the Skylab mock-up in the Mission Simulation and Training Facility at Houston.

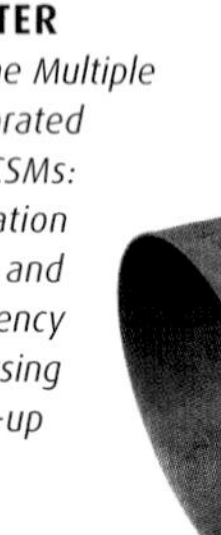

acquisition Sun sensor
aperture doors
solar array
sunshade
solar array for ATM
Apollo Telescope Mount
solar array mounting
ATM support strut
docking port
Apollo Command and Service Module
reaction-control thruster assembly
oxygen tank
Airlock Module
nitrogen tank
Multiple Docking Adapter
multispectral scanner
L-band antenna
propulsion engine nozzle

TECHNOLOGY

AMERICA'S FIRST SPACE STATION

Skylab

Originally conceived as part of the Apollo Applications Program in the late 1960s, the first US space station was all that remained after cuts in NASA's budget. It began life as the Orbital Workshop project, a plan to launch a Saturn IB rocket into orbit with a specially prepared S-IVB upper stage. This stage would enter orbit, and an Apollo crew would then dock with it, drain its remaining fuel, and begin to fit it out as a laboratory. The Skylab that ultimately flew was carried by a Saturn V, allowing a more ambitious design that was fully fitted out on the ground before launch.

CREW	3
LENGTH (INCLUDING CSM)	36.1m (118ft 6in)
MAXIMUM DIAMETER	6.6m (21ft 7in)
TOTAL MASS	34,473kg (76,000lb)
HABITABLE VOLUME	283 cubic m (9,985 cubic ft)
NUMBER OF DOCKING PORTS	2
DATE OF LAUNCH	14 May 1973
DATE OF RE-ENTRY	11 July 1979
MAIN CONTRACTOR	McDonnell Douglas

AIRLOCK MODULE AND MDA

Skylab's Airlock Module and Multiple Docking Adapter were built together at the McDonnell Douglas plant in St. Louis, then combined with the orbital laboratory fitted out elsewhere at the facility.

THE COMPLETE SKYLAB

This illustration shows Skylab as it would have looked had deployment gone smoothly. As it was, the launch cost the station its micrometeoroid shield and one of its two larger solar arrays.

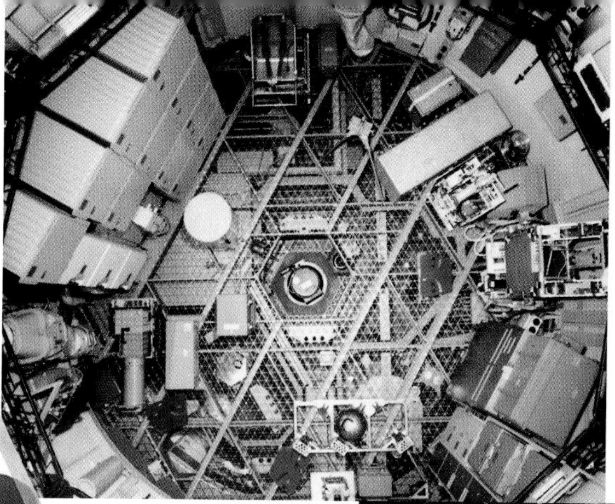

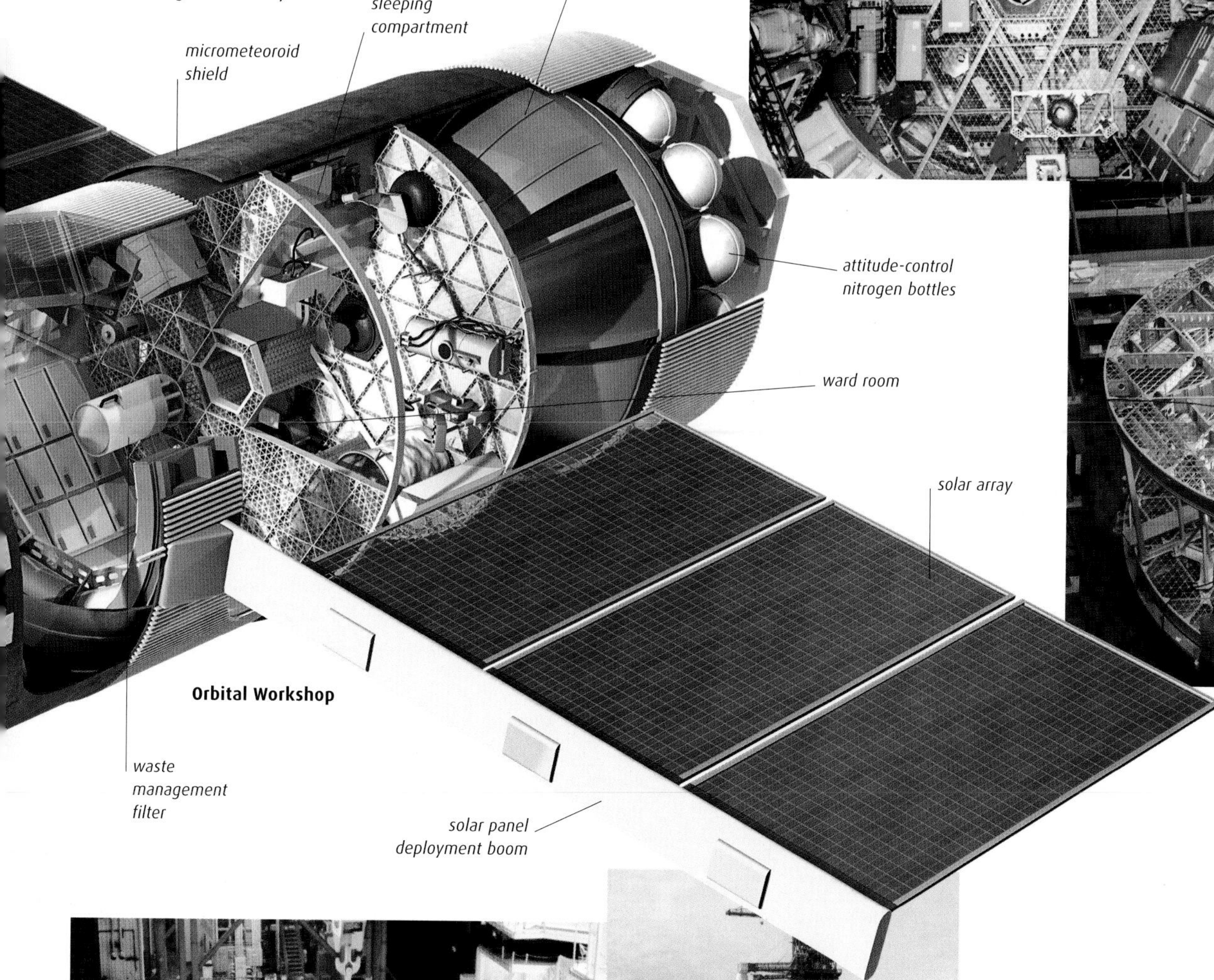

THE ORBITAL WORKSHOP

Skylab's main section appears spacious in this view from the Airlock Module towards the aft end of the station. The upper section held a food freezer and water tanks in addition to the experimental equipment. The waste airlock can be seen in the centre of the far wall.

PUTTING IN THE FLOOR

An important early stage in the transformation of an empty S-IVB shell into an Orbital Workshop was the addition of a two-storey lightweight floor grid. This would divide the finished station into an upper laboratory and a lower living area, with a hexagonal access hole between them.

LAUNCH ODDITIES

Skylab itself was launched on a special two-stage Saturn V – here (left), the Orbital Workshop section is being lowered onto the S-II second stage. The Apollo spacecraft, meanwhile, was launched on a far smaller S-IB rocket. In order to lift off from the massive Saturn V launch structure, the rocket was hoisted onto a pedestal called the Milk Stool (above).

HISTORY FOCUS

SKYLAB'S FATE

When Skylab was abandoned in 1974, NASA believed the station would remain in orbit until the 1980s. They planned an early Shuttle mission to attach a manoeuvring engine to the station, either boosting its slowly decaying altitude or pushing it into the atmosphere for a controlled re-entry. But high drag from the atmosphere sealed Skylab's fate, and it crashed back to Earth in July 1979. Most of the station landed in the Indian Ocean, but several large fragments (such as the one seen here) came down over western Australia.

The Apollo–Soyuz project

The decision to launch a joint Soviet–American spaceflight was largely a political one, but making it a reality required engineers and astronauts on both sides to overcome a number of technical and communication problems.

15 July 1975
Apollo and Soyuz 19 launch from opposite sides of the Earth within hours of each other.

16 July 1975
Orbital manoeuvres steer the two spacecraft towards a rendezvous.

17 July 1975
The Apollo CSM docks with Soyuz 19, and Americans and Russians shake hands in space for the first time.

18 July 1975
Astronauts and cosmonauts conduct joint experiments in orbit.

19 July 1975
Apollo and Soyuz carry out docking and undocking tests. After separating for a second time, they retreat to a safe distance to continue their scientific missions.

21 July 1975
Soyuz 19 returns to Earth.

24 July 1975
The Apollo CM returns to Earth. During re-entry, a series of mishaps results in the cabin flooding with propellant gases, almost suffocating the astronauts.

The years around 1970 saw a thawing of relations between the two sides in the Cold War – an interlude called *détente*, in which the rivalry between the superpowers briefly receded. At a summit in May 1972, US President Richard Nixon and Soviet premier Alexei Kosygin brought a formal end to the Space Race with the announcement that an Apollo and a Soyuz spacecraft would rendezvous and dock in space during 1975.

The announcement had been preceded by many months of patient diplomacy at all levels. Almost as soon as he took over from James Webb in October 1968, NASA Administrator Thomas O. Paine had begun planning for a future of cooperation, rather than competition, in space. He began sounding out leading Soviet figures about the possibilities for a collaborative spaceflight, beginning with Mstislav Keldysh, President of the Soviet Academy of Sciences. Discussions continued through the Apollo moon landings and the first Salyut launches, under both Paine and his successor James C. Fletcher, until finally a plan was agreed. The two-man Soyuz 19 spacecraft would rendezvous with an Apollo CSM in Earth orbit, on a mission called the Apollo–Soyuz Test Project (ASTP).

Although the main benefits of the project would be political, there were also practical implications. A system for docking Soviet and American spacecraft would open up new options for potential rescue missions should a spacecraft become stranded in Earth orbit. ASTP would also keep NASA in the manned spaceflight game during the long development of the Space Shuttle (see following chapter). In the longer term, the potential for technical advances and cost savings from pooled expertise was irresistible.

While the engineers worked feverishly on developing a system for uniting two defiantly incompatible spacecraft (see panel, above), the astronauts had their own barriers to overcome. NASA's crew consisted of Thomas Stafford, Vance Brand, and Deke Slayton, effectively NASA's chief astronaut (see p.94). The Soyuz 19 crew were to be Alexei Leonov (see panel, below) and Valery

TECHNOLOGY

THE ASTP DOCKING MODULE

Docking the Apollo and Soyuz modules would be a major challenge – the docking attachments on each spacecraft were incompatible, as were the atmospheres inside them. To solve these problems, the engineers designed an adaptor with suitable docking points at either end, and an airlock system in between, to allow the crews to make a gradual transition from one atmosphere to the other. During launch, the adaptor module – which was 3.15m (10⅓ft) long with a diameter of 1.40m (4½ft) – was stowed beneath the Apollo CSM, just like a lunar module. After reaching orbit, the CSM turned round, linked with the docking module, and pulled it free.

MISSION INSIGNIA
The Apollo–Soyuz mission logo, with its stylized representation of the docked spacecraft, was a Soviet design.

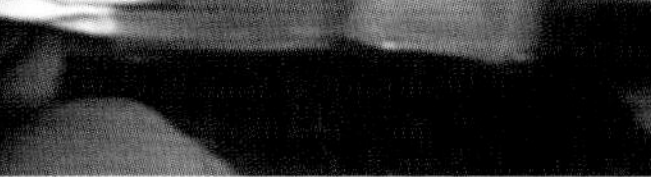

MEETING IN SPACE
(Left) Soyuz 19, photographed from the Apollo CSM, as it approaches for docking. (Above) Alexei Leonov and Thomas Stafford hold the flags of their respective countries during a broadcast for Soviet television.

PLANNING ON EARTH
NASA built a mock-up Soyuz at Houston for use in training. Here cosmonaut Leonov and astronaut Stafford pore over procedures for the orbital rendezvous.

APOLLO–SOYUZ
Joined together, the Apollo and Soyuz spacecraft clearly show their origins on rival sides of the Space Race. The block in the centre is the docking module, which allowed the two incompatible spacecraft to join together.

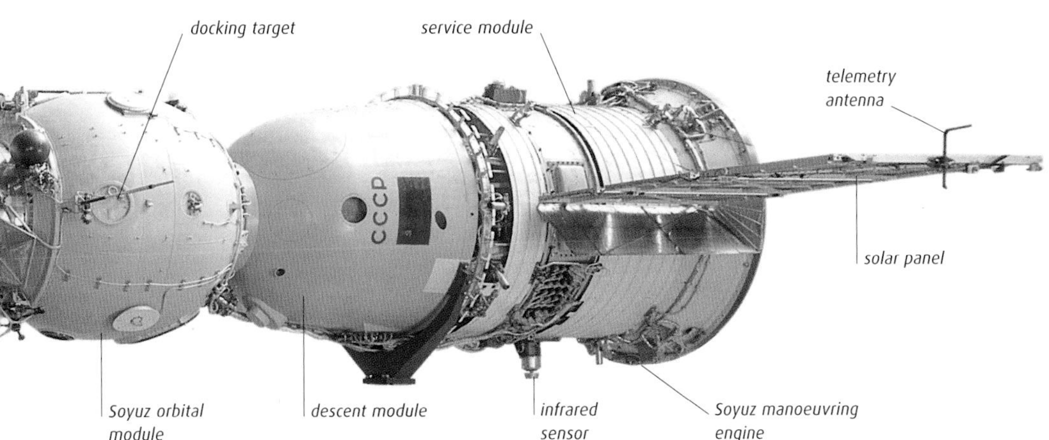

Kubasov. All five crew members, as well as their backups, had to go through intensive language training alongside their other studies. The crews, and many other members of the technical staff, also visited each other's training facilities and familiarized themselves with the other spacecraft's systems.

Orbital union

Soyuz 19 took off from Baikonur Cosmodrome on 15 July 1975, with the Apollo spacecraft (it had no call sign or other designation and was known simply as *Apollo*) joining it from Cape Canaveral about seven hours later. A day of orbital manoeuvres brought the spacecraft to their rendezvous, and they docked at 16:10 GMT on 17 July (see pp. 176–77). Astronauts and cosmonauts transferred from one spacecraft to the other several times, and though the tasks they carried out were largely related to ceremony and public relations, there were also practical tests of the docking module, with the spacecraft briefly undocking and then uniting once again, before finally going their separate ways.

After a successful return to Earth for both spacecraft, the crews did the usual round of press conferences, photocalls, and ceremonials. But the political mood was changing again, and as relations cooled once more between the superpowers, there would be no repeat of the mission. It would be another two decades before a second handshake in orbit would take place.

BIOGRAPHY

ALEXEI LEONOV

Born in Siberia, Alexei Arkhipovich Leonov (b.1934) showed early interest in both art and aviation but eventually joined the Soviet Air Force, where he was selected for the first cosmonaut trainee group in 1960. He first flew aboard Voskhod 2 in 1965, becoming the first man to walk in space (see p.102). Although he did not fly again until Soyuz 19, he had been scheduled for several other missions, including a proposed flight around the Moon and the cancelled second mission to Salyut 1. After his second flight, he became head of the cosmonaut team at Star City, eventually retiring with the rank of General in 1991 to concentrate once again on painting.

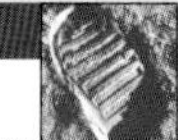

EXPERIENCE

THE SUPERPOWERS MEET IN SPACE

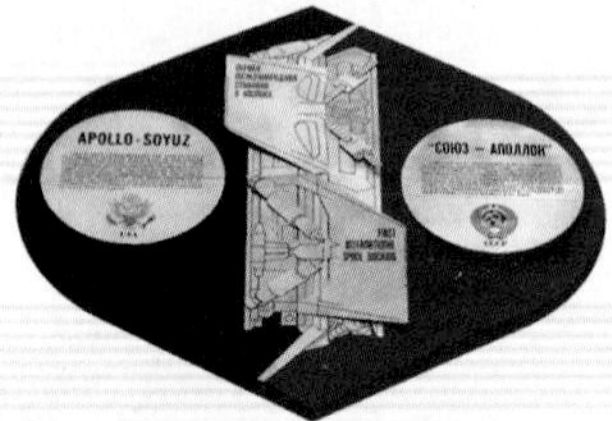

MISSION MEMENTO
Each crew carried two identical halves of this plaque – after exchanging sections, each took a complete version back to Earth.

A handshake in orbit

The Apollo–Soyuz Test Project was a triumph of early cooperation between the Soviet and American space programmes, with engineers working for many years to make two separately evolved spacecraft operate together. In July 1975, however, it ultimately came down to the skill of five people in orbit.

The numberless Apollo soared into the Florida sky atop its Saturn IB launcher on the afternoon of 15 July 1975, racing in pursuit of the Soviet Soyuz 19 that had launched earlier in the day. On reaching space, Command Module Pilot Vance Brand cried in Russian "Miy Nakhoditsya na orbite!" ("We are in orbit!"). A little under an hour later, the Apollo CSM separated from the S-IVB upper stage, and turned around to pull the Docking Module (DM) out of the stowage space normally reserved for a Lunar Module.

On 16 July, Soyuz performed several engine burns to help modify its orbit. Meanwhile, Apollo, after a perfect docking and extraction manoeuvre, made a series of its own burns to gradually take it up with the Soyuz, bringing the two craft together on the cosmonauts' 36th orbit of Earth.

Awaking early on 17 July, the Apollo crew sighted Soyuz at 1pm Houston time, and made radio contact moments later. The cosmonauts greeted their American counterparts enthusiastically:

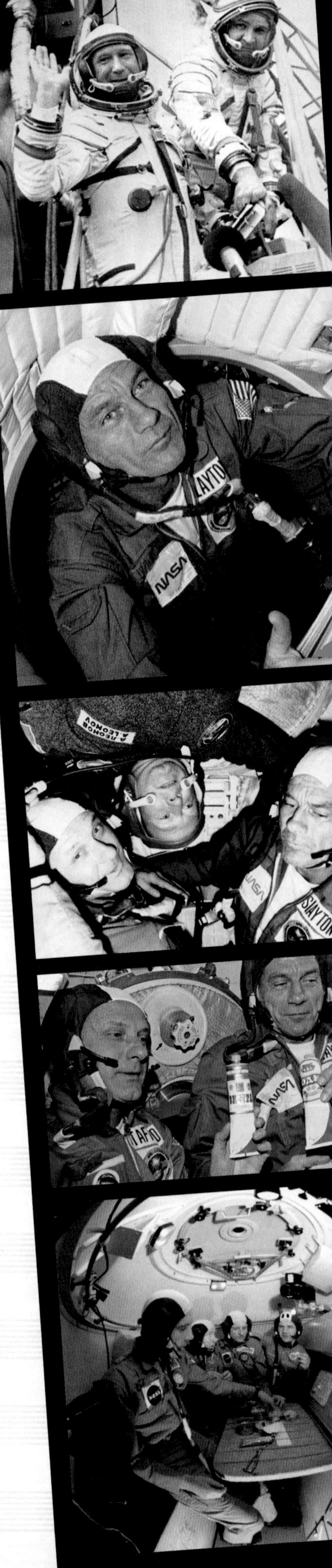

TOGETHER IN SPACE
The crew of Soyuz 19 left Earth 7½ hours before Apollo. Over the next two days of flight, both crews adjusted their sleep to synchronize with each other better. The spacecraft stayed locked together for a little under two days, during which time the Americans got to enjoy a banquet of Soviet space food aboard the Soyuz.

Deke Slayton: Soyuz, Apollo. How do you read me?
Valery Kubasov: Very well. Hello, everybody.
DS: Hello, Valery. How are you? Good day, Valery.
VK: How are you? Good day.
DS: Excellent ... I'm very happy. Good morning.
Alexei Leonov: Apollo, Soyuz. How do you read me?
DS: Alexei, I hear you excellently. How do you read me?
AL: I read you loud and clear.
DS: Good.

A first measure of separation revealed that the two craft were 222km (138 miles) apart. Over the next three hours, a series of manoeuvres by Apollo closed the gap. Given approval for docking, Leonov rolled Soyuz 19 towards the approaching Apollo, while US Commander Tom Stafford guided the American spacecraft in to a perfect rendezvous at 16:10.

IN TRAINING
The crew of Apollo and Soyuz 19 (left to right: Brand, Leonov, Stafford, Kubasov, and Slayton) formed close ties during their months of preparation.

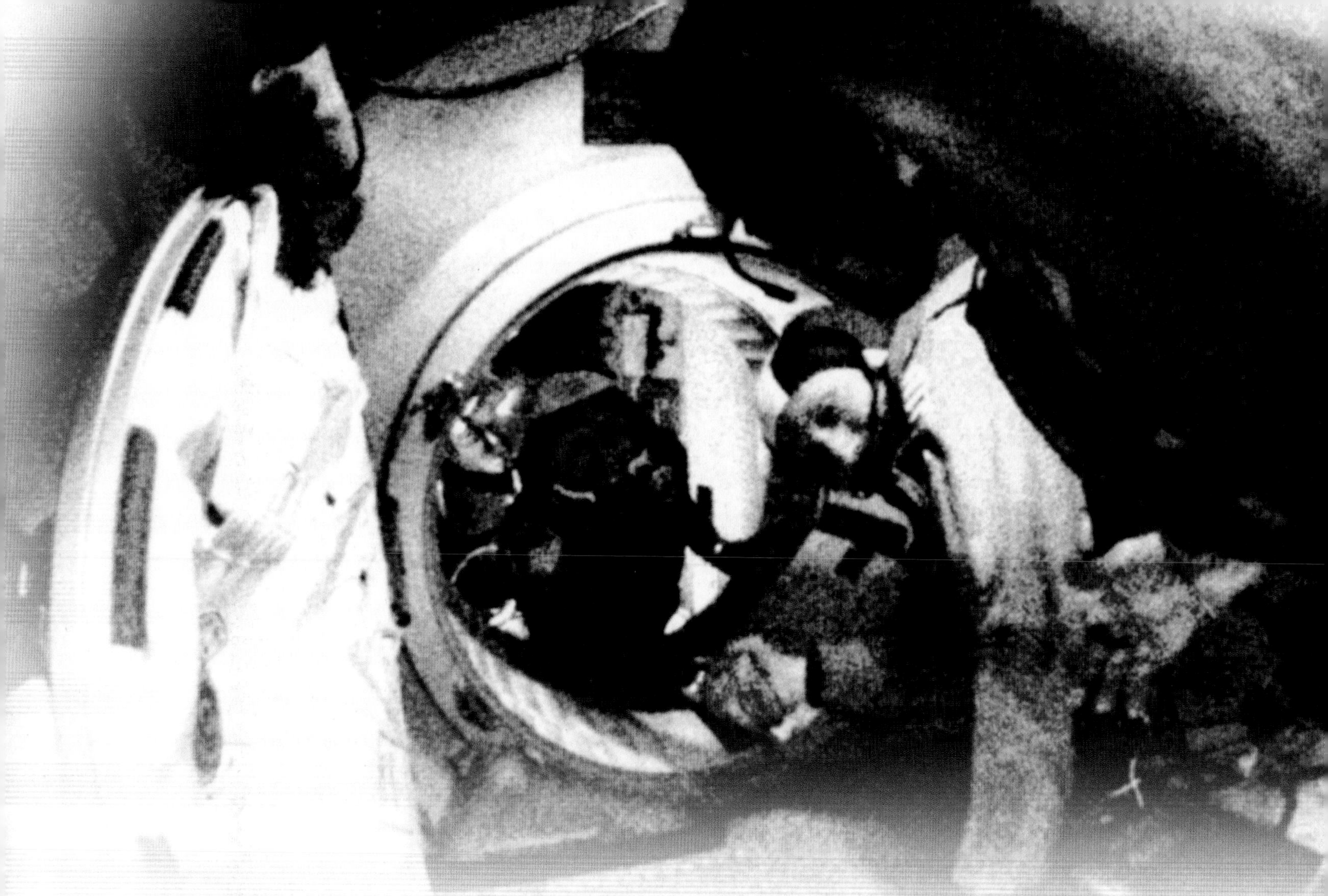

HISTORIC HANDSHAKE
Looking into the cable-strewn Soyuz Orbital Module, Stafford's first words were "Looks like they got a few snakes in there, too." Then he called "Alexei, our viewers are here – come over here, please."

"Man, I tell you, this is worth waiting **16 years** for!"

Deke Slayton catching his first view of the Earth from space, 15 July 1975

The Apollo crew had closed the hatch to the DM in case anything went wrong on final approach, while the cosmonauts had retreated to their Descent Module. As Deke Slayton reopened the DM hatch, a smell of burning glue from inside caused brief alarm and a delay while they waited for the air to clear. Slayton and Stafford now entered the DM and sealed themselves in as the atmosphere slowly adjusted to match that on Soyuz, while a Soviet announcer broadcast a message from premier Leonid Brezhnev:

"To the cosmonauts Alexei Leonov, Valery Kubasov, Thomas Stafford, Vance Brand, Donald Slayton. Speaking on behalf of the **Soviet people**, and for myself, I congratulate you ... The whole world is watching with rapt attention and admiration of your joint activities in fulfillment of the complicated programme of scientific experiments. The docking has confirmed the correctness of the technical decisions developed and realized by **cooperative friendship**."

Leonov and Kubasov opened the hatch on their side, then Stafford opened the final hatch and looked out into the Soviet spacecraft. Leonov was waiting to greet him, and the two commanders shook hands high above the French city of Metz.

MEETING PRESIDENT FORD
The US President took a keen interest in the Apollo-Soyuz rendezvous, questioning the astronauts and cosmonauts closely during a radio link-up. He later welcomed the US and Soviet crews to the White House.

An outpost in orbit

Recognizing that it had lost the race for the Moon, the Soviet Union rapidly switched its space programme to focus on the exploitation of Earth orbit.

19 April 1971
The unmanned Salyut 1 DOS station is launched from Baikonur by a Proton rocket.

23 April 1971
Soyuz 10 is launched, carrying the intended first Salyut crew of Vladimir Shatalov, Alexei Yeliseyev, and Nikolai Rukavishnikov (one of the design team). Later in the day, an attempt to board Salyut 1 fails when the docking system will not fully engage.

24 April 1971
Soyuz 10 returns to Earth.

10 May 1971
An investigation concludes that the Soyuz 10 docking failure was probably due to a fault on the spacecraft, not the station.

6 June 1971
Soyuz 11 is launched with an improved docking mechanism, and successfully docks with Salyut 1 a day later. The three-man crew begin a 23-day mission aboard the station.

29 June 1971
The crew of Soyuz 11 are killed returning to Earth.

10 October 1971
Salyut 1 breaks up on re-entry.

While the engineers of OKB-1 had spent most of the late 1960s developing technology for a Soviet trip to the Moon, by the end of the decade they had little to show for it but the Soyuz spacecraft. Meanwhile, the rival design bureau led by Vladimir Chelomei had been garnering political support for a series of manned military space stations known as Almaz. The Almaz station would be launched using Chelomei's powerful Proton rocket (see p.210), and cosmonaut crews and supplies would travel to and from it using the almost equally massive TKS ferry vehicle.

Following the decision to abandon manned lunar efforts and pretend there had never been a race with Apollo, launching a space station suddenly became a priority. Of course, there was still the question of whether a cosmonaut could survive the proposed month-long missions in weightless conditions – at the time, no cosmonaut had flown for longer than five days. The Soyuz 9 mission was to change all that, with Andrian Nikolayev and Vitaly Sevastyanov spending some 18 days in orbit.

A change of plan

Despite this success, other elements of the project were not going well. Both the station and the TKS ferry were taking a long time to develop, and there were strong reservations about using the new and unreliable Proton to launch manned missions. Despite intense hostility between Chelomei and Vasili Mishin, engineers from their two bureaux conspired to find a solution to the problem, developing the long-duration orbiting station (DOS is its Russian acronym), a hybrid combining elements of Almaz with ideas from OKB-1's own ambitious plan for a modular station. Soyuz spacecraft, launched on an R-7-derived rocket, would service the hybrid station.

The DOS plan helped speed up development, and by April 1971, the first of the hybrid stations was ready for launch. Salyut 1, as it was called (in "salute" to Yuri Gagarin's historic first flight a decade earlier), was launched by a Proton rocket and entered orbit some 264km (164 miles) above the Earth. With the station safely deployed, its intended first crew took off aboard Soyuz 10, rendezvousing on 24 April. The spacecraft managed to dock with Salyut 1, but an electrical fault prevented the cosmonauts swinging aside the bulkhead between the two spacecraft and gaining access to the station. After several attempts, they undocked and returned

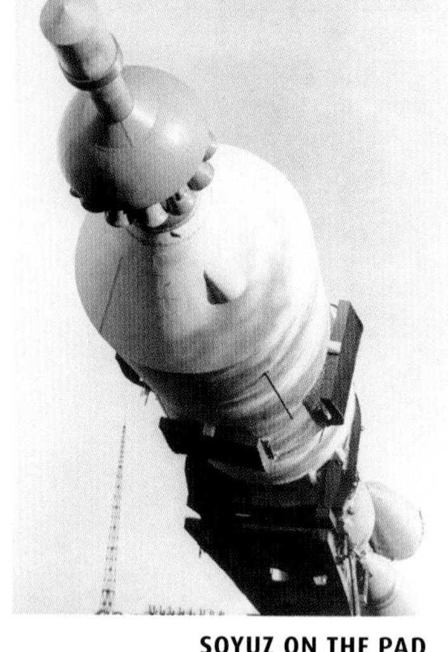

SOYUZ ON THE PAD
A Soyuz rocket with Soyuz 9 spacecraft, shroud, and escape tower in place is raised into an upright position prior to launch.

IN THE FACTORY
Engineers work on fitting the Soyuz-compatible docking port to the Almaz-derived components of Salyut 1. The 2m (80in) diameter compartment fitted to the narrower end of the station's main body.

LOST CREW
The crew of Soyuz 11 pose for a photo during training, Left to right are Commander Georgi Dobrovolsky, Test Engineer Viktor Patsayev and Flight Engineer Vladislav Volkov.

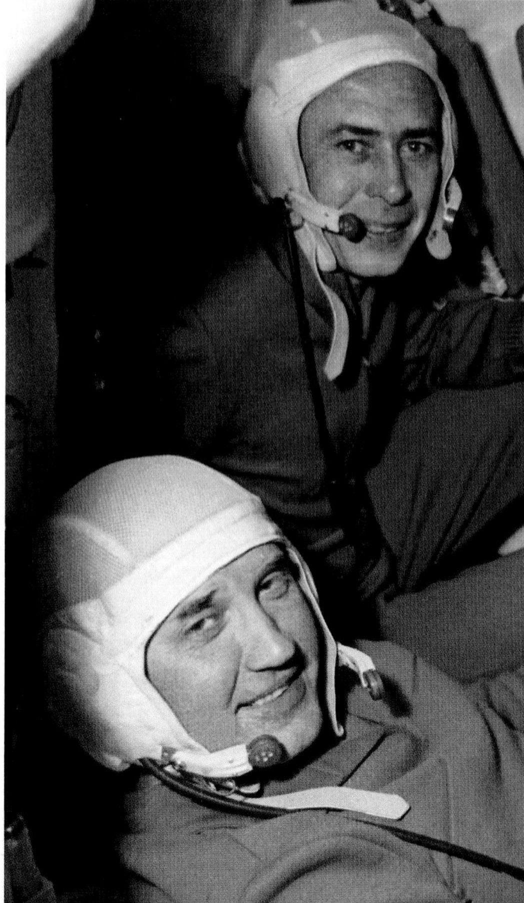

HISTORY FOCUS

STAR CITY

Since 1960, Soviet cosmonauts have lived and worked at the specially built town of Zvyozdny Gorodok near Moscow. The name means "little town of stars" in Russian, but in the West it is usually known as Star City. The town is home to the Gagarin Cosmonaut Training Centre, where facilities include full-size mock-ups and simulators of all the major Soviet and Russian space vehicles (including the Salyut stations), centrifuges for *g*-force training, and large water tanks for simulating weightless operations. A nearby airfield hosts aircraft that fly parabolic paths to simulate weightlessness. In the Soviet era, Zvyozdny Gorodok was a restricted area, but today foreign visitors include astronaut trainees, space enthusiasts, and tourists. Many cosmonauts and their families still live in the town.

TECHNOLOGY

SALYUT DESIGN

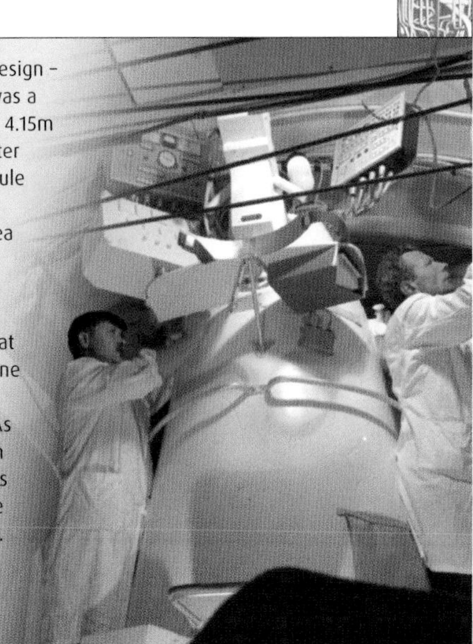

Vladimir Chelomei's original Almaz design – which formed the basis of Salyut – was a stepped cylinder with two segments 4.15m (13ft 7in) and 3m (9ft 10in) in diameter and 11.6m (38ft) long. A conical capsule at the forward end would allow it to be launched with a crew, but this idea was later abandoned. At the rear, a docking port would allow Chelomei's manned TKS ferries to approach. The Salyut DOS configuration developed at OKB-1 replaced the ring-shaped engine around the aft docking port with a Soyuz-based unit that sat centrally. As a result, a new docking system (with an airlock transfer compartment) was added to the forward end, giving the station a total length of 14.6m (48ft). The flown Almaz stations had fewer changes – the rear docking port was modified to accommodate Soyuz.

to Earth – official Soviet press releases claimed the mission was only ever intended as a docking test. Six weeks later, Soyuz 11 tried again – and this time, everything went smoothly.

After docking and moving into the station, the three-man crew established a round-the-clock schedule. While one person slept, another was off duty, relaxing, eating, and using various exercise devices to stave off the effects of weightlessness. The third, meanwhile, manned the station's array of scientific experiments, which included remote-sensing instruments to study the Earth, telescopes and other detectors for astronomy, and various biomedical experiments that frequently saw the cosmonaut as both scientist and guinea-pig.

Tragedy strikes

Things went well, and daily television broadcasts were sent back to Earth, until on 16 June a minor fire broke out. It was soon extinguished and was never a serious threat, but the scare led to the mission being cut short. Salyut 1 was put into automatic mode and its crew prepared to return to Earth, but then tragedy struck. As the re-entry module separated from the rest of the Soyuz 11 spacecraft, an explosive bolt forced open a pressure valve, allowing air to escape from the cabin and suffocating the cosmonauts before they re-entered the atmosphere.

The loss stunned the Soviet public, who had grown familiar with the cosmonauts through their broadcasts. It also grounded the Soyuz spacecraft, and with no other way of servicing the Salyut 1 station, its operators reluctantly allowed its orbit to decay, until it broke up on re-entry in October 1971.

ABOARD SALYUT 1
During more than three weeks on the station, the crew adapted well to life in orbit. Volkov and Dobrovolsky decided to grow beards, but Patsayev shaved regularly.

3 April 1973
Salyut 2, the first Almaz station, reaches orbit, but control is lost 22 days later, before it can be occupied.

24 June 1974
Salyut 3 is successfully launched into orbit.

4 July 1974
The crew of Soyuz 14 join Salyut 3, staying onboard for 14 days.

26 December 1974
A Proton rocket launches the DOS station Salyut 4.

11 January 1975
Soyuz 17 is launched, carrying the first crew to Salyut 4.

25 January 1975
Salyut 3 breaks up on re-entry after only one spell of occupation.

24 May 1975
Soyuz 18 takes a second crew to Salyut 4.

22 June 1976
The Almaz-based Salyut 5 launches successfully.

24 February 1977
Salyut 5 is abandoned following two successful manned missions.

Reinventing Salyut

The hiatus created by the Soyuz 11 tragedy gave Vladimir Chelomei's military Almaz project a chance to catch up with the hybrid space station under cover of the Salyut name.

The Soviet Union's next attempt at a space station launch, when it came, was to be another hybrid similar to Salyut 1. Launched in July 1972, the station was lost after a malfunction in the second stage of its Proton launch vehicle. Had it survived to orbit, this would have been Salyut 2 – but, as a failure, the Soviets simply ignored it.

Doubts about the reliability of the Proton rocket had forced Chelomei to adapt his original Almaz design so that it could dock with Soyuz, but by early 1973 the first Almaz and a new, upgraded hybrid were both ready. The Almaz station was launched first, and once in orbit on 3 April was officially announced as Salyut 2. Official press releases made no mention of the fact that this was a different type of station, intended largely as a manned spy satellite, but the difference in telemetry signals offered Western experts a clue to its true purpose. All seemed to be going well, and the first crew were preparing for launch, when Salyut 2 suddenly fell silent, victim of a catastrophic break-up in orbit.

As Soviet investigations soon traced the problem back to a fire that started in the Almaz propulsion unit, there was no reason to delay the hybrid station ready on the ground, and this was launched a month later, on 11 May. However, gremlins struck again as a fault in the propulsion system left the station spinning out of control, beyond hope of recovery. This time, the Soviets attempted to disguise the station as a failed satellite launch, named Cosmos 557.

Some success at last

Following these embarrassments, manufacture of both types of station was delayed for almost a year (in the interim, a pair of short-duration Soyuz-only missions kept up the Soviet presence in orbit). The next Almaz-type station was finally ready for launch in June 1974, and after a thorough orbital check-up this officially became Salyut 3. In early July, Pavel Popovich and Yuri Artyukhin docked with the station aboard Soyuz 14. As well as operating the reconnaissance camera, they conducted a number of remote-sensing experiments (see p.244) during their 16-day mission. Rigorous exercise meant that

IMPROVED HYBRID
The Salyut 4 design used three large solar panels that could rotate to face the Sun. It also had a new automatic docking system and improved water reclamation.

ALMAZ CONSTRUCTION
A military Almaz station nears completion at the Khrunichev factory near Moscow. Note the payload shroud, ready to be put in place before launch.

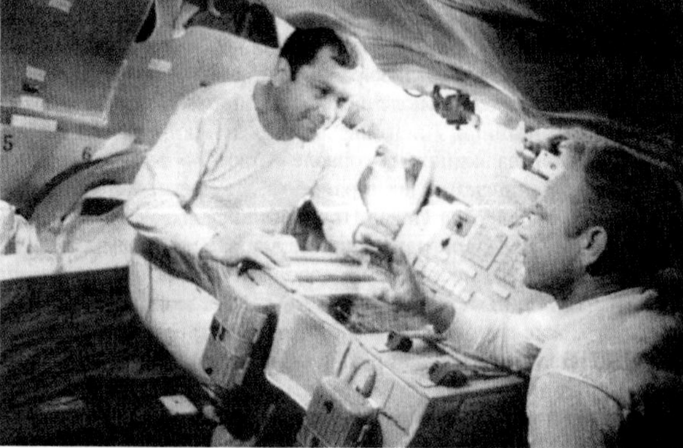

INSIDE AND OUTSIDE SALYUT 3
(Left) Cosmonauts Pavel Popovich and Yuri Artyukhin pose in front of the Almaz station a few days before launch. Above them is the docking port, surrounded by the manoeuvring engines.
(Above) A rare television picture from onboard a military Salyut shows Popovich and Artyukhin relaxing during free time. Even though it would operate in weightlessness, Salyut 3's interior was given a distinct floor and ceiling to help its crew adapt.

they returned to Earth in far better shape than previous long-duration cosmonauts. In late-August, Soyuz 15 set off, carrying a second crew intended for Salyut 3. But bad luck struck again, when the spacecraft was unable to make a successful rendezvous. Later in September, the station released a capsule containing the film from its cameras, and then it was left to its fate, breaking up in early 1975.

By this time, a new hybrid station (similar to Cosmos 557) was already in orbit. Salyut 4 was a significant advance for the programme, with a range of new features that made it more reliable and comfortable for the crew and a higher, more stable orbit. The first crew launched aboard Soyuz 17 on 11 January 1975 and stayed in space for a new record of 30 days, during which they adapted well to conditions and carried out a variety of experiments.

The launch of a second crew, on 5 April, ended in a dramatic escape after a rocket stage shut down prematurely and the Soyuz spacecraft had to make an emergency re-entry from an altitude of 180km (112 miles). A new replacement crew took off aboard Soyuz 18 on 24 May. Pyotr Klimuk and Vitali Sevastyanov remained in orbit for 63 days, but by the time they completed their planned tour of duty, the station was deteriorating, its windows fogged and mould growing on the walls.

TECHNOLOGY

THE SOKOL SPACESUIT

While Vostok-era cosmonauts wore pressure suits for safety reasons, by the Soyuz era the spacecraft had become a shirtsleeve environment. However, the loss of the Soyuz 11 crew changed all this, and a new suit, the Sokol, was introduced in 1973. Sokol was not intended for spacewalks – only for use in the event of an accident and during dangerous times such as re-entry. It is a two-layer suit, with a skin of rubberized synthetic material beneath an outer layer of canvas. The boots and hinged helmet are integrated into the suit, while the gloves are removable and lock into place on aluminium rings. Early versions of the suit came in two pieces that zipped together around the waist, but later variations (still in use today) are one-piece: the cosmonaut climbs into the suit through a V-shaped opening, an overlapping flap seals the inner skin, and zippers close the canvas layer. A ventilation system blows cabin air through the suit, but if pressure falls, it switches over to a bottled oxygen supply. The suit weighs 10kg (22lb) and is designed to keep the wearer alive for up to two hours in a vacuum. It is also intended to float in the event of an emergency splashdown.

pressure regulator
detachable glove
pressurized hood with plastic visor
connector for air and coolant lines
pressure gauge
utility pocket
pleated knee
boots are part of suit

Salyut 5 was the last Almaz station. Essentially the same as Salyut 3, this time the station carried materials science experiments as well as its main reconnaissance payload. The first crew, Boris Volynov and Vitali Zholobov, joined the station in July 1976, two weeks after its launch, and remained in space for 49 days. They left the station ahead of schedule after Zholobov in particular developed psychological problems. There were also suspicions of toxic gas in the cabin, but it was later reoccupied (after a failed attempt at docking by the crew of Soyuz 23), for a further 16 days by the crew of Soyuz 24.

The last Salyuts

Salyuts 6 and 7 combined the best elements of earlier military Almaz and civilian hybrid stations. They also added some innovations that allowed them to operate more effectively and for far longer.

29 September 1977
Salyut 6 reaches orbit.

10 December 1977
The crew of Soyuz 26 are launched on a 96-day mission to Salyut 6.

11 January 1978
Soyuz 27 docks to the spare port on Salyut 6, bringing two cosmonauts on a five-day visit.

22 January 1978
Progress 1, the first unmanned supply ferry, docks with Salyut 6.

19 June 1981
The Cosmos 1267 space tug docks with Salyut 6.

19 April 1982
Salyut 7 is launched.

14 May 1982
Salyut 7's commissioning crew arrive at the station aboard Soyuz T-5.

8 June 1985
Soyuz T-13 makes a manual docking with the deactivated Salyut 7 to bring it back online after a power failure.

17 November 1986
The final Salyut 7 mission is curtailed when Commander Vladimir Vasyutin falls ill.

When Salyut 6 was launched into orbit in September 1977, it represented a great leap forward for the Soviet space programme. The most significant change was that its engines had an off-axis design, allowing for a docking port at either end of the station. The ability to dock a spacecraft at either end of the station was a huge advance: unmanned supply ferries would now be able to come and go while the crew's own spacecraft remained safely docked; the station could welcome visitors or hand over from crew to crew without being uninhabited for long periods; and if a visiting crew left their new spacecraft with the station and returned in the older one already in orbit, Salyut missions would no longer be limited by the Soyuz capsule's relatively short operational life.

LIFE ON SALYUT 7
(Above) The crew of Soyuz T-12, Vladimir Dzhanibekov, Svetlana Savitskaya, and Igor Volk relax onboard the station. (Below) Dzhanibekov with Viktor Savinykh during their mission to revive the station in 1985.

The career of Salyut 6

If all these plans were to become a reality, though, Salyut 6 would need more luck than its predecessors, and things did not start well. The first crew, launched aboard Soyuz 25, had to turn back frustrated when the docking mechanism between station and spacecraft failed to lock. Things went better for Soyuz 26, and Georgi Grechko and Yuri Romanenko became the first full-time crew. After commissioning the station and beginning its astronomy and Earth science programmes, they received a visit from Soyuz 27, and later unloaded the first cargo from a Progress supply ferry (see p.210). Later in the 96-day flight, a second guest crew visited, including Czech cosmonaut Vladimir Remek (see p.240).

When Grechko and Romanenko finally returned to Earth, it was aboard the Soyuz 27 spacecraft, left in orbit after its original crew had returned on Soyuz 26. This marathon formed a template for later Soviet space station operations – the only thing it did not attempt was a crewed handover. Instead, Salyut 6 was returned to automatic mode until the arrival of Soyuz 29 in June 1978. The new crew were Vladimir Kovalyonok and Alexander Ivanchenkov. During their marathon 140-day mission, they welcomed three Progress ferries and two visiting crews, with Polish and East German guest cosmonauts, and expanded the station's scientific programme into materials science. Two six-month tours then followed, in 1979 and 1980. By the time the final 75-day mission ended, with the return of Soyuz T-4 in May 1981, preparations were well underway for the next station. Before Salyut 6 was decomissioned in July 1982, it received one final visitor, the unmanned spacecraft Cosmos 1267, which was a test of Chelomei's TKS space tug, flying at last.

SALYUT 6
The latter generation of Salyut space stations incorporated the best features of the Almaz design – such as the environmental systems and gyrodynes (electrically powered stabilisers that oriented the station in orbit without wasting propellant) – together with successful elements from Salyut 4, most notably the improved power and navigation systems.

SALYUT 7 IN ORBIT
The layout of Salyut 7 surprised those who had expected the next station to have a modular design (like the later Mir). Nevertheless, the station proved that Salyut 6's four-year lifetime was no fluke and that stations could operate for long periods in orbit.

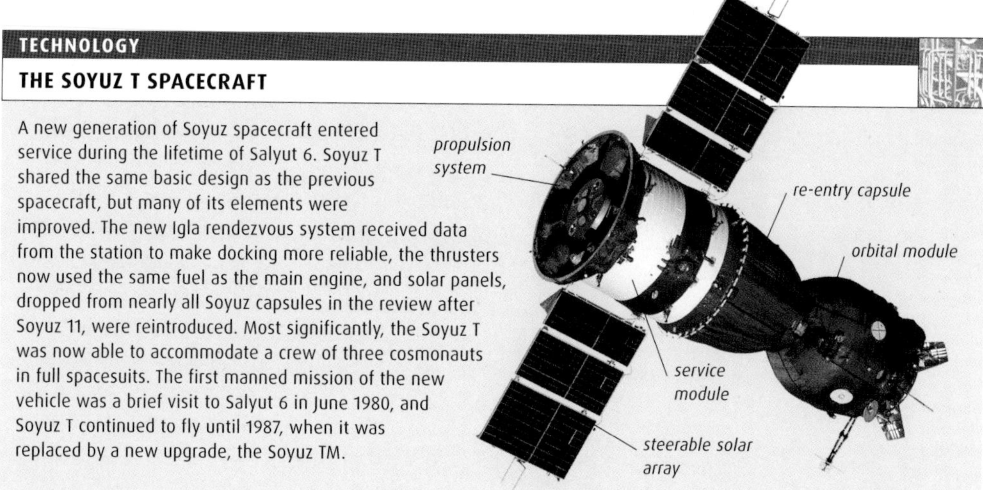

TECHNOLOGY

THE SOYUZ T SPACECRAFT

A new generation of Soyuz spacecraft entered service during the lifetime of Salyut 6. Soyuz T shared the same basic design as the previous spacecraft, but many of its elements were improved. The new Igla rendezvous system received data from the station to make docking more reliable, the thrusters now used the same fuel as the main engine, and solar panels, dropped from nearly all Soyuz capsules in the review after Soyuz 11, were reintroduced. Most significantly, the Soyuz T was now able to accommodate a crew of three cosmonauts in full spacesuits. The first manned mission of the new vehicle was a brief visit to Salyut 6 in June 1980, and Soyuz T continued to fly until 1987, when it was replaced by a new upgrade, the Soyuz TM.

The last Salyut

Launched in April 1982, Salyut 7 was a slightly upgraded version of its predecessor. The improved orbital life of the Soyuz T capsule (see panel, above) reduced the need for regular visits by guest crews to exchange the docked spacecraft, but visits were still needed from time to time, as missions grew longer.

The new station was launched in an almost empty state – once Anatoli Berezovoi and Valentin Lebedev were aboard, scientific equipment and consumables were supplied by Progress (and later TKS) ferries. The crew soon welcomed visitors from Soyuz T-6, including French cosmonaut Jean-Loup Chrétien. In August, the visit of Soyuz T-7 brought with it the second female cosmonaut, Svetlana Savitskaya. Berezovoi and Lebedev remained in orbit for 211 days, smashing previous records. The second long-stay crew, who arrived the following April, stayed for only five months, but a three-man mission in 1984 saw Leonid Kizim, Vladimir Solovyov, and Oleg Atkov extend the record to 236 days, and included a visit by Indian cosmonaut Rakesh Sharma.

Early in 1985, with the unmanned station in automatic mode, an electrical fault drained the batteries, and it seemed as if Salyut 7 was to be abandoned. But a rescue mission by Soyuz T-13 brought it back to life for two more long-duration missions – and this time the cosmonauts did achieve the orbital handover between crews that would be key to the next, and final, Soviet space station.

THE FIRST FEMALE SPACEWALKER
During a second visit to Salyut 7 in July 1984, cosmonaut Svetlana Savitskaya became the first woman to walk in space, during a three-hour, 35-minute EVA.

USA – NASA achievements

USA – Apollo-Soyuz

USA – Ten years of manned spaceflight

USA – Gemini 4

USA – Pioneer

USA – Echo 1

USA – Apollo 8

USSR – Vostok 1

USSR – Soyuz 4 and 5, first-day cover

USSR – Mars 1

USSR – Vostok 6

USSR – Gagarin 1964

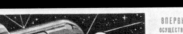

USSR – Voskhod 2

USSR – Vostok 5

USSR – Salyut 6

Rwanda

Bulgaria

Czechoslovakia

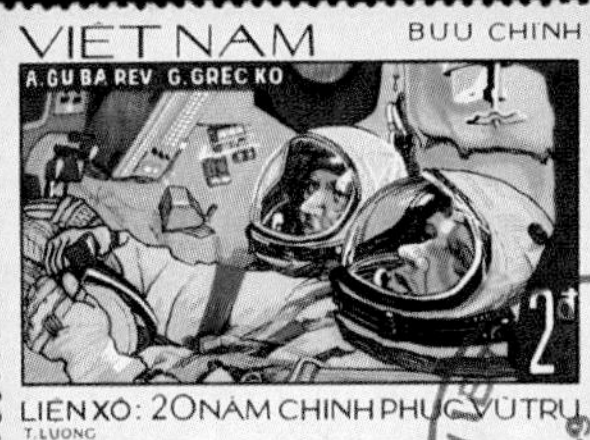

Vietnam

Cambodia

North Korea

Space Age stamps

Almost as soon as the Space Race began, its associated propaganda was put to use in ephemera of all kinds. The gallery of stamps shown here shows how space triumphs were used to shape the national image.

For all its failings, the former Soviet Union was an undoubted master of propaganda – within days of Sputnik 1's launch, the communist nation's triumph was commemorated on stamps issued not only in the USSR itself, but also across Eastern Bloc countries such as Romania and Czecholslovakia. The Soviet command economy was soon turning out other varieties of space-related ephemera, but the stamps are particularly evocative of the modernist dreams that accompanied the birth of spaceflight – representations of cosmonauts and spacecraft ranged from the self-consciously stylized and heroic to the photographic and futuristic.

While other communist countries toed the party line, they soon had their own space achievements to commemorate – as the Soviet Intercosmos programme put astronauts from countries as diverse as Vietnam, Mongolia, and Cuba into orbit.

NASA also found an eager supporter in the United States Postal Service, and stamps were issued to commemorate most US manned and unmanned space achievements. Many other nations also seized on the Space Race as a fitting subject for their stamps, and the decision over whether to depict US or Soviet missions frequently reflected the political complexion of their governments.

Mongolia

Mongolia

Nicaragua

Romania

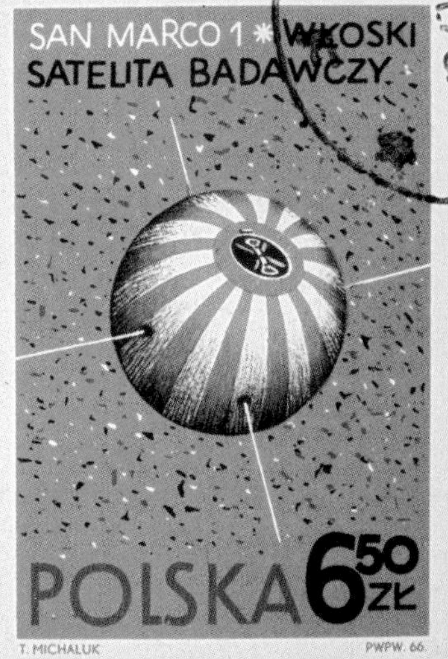

Hungary

MAY 1992: SPACE BUILDING
Astronauts Kathryn Thornton (foreground) and Thomas Akers rehearse assembly procedures for the projected, but never completed, Freedom Space Station in the open cargo bay of the Space Shuttle Endeavour*. In the background hangs the pale blue globe of the Earth.*

WORKING IN SPACE

FACED WITH FALLING BUDGETS in the wake of Apollo, NASA turned its attention to what it hoped would be cheaper methods of exploring space. The benefits of winged spacecraft had long been recognized, but such vehicles were sidelined in favour of the ballistic approach as the Space Race took hold. Now the idea was reborn in NASA's Integrated Space Transportation System – a Space Shuttle that would make routine trips to and from orbit, and a large space station that would be serviced by the new spacecraft. But there were insufficient funds for both projects, so the station was soon abandoned to concentrate on the Shuttle, which would now take on the role of orbiting laboratory as well as launch vehicle.

In the Soviet Union, meanwhile, space stations went from strength to strength, still supported by the reliable Soyuz spacecraft family. The last Soviet station, Mir, was a great leap forward with a complex, modular design. And political changes on Earth dictated that this was where the two nations would finally come together in space.

8 June 1959
NACA's X-15 hypersonic research aircraft makes its first flight.

10 December 1963
USAF cancels its X-20 Dyna-Soar programme.

12 July 1966
The first test of NASA's M2-F2 lifting body prototype is a success.

24 October 1968
The X-15 makes its 199th and final flight.

17 April 1969
The X-24A, prototype of a lifting body planned by the USAF for potential launch on a Titan rocket, makes its first test flight.

5 January 1972
President Nixon announces that NASA's next major space project will be the development of a winged reusable launch vehicle – the Space Shuttle. Shortly afterwards, USAF signs up as a partner in the Shuttle programme, and eventually abandons its research into rocket-launched lifting bodies.

Early spaceplanes

The idea of a winged space vehicle was considered as an alternative to the ballistic capsule approach many times before NASA finally decided to develop its reusable Space Shuttle system.

The basic idea of a winged, rocket-powered space vehicle that can use aerodynamic lift to travel at least part of the way into space was first proposed in the 1930s by Eugen Sänger, a member of the German VfR rocket society. His concept involved an aircraft launched at supersonic speeds on a rocket-powered sled (see p.300). Other ideas often used some combination of a ballistic launch and a plane-like descent, such as the British MUSTARD concept (see panel, below).

The true ancestors of the Space Shuttle, however, were the hypersonic X-craft. Close relations of the Bell X-1 (see p.34), these craft could fly at more than five times the speed of sound and were tested by NACA and NASA throughout the 1950s and 1960s. Several of the later designs had stubby wings and relied on the shape of the fuselage itself to generate most of their aerodynamic lift – they are often referred to as lifting bodies. Normally carried to high altitude while attached to a larger aircraft, they fired their rocket engines on release, accelerating rapidly and following trajectories that looped up to 28km (17 miles) above the Earth, before plummeting back in a hair-raising unpowered, supersonic glide. Although NASA did not pursue the lifting body concept at the time, it has proved highly influential in plans for the next generation of spaceplanes (see p.300).

Another type of X-craft was typified by the X-15, a rocket-powered research plane piloted by Neil Armstrong and Scott Crossfield (see panel, opposite) among others. This long, slender aircraft still relied on a lift to high altitude, but with larger wings it generated more lift as it fell back to Earth. Perhaps the closest parallel of all to the Space Shuttle was the X-20 Dyna-Soar – an abandoned USAF plan for a winged aircraft that would have been launched into orbit on top of a rocket.

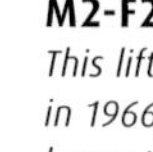

M2-F2
This lifting body first flew in 1966, dropped from beneath a B-52 bomber. It tested techniques for high-speed gliding that would eventually be used by the Space Shuttle.

SPIRAL
The Soviet Spiral was an orbiter to be launched from a hypersonic aircraft. It was abandoned in 1971.

Birth of the Space Shuttle

Throughout the early decades of space exploration, there were no reusable launch vehicles, and so each launch was extremely expensive. NASA's plans for a Space Shuttle, developed in the early 1970s under the administration of James C. Fletcher and announced by President Nixon on 5 January 1972, were supposed to drastically cut the costs of reaching orbit, and make routine spaceflight a reality.

With budget and political backing secured, NASA invited concepts from contractors and was deluged with a huge variety of concepts. The most popular early plan envisaged a two-stage vehicle, with the spaceplane carried most of the way to orbit by a large rocket-powered, piloted carrier aircraft – a larger equivalent of the X-15 system. When this concept was abandoned as too costly, NASA reluctantly recognized that not all of the system, now known formally as the Space Transportation System (STS), could be completely reusable.

ARMSTRONG THE PILOT
Neil Armstrong poses with the North American X-15 following a successful test flight. The hypersonic plane flew 199 flights between 1959 and 1968.

TECHNOLOGY

MUSTARD

Of the many great missed opportunities of the Space Age, the Multi-Unit Space Transport And Recovery Device (MUSTARD) is one of the most intriguing. First proposed by the British Aircraft Corporation in 1965 but abandoned a few years later, MUSTARD would have used three identical "stacked" lifting bodies, launched like a conventional rocket. The lower stages would separate and glide back to Earth at altitudes of 45–60km (28–38 miles), after pumping their remaining fuel into the orbiter stage. This would in theory allow the orbiter to reach space with full fuel tanks, potentially allowing it to continue to the Moon.

X-PLANE

The X-15 was 15m (50ft) long, with a wingspan of just 6.7m (22ft). Rear rockets powered it to a height of 108km (67 miles) and speeds of up to Mach 6.7.

BIOGRAPHY

SCOTT CROSSFIELD

Albert Scott Crossfield (1921–2006) was widely viewed as America's top test pilot – the man who often took experimental aircraft like the X-15 into the sky for the first time. After serving as a pilot and flight instructor during the Second World War, he joined NACA in 1950 but left to work for North American Aviation in 1955. In the late 1950s, he and several of his colleagues were briefly considered as pilots for a manned spaceflight.

TECHNOLOGY

SPACE SHUTTLE LAUNCH AND RETURN

The Shuttle system

The final version of the Space Transportation System (STS) was a compromise born of necessity. The US military's need for a sizeable orbiter to launch its largest payloads made it impossible to design a fully reusable system, but the Shuttle is at least largely reusable. In addition to the Shuttle orbiter itself (see pp.196–97), there are three other elements – a large External Tank (ET) and two side-mounted Solid Rocket Boosters (SRBs). Each of these plays an important role in getting the Shuttle orbiter into space.

HEIGHT (TO TOP OF ET)	56.14m (149ft 7in)
MASS AT LAUNCH	2,029,203kg (4,474,574lb)
LAUNCH PROPULSION	2 x SRBs, 3 x SSMEs
TOTAL THRUST AT LIFT-OFF	3.55 million kgf (7.8 million lbf)
PAYLOAD TO LEO	24,400kg (53,700lb)
MAIN CONTRACTORS	SRB: Thiokol; ET: Lockheed Martin; SSME: Rocketdyne

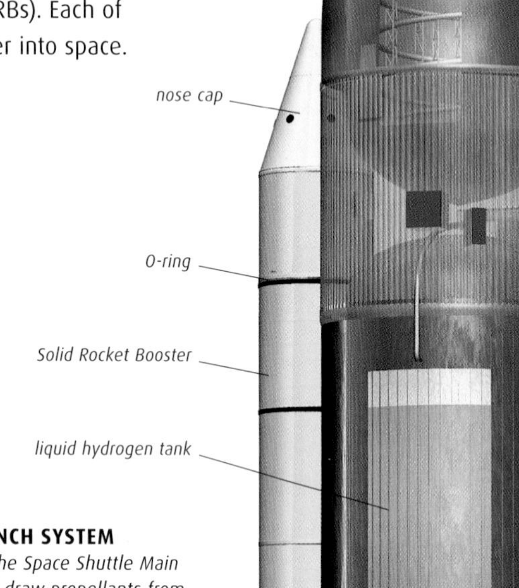

SHUTTLE LAUNCH SYSTEM

During launch, the Space Shuttle Main Engines (SSMEs) draw propellants from the External Tank, and two Solid Rocket Boosters help to get the Shuttle moving. The boosters fall away first, then as the Shuttle nears orbit, it also discards the External Tank.

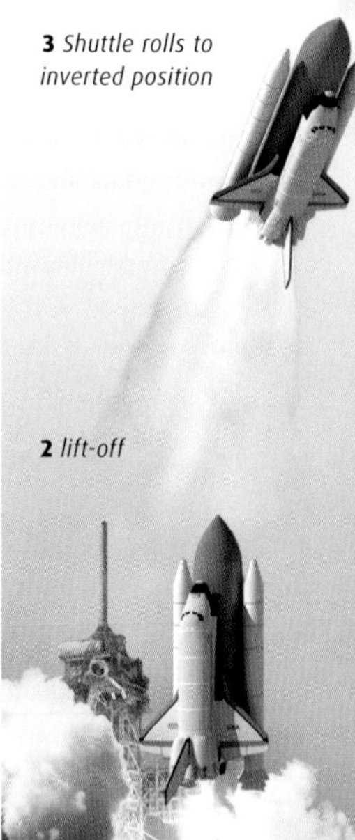

STACKING THE SHUTTLE

The Shuttle system is put together at Kennedy Space Center's Vehicle Assembly Building. The boosters and External Tank are mounted on top of the Mobile Launch Platform (see opposite). The orbiter arrives from its separate processing facility and is hoisted high into the air and lowered into place.

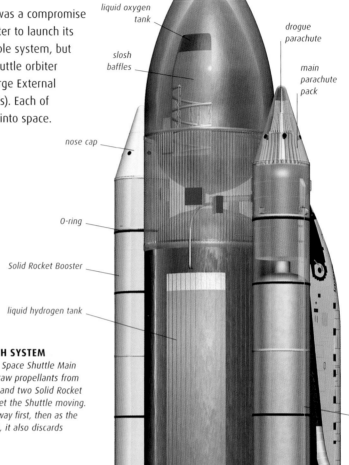

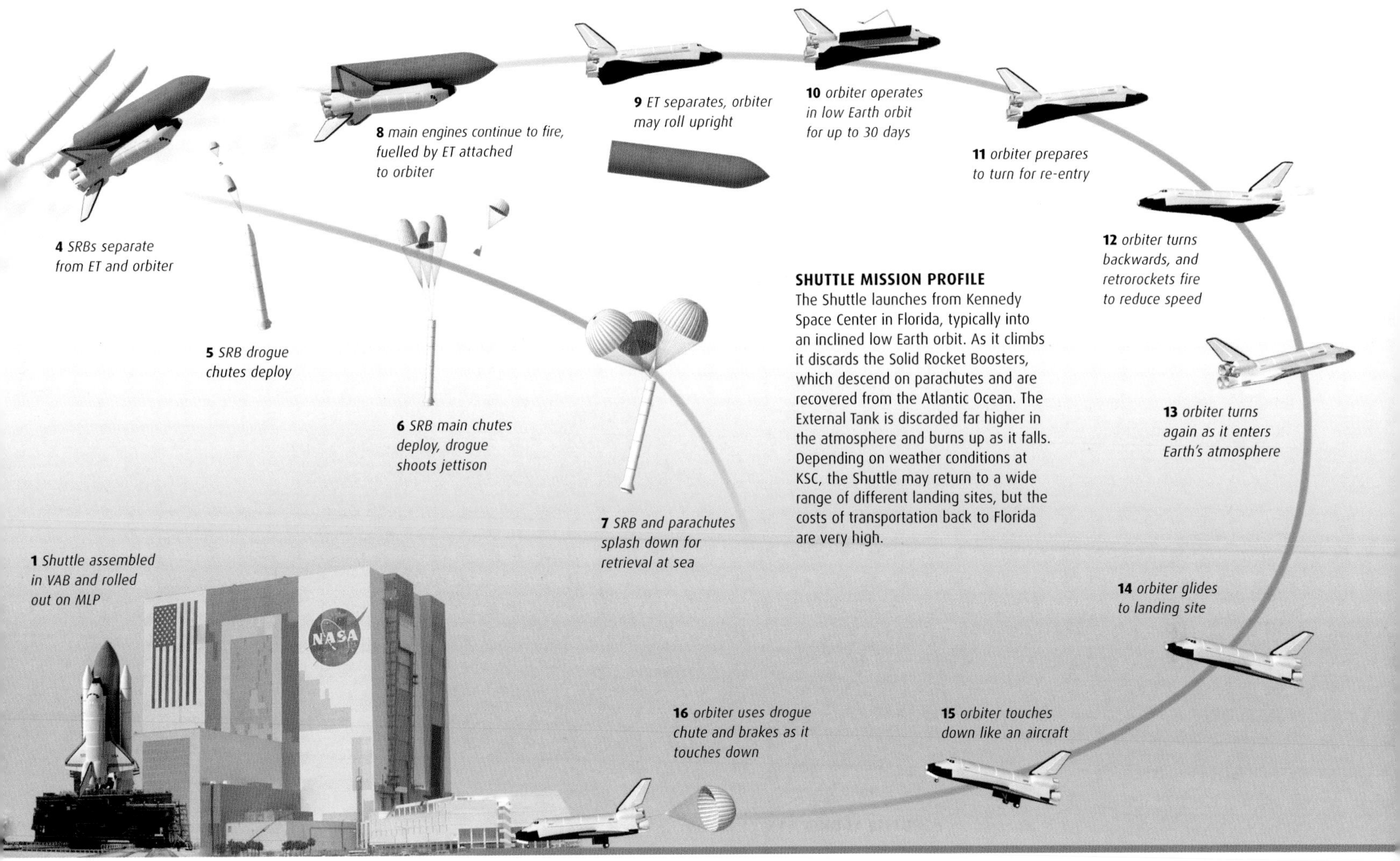

1 *Shuttle assembled in VAB and rolled out on MLP*

4 *SRBs separate from ET and orbiter*

5 *SRB drogue chutes deploy*

6 *SRB main chutes deploy, drogue shoots jettison*

7 *SRB and parachutes splash down for retrieval at sea*

8 *main engines continue to fire, fuelled by ET attached to orbiter*

9 *ET separates, orbiter may roll upright*

10 *orbiter operates in low Earth orbit for up to 30 days*

11 *orbiter prepares to turn for re-entry*

12 *orbiter turns backwards, and retrorockets fire to reduce speed*

13 *orbiter turns again as it enters Earth's atmosphere*

14 *orbiter glides to landing site*

15 *orbiter touches down like an aircraft*

16 *orbiter uses drogue chute and brakes as it touches down*

SHUTTLE MISSION PROFILE
The Shuttle launches from Kennedy Space Center in Florida, typically into an inclined low Earth orbit. As it climbs it discards the Solid Rocket Boosters, which descend on parachutes and are recovered from the Atlantic Ocean. The External Tank is discarded far higher in the atmosphere and burns up as it falls. Depending on weather conditions at KSC, the Shuttle may return to a wide range of different landing sites, but the costs of transportation back to Florida are very high.

ATLANTIS STACK ROLL-OUT
A spectacular view from the roof of the Vehicle Assembly Building shows the fully assembled Space Shuttle Atlantis *on top of its Mobile Launch Platform and ready to move to the launch pad.*

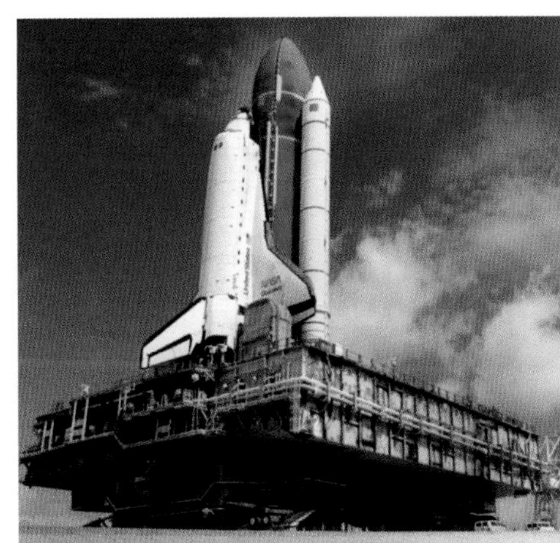

CRAWLER TRANSPORTER
The Shuttle is assembled on top of a Mobile Launch Platform (MLP), which is then picked up by a powerful mobile tractor unit for transport to the pad. The entire assembly weighs 8,170 tonnes (18 million lb) and moves at a top speed of 1.6kph (1mph).

FIERY LAUNCH
The pillar of flame on which the whole Shuttle assembly rises into the sky comes largely from the two Solid Rocket Boosters – the exhaust from the Space Shuttle Main Engines is much hotter, blue, and almost transparent.

Wings into orbit

Developing a spacecraft as complex as the Space Shuttle inevitably took longer than planned, but after a variety of tests, the Shuttle was finally ready for its maiden launch in 1981.

12 April 1981
Columbia launches from Cape Canaveral on its maiden flight.

14 April 1981
Columbia returns to Edwards Air Force Base, landing on a dry lake bed.

28 April 1981
The Shuttle orbiter returns to Cape Canaveral for processing and repairs.

12 November 1981
Columbia is launched on its second test flight.

14 November 1981
Columbia returns to Earth early, landing again on the lake bed at Edwards.

22 March 1982
Columbia is launched on its third test flight.

30 March 1982
The Shuttle touches down on a reserve strip at White Sands, New Mexico.

27 June 1982
Columbia launches on its final test flight, STS-4.

4 July 1982
Columbia returns to Edwards, landing for the first time on the runway.

Although NASA was given the go-ahead to build the Space Shuttle in the early 1970s, a host of issues needed to be solved before construction work could even begin, and optimistic hopes of a first flight in 1977 were soon being revised.

One significant cause of delay was the question of military use. In order to provide the predicted savings in launch costs, the Shuttle would have to fly regularly, with perhaps one mission every two weeks. The demand for such frequent flights would only be there if the US totally abandoned unmanned launches, including those for military payloads. For this reason, the Air Force agreed to contribute to development costs – but it extracted a heavy price in return. The entire spacecraft would have to be larger to carry military payloads. It would have to be capable of operating from the USAF's launch site at Vandenberg Air Force Base. And the USAF would be guaranteed a number of launch slots every year.

With all these issues taken into account, work on the prototype Shuttle orbiter, originally to be known as *Constitution*, got under way in June 1974.

BIOGRAPHY
JOHN YOUNG

Californian John W. Young (1930–2018) had one of the most varied careers of any astronaut. Joining NASA in 1962, he flew on Geminis 3 and 10 and Apollos 10 and 16. Taking over the running of the Astronaut Office from Deke Slayton in 1974, he was in charge of NASA's astronaut selection through to 1987 and commanded STS-1 and STS-9, the first Spacelab mission. He retired from NASA in 2004.

Ready for flight

The Shuttle's complexity made it difficult to test its elements separately. A dummy called *Pathfinder* was built at Marshall Spaceflight Center to test some support systems, scale wind-tunnel models were analyzed at Ames Research Center, and the prototype (now renamed *Enterprise*) was launched from a Boeing 747 in flight tests. But the nature of the orbiter meant there was no way to launch a boilerplate model into space. Similarly, while there were many test firings of the engines, they could not be flown alone. So the first launch of the Space Shuttle became the ultimate "all-up" test – after nine years of development and many billions of dollars of investment, the maiden flight of the fully functional orbiter *Columbia*, designated STS-1, on 12 April 1981, was one of those occasions where a single fault might cause a critical failure, even if everything else worked.

The crucial mission was to be helmed by experienced astronaut John Young (see panel, above), with rookie Bob Crippen at his side. To minimize the risk to their lives, they wore specially designed pressure suits modified from those

GLIDE TESTING
The Shuttle prototype Enterprise *flies free during a test flight in late 1977. Launched from the back of its 747 carrier,* Enterprise *was fitted with instruments to record its flight characteristics at subsonic speeds – there was no way to test the Shuttle's supersonic performance until a real launch.*

MAIDEN FLIGHT
Columbia *arrives for duty at the Kennedy Space Center, Cape Canaveral, piggybacked on a converted 747 carrier aircraft. The aerodynamic tailcone over the Shuttle's main engines reduces turbulence.*

LIFT-OFF!
With its three SSMEs and two SRBs firing simultaneously, Columbia *blasts free of Pad A at Launch Complex 39 on its maiden mission. The date, 12 April 1981, marked the 20th anniversary of Yuri Gagarin's first manned spaceflight.*

FIRST IMPRESSIONS
John Young (foreground) and Bob Crippen monitor controls on Columbia *during STS-1. On their return to Earth, an enthusiastic Crippen dubbed the Shuttle "a superb flying machine".*

worn by test pilots. As things turned out, the mission went entirely to plan, and the astronauts spent two days giving *Columbia* her orbital shakedown before returning to Earth at Edwards Air Force Base in California.

Despite its overall success, the maiden flight still encountered some problems – most significantly with the Shuttle's thermal tiles (see panel, below), and these were partly responsible for pushing back the date of the second test flight to 12 November 1981. Although STS-2's launch went smoothly and Joe Engle and Dick Truly were able to test the Remote Manipulator System (RMS) for the first time, the planned five-day mission was cut down to just three days due to a problem with *Columbia*'s power-generating fuel cells.

STS-3 launched as planned in March 1982. One of its aims was to study conditions around the Shuttle in space, and to do this the RMS hoisted an instrument pallet out of the cargo bay. The flight was planned to last seven days, but poor weather forced NASA to switch landing sites, and the Shuttle ultimately came back, a day late, to White Sands in New Mexico.

One final test flight was scheduled for June 1982, carrying a classified military cargo. Although a fault with the payload caused problems, the Shuttle itself once again performed well, and after it touched down at Edwards on 4 July, President Reagan announced that the Shuttle was now operational.

TECHNOLOGY

TILE DAMAGE

Throughout the early flights, damage to the Shuttle's fragile ceramic tiles was a recurrent issue. Problems with the delicacy of the tiles themselves, and the adhesive that fixed them to the hull, delayed *Columbia*'s maiden launch, and the Shuttle suffered from lost or damaged tiles on several occasions. Fortunately for the astronauts, their luck held – the damaged areas were not those that took the brunt of heat on re-entry.

1982

11 November 1982
Columbia launches on its fifth mission to deploy two commercial communications satellites.

4 April 1983
The second Space Shuttle, *Challenger*, enters service on a mission to launch the first Tracking and Data Relay Satellite.

18 June 1983
Challenger's second flight, sees the launch of two more communications satellites and also rehearses satellite retrieval.

30 August 1983
Challenger launches on her third mission, deploying an Indian communications satellite and carrying out more rehearsals for satellite retrieval operations.

28 November 1983
Columbia launches on the STS-9 mission, carrying the European-built Spacelab module.

3 February 1984
Challenger sets out on the eight-day STS-41B mission, which involves satellite deployments and tests of the MMU.

Early Shuttle missions

Following the fourth orbital test flight of *Columbia*, President Reagan declared the Space Shuttle operational on 4 July 1982. The Shuttle formally entered service with a variety of satellite-related missions.

The fifth flight of *Columbia*, STS-5 in November 1982, was the first to carry a commercial satellite onboard – a mission that was seen as vital to establishing the Shuttle as a commercial launch vehicle. In addition to the flight crew, STS-5 therefore became the first mission to carry "mission specialists" – highly trained non-pilot astronauts. The payload of two communications satellites deployed perfectly, and the only major problem on the mission was a spacesuit flaw that prevented the first spacewalk of the Shuttle programme.

Nearly five months later, *Challenger* took to the skies on her maiden flight, STS-6. This mission saw Story Musgrave and Donald Peterson test NASA's new spacesuits on tethered spacewalks, but its main aim was to launch TDRS-A, the first of the Shuttle's Tracking and Data Relay Satellites (see p.201). Deployment was flawless, but a problem in the Inertial Upper Stage (the powerful rocket intended to push the satellite into its final orbit), left it stranded. Controllers on the ground eventually found a way to edge the satellite towards its intended final orbit, but the time this took – coupled with delays to heavy satellite launches while the IUS problem was solved – put the TDRS programme behind schedule.

Altered priorities

The mission roster was hurriedly rejigged to work around the TDRS problems, and *Challenger's* STS-7 flight launched a pair of communications satellites in April. It also released the German-built Shuttle Pallet Applications Satellite (SPAS), a temporary satellite designed to float free of the Shuttle. Recapturing the SPAS at the end of the mission provided experience that would be useful when the Shuttle needed to retrieve larger satellites from orbit for servicing (see p.201). The same problems were addressed on the next mission, STS-8, where alongside the Indian satellite Insat-1B, *Challenger* carried a heavyweight dummy satellite to allow testing of the manipulator arm under stress.

While most of these flights carried small-scale experiments in the orbiter's mid-deck, the first full-scale science mission was Spacelab 1, carried aboard *Columbia* for the STS-9 launch in November 1983 (see p.198). A change in the numbering system meant that the tenth Shuttle mission was *Challenger's* STS-41B. Two satellites were successfully deployed, only for their Payload Assist Modules (smaller, lower-powered equivalents of the IUS) to malfunction. However, the Manned Maneuvering Unit (MMU) was successfully tested, permitting the first untethered spacewalk.

SATELLITE RELEASE
TDRS-A sits atop a pivoting turntable in the Shuttle cargo bay just prior to its release into orbit on Challenger's STS-6 flight. The problematic IUS rocket stage is shrouded in gold foil beneath the satellite.

FLYING FREE
Bruce McCandless somersaults above Challenger's cargo bay during a test of the MMU. In his hands he carries the Trunnion Pin Attachment Device (TPAD), a piece of equipment that, when attached to a satellite, allowed it to be grasped by the Shuttle's robot arm.

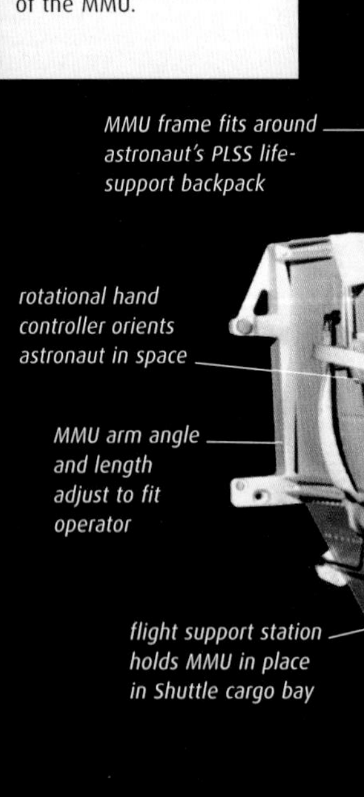

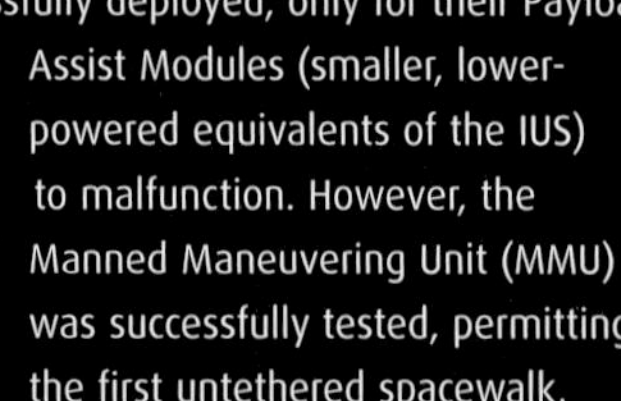

MANNED MANEUVERING UNIT
The MMU uses the controlled release of gaseous nitrogen propellant (GN2) to allow an astronaut to fly free in space. The pack contains two independent propulsion systems, each with its own GN2 tanks and set of four "triad" thrusters (each a set of three nozzles allowing the escaping gas to push the astronaut around three axes).

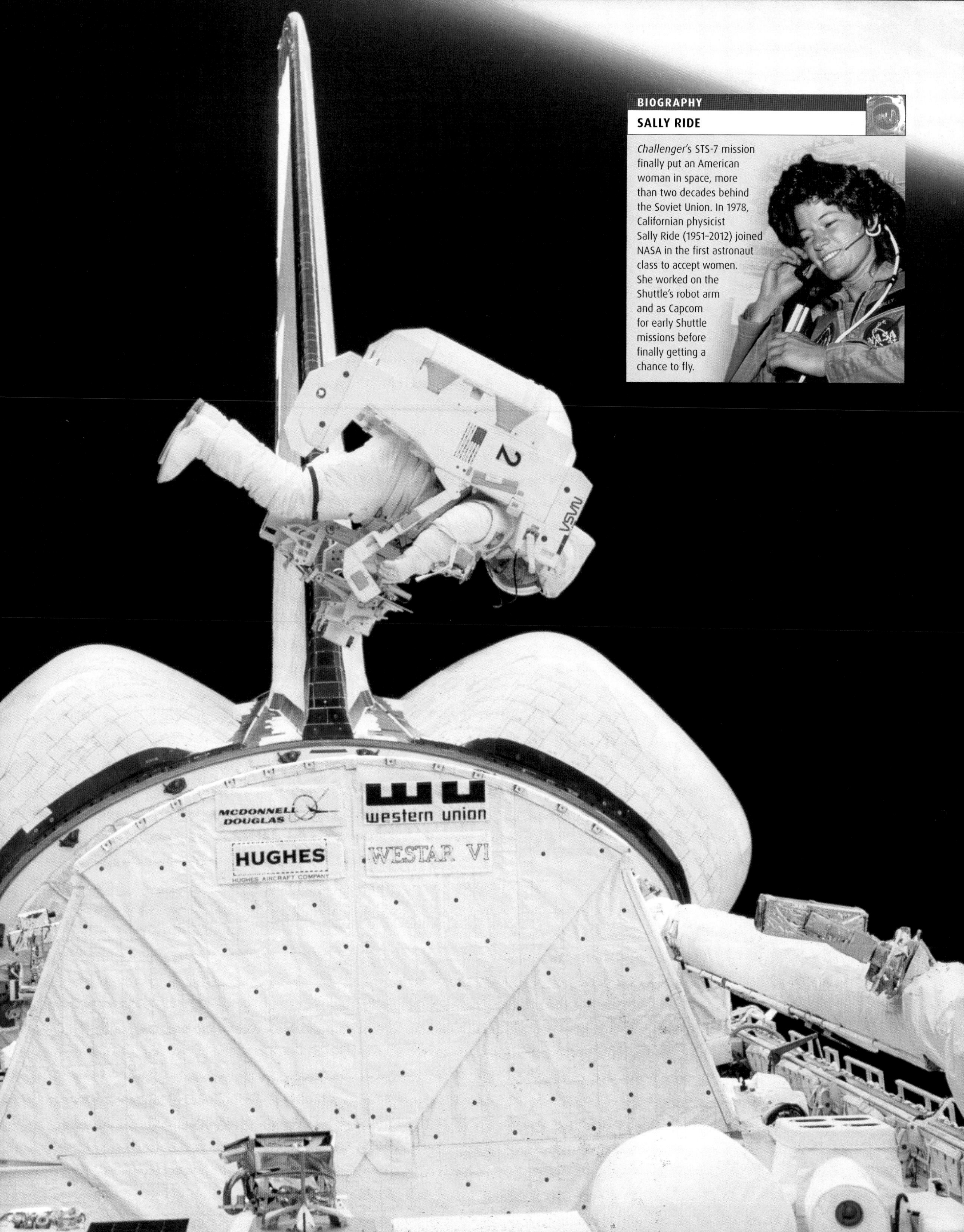

BIOGRAPHY

SALLY RIDE

Challenger's STS-7 mission finally put an American woman in space, more than two decades behind the Soviet Union. In 1978, Californian physicist Sally Ride (1951–2012) joined NASA in the first astronaut class to accept women. She worked on the Shuttle's robot arm and as Capcom for early Shuttle missions before finally getting a chance to fly.

Shuttle orbiter

The size of a short-haul jet airliner, the Space Shuttle orbiter is by far the largest spacecraft ever launched into orbit. Although its main engines are used only during launch, a complex system of secondary thrusters and manoeuvring engines gives the orbiter mobility and versatility in orbit. The payload bay can carry up to two satellites or the Spacelab laboratory into orbit, while the Remote Manipulator System is used for satellite deployment or retrieval and for orbital construction tasks.

CREW	A maximum of seven
LENGTH	37.24m (122ft 2½in)
HEIGHT	17.27m (56ft 8in)
WINGSPAN	23.79m (78ft ½in)
WEIGHT AT LAUNCH	99,318kg (218,958lb)
ENGINES (ORBITAL)	2 x OMS main engines (hydrazine/N204) 44 x RCS thrusters (hydrazine/N204)
OPERATING ALTITUDE	300–620km (185–385 miles)
MAIN CONTRACTOR	North American Rockwell

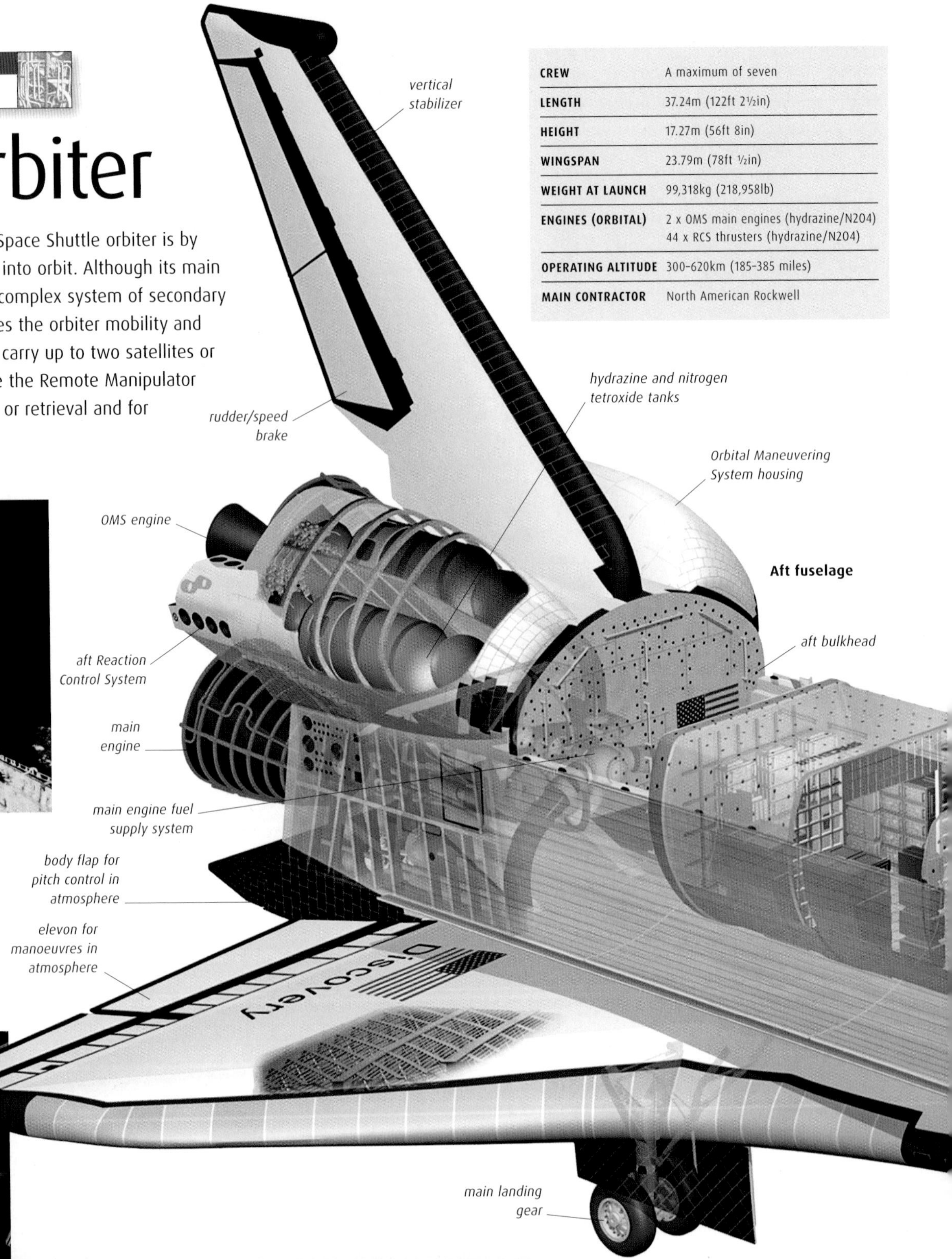

vertical stabilizer

rudder/speed brake

hydrazine and nitrogen tetroxide tanks

Orbital Maneuvering System housing

OMS engine

Aft fuselage

aft bulkhead

aft Reaction Control System

main engine

main engine fuel supply system

body flap for pitch control in atmosphere

elevon for manoeuvres in atmosphere

main landing gear

ORBITAL THRUSTERS

The two pods at the back of the Shuttle carry its Orbital Maneuvering System (OMS) engines – a pair of engines derived from the Apollo Service Module engine that allow the Shuttle to make orbital corrections and the retroburn for re-entry. They also contain thrusters for the Reaction Control System (RCS), which orients the Shuttle in space.

SHUTTLE ENGINE TEST

The Space Shuttle Main Engines are extremely complex and are fuelled by high-speed turbopumps that draw propellants from the External Tank. Temperatures in the combustion chamber can reach 3,300°C (6,000°F).

LANDING PROCEDURE

The Shuttle begins its descent by turning backwards in its orbit and using the OMS engines as retrorockets. It then turns again for re-entry, angling itself so that the heat-resistant ceramic tiles on the underside take the brunt of friction. Once it has slowed to merely supersonic speeds, it is effectively the world's heaviest glider, and gets only one chance at landing. Landing at high speed, it needs a very long runway for its parachute-assisted braking.

THE VIEW FROM ABOVE

The orbiter always opens its payload bay doors after reaching space – their interiors function as radiators that help regulate the spacecraft's temperature. In this photograph from August 2001, Discovery *is approaching the International Space Station, carrying the Italian-built Multi-Purpose Logistics Module – a cargo container for the ISS.*

ORBITER DOCKING SYSTEM

The Orbiter Docking System (ODS) was installed above the orbiter's external airlock in the mid-1990s, for use on Shuttle–Mir and ISS missions. Once the docking ring has made initial contact, an automatic docking sequence is started, which nudges the orbiter towards the other vehicle and extends latches to form an airtight seal.

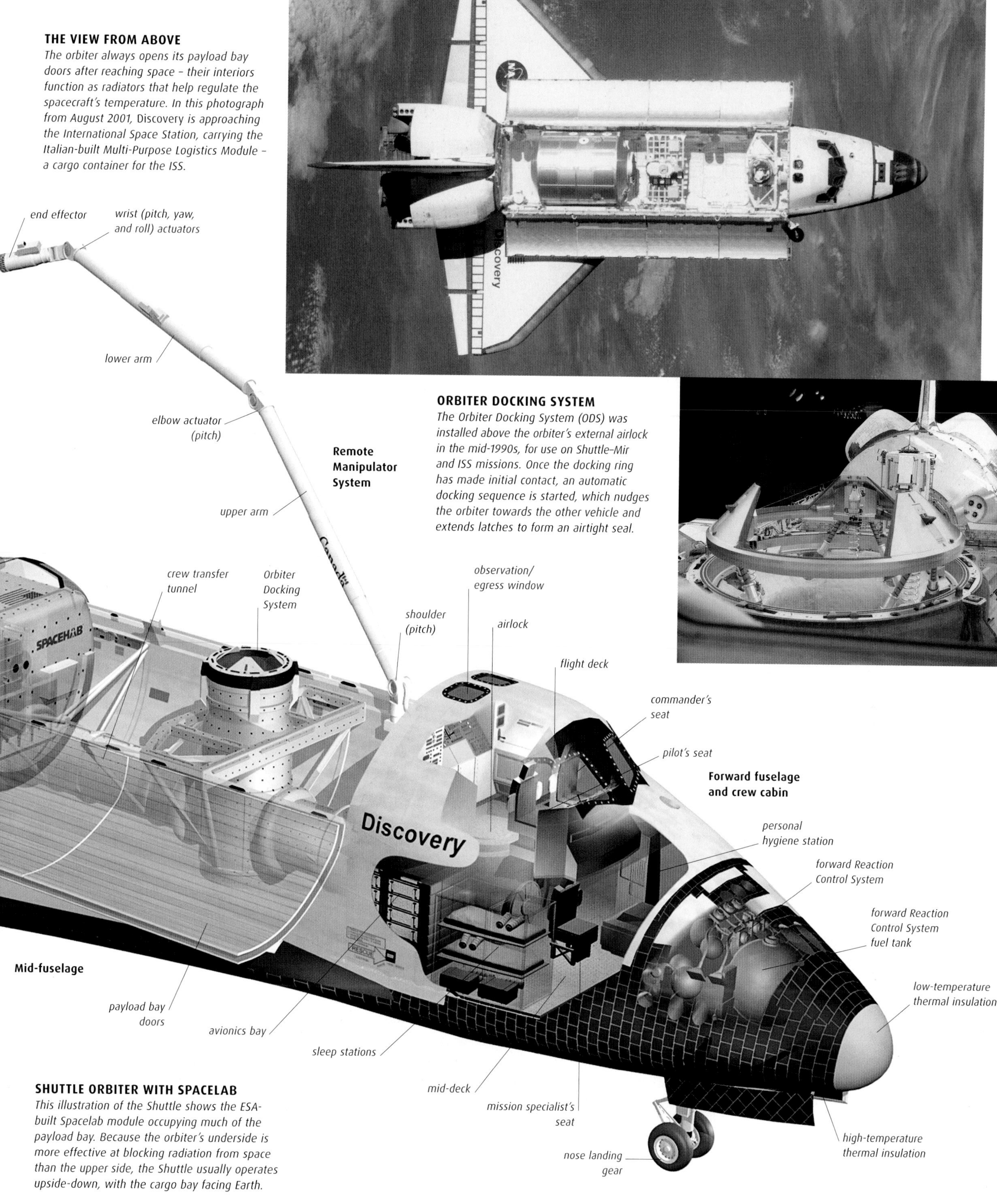

SHUTTLE ORBITER WITH SPACELAB

This illustration of the Shuttle shows the ESA-built Spacelab module occupying much of the payload bay. Because the orbiter's underside is more effective at blocking radiation from space than the upper side, the Shuttle usually operates upside-down, with the cargo bay facing Earth.

1983

Spacelab

28 November 1983
The first Spacelab mission, to test scientific techniques in orbit, begins.

29 April 1985
Spacelab 3 carries a variety of life sciences and microgravity experiments.

29 July 1985
Spacelab 2 carries instruments to study the physics of the Sun.

30 October 1985
Two German scientists and a Dutch ESA astronaut fly on the German-sponsored Spacelab D1.

2 December 1990
ASTRO-1 becomes the first astronomy-centred Spacelab mission.

12 September 1992
Spacelab J carries Japanese life-science and microgravity experiments.

27 June 1995
The Shuttle–Mir Spacelab studies the effect of long-duration spaceflight on the Mir astronauts.

1 July 1997
MSL-1R carries specialist materials processing equipment into orbit.

Built by the European Space Agency (ESA) to fit into the Shuttle cargo bay, the Spacelab laboratory module was first flown in 1983 and helped to address NASA's lack of a permanent space station.

NASA and the forerunner of the ESA, ESRO (see p.228), agreed to develop a laboratory module for the Shuttle's cargo bay in 1973. Europe would provide the module for free, and in exchange their astronauts would fly aboard the Shuttle as payload specialists. STS-9 was scheduled as the first of these international missions, but Spacelab would make heavy use of the TDRS satellite system in order to relay experimental data back to Earth, and so the schedule was put at risk by early problems with the satellites (see p.201). However, once TDRS-A had crawled into its proper orbit and been activated in autumn 1983, the mission was given the go-ahead – the political benefits of taking an international crew aboard the Shuttle outweighed the possible loss of data due to reliance on only one satellite.

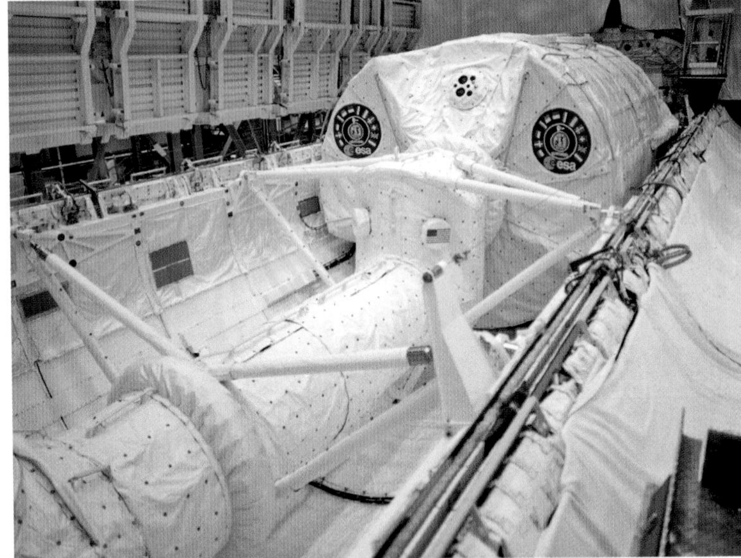

SPACELAB INSTALLED
Spacelab and its access tunnel are fitted in Columbia's cargo bay during preparations for the STS-9 mission.

Double success

In the end, Spacelab 1 was a huge success, both politically and scientifically. The laboratory carried a wide range of experiments investigating everything from physics and materials science to space biology and astronomy – its overall aim was to demonstrate the feasibility of carrying out such research in orbit. The six-man crew, including West Germany's Ulf Merbold (see panel, below) worked in groups of three for 12-hour shifts, sending a steady stream of data back to Earth. By the end of the ten-day mission, more scientific data had been returned to Earth than during the entire Skylab programme.

Spacelab was built for use in various configurations – as well as the pressurized lab, it had a system of external pallets to carry telescopes and experiments requiring exposure to vacuum. Spacelab modules were flown on 15 more Shuttle missions through to 1998 (NASA had been so impressed with the original European version that it paid ESA to build a second one). Pallets were flown on a further nine missions, up to *Endeavour*'s STS-99 in 2000, which carried a radar that mapped the elevation of 80 per cent of the Earth's surface.

Most later Spacelab missions concentrated on specific areas of science, such as astronomy, life sciences, medicine, geophysics, and materials science. However, science aboard the Shuttle was not confined to Spacelab missions – most flights carried smaller experiments on the orbiter's mid-deck, and other scientific instruments have been flown on the Shuttle outside of the Spacelab programme.

TEMPORARY SATELLITE
Part of the Spacelab system, the Long-Duration Exposure Facility (LDEF) was released into orbit in 1984 and retrieved in 1990.

BIOGRAPHY

ULF MERBOLD

Germany's Ulf Merbold (b.1941) was the first non-American to fly aboard the Space Shuttle, joining the first Spacelab mission in 1983. A physicist by training, he was selected by ESA as a potential payload specialist for the mission in 1977. He also flew on the STS-42 International Microgravity Laboratory mission in 1992, and became the first ESA astronaut to work on the Mir space station, during the 32-day "Euromir" mission of 1994.

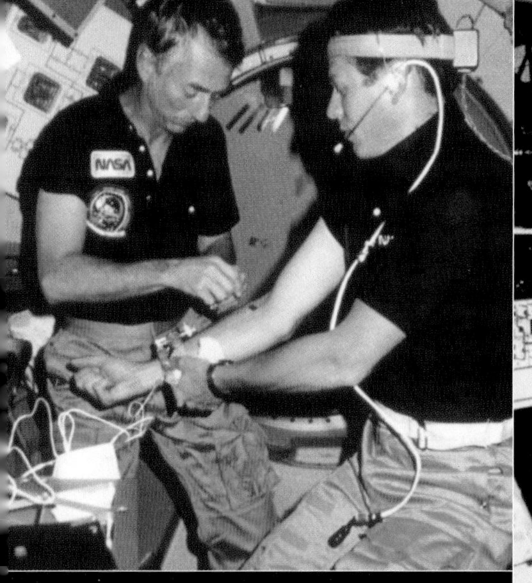

BLOOD TESTING
Owen Garriott takes a blood sample from Byron Lichtenberg during the first Spacelab mission. Blood analysis during and after spaceflight reveals that weightlessness affects the production of red blood cells.

MULTITASKING
(Above) Robert Parker (on the left) and Ulf Merbold were able to work on experiments while "wired up" to biometric sensors on Spacelab 1. An ESA experiment, the fluid physics module, can be seen on the left.

LIVE AND DIRECT
(Left) The TDRS satellite network allows huge amounts of experimental data to be sent back from the Shuttle, including near-continuous live television. Here Gregory Linteris (foreground) and Donald Thomas are seen during the MSL-1R Spacelab of 1997.

MICROGRAVITY SCIENCE
(Above) Taylor Wang displays part of a "drop dynamics" experiment aboard Spacelab 3. The weightless environment of the Shuttle allows physicists to see how materials behave away from the influence of gravity.

CAPTURING INTELSAT VI

Retrieving a giant Intelsat comsat during STS-49 required the first ever three-person spacewalk, by (left to right) astronauts Richard Hieb, Tom Akers, and Pierre Thuot. Beyond Akers lies the new motor that would finally boost the satellite to its correct orbit.

Satellite servicing

The arrival of the Space Shuttle made it possible for the first time to retrieve valuable satellites – either repairing them in orbit or returning them to Earth in the cargo bay for more complex refurbishment.

The most eyecatching of the Shuttle's early flights came in April 1984, when *Challenger* launched into an unusually high orbit at the start of the STS-41C mission. After deploying the LDEF satellite (see p.198), the Shuttle's main goal was to rendezvous with a crippled satellite 556km (345 miles) above the Earth and restore it to working order.

The target was the Solar Maximum Mission satellite (SolarMax for short), launched in 1980 to study activity on the Sun but unable to target its instruments thanks to a failure in its control systems. A first attempt to grapple with the satellite, using an MMU piloted by George D. Nelson, failed, and a second try at grabbing SolarMax with the Shuttle's manipulator arm simply sent it into a chaotic spin. The following day, things went better, and the satellite was safely brought into the cargo bay, where it was anchored to a special repair platform. The repairs themselves were lengthy but successful, upgrading the satellite's scientific payload as well as replacing its attitude control system. After two spacewalks, SolarMax was released back into orbit, where it continued to function until it burned up on re-entry to the Earth's atmosphere in 1989.

Chasing satellites

The next repair mission was a case of tidying up unfinished business – rescuing the pair of mis-deployed satellites released during the STS-41B mission in February 1984 (see p.194). Both had fallen far short of their intended orbits after their Payload Assist Modules (PAMs) had failed. After a brief pause in commercial launches while the problem was sorted out, the maiden flight of *Discovery* in August had flawlessly deployed another pair of communications satellites. With the PAM problem apparently resolved, *Discovery*'s second mission, STS-51A, in November 1984, was to retrieve the stranded satellites for a refit on the ground. A difficult mission, requiring the chase-down and capture of two separate satelllites, was made harder still by the lack of grips on the satellites – unlike SolarMax, they had not been built with in-flight servicing in mind.

The STS-51D mission offered a great demonstration of the Shuttle's flexibility. In April 1985, *Discovery* released a US Navy Leasat communications satellite from the cargo bay, only to see it spin away without power as a switch intended to trigger its built-in kick motor failed to activate. Rather than abandon the satellite, the mission was extended – the satellite was chased down, and a spacewalk was improvised in which the astronauts attempted to trigger the switch manually. In the end, the switch failed to wake the satellite but the astronauts' assessment paved the way for repair by *Discovery* a few months later.

Following the hiatus caused by the *Challenger* disaster in January 1986 (see p.202) it was some time before NASA attempted any further satellite repairs (see p.207). When it came, however, it was truly spectacular, as *Endeavour* caught up with and retrieved the enormous Intelsat VI communications satellite. Experience in handling this giant in space paved the way for the even more ambitious repair missions that would be needed to help maintain the Hubble Space Telescope (see pp.252–53).

TECHNOLOGY

THE TDRS SATELLITE SYSTEM

Situated 35,786km (22,240 miles) above the Earth's equator, NASA's Tracking and Data Relay Satellites orbit the Earth once each day, acting as relay stations for data from faster-orbiting satellites and spacecraft closer to Earth. Seen from Earth, each TDRS remains stationary in the sky – a fixed platform through which data can be sent and received from satellites at lower altitude. Moving a small antenna on a vehicle such as the Space Shuttle in order to track each TDRS is far easier than tracking the Shuttle from the ground as it speeds across the sky.

helmet with Sun visor down

helmet lights

PLSS life-support unit

cuff checklist

Hard Upper Torso (HUT)

two-part glove assembly

Lower Torso Assembly (LTA) including legs

harness attachment

integrated boots

EXTRAVEHICULAR MOBILITY UNIT

The Extravehicular Mobility Unit (EMU) spacesuits of the Shuttle era are very different from those used on Apollo. Designed for work in weightless conditions, they have a hard upper torso unit and an integrated PLSS backpack.

3 February 1984
Two communications satellites deployed from *Challenger* fail to reach their intended orbit after a rocket motor failure.

8 April 1984
Challenger captures and repairs the SolarMax satellite – the first ever satellite maintenance in orbit.

13–15 November 1984
Discovery retrieves the Westar-VI and Palapa B-2 satelllites, deployed by *Challenger* in February, for a refit back on Earth.

16 April 1985
Discovery chases down and attempts to repair the faulty Leasat-3 communications satellite launched earlier in this STS-51D mission.

31 August 1985
Discovery's STS-51I mission successfully recaptures and "hotwires" the Leasat-3 satellite.

13 May 1992
Endeavour captures and repairs the 4.2-tonnne Intelsat VI satellite as part of the STS-49 mission.

2 December 1993
Endeavour is launched on the first of several servicing missions to the Hubble Space Telescope.

FATAL LIFTOFF
Cameras at pad 39-B photographed Challenger *on its way to catastrophe. Residue from the SRB fuel formed a temporary seal over the cracked O-ring, delaying the disaster.*

> "Reality must take precedence over public relations, for **nature cannot be fooled**."
>
> **Richard Feynman's appendix to the Rogers Report, 1987**

BIOGRAPHY

CHRISTA MCAULIFFE

NASA set up its Teacher in Space project in 1984, in an attempt to reignite public interest in space. Out of 11,500 applicants, Christa McAuliffe (1948–86), a Social Studies teacher at Concord High School, New Hampshire, was chosen for her inspirational teaching style. She began training in autumn 1985 and was welcomed to the team of the doomed STS-51L, dying with her six crewmates on 28 January 1986. In 2004, NASA began a new "educator astronaut" programme. McAuliffe's original backup, Barbara Morgan, flew aboard STS-118 in August 2007.

The *Challenger* disaster

By the end of 1985, the Space Shuttle programme finally seemed to be getting into gear, with shorter intervals between flights and faster turnaround times for individual spacecraft – but then disaster struck.

BLOWTORCH EFFECT
Fifty-eight seconds after launch, ground-based cameras capture a superheated jet of flame emerging from a joint on Challenger's *right SRB, burning through a vulnerable support strut.*

The launch of *Challenger*'s STS-51L mission on 28 January 1986 was the focus of unusual attention. It was the 25th flight, a minor landmark in itself, and was also carrying a notable passenger – Christa McAuliffe, from New Hampshire, had been selected to become the first teacher in space and would be delivering lessons over a live television link from the Shuttle to schools across the US.

But just 73 seconds after launch, both NASA and the general public were left reeling as *Challenger* exploded in mid-flight, instantly killing all seven astronauts onboard, showering debris across the Atlantic Ocean, and fatally undermining America's dreams of routine manned spaceflight.

BIOGRAPHY

RICHARD FEYNMAN

Physicist Richard P. Feynman (1918–88), best known for his work on fundamental atomic forces, sat on the Rogers Commission into the *Challenger* disaster. He famously revealed the vulnerability of the SRB O-rings by dipping a sample of material into a jug of iced water, then snapping it with his hands. Feynman's conclusions about the Shuttle's safety were even more damning than those of the official report, and were added as an appendix – his estimate of one failure in roughly every 50 missions was to prove tragically accurate.

Accident investigation

Future Shuttle flights were immediately grounded while an investigation got under way. Television pictures soon revealed the direct cause of the accident – a jet of scalding flame emerging from the side of one of the SRBs and burning through one of the struts that held it to the rest of the launch assembly. As the strut gave way, the SRB had swung round, slamming into the external tank and rupturing it in a huge explosion that completely destroyed the spacecraft, instantly killing the crew.

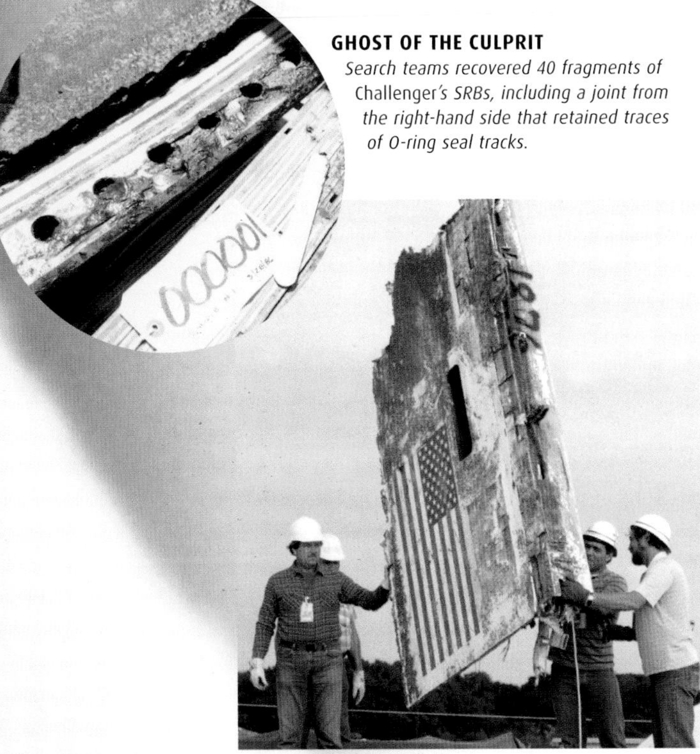

GHOST OF THE CULPRIT
Search teams recovered 40 fragments of Challenger's *SRBs, including a joint from the right-hand side that retained traces of O-ring seal tracks.*

ENTOMBMENT
Following the investigation, the recovered Challenger *wreckage was buried in empty missile silos on Canaveral Air Force Station.*

Tracing the accident back to its origins, however, took longer. President Reagan appointed an investigation board, which included Neil Armstrong and Richard Feynman (see panel, above). They tracked the jet of flame back to a crack in one of the joints between sections of the right SRB – the night before launch, frost had formed on the Shuttle, and a rubber O-ring seal had become brittle and unable to fill the joint properly under the stress of launch.

However, the investigation did not stop there – it comprehensively tore apart NASA's managerial structure, attacking a culture of complacency in which the concerns of several engineers about the launch had become lost in bureaucracy and never reached the ears of key safety managers.

Although there was probably nothing that could have saved *Challenger*'s crew in this exact chain of events, the board also recommended extensive changes to Shuttle hardware that would keep the spaceplane grounded for a total of 32 months. *Challenger* would be replaced – NASA had enough "spare parts" to put together a new Shuttle, *Endeavour*. But a more conservative launch policy would result in far longer turnaround times, putting an end to the optimistic estimates of launch rates on which the entire Shuttle concept was founded.

22 January 1986
STS-51L is scheduled for an afternoon launch but is delayed.

24 January 1986
More delays are caused by poor weather at an emergency landing site in Senegal. The abort site is moved to Casablanca, but this runway cannot handle a night landing, so the launch is moved to local morning. Later, poor weather forecasts at Cape Canaveral put the launch back still further.

27 January 1986
High winds force another delay. Forecasts of cold weather for next morning prompt engineers at SRB maker Morton Thiokol to call a teleconference with NASA managers regarding a possible problem with the O-rings. In the end, Thiokol's own management overrule the engineers and recommend that the launch should proceed.

28 January 1986
Challenger launches at 11:38 EST.

1956 1957 1958 1959 1960 1961 1962 1963 1964 1965 1966 1967 1968 1969 1970 1971 1972 1973 1974 1975 1976 1977 1978 1979 1980 1981 1982 1983 1984 1985 1986 1987 1988 1989 1990 1991 1992 1993 1994 1995 1996 1997 1998 1999 2000 2001 2002 2003 2004 2005 2006 2007 2008 2009 2010 2011 2012 2013 2014 2015 2016 2017 2018 2019 2020

EXPERIENCE

THE CHALLENGER DISASTER

"Go at throttle up"

For those on the ground or watching the televised launch coverage, the explosion that destroyed *Challenger* came without warning. But the decision to launch had hung in the balance almost until the last moment.

THE CREW
Photographed following a launch rehearsal on 8 January, Challenger's crew (left to right: McAuliffe, Jarvis, Resnik, Scobee, McNair, Smith, and Onizuka) pose in the White Room next to the Shuttle hatch.

On the morning of 28 January 1986, the crew of *Challenger* prepared for the launch from Cape Canaveral. Out on Pad 39B, the ice team had been working through the night to tackle frost forming on the Fixed Service Structure supporting *Challenger*. After a week of delays, there was an eagerness to get under way, although Rockwell, the Shuttle's main contractor, was now expressing concerns about launch safety. A hardware failure led to a two-hour delay, during which Program Manager Arnold Aldrich agreed to a further ice inspection at 10:30, but the engineers' worries about the O-rings had not reached him, and when the ice appeared to be melting, he gave the go-ahead. The Shuttle launched at 11:38 EST, but a minute later disaster struck:

SHOCK IN THE CROWD
At first, spectators watching from the Launch Complex 39 Observation Gantry were unsure of the explosion's meaning – those who had never seen a Shuttle launch before had been told that SRB separation could be spectacular, and it took several moments to realize something had gone badly wrong.

Dick Scobee: Point nine.
Michael Smith: There's Mach one.
DS: Going through nineteen thousand. Okay we're throttling down. Throttling up.
MS: Throttle up.
DS: Roger.
MS: Feel that mother go.
DS: Woooohoooo.
MS: Thirty-five thousand going through one point five.
DS: Reading four eighty six on mine.
MS: Yep, that's what I've got, too.
DS: Roger, go at throttle up.
MS: Uhoh [The line then goes dead].

COUNTDOWN TO DISASTER
The Challenger crew had spent four months in training for their STS 51-L mission, but on the morning of 28 January, long icicles had formed on the Shuttle's service structure. Debate about the effects of the ice continued even as the crew made their way to the Shuttle, and the final decision to go ahead came just 20 minutes before launch.

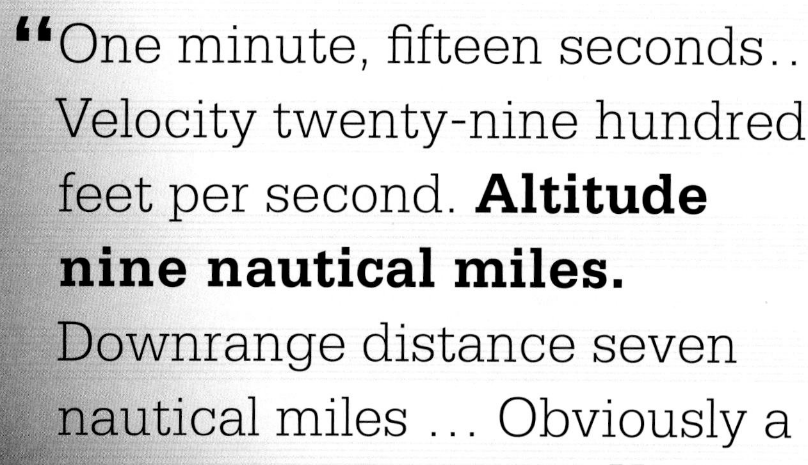

"One minute, fifteen seconds... Velocity twenty-nine hundred feet per second. **Altitude nine nautical miles.** Downrange distance seven nautical miles ... Obviously a **major malfunction."**

NASA launch commentary, 28 January 1986

From the ground, no one saw the puffs of smoke that emerged from an O-ring seal on the right-hand SRB. Within two and a half seconds of ignition, they had disappeared anyway, as the expanding SRB casing temporarily closed the gap formed by the cracked O-ring. It was not until 59 seconds into the flight that flame suddenly spurted from the joint, growing rapidly into a white-hot blowtorch that was blown backwards onto the central tank and the SRB's own support strut. At 72 seconds, the strut gave way, and a second later, the external tank ruptured, its exposed propellants exploding in a blast that tore the *Challenger* orbiter to pieces, killing all on board.

STATE OF MOURNING
President Ronald Reagan had been due to give his State of the Union address to the nation on the night of 28 January. Instead, he delivered a moving address to a shocked and grieving nation.

"We mourn seven heroes: **Michael Smith, Dick Scobee, Judith Resnik, Ronald McNair, Ellison Onizuka, Gregory Jarvis, and Christa McAuliffe** ... The crew of the Space Shuttle *Challenger* honoured us by the manner in which they lived their lives. **We will never forget them** nor the last time we saw them this morning, as they prepared for their journey and waved goodbye and 'slipped the surly bonds of Earth' to 'touch the face of God'."

President Ronald Reagan, 28 January 1986

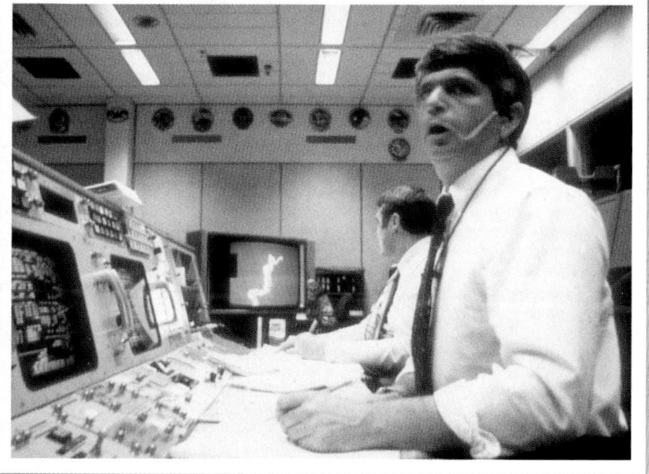

MISSION CONTROL
Staff at Johnson Space Center in Houston watch in stunned disbelief as their communications and telemetry disappear and the television monitors reveal the awful reason why (main picture).

RETRIEVAL OF LDEF
During the STS-32 mission of January 1990, Columbia *retrieved an experiment called the Long Duration Exposure Facility, which had been released by* Challenger *back in 1984.*

After *Challenger*

The Space Shuttle's return-to-flight programme began falteringly, but within a couple of years, the Shuttle and its crews were accomplishing some of its most ambitious missions yet.

When *Discovery* returned to space in late September 1988, the eyes of the world were watching what would otherwise have been a relatively mundane mission. As well as testing the modified hardware, NASA was using the flight to finally continue its deployment of the TDRS data satellite network. TDRS-C would supplement the earlier TDRS-A, which was now beginning to fail. Over the next two years, the Shuttle would launch a further three satellites, improving and reinforcing its orbital communications.

Many of the other Shuttle missions in this period were launches of classified, defence-related satellites. With the Shuttle grounded for so long, the US military had been without a heavy-lift launch vehicle (the USAF's own Titan IV was not yet ready), and they did not hesitate to demand their share of launch slots. Aside from a couple of commercial satellite launches, other significant missions of the time included much-delayed spaceprobe launches (of the Galileo, Magellan, and Ulysses probes) and deployment of the Hubble Space Telescope (HST), delayed since 1986 and finally launched in April 1990. Spacelab missions resumed in 1990 with the ASTRO-1 astronomy mission. Throughout 1991, the rate of missions gradually ramped up again, and some limited spacewalking took place in the cargo bay.

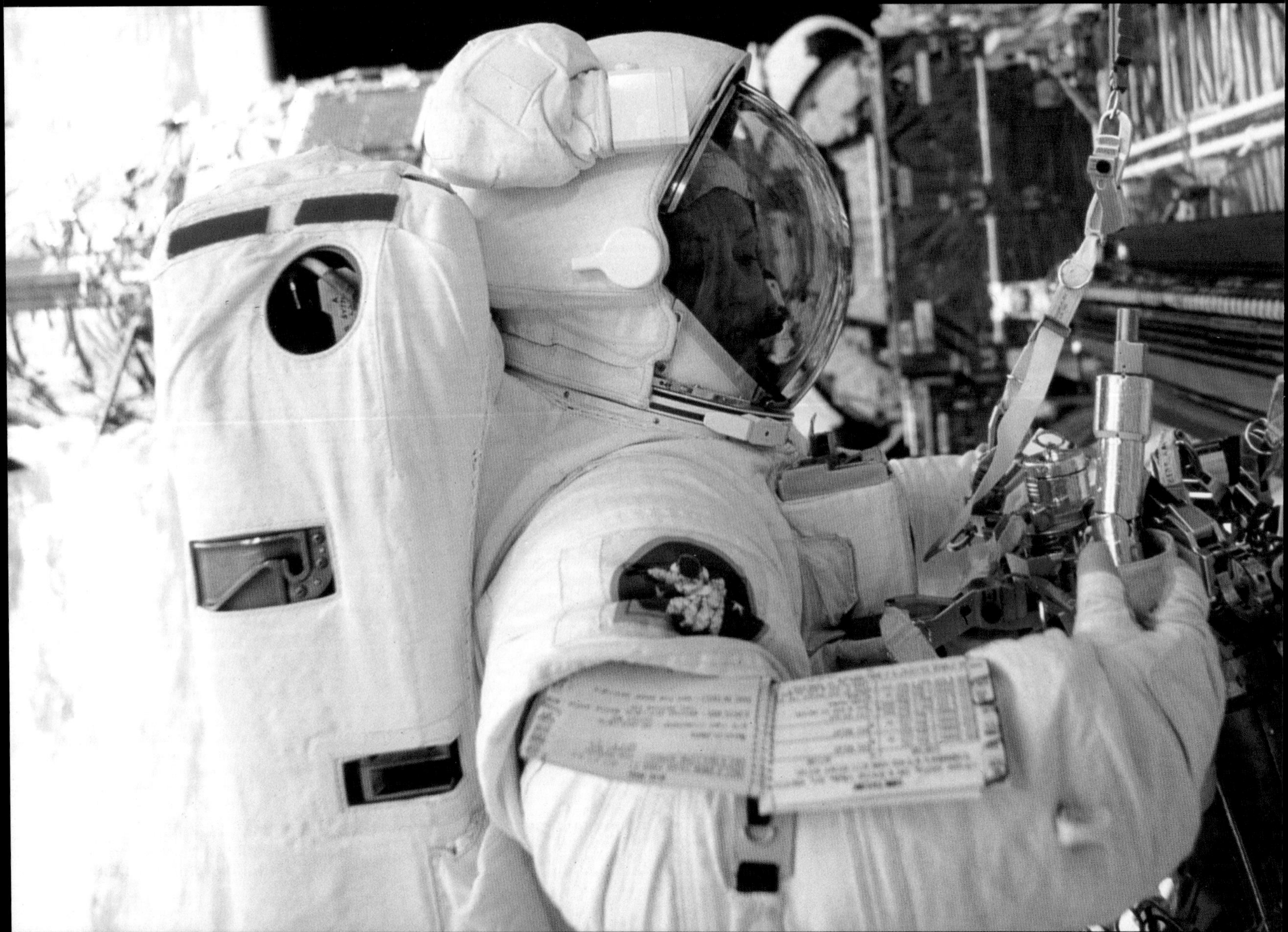

DEPLOYING GALILEO
*The post-*Challenger *hiatus meant that when this Jupiter probe was finally deployed, it had missed its best launch window and had to take a circuitous route to the giant planet.*

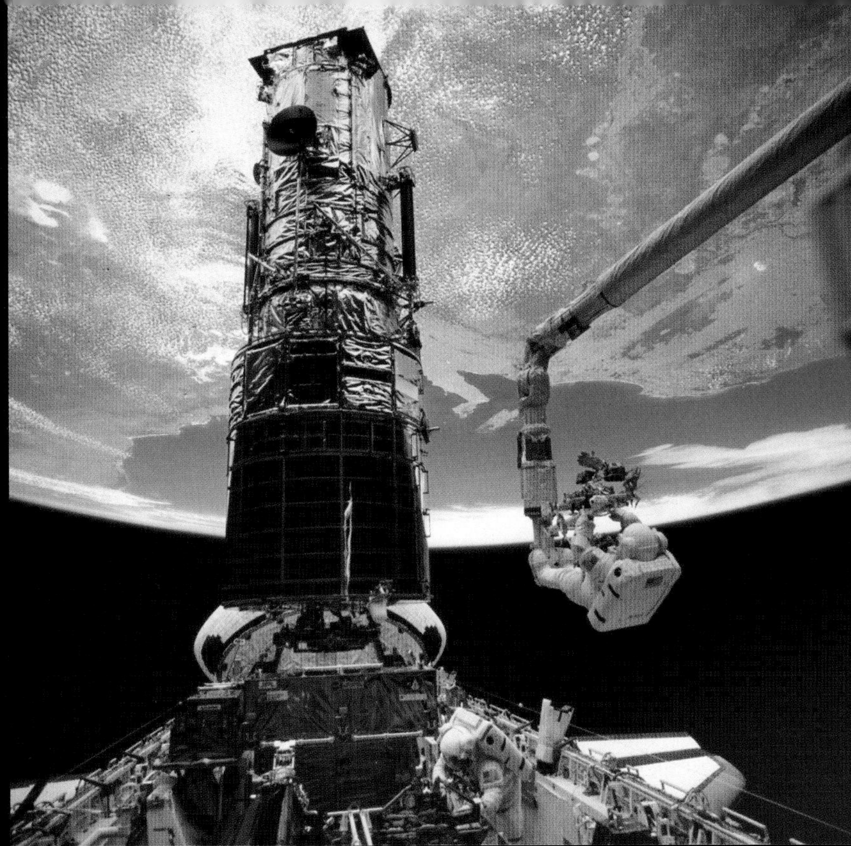

HITCHING A LIFT
Astronaut Story Musgrave rides to the top of the HST aboard Endeavour*'s robot arm, preparing to fit covers to some instruments during the last of STS-61's five spacewalks. Jeffrey Hoffman works below him in the payload bay.*

29 September 1988
Discovery launches on the first Space Shuttle mission since 1986. A decision to simplify the mission codes sees *Discovery*'s flight named STS-26. However, schedule delays will frequently lead to later missions launching out of order.

4 May 1989
Atlantis deploys the Venus-bound Magellan spaceprobe.

18 October 1989
Atlantis launches the Galileo probe to Jupiter.

25 April 1990
Discovery deploys the Hubble Space Telescope.

6 October 1990
The Ulysses solar probe is launched from *Discovery*.

21 June 1993
Endeavour's STS-57 is the first mission to carry the Spacehab experimental module.

2 December 1993
Endeavour launches on its epic STS-61 mission to repair the HST.

3 February 1995
The launch of *Discovery* on STS-63 marks the beginning of the Shuttle–Mir programme.

Endeavour debuts

Challenger's replacement, *Endeavour*, finally took to the skies in May 1992 with the STS-49 mission. Its aim was the most ambitious satellite repair yet attempted – servicing an enormous Intelsat communications satellite that had been stranded when the upper stage of its Titan rocket failed back in 1990. The successful mission provided valuable experience that helped in planning an even more ambitious mission to the Hubble Space Telescope in 1993 (see panel, right).

Throughout the early 1990s, NASA's four Shuttles carried out a successful and varied series of missions. Satellite and spaceprobe deployments continued, along with the release and retrieval of short-term independent satellites, or free fliers, developed with the European and Japanese space agencies. There were also further Spacelab missions – the experiments carried in the mid-deck storage lockers inside the spacecraft were supplemented by the commercially built Spacehab module, which fitted behind the main cabin while leaving most of the cargo bay empty, and contained additional experiment lockers for paying customers.

Alongside a programme of joint missions with the Russian space station Mir (see p.217), a number of missions of the early and mid-1990s also developed techniques for construction of the International Space Station (ISS). Spacewalking astronauts practised building structures that might be used in a space station, including long truss sections and frameworks (see p.186). There were even opportunities to test rather more speculative ideas – *Atlantis* carried an experimental trolley-car for astronaut transport in its cargo bay in April 1991, while two other missions unwound a tethered satellite that generated electricity as its 1km- (⅔-mile-) long tether sliced through the Earth's magnetic field.

REPLACING HUBBLE'S SOLAR PANELS
Mission specialist Kathryn Thornton fixes replacement solar arrays to the Hubble Space Telescope during STS-61. Spacewalk activities are listed in the checklist on her right cuff.

TECHNOLOGY

HUBBLE SPACE TELESCOPE REPAIR MISSIONS

Shortly after the Hubble Space Telescope's much-delayed deployment, NASA was highly embarrassed at the discovery that its great orbital observatory was delivering out-of-focus images (see p.252) – a major flaw in the mirror design had not been spotted before launch, and the telescope was crippled. Fortunately, the HST was designed with orbital servicing in mind – and each of its major instruments could be removed and replaced with relative ease. Engineers on the ground soon devised a system of corrective optics that could fit into one of the instrument bays and bring Hubble's vision into focus for the remaining instruments. The capture and repair of the HST by *Endeavour* in 1993 ultimately took five spacewalks, but it resurrected the HST. Further maintenance missions took place in 1997, 1999, 2002, and 2009.

ON THE FLIGHT DECK
The Shuttle upper deck is a maze of instrumentation that clearly shows its origins in the 1970s. From their seats, Commander and Pilot could control every aspect of the Space Shuttle's flight.

OMS PRPLT QTY
LEFT
FWD/KIT
RIGHT
MISSION TIME
DAY
HR
MIN
SEC
TK 5
TK 4
02
03
HeXIO
PRPLT
OMS PRPLT
RCS
OXID
FUEL
LOWEST
OMS
GMT
TRIM
RHC/PNL ENABLE
PANEL ON
HUD POWER ON
INHIBIT
OFF
PLT1
KSC 15
MEDS

1986

The Mir space station

Less than a month after the American space programme had suffered the loss of *Challenger*, the Soviet Union launched what turned out to be its last great space project – the space station Mir.

19 February 1986
The empty Mir core module is launched from Baikonur Cosmodrome into an orbit 390km (242 miles) above the Earth.

13 March 1986
The first crew of two cosmonauts is launched on an 18-week mission to both Mir and Salyut 7.

5 February 1987
After a seven-month hiatus caused by delays to Kvant-1, Mir's second crew is launched aboard Soyuz-TM 2.

9 April 1987
Kvant-1 docks with Mir.

22 July 1987
Soyuz-TM 3 brings a crew including Mohammad Faris, the first Syrian astronaut, to Mir. During the visit, the first in-orbit crew exchange is made.

21 December 1988
Cosmonauts Musa Manarov and Vladimir Titov return to Earth after spending 366 days in orbit.

6 December 1989
Kvant-2 is added to the Mir core.

10 June 1990
The Kristall materials science module docks with Mir.

Development work on a more complex modular space station – one that would have more than two docking points, allowing the attachment of extra individual elements to the basic core – began in the mid-1970s. The station's design and construction was undertaken at NPO Energia (the former OKB-1) under Valentin Glushko, and initially it was supposed to use only the relatively lightweight, Soyuz-derived elements of the bureau's own Salyut stations. But the cancellation of Vladimir Chelomei's military Almaz programme (see p.179) led to a decision that the station should also accommodate much heavier modules developed from Chelomei's TKS space ferry. Development progressed throughout the early 1980s but was sidetracked by the pressure of other spacecraft programmes, such as the Progress cargo ferry (see panel, right) and Buran (see p.214). It only took priority once the bureau was given a launch deadline of spring 1986 – to coincide with the 27th Communist Party Congress.

Building in orbit

A lot of information about Mir (the name means "peace" and also "world") was available from the start – part of the new Soviet leader Mikhail Gorbachev's policy of *glasnost*, or "openness". The nature of the station was revealed within a few weeks of its launch – scientific equipment would be added piece by piece in additional modules, keeping the core of the space station comparatively uncrowded. This was just as well, since Mir was intended to be the first space station with more-or-less continuous occupancy. Once the first crew got Mir up and running on 13 March, they then travelled to the abandoned Salyut 7, revived it from deep-frozen slumber, and ultimately transferred a large amount of experimental material to the new station.

The second major element of Mir, the Kvant-1 astrophysics module, docked with the station just over a year later. Originally designed for Salyut 7, Kvant-1 also contained Mir's first set of gyrodynes – electronically controlled stabilizing wheels that allowed the station's attitude to be altered in orbit without wasting valuable thruster fuel.

In late 1989, Kvant-2 brought an improved life-support system. In the meantime, various crews had come and gone – in December 1988, Vladimir Titov and Musa Manarov had become the first people to spend a year in space and had recovered well. Mir's multiple ports allowed two spacecraft to be docked at the same time, so a new crew could arrive without the old one having to leave. Because Soyuz could carry three people and Mir had a typical crew of two, foreign "guest cosmonauts" could also visit, even if an entire crew was being replaced at the same time.

Kristall, a materials science and geophysics laboratory, was added to the station in June 1990 – but by the time Mir expanded further, the Soviet Union would have slipped into history.

TECHNOLOGY

THE PROGRESS CARGO SPACECRAFT

In order to keep a crew in orbit for long periods of time, a new type of spacecraft was needed – an automated supply vehicle that could fly to orbit, rendezvous, and dock with a space station almost automatically. This was the purpose of the Progress ferry. The original Progress evolved from Soyuz manned ferries (type 7K-T) used in the Salyut 6 mission and made its debut in August 1978. An upgraded version, Progress M, incorporated technology developed for the Soyuz T and was first used in August 1989. This newer version could also transport a capsule capable of returning to Earth with samples from Mir.

RUSH HOUR IN ORBIT
By late 1990, Mir was in an upside-down T-shaped configuration. Kvant-1 is mounted above the core module at the end of Mir's long axis, while Kvant-2 and Kristall form the cross-bar of the "T".

TECHNOLOGY

THE PROTON LAUNCH VEHICLE

All of Mir's major components were launched with the Soviet Union's most reliable heavy launch vehicle, the Proton. Formally known as the UR-500, the Proton got its name from a series of heavy satellites that it launched early in its career, which began in 1965 and continues today. The rocket was designed for launching the Almaz station and various lunar missions, and was a rival to Korolev's ambitious N1. Despite appearances, the "boosters" around the base are actually integral to the first stage – they hold the rocket engines and the UDMH fuel (see p.45), while the core carries the nitrogen tetroxide oxidizer.

FIRST CREW
Leonid Kizim (left) and Vladimir Solovyov were chosen for the Mir-Salyut mission because they had already spent a long tour of duty aboard Salyut 7.

TECHNOLOGY

MODULAR SOVIET SPACE STATION

Mir space station

The world's first modular space station used elements of the earliest civilian (DOS) Salyut stations, with additional laboratories and modules that were often based on Vladimir Chelomei's military TKS ferry design. The core module was based on the DOS design used in Salyuts 6 and 7, with a docking module offering five attachment points at one end and a single docking point at the other. The station grew in fits and starts, and the final modules, Spektr and Priroda, were only completed and docked to the station following an injection of NASA cash in the early 1990s.

CREW	3
LENGTH	32.9m (108ft)
MAXIMUM DIAMETER	4.35m (14ft 3in)
TOTAL MASS	117,205kg (258,380lb)
HABITABLE VOLUME (CORE)	90 cubic m (3,175 cubic ft)
NUMBER OF DOCKING PORTS	2
DATE OF LAUNCH	19 February 1986
DATE OF RE-ENTRY	23 March 2001
MAIN CONTRACTOR	Energia/Chelomei

LAST SOVIET STATION
In its completed form, Mir incorporated six major modules, plus a docking module for Space Shuttle visits. One or two Russian spacecraft were also usually docked to the station.

KRISTALL AT KRUNICHEV
Although Mir was designed by the Energia bureau (formerly OKB-1), some of its modules, such as Kristall, were built at the Krunichev factory in Moscow.

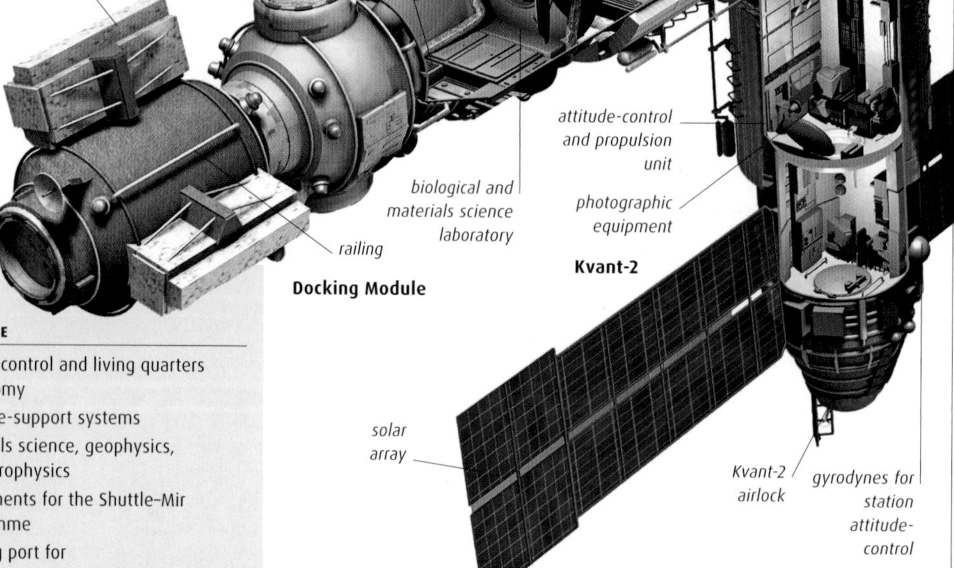

CONSTRUCTION HISTORY

MODULE	DOCKING DATE	PURPOSE
Core	n/a	Central control and living quarters
Kvant-1	April 1987	Astronomy
Kvant-2	December 1989	New life-support systems
Kristall	June 1990	Materials science, geophysics, and astrophysics
Spektr	June 1995	Experiments for the Shuttle–Mir programme
Docking Module	November 1995	Docking port for the Space Shuttle
Priroda	April 1996	Remote sensing module

MIR CORE MODULE
The core module contained living quarters and the station's main control console. Designed for use in weightless conditions, handles run along the walls and hatches are high up on the bulkheads. The unseen ceiling has exercise equipment attached. Cosmonauts could secure themselves to the chairs by folding their feet back under the seats.

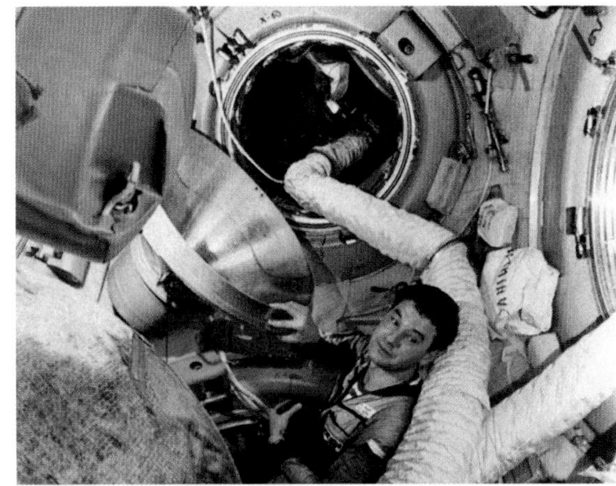

INSIDE THE CONNECTING NODE
Valeri Korzun negotiates air-conditioning hoses snaking their way through the main connecting node, which linked the core module to five other areas of the station.

MIR IN USE
After a decade of use, the station's interior was a jumble of cables, experiments, and the ephemera of everyday life. However, the cosmonauts were still careful to make sure that floating objects were secured onto surfaces.

urinal funnel

waste storage tank

pump

solid waste container

SPACE TOILET
Cosmonauts travelling to and from Mir had to rely on some basic toilet facilities aboard their Soyuz taxis.

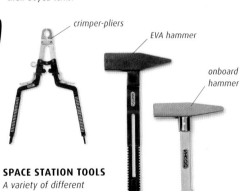

crimper-pliers

EVA hammer

onboard hammer

TIMEKEEPERS
Although the station had an orbital period of around 90 minutes, and so saw 16 sunrises every day, a normal daily rhythm was maintained with the help of onboard clocks.

SPACE STATION TOOLS
A variety of different tools were devised for the Mir astronauts to use either inside or outside the station. Frequent spacewalks (EVAs) were used to service the station or install scientific experiments in open space.

Buran – the Soviet shuttle

The Soviet Union's attempt to match NASA with a partially reusable spaceplane system pushed the country's space programme to the limit and ultimately foundered under pressure from technical difficulties and a faltering economy.

With hindsight, the decision to develop a Soviet equivalent to NASA's Space Shuttle was a disastrous mistake – some have even suggested that the project's crippling development costs helped push the entire Soviet economy over the edge, ultimately leading to the country's collapse.

Of course, when the decision was announced in 1974, it seemed quite logical – NASA's Shuttle, partly financed by the US Air Force, was expected to play a key strategic role in the military exploitation of space, and naturally the Soviets felt they had to have a similar system to maintain parity – even if the exact applications for it remained uncertain.

Design issues

Development of this major new programme, known as the Reusable Space System (MKS in its Russian acronym) was entrusted to NPO Energia. While the spacecraft itself ended up looking strikingly similar

ON THE PAD
Buran awaits its maiden flight in the Kazakh desert. The orbiter's resemblance to the US Space Shuttle is obvious, but the rest of the assembly – the long Energia rocket with its four boosters – is unique.

UNDER CONSTRUCTION
The MKS orbiters were mainly built at the NPO Energia factory in Kaliningrad, near Moscow (shown here). They were then flown to Baikonur Cosmodrome piggybacked on a converted Antonov An-225 aircraft.

1 August 1974
Valentin Glushko, head of the newly merged NPO Energia design bureau, orders work to begin on a new heavy-lift launcher and a reusable orbiter spacecraft.

12 February 1976
Work on the MKS system is formally approved by the Soviet government.

1 January 1986
Constant delays lead to an extensive shake-up in the Buran project's management.

15 May 1987
The first launch of the Energia rocket puts a Polyus military satellite into space, though it is classed as a failure when a guidance problem prevents the Polyus reaching its proper orbit.

29 October 1988
A first attempt at launching Buran is aborted 51 seconds before launch due to a software fault.

15 November 1988
Buran makes a successful automatic flight.

30 June 1993
President Yeltsin of Russia cancels the MKS project.

to its US rival, its operating principle and launch hardware were quite different. Rather than build large solid rocket boosters, of which they had no experience, the design team soon settled on the idea of a new, entirely liquid-fuelled, rocket system, ultimately called the Energia (see panel, below).

The MKS orbiter itself, usually known as Buran ("snowstorm") after the first of three ultimately built, was externally an almost exact twin of the Space Shuttle. Although the Soviet designers tried to come up with an alternative look, wind tunnel tests soon revealed that NASA had done its job well and the Shuttle was already the best shape for the job. The main difference between the two vehicles was that the Buran orbiter did not carry main engines inside its body, relying instead on the Energia launcher's own powerful engines. This made the system more wasteful than NASA's, but it potentially allowed Buran to carry an extra five tonnes of payload compared to the Shuttle and also meant that the Energia could function as an independent launcher, carrying payloads other than the orbiter.

Brief career

As with the Shuttle a variety of more or less complex dummy orbiters were built during MKS development, some of which were used in glide tests. However, the real Buran orbiter flew just once, on 15 November 1988. Like other Soviet spacecraft, the entire system had been designed for automatic as well as piloted flight, and Buran's first and only mission was controlled entirely from the ground. The flight lasted 206 minutes, during which the spacecraft orbited the Earth twice and executed a perfect landing. The system had proven itself – but too late. By now, the Soviet system was in its death throes, and the programme was suspended as a cost-cutting measure. Two half-finished orbiters were left in limbo while Buran was reduced to touring air shows as a display of ingenious Soviet technology. By the time the programme was formally cancelled in 1993, the country and system that produced it were themselves things of the past.

SAD END
After the project's cancellation, the Buran orbiter remained stuck in Kazakhstan, sheltered in a deteriorating hangar at Tyuratam. In May 2002, the roof of the hangar collapsed, destroying Buran and its Energia launcher and killing eight people.

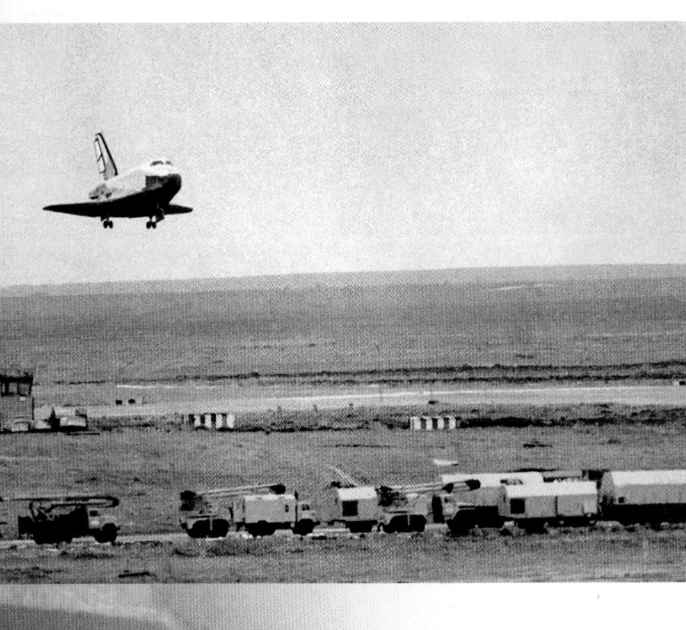

APPROACH AND LANDING
Buran glides towards the landing strip under automatic control at the end of its first flight around the Earth. Longer automatic flights should have followed over the next few years, before manned flights in the mid-1990s.

TECHNOLOGY

THE ENERGIA LAUNCHER

With a central core surrounded by strap-on boosters, Energia resembled many past Soviet rockets, but it was also a modular system – different configurations of boosters, and even extra stages stacked on the rocket, could turn it into a general-purpose heavy launch vehicle. Even without Buran, Energia could have had a useful role as a launcher, but it was ultimately cancelled with the rest of the MKS, leaving the Proton as Russia's heavy-lift rocket of choice.

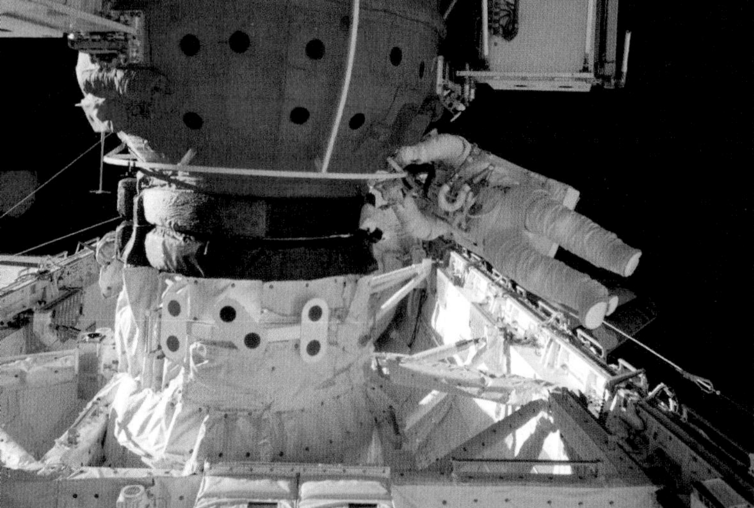

NEAR MIR
The crew of Discovery *photographed cosmonaut Valeri Polyakov watching their close approach in February 1995. Polyakov, an expert in space medicine, was nearing the end of a record-breaking 14-month endurance mission aboard Mir.*

ORBITAL UPGRADE
Astronaut Rich Clifford works on the Mir docking module during STS-76 in March 1996. Although tethered to the Shuttle, Clifford also tested a new emergency manoeuvring unit called SAFER.

APPROACHING MIR
Atlantis *photographed Mir hanging above the Earth shortly before the third docking of the spacecraft, in March 1996.*

STS-86 ROLLOUT
Atlantis *made seven flights to Mir from 1995–97. A refit in late 1997 led to* Endeavour *and* Discovery *making the last two flights.*

Working together

The collapse of the Soviet Union and the chaos that followed in its successor states finally brought the Russian and American space programmes together for a series of daring Shuttle–Mir missions.

JOINT MISSION
The STS-71 mission patch reflected Atlantis's historic link-up with Mir. As was as helping to create the then largest man-made object in space, the mission was also the USA's 100th manned spaceflight.

ATLANTIS DOCKED
Cosmonauts Solovyov and Budarin photographed the united Shuttle and space station during STS-71 on 4 July 1995. The cosmonauts had boarded a Soyuz spacecraft for an inspection "flyaround" prior to the departure of Atlantis. *Together, station and Shuttle comprised by far the largest structure ever put into space.*

The Soviet Union's dramatic disintegration at the end of 1991 left its space programme in chaos. While much of its infrastructure was inside Russia itself, some vital assets were beyond Russian borders in suddenly independent states. Kazakhstan, for instance, was soon demanding payment for use of its launch facilities at Tyuratam.

Meanwhile Mir remained in orbit, its politically stranded cosmonauts Alexander Volkov and Sergei Krikalev dubbed "the last Soviet citizens". Cash shortages meant the station was still only half-finished, with two large modules still waiting on the ground. Eager to extend the hand of friendship and help stabilize the fragments of their former rival, American politicians saw the possibility of once more turning space exploration to political ends.

In June 1992, US President George Bush and Russia's new President Boris Yeltsin announced their intention to cooperate in space. NASA Administrator Dan Goldin and Yuri Koptev, director of the newly formed Russian Space Agency (RSA), soon agreed a tentative programme of astronaut exchanges. With the aid of a substantial cash injection from NASA, normal service on Mir was resumed, and in February 1994 Sergei Krikalev became the first cosmonaut to fly aboard the Shuttle. By this time, the new Clinton administration had broadened the scope of cooperation with the Russians, who were to become major partners in the International Space Station (ISS – see p.288). A series of joint Shuttle–Mir missions would help keep the Russian station operating and give NASA astronauts experience of long-duration spaceflight before construction work on the ISS started in the late 1990s.

BIOGRAPHY

HELEN SHARMAN

Briton Helen Sharman (b.1963) became her nation's first astronaut with an eight-day stay aboard Mir in May 1991. Food scientist Sharman was selected in 1989 – one of thousands who applied for the Project Juno mission, intended to be sponsored by British business. She spent 18 months in training at Star City, and when the fee for her trip could not be raised, the Soviet Union agreed to let her fly in exchange for her assistance in their own experiments.

Shuttle–Mir

In order to dock with Mir, one Shuttle would have to be fitted with a special adaptor, and *Atlantis* was chosen for the upgrade. Following a close orbital rendezvous by *Discovery* in early 1995, *Atlantis* and Mir docked in space on 29 June. In the meantime, one of Mir's remaining elements, the Spektr Earth-sciences module, had finally been launched, now fitted out with additional equipment needed for NASA's own programme of experiments.

An American astronaut, Norman Thagard, had also been aboard Mir for more than three months. For this first visit, *Atlantis* brought with it a Spacelab module fitted with equipment to assess the health of the station's outgoing crew. The incoming crew also included two Russian cosmonaut passengers who would take over on Mir – Anatoli Solovyov and Nikolai Budarin.

Later visits brought new equipment and supplies to Mir, as well as ferrying personnel to and from the station. A new docking adapter was fitted that allowed the Shuttle to dock in a more convenient configuration, and in April 1996 the Priroda module was finally added to Mir. This last element of the station contained remote-sensing and materials-science equipment.

TECHNOLOGY

EXERCISE IN SPACE

Long stays on Mir meant that American astronauts had to get used to the exercise regimes cosmonauts had been practising for years. While it was relatively easy to recover from a few days of weightlessness aboard the Shuttle, months in orbit had greater physiological effects, and astronauts such as Shannon Lucid used Russian equipment to keep fit in space. With no gravity to work against, most of the exercise devices forced the user to work against tension or interia. These included a cycling machine and a treadmill with elastic attachments to strap down the user.

3 February 1995
Discovery comes within 10m (33ft) of the Russian space station during the STS-63 "Near Mir" mission.

1 June 1995
The Spektr module docks successfully with Mir.

29 June 1995
Atlantis and Mir dock for the first time during the STS-71 mission.

14 November 1995
Atlantis STS-74 adds a docking module to Mir.

24 March 1996
STS-76 ferries equipment to and from Mir and leaves a new American crew member, female astronaut Shannon Lucid.

26 April 1996
Priroda, the last of Mir's main modules, is joined to the station.

15 January 1997
Atlantis STS-81 collects John Blaha and leaves Jerry Linenger as Mir's new long-term guest.

17 May 1997
Atlantis STS-84 docks with Mir. Jerry Linenger is replaced by British-born astronaut Michael Foale.

27 September 1997
Atlantis STS-86 arrives at Mir to collect Michael Foale and leave David A. Wolf onboard.

FACE TO FACE

The five astronauts of Atlantis's STS-74 mission look out from the Space Shuttle's aft flight deck overhead window to greet their colleagues aboard Mir. Moments earlier, the two spacecraft had docked for only the second time.

The end of Mir

Throughout the later Shuttle–Mir missions, the Russian space station was showing its age. When ambitious schemes to keep Mir running as a private concern came to nothing, the station was doomed to a fiery end.

FINAL MISSION
Cosmonauts Sergei Zalyotin (centre) and Alexander Kaleri set out on what would prove to be the last visit to Mir, in April 2000.

1976 1977 1978 1979 1980 1981 1982 1983 1984 1985 1986 1987 1988 1989 1990 1991 1992 1993 1994 1995 1996 1997 1998 1999 2000 2001 2002 2003 2004 2005 2006 2007 2008 2009 2010 2011 2012 2013 2014 2015

24 January 1998
Endeavour STS-89 arrives to retrieve David A. Wolf and deliver Andy Thomas, Mir's last American resident.

4 June 1998
Discovery docks with Mir during its STS-91 mission.

8 June 1998
Discovery undocks from Mir, marking the end of the Shuttle–Mir programme.

22 February 1999
French spationaut Jean-Pierre Haigneré arrives at Mir aboard Soyuz-TM 29, to work alongside the last Russian crew.

28 August 1999
Mir's final working crew mothball the station and return to Earth at the end of almost ten years of continuous occupation.

6 April 2000
Cosmonauts Alexander Kaleri and Sergei Zalyotin return to Mir and revive it for a possible new life as a commercial venture.

23 March 2001
Mir re-enters the Earth's atmosphere and burns up over the Pacific Ocean.

American astronauts visiting Mir found the station ran along unfamiliar lines. While NASA controllers micromanaged each Shuttle astronaut's day in blocks of activity, Mir cosmonauts were allowed to organize much of their own time. Although partly an inevitable consequence of Mir's limited contact with the ground (the Russians had no equivalent of NASA's TDRS satellite system to maintain constant contact), the Americans soon found they enjoyed their independence during such long missions – a lesson that would be applied when it came to managing the ISS.

The coming together of space programmes also emphasized other differences between them – most significantly, how rapidly Mir was aging compared to the frequently serviced, constantly refurbished Shuttle. Every month seemed to bring a new snag, ranging from leaks to power failures and even bumps from Soyuz and Progress spacecraft flying close to the station. However, the worst crises were undoubtedly the fire of February 1997 (see over), and the collision of Progress-M 34 with the Spektr module four months later (see panel, opposite).

The final phase

Discovery's departure in June 1998 marked the end for a chapter in US–Russian space co-operation. The first segments of the ISS would soon be in orbit, and NASA was keen for the RSA to dedicate its limited resources to the new project rather than sustaining the crumbling Mir. The RSA duly announced that the station would be deliberately taken out of orbit the following year.

Nevertheless, the end of Shuttle visits saw a temporary return to business as usual on Mir, as Talget Musabayev and Nikolai Budarin resumed work. In the following months crew rotations continued, and Mir welcomed a variety of guests, including Yuri Baturin, a distinguished space physicist and former adviser to President Yeltsin, who opined that Mir should be funded for another two years. Delays to the ISS brought more calls to extend Mir's life, but without money there was little the RSA could do.

Meanwhile RSC Energia (the company that now actually owned Mir) sought private funding to continue operation. One important element of this was the ongoing guest cosmonaut programme, which culminated in a six-month tour for France's Jean-Pierre Haigneré from February 1999. In January, Yuri Semenov (see panel, above) announced a potential investor, but by the end of February, that hope had evaporated. Viktor Afanasayev and Sergei Avdeyev closed down much of the station before departing with Haigneré on 28 August 1999. The RSA now planned to bring Mir down early in 2000, while they could still guide it to a safe crash site – but there was one more twist to come.

In January 2000, Energia announced investment from MirCorp, a company that would run the station for research and tourism. To NASA's annoyance, the RSA supported the venture, launching a ferry to boost Mir's orbit in February, and a ten-week Soyuz mission in early April to carry out repair and servicing work.

Throughout 2000, MirCorp announced various schemes – plans to launch an actor into space, a gameshow contest for a flight aboard Mir, and even the first space tourist, one Dennis Tito (see p.308). But launch dates continued to slip, and Energia eventually accepted the inevitable, terminating the deal in December. Mir's impressive career finally ended with a spectacular demise over the Pacific in March 2001.

BIOGRAPHY

YURI SEMENOV

After engineering studies and work for another missile design bureau, Yuri Pavlovich Semenov (b.1935) transferred to Korolev's OKB-1 in 1964. He soon became Deputy Leading Designer on the Soyuz project and then Leading Designer of the abandoned L1 Moonship. Through the 1970s and 1980s, he was Chief Designer (under Designer General Valentin Glushko) for the Soyuz, Salyut, Mir, and Buran projects. He took over at the head of NPO Energia in 1989 and led it in its various subsequent incarnations until his retirement in 2005.

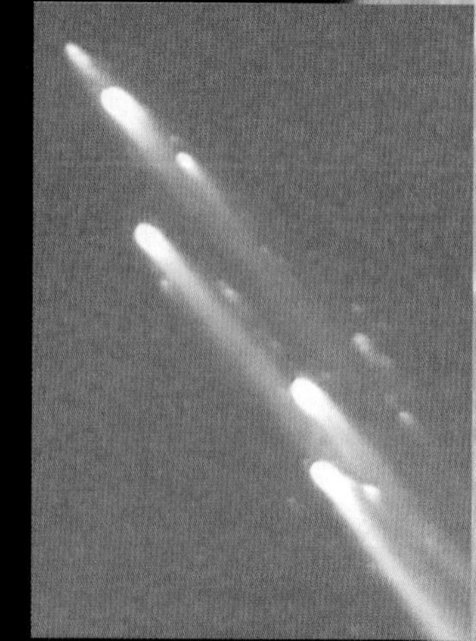

LAST MOMENTS OVER FIJI
On 23 March 2001, the Progress-M1 ferry attached to Mir fired its engines to lower the station's orbit and plunge it into the Earth's atmosphere.

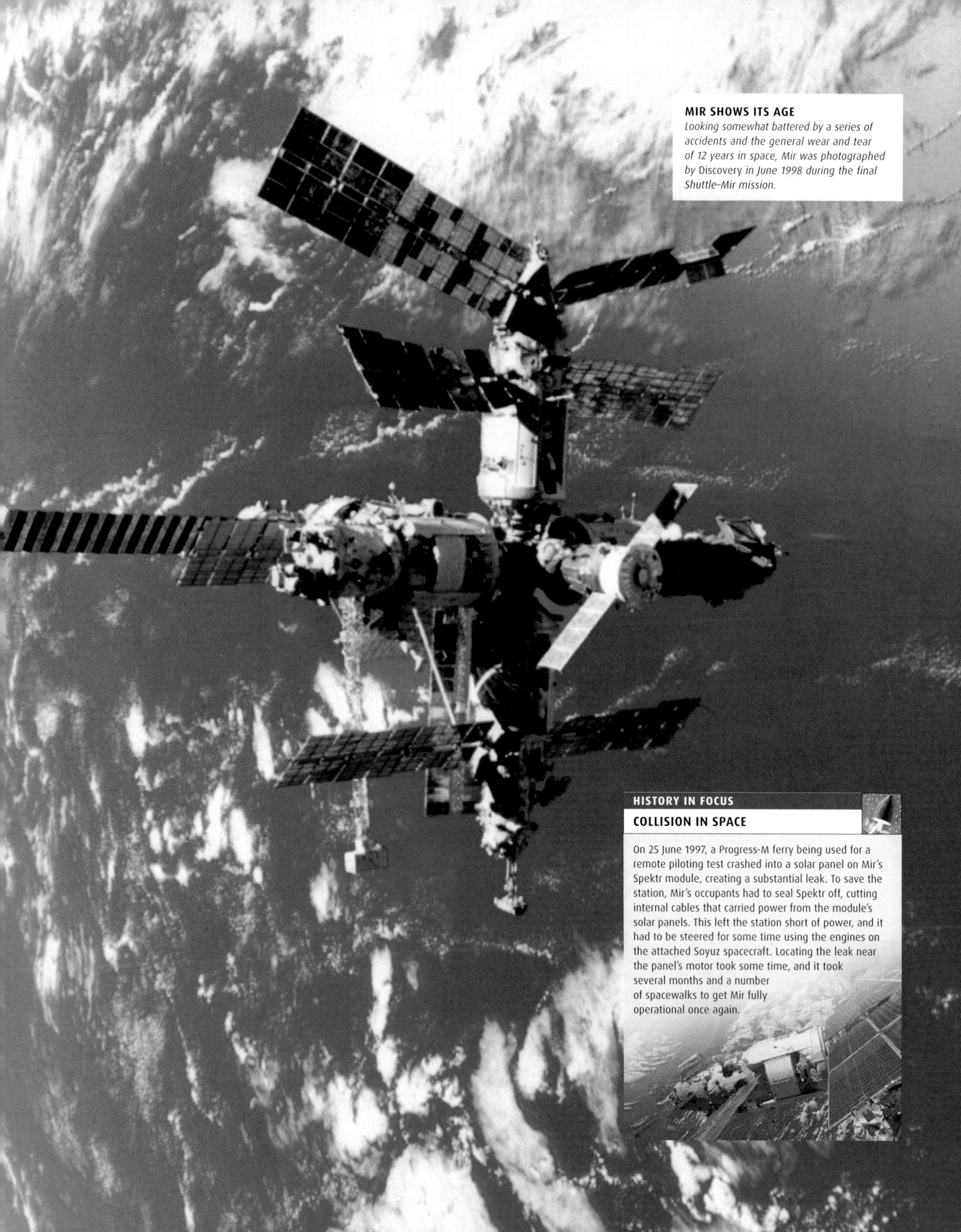

MIR SHOWS ITS AGE
Looking somewhat battered by a series of accidents and the general wear and tear of 12 years in space, Mir was photographed by Discovery *in June 1998 during the final Shuttle–Mir mission.*

HISTORY IN FOCUS

COLLISION IN SPACE

On 25 June 1997, a Progress-M ferry being used for a remote piloting test crashed into a solar panel on Mir's Spektr module, creating a substantial leak. To save the station, Mir's occupants had to seal Spektr off, cutting internal cables that carried power from the module's solar panels. This left the station short of power, and it had to be steered for some time using the engines on the attached Soyuz spacecraft. Locating the leak near the panel's motor took some time, and it took several months and a number of spacewalks to get Mir fully operational once again.

EXPERIENCE

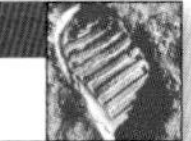

EMERGENCY IN ORBIT

Fire on Mir

JERRY LINENGER
NASA astronaut Linenger was the fourth American aboard Mir. As the station's physician, he had special duties in the emergency – watching in case anyone fell ill and keeping a particular eye on Korzun as he fought the fire.

By the mid-1990s, Mir was approaching a decade of continuous operation and had started showing its age. In addition, a series of accidents and mishaps threw the station's future into doubt. Of these, the most dangerous was the fire of February 1997.

The fire broke out during one of Mir's crowded handover periods – Soyuz TM-25 had arrived 12 days earlier, bringing Vasily Tsibliev, Aleksandr Lazutkin, and German visitor Reinhold Ewald. The American Jerry Linenger had been aboard for several weeks, while Valeri Korzun and Alexander Kaleri were near the end of their mission. The six residents were supplementing the station's air supply with solid-fuel oxygen generators (SFOGs) that produced oxygen through a slow chemical reaction. After dinner on 24 February, Lazutkin went to activate another SFOG cylinder in the Kvant module when it erupted in a shower of sparks – a "baby volcano" as Lazutkin described it. Ewald spotted the fire and alerted the rest of the crew, with Korzun clambering through the hatch to pull Lazutkin from the flames. A wet towel had little effect and Korzun called for fire extinguishers – but the first one he tried failed to work. As the smoke grew worse, and the crew put on oxygen masks, the fire alarm alerted Linenger, who had retired early to bed:

" **The smoke was immediate.** It was dense … I could see the five fingers on my hand, I could see a shadowy figure of the person in front of me who I was trying to monitor to make sure he was doing okay … Where he was standing he **could not see his hands in front of his face**. In the distant modules … the smoke was still dense, so it was very surprising how fast and rapid the smoke spread throughout the complex … I did not inhale anything, and I don't think anyone else did because the thickness of the smoke told you that **you could not breathe**. So, everyone immediately went to the oxygen ventilators. They worked very [well], and they protected us from inhalation injury. "

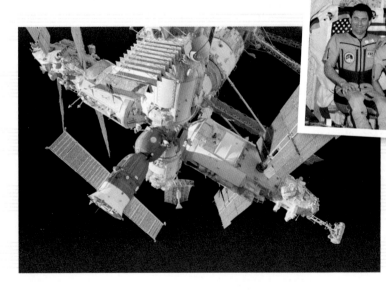

AFTER THE FIRE
Jerry Linenger would continue his mission alongside Tsibliev and Lazutkin (above, left and right of Linenger) until May, when Atlantis arrived to collect him and deliver Michael Foale.

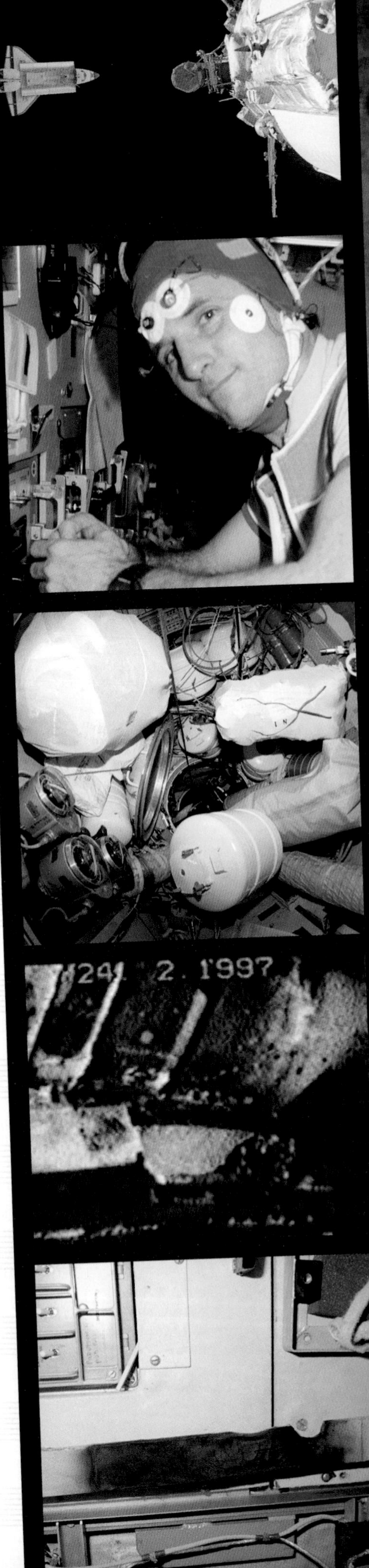

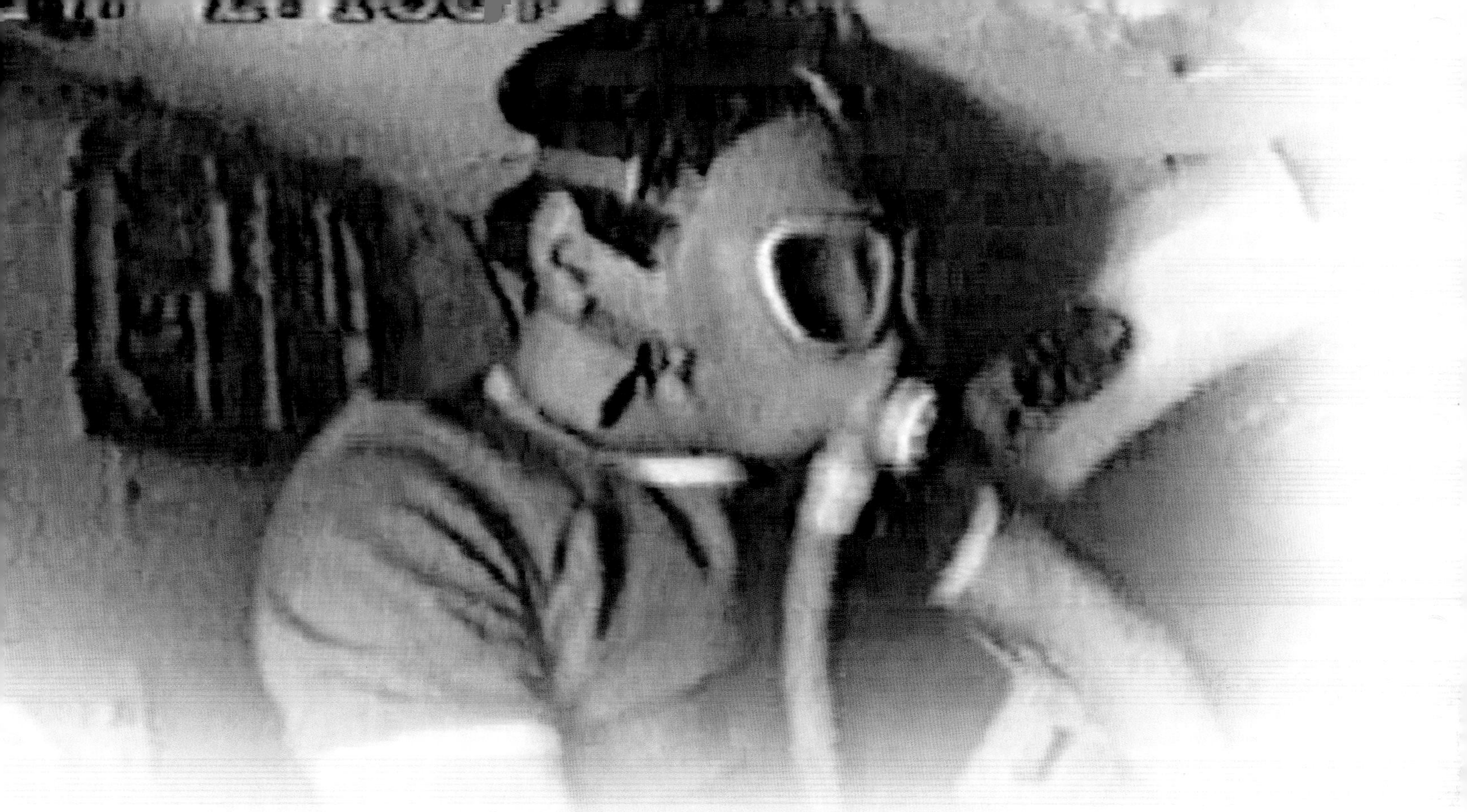

SITTING IT OUT
With the fire out, the crew had to wear their oxygen masks for several hours while Mir's air filters did their work and the smoke cleared. With so many oxygen masks used, they had to preserve those remaining if all six crew were to remain on the station.

"When I saw the ship was full of smoke, my natural earthly reaction was to want to open a window. And then, **I was truly afraid** for the first time."

Aleksandr Lazutkin, interviewed by Nova TV, 1998

Korzun told Lazutkin to prepare one of the two Soyuz spacecraft for evacuation. But there was a problem – the other Soyuz lay on the other side of the fire. As Linenger arrived, the Russian commander ordered the crew to work in pairs, in case one was overcome by smoke. Ewald fetched more oxygen masks (some did not seem to be working), while Tsibliev and Linenger collected fire extinguishers from around the station, before returning to the base block. Eventually, the fire was out – but smoke lingered in the station for some time. Oxygen masks were also running short, and it was some hours before the risk of an evacuation was over.

INFERNO IN SPACE
Linenger had arrived on the Shuttle Atlantis *in mid-January and begun a programme of biomedical science. The fire broke out in the crowded Kvant-1 module and, though brief, coated surfaces throughout the station in a thick layer of grime. Thankfully, aside from the destroyed SFOG canister, there was no serious damage.*

"I grabbed the respirator off the wall, activated it, took a breath, and I **didn't get any oxygen**. At that point, there was a lot of smoke. I took the mask off. Again, Earth instinct made me look low to **try to find a clear spot** where I could get a quick breath because I was getting very short of breath at that time. But it was solid smoke. **Smoke does not rise in space** like it does on the ground. It's just everywhere. I went to the other respirator on the other wall. Opened it up. At that point, Vasily was there. He saw **I was getting into trouble.** He helped me get the thing out. I activated it again. Put it on. Breathed in, and luckily got oxygen at that point."

Jerry Linenger, interviewed by Nova TV, 1998

EXPERIMENTAL PROGRAMME
During Linenger's 132-day stay on Mir, he concentrated on life sciences and biotechnology, using his scientific background. He is shown here in Mir's Priroda module.

Columbia and after

The tragic loss of a second Space Shuttle in 2003 led to another hiatus in the launch programme, delaying the effort to build the International Space Station and ultimately forcing the Shuttle into retirement.

While the Space Shuttle *Challenger* had been lost at the start of her flight, *Columbia* was minutes from home when disaster struck on 1 February 2003. The first signs that something was seriously wrong came with an apparently faulty indication that one of the tyres on the Shuttle's left wing had been deflated. Within seconds, however, temperature sensors both on and within the wing began to rise. The last communication from the Shuttle gave no indication that the crew knew they had a major problem, but at 8:59am EST, telemetry data was abruptly lost. Then reports started to come in of fireballs seen in the sky over Texas and smoking debris falling to the ground. It was soon clear that *Columbia* had broken up during re-entry.

Fatal impact

Tragically, the cause of this disaster had been noticed earlier, but a series of mistakes led managers and engineers to underestimate the danger it presented. Stress and vibration during *Columbia*'s launch 16 days earlier had shaken loose a large chunk of lightweight insulating foam from the bipod struts that supported the Shuttle's nose above the external tank. As the foam flew off, ground-based video cameras saw it strike *Columbia*'s wing.

NASA immediately began to assess the risk, but managers felt that comparisons with similar impacts in the past (often noticed only when the orbiter returned to Earth with scars) suggested there was little to worry about. However, several key factors in this case had been missed – the lost chunk of foam was much larger than on any previous mission, and had struck the Shuttle at a different angle and at much higher speed. Furthermore, although no one could have known it, the debris had hit a very vulnerable spot – the leading edge of the wing, which received most of the heat during re-entry. As *Columbia* entered the upper atmosphere at 24 times the speed of sound, the usual sheath of hot gases developed around it, and found their way inside the Shuttle through a hole. At an altitude of 70km (44 miles), the orbiter's wing broke up, taking the rest of *Columbia* with it.

The aftermath

Following guidelines set up after the *Challenger* disaster, an expert committee, the Columbia Accident Investigation Board, was immediately estabilshed to discover what had gone wrong. It took some time to conclude that a piece of lightweight foam had indeed doomed the spacecraft, but debris collected from along the re-entry path soon revealed the story of *Columbia*'s break-up in detail, and when the Shuttle's black box flight recorder was found, it confirmed that the trouble had started on the left wing.

Once again the Shuttle fleet was grounded for modifications, and once again the verdict on NASA management was damning – but this time the effects went further. Construction of the International Space Station (ISS) ground to a halt, and NASA was forced to rely on Russian launches simply to keep the station crewed and supplied.

In the longer term, the loss of a second orbiter was too much for the Shuttle programme to bear – it was now clear that a large spacecraft strapped alongside its fuel tank and boosters was vulnerable

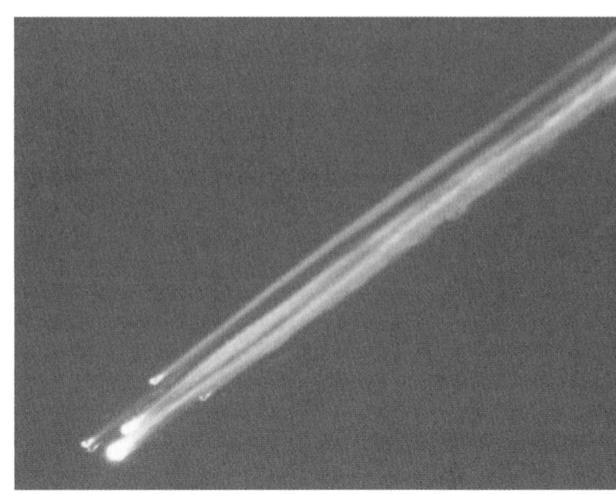

ALL LOST
(Left) The crew shortly before setting off on the STS-107 mission. Tragically they all perished when Columbia *broke up in the skies over Texas (right).*

> "Their mission was almost complete, and **we lost them so close to home**."
>
> **President George W. Bush, 4 February 2003**

POST MORTEM
The break-up of Columbia *scattered debris across a large swathe of the southern United States. Accident investigators collected as much as possible and, applying the same technique used with other air crashes, laid the pieces onto an outline of the orbiter marked out inside one of the hangars at Kennedy Space Center.*

to accidents that would not affect other vehicles. In economic terms, the problems were even clearer – the development of rival rockets and commercialization of the launch market, coupled with the Shuttle's relatively slow launch rate, had already made it clear that the Shuttle would never undercut the competition. After *Challenger,* most routine satellite launches, even for NASA missions, had returned to unmanned rockets, with the Shuttle reserved for more ambitious tasks. *Columbia* decided the issue, and in January 2004, President Bush announced that the Shuttle would be retired as soon as its construction work on the ISS was complete.

Final missions

The go-ahead for a return to flight by the remaining Shuttles was made only after stringent new safety precautions were put in place. Absolute priority was given to completion of the ISS in as few missions as possible, and eventually a revised schedule was agreed in March 2006. By this time, *Discovery* had already made an initial return-to-flight mission, docking with the ISS to deliver supplies and exchange crew in late July 2005. The mission involved a thorough inspection of the orbiter hull from space, and some minor repairs were carried out. More problematically, a large piece of insulation foam was found to have broken away from the external fuel tank. While it did not strike the orbiter, this led to the Shuttle being grounded for another 11 months while further safety measures were put in place. Finally, following a second return-to-flight mission in July 2006, approval was given for the remaining launches.

There were more twists to come, however. In particular, one safety edict had been that remaining missions should have the option of "abandoning ship" to the ISS in the event that an orbiter was badly damaged during launch. This initially forced the cancellation of a final servicing mission to the Hubble Space Telescope (whose orbit was incompatible with this option), but pressure from the public and scientific community eventually led to the mission being reauthorized and taking place in May 2009.

The final Shuttle mission, STS-135, ended with the touchdown of *Atlantis* on 21 July 2011. It was the end of an era for manned spaceflight.

TECHNOLOGY

LESSONS LEARNED

When the Shuttle returned to space again in July 2005, NASA had a far more cautious attitude. Tiles in vital areas would now be inspected in space, either during a spacewalk, or using a new camera attached to the robot arm. Despite all the safety improvements, though, checks during *Discovery*'s initial return-to-flight mission revealed that there were still minor problems with the tiles and gave the astronauts onboard a chance to test new techniques for carrying out in-orbit repairs.

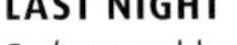

LAST NIGHT
Endeavour blasts free from pad LC-39A on 8 February 2010, the last night launch of the Shuttle programme. During the 13-day STS-130 mission, it deployed the Tranquility module and Cupola (see p.295) to the International Space Station.

16 January 2003
With its planned launch delayed 18 times over a two-year period, *Columbia* finally lifts off on the STS-107 mission to conduct microgravity and Earth science research.

17 January 2003
Routine analysis of launch images reveals a large chunk of debris from the ET support strut potentially striking the Shuttle's left wing.

19 January 2003
Engineers' requests for an emergency spacewalk to inspect the Shuttle's wing are ignored by NASA.

22 January 2003
Requests from NASA engineers for Department of Defense technology to image the Shuttle in orbit are withdrawn by NASA management.

1 February 2003
Columbia disintegrates on re-entry to Earth's atmosphere. The crew of Kalpana Chawla, Ilan Ramon, William McCool, Michael Anderson, Laurel Clark, David Brown, and Rick Husband are all killed. All Shuttle flights are suspended.

26 July 2005
The Shuttle returns to flight with *Discovery*'s STS-114 mission to the ISS, but further flights are delayed due to safety reasons.

4 July 2006
Discovery launches on a second "return-to-flight" mission.

21 July 2011
The final touchdown of *Atlantis* at Kennedy Space Center brings the Shuttle era to an end.

ENGINES OVER EARTH

This spectacular close-up of the Space Shuttle Discovery*'s tail section was taken from the International Space Station. The rear of the Shuttle is covered in rocket nozzles, including the main engines, the smaller OMS engines used in orbit, and the RCS attitude adjustment thrusters.*

Discover

20 March 1964
ESRO and ELDO come into existence.

4 June 1964
The first test launch of the Europa project, firing a Blue Streak missile at Woomera, is successful.

14 November 1966
A test launch of the modified Europa first stage with dummy upper stages succeeds.

4 August 1967
In the first test of Europa with an active second stage, the upper stage fails to ignite.

3 October 1968
The first European satellite, ESRO 1A, reaches orbit on a NASA Scout B rocket.

12 June 1970
After a string of accidents, the last Europa launch from Woomera fails to deploy a satellite when the payload shroud jams.

5 November 1971
A final Europa launch from Kourou fails due to a guidance problem in the third stage. The project is suspended for review, then cancelled.

European space efforts

The smaller nations of Europe could not hope to compete individually with the massive space efforts of the superpowers, but by banding together they ultimately became a significant space power.

Britain and France began to develop their own ballistic missile programmes well before the launch of Sputnik 1, and scientists and engineers on both sides saw the potential for turning these weapons into satellite launchers (see p.56), but it was soon clear that neither nation had the resources to pursue a major space programme on its own. The drive towards closer political cooperation in western Europe, however, meant that a Europe-wide space effort would be a natural progression.

The first sign of this new policy was the formation of the European Space Research Organization (ESRO) in 1964. France, Britain, and Germany were its leading players and contributed most of the budget, but there were seven other initial members – Italy, Belgium, the Netherlands, Sweden, Denmark, Spain, and Switzerland.

ESRO's main purpose was to coordinate European space policy and direct research efforts for the peaceful use of space. Over the next ten years, the organization developed seven scientific satellites – four that studied the Earth's upper atmosphere and aurorae (the northern and southern lights), two that studied the Earth's magnetic field and the solar wind, and an orbiting ultraviolet observatory.

EUROPEAN UNION
Representatives of the ten ESRO member states sign the new European Space Agency into being. ESA formally took over ESRO and ELDO operations on 31 May 1975.

A European launcher

Each of ESRO's satellites ultimately had to rely on an American rocket to put it into space, but it seemed obvious that Europe should have its own launch vehicle. After Britain's attempts to develop the three-stage Black Prince launcher with partners from the British Commonwealth came to nothing, the project was taken over by the European partnership. A new agency, the European Launcher Development Organization (ELDO), was set up to coordinate its

DUMMY LAUNCH
In this test launch of the Europa launch vehicle at Woomera in 1966, only the lower stage (modified from Blue Streak) is active – the rest of the rocket and a satellite onboard are dummies.

development – it included six members of ESRO and also Australia. The new Europa launch vehicle, as it was known, was to use Britain's Blue Streak missile as its base, with upper stages provided by France and Germany. Italy took responsibility for the payload fairing, the Netherlands for telemetry systems, and Belgium for the ground tracking, while the launch site would be at Woomera, Australia.

With so many contributions from isolated teams in various countries and no overall firm leadership, the Europa project proved the impossibility of launch-vehicle design by committee. While the Blue Streak-based stage performed reliably throughout a series of launches from 1966 onwards, either the French Coralie second stage or the German Astris upper stage failed in each case. The British decision to discontinue all work on Blue Streak in 1968 finally sealed the project's fate.

A new beginning

Inquests into the death of Europa laid much of the blame at the feet of ELDO itself. It was clear that Europe would not step back from space completely – even in the darkest hours of the Europa project, France was using its Diamant launcher for successful launches from both Algeria and its new facility in Kourou, French Guiana. But Britain was rapidly losing confidence in the future of space exploration, and after the cancellation of the Black Arrow project in 1971, it largely restricted itself to scientific contributions to various satellites. ELDO had been critically undermined by the Europa failure, and while France was eager to develop a new pan-European launcher as a successor to Diamant, it wanted the largest share of the work and the final say in many elements of the new programme. The other nations were only too happy for France to shoulder the burden, and so in May 1975 the European powers united all their joint space efforts in a single new organization – the European Space Agency.

The fledgling agency would inherit all of ESRO's projects and use many of the facilities of the French CNES space agency, including its launch centre at Kourou. One of the first satellites launched under the new ESA regime was the gamma-ray astronomy observatory Cos-B, though this mission, like previous European satellites, was carried into orbit by an American Delta rocket. The situation could not last, however – the development of microelectronics was already paving the way for the commercial satellite boom of the late 1970s, and ESA wanted its share of what would surely become a lucrative market.

ESA'S FIRST SATELLITE
The Cos-B gamma-ray observatory, launched for ESA by a US rocket on 9 August 1975, was largely developed under ESRO. Like many of their satellites, it was a great success.

ROCKET ASSEMBLY
ESA's launcher systems are all tested at the 58m- (190ft-) high Launcher Integration Building at Kourou. From here, rockets are transferred along a rail track to the Final Assembly Building, and then on to the launch pad.

GUIANA SPACE CENTRE
(Left) France selected Kourou for its new launch centre because of its near-ideal location. At latitude 5°N, rockets get a large speed boost from the Earth's rotation as they are launched towards the Atlantic. (Below) An ESA Ariane 5 rocket is being readied at Kourou's Launch Complex 3.

The European Space Agency

31 May 1975
ESA is established by the ten ESRO member states.

9 August 1975
A US Scout rocket launches ESA's first satellite, Cos-B.

24 December 1979
The first test flight of Europe's Ariane launch vehicle is a success.

24 June 1982
Jean-Loup Chrétien becomes the first western European in space, aboard Soyuz T-6.

14 March 1986
ESA's first spaceprobe, Giotto, makes a successful flyby of Halley's Comet.

21 October 1998
After a pair of failures, Ariane 5 enters service with a successful launch.

25 December 2003
Mars Express enters orbit and begins a 15-year survey of Mars.

12 November 2012
ESA announces it will supply the service module for the Orion spacecraft.

6 August 2014
After a 10-year journey, ESA's Rosetta probe enters orbit around Comet 67P.

Since its formation in 1975, ESA has become a major player in the commercial launch industry. It has also developed many groundbreaking satellites and probes and sent astronauts into space in collaboration with other space powers.

A major pillar of ESA's success has been the Ariane launcher, developed largely by the French CNES space agency. After the problems with the Europa project, France contributed 60 per cent of the development costs in return for control of the project. Germany put in another 20 per cent, with the other ESA nations making up the remainder. Ariane I, with its first and second stages powered by French Viking engines, was low-tech but reliable. A successful maiden flight in 1979 paved the way for 11 more launches, with only two failures early in the programme.

Arianes 2 and 3, which entered service in the mid-1980s, were essentially similar rockets, though Ariane 3 had strap-on boosters. They and their successor, Ariane 4, had very good success rates, allowing Arianespace, a company formed to run the commercial launch operation, to take a substantial segment of the satellite launch market. Today, ESA operates Ariane 5 (see over) alongside Vega (a solid-rocket launcher for smaller payloads, built with the Italian Space Agency). A new Ariane launcher is under development (see panel, opposite) and the agency has grown from its initial 10 member states to 22.

BIOGRAPHY

JEAN-LOUP CHRÉTIEN

Europe's first astronaut (or spationaut) was French air force pilot Jean-Loup Chrétien (b.1938). He was first selected as a candidate for cosmonaut training when the Soviet Union offered France a place in its Intercosmos programme (see p.240) and flew to Salyut 7 aboard Soyuz T-6 in June 1982. Appointed head of the CNES Astronaut Office, he was backup for the first spationaut to join the Space Shuttle in 1985 and flew on another Soviet mission, this time to Mir in 1988. Moving to America, he retrained with NASA and returned to Mir aboard *Atlantis* in 1997. He retired in 2001.

Probes and satellites

ESA inherited a number of satellite projects from ESRO and has since expanded into nearly all the main satellite applications. Astronomical observatories have included Cos-B, Hipparcos (a survey telescope that measured the properties of half a million stars) and the Infrared Space Observatory (ISO). The Agency has developed a particularly successful series of remote-sensing satellites: its Earth resources satellite ERS-1 was among the first to carry synthetic aperture radar (see p.267) and was followed up by ERS-2 and Envisat, which carries an array of instruments to monitor climate change.

Further afield, Europe made a spectacular space-exploration debut when its Giotto probe flew past the nucleus of Halley's Comet in 1986 (see p.272). Since then, ESA has contributed the Huygens Lander to the Cassini Saturn mission, launched its own planetary orbiters – Mars Express and Venus Express – and built a number of smaller probes, including the highly successful SMART-1, an experimental spacecraft powered by an ion drive (see p.285). Perhaps ESA's biggest exploration triumph so far has been the Rosetta comet orbiter, which accompanied Comet 67P/Churyumov-Gerasimenko for two years as it passed close to the Sun in 2014–16. However, the BepiColumbo mission to Mercury, ExoMars lander, and JUICE Jupiter orbiter promise to be even more ambitious.

ASTRONAUT TRAINING
ESA astronauts Pedro Duque (Spanish, seated left) and Paolo Nespoli (Italian, seated right) train for flights aboard the ISS. Training at the European Astronaut Centre in Cologne is supplemented with training at Houston and the Gagarin Cosmonaut Training Centre.

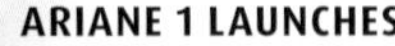

ARIANE 1 LAUNCHES
The successful launch of Ariane 1 in December 1979 pointed the way to Europe's future in space. Although there were still some teething troubles to come, ESA's rockets have established an almost unrivalled record of reliability.

ARIANE 5 ECA
An upgraded version of Ariane 5 waits on the pad at Kourou prior to a (failed) launch in December 2002. The ECA variant has the capacity to take payloads of up to 10,500kg (23,100lb) into geostationary transfer orbit.

European astronauts

ESA maintains its own astronaut corps, based at the European Astronaut Centre in Cologne, Germany. Attempts to develop a small manned spacecraft called Hermés stalled in the early 1990s, but ESA astronauts have flown as guest cosmonauts on Soyuz launches (see panel, opposite, and p.240) and as payload specialists on a number of Space Shuttle missions. The agency supplied several components for the ISS (see p.288), including the Columbus laboratory module, launched in 2008. European astronauts play an important role in crewing the station expeditions, and ESA also developed the Automated Transfer Vehicle (ATV) for resupplying the ISS. Five of these vehicles were launched by Ariane 5 between 2008 and 2014, each remaining attached to the ISS for several months before being detached and deorbited.

Since 2011 ESA has worked with the Russian Federal Space Agency (RSA) to launch Soyuz rockets from its Kourou spaceport. Kourou's equatorial location offers better payload capacity for Russian launches, while ESA gained another reliable launch vehicle. In 2012, the agency joined NASA's Orion programme, agreeing to develop a service module based on the ATV, which will serve as a major component of the new spacecraft.

TECHNOLOGY

ARIANE 6

The next generation of European launch vehicle, Ariane 6 is currently under development and scheduled for its first launch in 2020. After considerable debate over the design, the approved final version has two variants, known as Ariane 62 and Ariane 64. Both combine a core stage powered by the reliable Vulcain 2 engine with the upper stage powered by the new Vinci engine that can be stopped and restarted for precision payload deployment. The core stage is assisted by either two or four solid-fuel boosters, depending on the cargo. Other innovations aimed at reducing costs and improving reliability include 3-D printing of components and a laser ignition system.

EVOLUTIONARY LAUNCHER
The Ariane 4 (right) was the final stage in the development of the original Ariane launcher. Based on Viking engines made by the French company Societé Européenne de Propulsion, Ariane was upgraded throughout the 1980s, with the addition of longer stages, liquid- and solid-fuelled boosters, and more powerful rocket engines.

HEIGHT	58.4m (191ft 7in)
CORE DIAMETER	3.8m (12ft 5in)
TOTAL MASS	240,000kg (520,000lb)
ENGINES	4 +1 x Viking 2B (N204/UDMH) 1 x Viking 4B 1 x Viking 2B (upper stages)
LAUNCH THRUST	276,586kgf (608,490lbf)
MANUFACTURER	Aerospatiale

ARIANE 4 SECOND STAGE
The 11.5m (37ft 8in) second stage remained externally the same throughout Arianes 1 to 4. The single rocket engine, however, was upgraded to a Viking 4B.

fairing
payload
independent upper stage
upper-stage engine
third-stage propellant tank
third-stage engine
second-stage propellant tank
interstage

LAUNCH PREPARATIONS
The first stage of an Ariane 5 rocket is hauled upright within ESA's Launcher Integration Building at Kourou, French Guiana. Two solid rocket boosters will then be attached to the sides, and the upper stages and payload mounted on top.

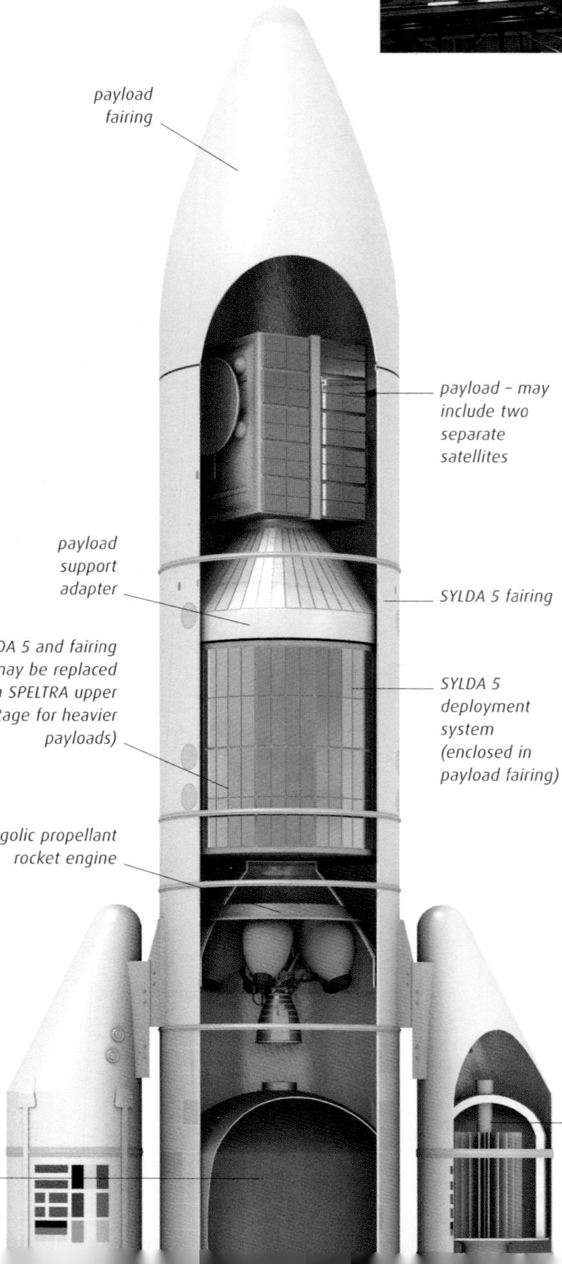

payload fairing
payload – may include two separate satellites
payload support adapter
SYLDA 5 fairing
SYLDA 5 and fairing (may be replaced with SPELTRA upper stage for heavier payloads)
SYLDA 5 deployment system (enclosed in payload fairing)
hypergolic propellant rocket engine
liquid oxygen storage tank
solid propellant engine

VULCAIN ENGINE
The new version of Ariane needs a different engine design to burn its cryogenic fuels, partly because they need an ignition trigger in order to burn. The new engine is called Vulcain.

Ariane launchers

ESA's Ariane launchers showcase two different approaches to launch-vehicle design. Ariane 4 was the final stage in the evolution of the original Ariane rocket, with a slender design harking back to the French Diamant of the 1960s. Its Viking rockets were also developed from the Vexin that powered Diamant, and it used traditional hypergolic propellants. Ariane 5, in contrast, was a new design built by the European Aeronautic Defence and Space Company (EADS) – the first European rocket to use cryogenic propellants.

ARIANE 4 FIRST STAGE
The lower stage of Ariane 4 used four Viking 2B engines. These highly reliable engines, first introduced on the Ariane 2 and 3 variants, allowed the vehicle to make 113 successful launches throughout its career. The propellants, dinitrogen tetroxide (N_2O_4) and unsymmetrical dimethylhydrazine (UDMH), are hypergolic – they combust on contact and do not require low-temperature storage.

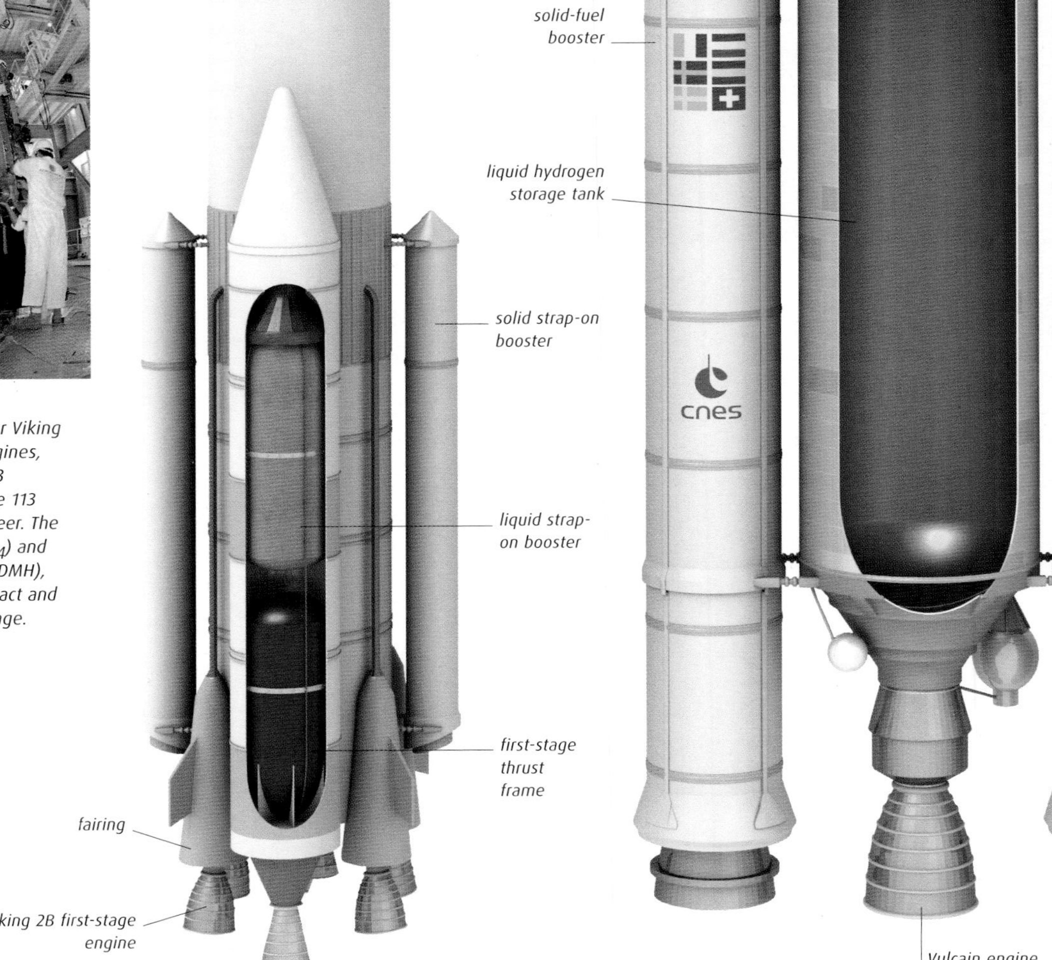

CRYOGENIC LAUNCHER
The Ariane 5 (left) is a completely new vehicle, developed in the 1990s. Unlike previous Ariane rockets, it uses liquid oxygen (LOX) and liquid hydrogen (LH2) as fuel for its central stage. These more powerful propellants need to be stored at low temperature and are termed cryogenic.

HEIGHT	54.05m (177ft 4in)
CORE DIAMETER	5.4m (17ft 8in)
TOTAL MASS	746,000kg (1,644,000lb)
ENGINES	1 x Vulcain (LOX/LH22) 2 x P230 SRBs
LAUNCH THRUST	1,162,531kgf (2,562,800 lbf)
MANUFACTURER	EADS

THE ARIANE SERIES
ESA's early rockets, from Ariane 1 (first launched in 1979) through to Ariane 4, were all evolutionary, developed from the basic design of Ariane 1. Ariane 5, conversely, was a completely new design, first launched in 1996 but not completely successful until its third flight in 1998. The new Ariane 6 combines an upgraded version of the Ariane 5 core stage with new solid-fuel boosters and an upper stage powered by a restartable engine called Vinci.

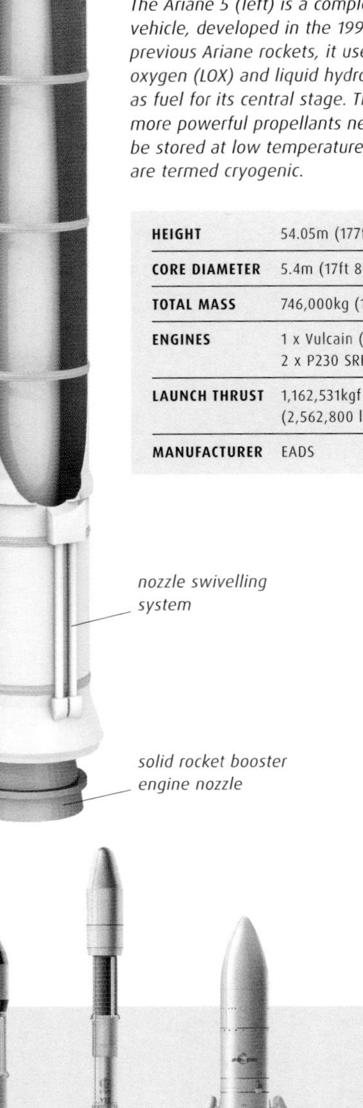

Japan in space

For a nation that did not fire even an experimental rocket until the mid-1950s, Japan rapidly established itself as a force to be reckoned with, launching a multitude of satellites and spaceprobes from 1970 onwards.

11 February 1970
Japan's first satellite, Ohsumi, is launched into orbit by an ISAS rocket.

7 April 1978
NASDA launches an experimental direct broadcast television satellite called Yuri.

March 1986
The ISAS probes Suisei and Sakigake fly past Halley's Comet.

2 December 1990
Reporter Toyohiro Akiyama becomes Japan's first cosmonaut.

12 September 1992
Mamoru Mohri becomes the first Japanese astronaut to fly on the Space Shuttle.

1 October 2003
NASDA, ISAS, and the NAL are merged to form a new agency, JAXA.

19 November 2005
The Hayabusa probe lands on asteroid Itokawa to collect samples.

20 May 2010
An H-II rocket launches the Venus probe Akatuski and a prototype "solar sail" spacecraft, IKAROS.

The beginnings of the Japanese space programme can be traced back to the enthusiasm of one man – aeronautical engineer and Tokyo University professor Hideo Itokawa, nicknamed Dr. Rocket. After conducting a number of successful launches of small rockets in the 1950s, Itokawa persuaded the Japanese government to fund his Institute of Space and Astronautical Science (ISAS).

In 1969 a second organization, the National Space Development Agency of Japan (NASDA) was established, charged by the government with turning Japan into a major space power. Although the presence of rival agencies led to occasional clashes, and split the available expertise, each body had its own specialist areas: ISAS concentrated on space research projects such as astronomy satellites and planetary probes, while NASDA focused on the development of commercial launch vehicles, other satellite applications, and manned spaceflight.

Both agencies had an impressive record of success throughout the 1970s and 1980s, but a series of expensive failures in the 1990s ultimately forced the Japanese government to merge them (along with NAL, Japan's National Aerospace Laboratory) into a new organisation, the Japan Aerospace Exploration Agency (JAXA) in 2003.

Launch vehicles and satellites

Under Itokawa's direction, ISAS developed a series of small solid-fuelled launchers known as the Lambda (L) and Mu (M) series from the 1960s onwards, and it was an L-4S that carried Ohsumi, Japan's first satellite, into orbit in February 1970. In keeping with the agency's main fields of interest, ISAS satellites focused on studying the environment around the Earth and on orbital astronomy. Successes include the Yohkoh solar observatory (a joint project with the United States and the United Kingdom) and the HALCA (Highly Advanced Laboratory for Communications and Astronomy) radio astronomy mission.

NASDA's launchers, in contrast, were larger and liquid fuelled – their first N-series rockets were effectively US Deltas (see p.245) built under license in Japan. However, continued improvements have seen the Japanese rockets evolve along their own route, to a point where JAXA's current H-IIA has an entirely Japanese design (see panel, opposite).

An N-1 rocket launched NASDA's first satellite, called Kiku, in 1975. Although this was only an engineering test, it paved the way for an array of different applications including a network of comsats, a direct broadcast satellite TV system, weather satellites, and Earth and ocean observers. Recently, Japan has followed China in developing a recoverable satellite system, the Unmanned Space Experiment Recovery System (USERS).

OHSUMI
The 24kg (53lb) Ohsumi was purely a test satellite, sending out a signal that revealed its location and the readings of simple instruments onboard.

N-1 LAUNCH
An N-1 rocket soars into the skies of Tanegashima in the mid-1970s. Its resemblance to the US Delta rocket is obvious.

BIOGRAPHY

TOYOHIRO AKIYAMA

The first Japanese person in space, Toyohiro Akiyama (b.1942) can also claim a place in the history books as the first journalist in space. His trip to Mir aboard Soyuz TM-11 in 1990 was paid for by the Tokyo Broadcasting System (TBS) network, and he made several live TV broadcasts during his time in space. Akiyama beat NASDA astronaut Mamoru Mohri into space after the launch of the Japanese-sponsored Spacelab-J Shuttle mission was delayed in the wake of the *Challenger* disaster.

SPACEPORT PANORAMA
JAXA's main launch site, at Tanegashima, is probably the world's most picturesque spaceport – although launches do have to be timed to fit around the local fishing industry.

1956 1957 1958 1959 1960 1961 1962 1963 1964 1965 1966 1967 1968 1969 1970 1971 1972 1973 1974 1975 1976 1977 1978 1979 1980 1981 1982 1983 1984 1985 1986 1987 1988 1989 1990 1991 1992 1993 1994 1995 1996 1997 1998 1999 2000 2001 2002 2003 2004 2005 2006 2007 2008 2009 2010 2011 2012 2013 2014 2015 2016 2017 2018 2019 2020

DR. ROCKET
Hideo Itokawa, musician, ballet dancer, and pioneer of the Japanese space programme, poses next to one of his experimental "Baby" rockets of the late 1950s.

Just as Europe's first venture beyond Earth orbit was the Giotto probe, so too Japan's Sakigake and Suisei 1986 missions to Halley's Comet paved the way for later missions to other worlds, though these missions met with mixed success.

Hiten, a technology test probe designed to relay signals from a small lunar orbiter, functioned well but was rendered useless when its partner mission, Hagomoro, failed. Nozomi, an orbiter intended to study the Martian atmosphere, missed entering orbit around Mars after running short of fuel. In contrast, the Hayabusa mission landed on the Near Earth Asteroid 25143 Itokawa in November 2005. Although a lander element failed to deploy, it successfully collected dust samples from the surface, returning them to Earth via a re-entry capsule in 2010. The SELENE lunar orbiter, launched in 2007 for a two-year mission (see p.259), was an unqualified success, but in 2010 the Akatsuki probe failed to enter its planned orbit around Venus. Fortunately, it was finally able to enter a different orbit five years later, becoming the first Japanese spacecraft in orbit around another planet. JAXA is also a partner in the BepiColombo Mercury mission with ESA (see p.306) and has plans for a sample-return mission to the Martian moon Phobos.

Japan's astronauts

A number of Japanese astronauts and cosmonauts have reached orbit. NASDA had agreements with NASA for astronauts to fly as payload specialists on the Space Shuttle, but delays caused by the loss of *Challenger* meant that journalist Toyohiro Akiyama beat NASDA's specialists to become the first Japanese person in space (see panel, opposite). The first NASDA astronaut, Mamoru Mohri, flew aboard the Japanese-sponsored Spacelab-J Shuttle mission in 1992. With the end of the Shuttle programme, the future of Japanese manned spaceflight was assured by NASDA's (and now JAXA's) role in the International Space Station – in particular, the complex Japanese Experimental Module (JEM, also known as Kibo, meaning "hope").

The largest module on the ISS, JEM consists of several components that were launched on various Shuttle flights in 2008–09. Japan has also developed an unmanned cargo spacecraft, the H-II Transfer Vehicle – or HTV – for resupplying the ISS. In the longer term, JAXA still has plans for a small rocket-launched spaceplane and single-stage-to-orbit (SSTO) launchers.

TECHNOLOGY

H-II LAUNCH VEHICLES

JAXA's current workhorse launcher is the H-IIA, a basic two-stage rocket fuelled by liquid hydrogen and oxygen, to which a variety of booster modules can be added if necessary, depending on the size of payload and its destination. The H-IIA was developed from NASDA's earlier H-II, which used similar fuels and had similar flexibility, but suffered major reliability problems that brought NASDA to a crisis. Since its first flight, in August 2001, the H-IIA has suffered just a single failure, and JAXA has subsequently developed a more powerful version, the H-IIB (right) – primarily used for launching HTV cargo ferries to the ISS.

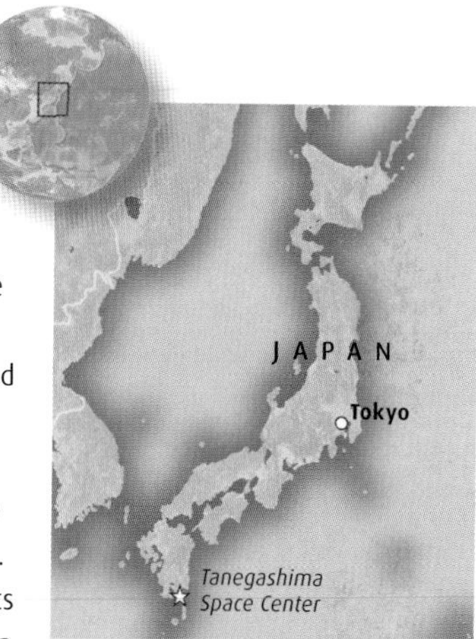

JAPAN'S SPACE CENTRE
Located on an island to the south of Kyushu in southern Japan, Tanegashima Space Center lies as close to the equator as possible, but is still at latitude 30°N.

发射流程
06 48

Long March into space

Although China was the birthplace of the rocket, it was slow to become a modern space power. Yet it now has its own military programme and operates commercial launch vehicles and satellites.

SPACE PROPAGANDA
Just like the rival powers of the Space Race, China today presents its space programme as a symbol of national pride and superiority, as seen in posters like this one.

The catalyst for the formation of a national space programme in China dates back to the mid-1950s, when the recently established communist government welcomed the return of Tsien Hsue-shen from the USA (see panel, right). At the time, relations between the People's Republic of China and the Soviet Union were at their warmest, and an agreement signed in 1956 kick-started the Chinese missile programme by providing access to Soviet technology and expertise. However, China was far from immune to the political machinations and dogmatism that often frustrated Soviet space scientists, and although missile development continued to be a priority, support for the space programme waxed and waned – particularly after relations with the Soviets began to deteriorate in 1960.

Nevertheless, in 1970 China launched its first satellite, Dong Fang Hong-1. This was an updated equivalent of Sputnik 1, orbiting the Earth while transmitting a recording of the communist patriotic song *The East is Red*. The launch vehicle was Tsien's own Chang Zheng ("Long March", abbreviated as either CZ or LM) design – which is still in use today, albeit in a different form (see panel, below). The success of the CZ rocket allowed China to enter the commercial launch market in 1985. Early customers included the comsat companies Asiasat in Hong Kong and Optus in Australia, as well as organizations in Sweden and Pakistan. Much of the Iridium satellite phone network has also been launched from China.

XICHANG CONTROL
Uniformed officers monitor a simulated launch from the control centre at Xichang Satellite Launch Centre, which is situated some 6km (4 miles) from the launch pad.

BIOGRAPHY

TSIEN HSUE-SHEN

Widely regarded as the founding father of China's space programme, Tsien Hsue-shen (1911–2009) also played a key role in the early US space effort. Born and educated in China, he won a scholarship to study in the US in 1935. After moving to the California Institute of Technology in 1936, Tsien began work on experimental rockets. During the Second World War, he helped establish the Jet Propulsion Laboratory (JPL), but in 1950 he was arrested at the height of US anti-communist paranoia. After deportation to China in 1955, he did indeed join the Communist Party, leading the Chinese effort to develop ballistic missiles and launch vehicles.

A range of satellites

Since the first launch, China has put more than 50 satellites in the Dong Fang Hong (DFH) series into orbit – though, rather like the Soviet Cosmos satellite series, the DFH designation is used as a catch-all for a wide range of satellites, including remote-sensing missions, atmospheric studies, and a system of regional comsats also known as Chinasat.

A separate series of spacecraft is the FSW (a Chinese acronym for recoverable test vehicle). These satellites carry a capsule that can be used to return the results of satellite experiments to Earth, giving China an ability matched only by the US and Russia. Re-entry technology was developed as part of China's secret military satellite programme (in order to return photographic film from orbiting spy cameras), but it has also been used to return remote-sensing data and a variety of experiments from orbit.

In keeping with the Chinese attitude to their space programme, the technology was soon made commercially available, and the first customer to take advantage of the service was the French company Matra in 1987.

RUSSIA
Jiuquan Satellite Launch Centre
Beijing
Taiyuan Satellite Launch centre
Xichang Satellite Launch Centre
CHINA
INDIA
Wenchang Satellite Launch Centre

CHINESE SPACE FACILITIES
China's four major spaceports each specialize in a different type of launch. Geostationary satellites launch from Xichang or Wenchang, polar satellites from Taiyuan, and satellites with inclined orbits from Jiuquan.

TECHNOLOGY

LONG MARCH CZ ROCKETS

China's launchers were originally based on the DF-5 ICBM, but soon developed along their own path. The CZ-2 is a family of launchers built around a common core, to which a variety of boosters and extra stages can be added. Despite the introduction of the CZ-3, which uses more powerful liquid hydrogen and oxygen as propellants, the older rocket is still in use. Development problems for the CZ-3 culminated in a 1996 disaster when a rocket hit a village near its launch site. Following a thorough overhaul, the CZ-3 has operated flawlessly. Two new vehicles, the CZ-5 and CZ-6, have debuted in recent years.

24 April 1970
China launches its first satellite, Dong Fang Hong-1 (DFH-1, sometimes also known as China-1), on a CZ-1 Long March rocket.

26 November 1975
The maiden flight of the CZ-2C rocket puts an FSW surveillance satellite, with a recoverable film capsule, into orbit.

8 April 1984
Dong Fang Hong-2 is launched by a CZ-3 rocket to become China's first geostationary comsat.

25 October 1985
China announces that it will make its launch vehicles and facilities commercially available.

7 September 1988
China launches its first weather satellite, Fengyun-1.

7 April 1990
China fulfils its first commercial contract, launching the Asiasat-1 comsat.

14 October 1999
The Ziyuan-1 remote-sensing satellite, a joint project of China and Brazil, is launched.

20 November 1999
China launches the first spacecraft in its new Shenzhou programme.

The spread of spaceflight

Since the 1960s, a handful of nations other than the major space powers have developed their own launchers and satellite programmes as commercial enterprise encourages access to space.

29 September 1962
The first non-superpower satellite, Canada's Alouette 1, is launched by a US rocket.

19 April 1975
India's first satellite, Aryabhata, is launched by the Soviet Union.

18 July 1980
India launches the Rohini 1B test satellite with its own SLV-3 rocket.

19 September 1988
Israel launches its first satellite, the prototype for a series of Ofek reconnaissance satellites.

18 April 2001
The first flight of India's Geosynchronous Satellite Launch Vehicle launches the GSAT-1 test satellite.

23 October 2004
Brazil successfully launches a VSV-30 rocket carrying a suborbital test satellite.

2 February 2009
Iran successfully orbits its Omid telecommunications satellite using the Safir launch vehicle.

21 January 2018
The US/New Zealand firm Rocket Lab successfully deploys satellites on the second flight of its Electron rocket.

Several countries have developed national satellite projects that have been sent into orbit with the help of launch vehicles supplied by the more established space nations. Canada was the first non-superpower country to have its own satellite in orbit: Alouette 1 was launched by a US Thor-Agena B rocket in September 1962 and led to a whole series of similar satellites, as well as later national projects such as the Anik comsats.

Other nations soon followed. NASA launched Italy's San Marco 1 atmospheric probe in 1964, Britain's Ariel 3 and Australia's Wresat in 1967, and Germany's Azur 1 to investigate the Van Allen Belts in 1969. A more ambitious German mission followed in 1974 – the first Helios probe, designed to orbit close to the Sun. In the 1970s, the growth of satellite applications, and the establishment of specialist organizations and businesses to build satellite networks, broadened the launch market beyond national governments. By the 1980s, NASA had competition for these hungry customers – at first from ESA, but then from China and even the Soviet Union.

But, while most nations seem content to pay for their launches or enter into collaborations with the major space powers, a few countries have worked hard to develop their own launch capability.

ALOUETTE 1
The first Canadian satellite, designed to study the ionosphere layer of Earth's upper atmosphere, established a close relationship between NASA and Canada.

India in space

India's space programme began in the mid-1960s with the foundation of ISRO, the Indian Space Research Organisation. ISRO developed the country's first satellites, including Aryabhata, launched by a Soviet rocket in 1975. India has tended to concentrate its satellite work in fields of national interest, producing remote-sensing satellites and comsats to aid the country's development. The first communications satellite, INSAT 1A, was launched in 1982 on a NASA Delta rocket, and India was also involved in the development of Direct Broadcast satellite television through the 1976 Satellite Instructional Television Experiment (SITE), which beamed educational programming to remote villages through a NASA satellite. Recently, ISRO revisited the concept with its EDUSAT mission, launched in September 2004. Indian remote-sensing satellites, meanwhile, have concentrated on hydrology and mineralogy, looking for vital water deposits and potentially valuable mineral resources.

Since the 1970s, ISRO has also worked to develop a series of SLV (Satellite Launch Vehicle) rockets. The four-stage, solid-fuelled SLV-3 launched its first payload, a test satellite called Rohini 1B, in 1980. For heavier payloads, a modified version called the ASLV (Advanced SLV) is assisted at launch by twin boosters.

The more powerful PSLV (Polar SLV) is a four-stage rocket with alternating solid- and liquid-fuelled stages. It made its first flight in 1994 and has become India's most widely used launcher. A Geosynchronous SLV (GSLV), able to lift heavier loads to higher altitudes, was introduced in 2001.

All ISRO vehicles are launched from the Satish Dhawan Space Centre, on the island of Sriharikota off India's southeastern coast. From here, rockets are launched over the Bay of Bengal at a latitude of about 14°N.

GEOSYNCHRONOUS SLV
India's latest launcher, the GSLV blasts off with EDUSAT aboard on 20 September 2004. The GSLV is a three-stage rocket with a solid first stage, a second stage based on traditional liquid fuels, and a new third stage fuelled by liquid hydrogen and oxygen.

STACKING EDUSAT
The EDUSAT satellite is winched into place on top of its GSLV launch vehicle. The satellite provides interactive satellite TV-based education in remote areas of India.

BIOGRAPHY

VIKRAM SARABHAI

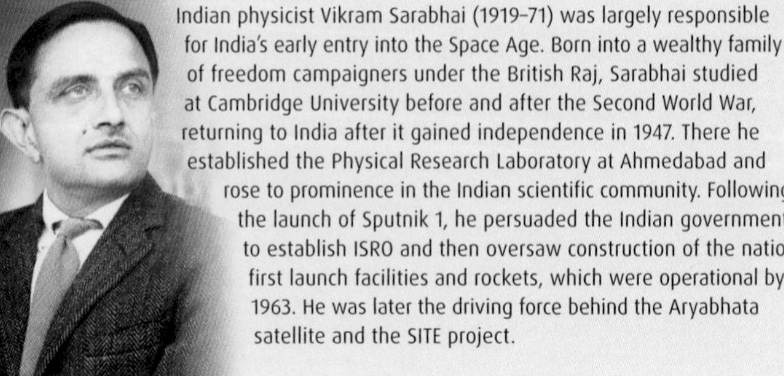

Indian physicist Vikram Sarabhai (1919–71) was largely responsible for India's early entry into the Space Age. Born into a wealthy family of freedom campaigners under the British Raj, Sarabhai studied at Cambridge University before and after the Second World War, returning to India after it gained independence in 1947. There he established the Physical Research Laboratory at Ahmedabad and rose to prominence in the Indian scientific community. Following the launch of Sputnik 1, he persuaded the Indian government to establish ISRO and then oversaw construction of the nation's first launch facilities and rockets, which were operational by 1963. He was later the driving force behind the Aryabhata satellite and the SITE project.

Israel in orbit

The Israeli space effort got underway in earnest in 1983 and is managed by ISA, the Israel Space Agency. ISA's launch vehicle, the Shavit, is based on the Jericho ballistic missile, solid-fuelled, and capable of launching only small payloads. Although it launched the first Israeli satellite, Ofek 1, in 1988, and has since launched several more of these reconnaissance satellites, other ISA satellites have been launched on European, Russian, and Indian rockets. Shavit's limitations are partly due to regional politics – in order to avoid overflying neighbouring Arab states, the rocket is launched westwards over the Mediterranean, so that it is effectively slowed down, rather than boosted, by Earth's rotation.

FIRST VLS-1 LAUNCH
The first prototype of Brazil's Veiculo Lancador de Satelites blasts off from the Alcântara launch centre close to the equator. The rocket was destroyed 65 seconds into its flight after veering off course due to a failed booster.

Brazil and beyond

For Brazil, meanwhile, the road to space has proved anything but easy. Two prototypes of its VLS-1 launch vehicle failed during tests, and a third exploded when one of its solid-fuel rockets ignited prematurely, killing 21 people and levelling the launch facility in August 2003. Although the country bounced back to make a suborbital spaceshot a year later, development has since faltered, with attempts at revival through partnerships with other space powers resulting in only sporadic advances and occasional reversals. Under the latest Southern Cross programme, it's hoped that assistance from the Russian space agency Roscosmos will see descendants of the VLS finally reach orbit in the early 2020s.

While Brazil has struggled openly, however, it has been beaten into space by two rather more secretive powers. In 2009, Iran launched its first satellite, a telecommunications test-bed called Omid, into orbit using the two-stage Safir launch vehicle. It has since launched other satellites, and sent a number of animal payloads, including monkeys, on suborbital flights using the Kavoshgar sounding rocket. North Korea launched its first satellite, the Earth observatory Kwangmyŏngsŏng-3 Unit 2, in December 2012.

New Zealand became the most recent nation to join the space club in 2017, when a privately funded spaceport was established by the commercial manufacturer Rocket Lab on North Island's remote Mahia Peninsula. An Electron rocket launched from here reached orbit and deployed three small CubeSat in January 2018.

NEW ZEALAND'S FIRST
Rocket Lab's first Electron launch vehicle, designated "It's a Test", is transported prior to launch in May 2017. This innovative rocket uses electric fuel pumps and 3-D-printed engine parts.

International astronauts

The vast majority of spacefarers have been citizens of the former Soviet Union or the United States, but politics and commerce have led to a variety of nations getting their own astronauts into space.

ILAN RAMON
Israel's first astronaut was killed returning to Earth during the Columbia Space Shuttle disaster of 2003.

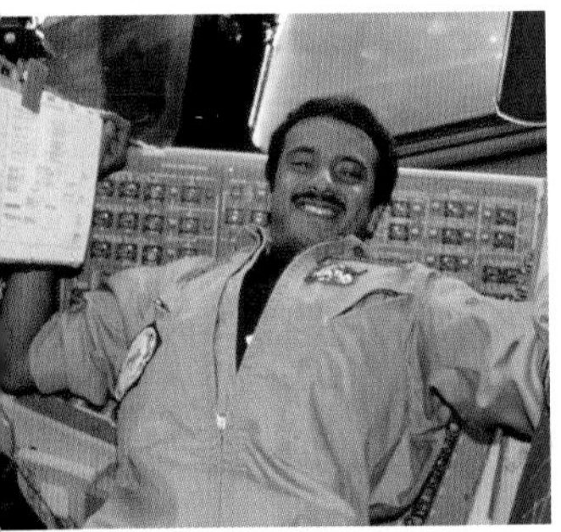

PRINCE SULTAN AL-SAUD
Saudi Prince Sultan Al-Saud flew aboard Discovery *during the 1985 launch of the Arabsat-1B comsat.*

TAKAO DOI
Astronaut Takao Doi flew aboard the Space Shuttle Columbia *on mission STS-87 in 1997 and became the first Japanese person to perform a spacewalk.*

The Soviet Union was first to recognize the political capital to be made by taking passengers into space. Since the Soyuz spacecraft servicing their space stations had an operational life shorter than the typical station mission, visiting cosmonaut crews were used to bring fresh spacecraft to the station – arriving in a new Soyuz and departing aboard the older one. These brief visits offered the opportunity to take along cosmonauts from other nations – a programme the Soviets called Intercosmos.

At first, invitations were dictated by the politics of the Cold War – guest cosmonauts came from members of the Warsaw Pact alliance or from nations with which the USSR wanted to maintain good relations. Beneficiaries included Czechoslovakia (Vladimir Remek, 1978), Poland (Miroslaw Hermaszewski, 1978), Vietnam (Tuan Pham, 1980), and India (Rakesh Sharma, 1984). Although the operating life of the spacecraft improved with the Soyuz T and TM models, the duration of the long-term residencies also grew greater, so there was still a need to bring new spacecraft. The reintroduction of three-man flights with Soyuz T also meant that there was almost always a spare seat onboard.

From the mid-1980s onwards, as relations between the West and the Soviet bloc finally thawed, the range of guest cosmonauts increased to include passengers such as Helen Sharman from the UK (see p.217). The eventual collapse of the Soviet Union in 1991 and the appearance of a newly capitalist, impoverished Russia led to further expansion of the programme. Mir was "open for business", and foreign visitors, now paid for by their own governments, space agencies, or even private companies, jostled on the station with American astronauts. Towards the end of Mir's operating life, France paid for its spationaut Jean-Pierre Haigneré to stay on the station for a full six-month research tour.

Shuttle visitors

NASA did not use the Space Shuttle for such overtly political purposes, but the nature of Shuttle crews meant that there were soon opportunities for foreign astronauts to take their place onboard. While NASA's own pilots and mission specialists (professional astronauts) are US citizens or naturalized Americans, Shuttle payload specialists were chosen in collaboration with mission sponsors, and as a result international astronauts from various partner agencies frequently joined Shuttle crews. The agreement to build Spacelab automatically assured ESA of a number of flights for its astronauts, and Japan and Germany both sponsored additional Spacelab missions in the 1980s and early 1990s. Involvement in the ISS has seen frequent European astronaut flights continue in the post-Shuttle era.

NASA also has a special relationship with the Canadian Space Agency, which supplied the robot-arm systems for both the Space Shuttle and the ISS. As a result, Canada was invited to nominate several people to fly aboard the Shuttle, kick-starting its own astronaut programme.

VLADIMIR REMEK
Czechoslovakia's first cosmonaut, and the first spacefarer from a non-superpower nation, Vladimir Remek (left) flew on the Soyuz 28 mission of 2–10 March 1978.

DOUBLE FIRST
The 2015 Soyuz TMA-18M mission to the ISS carried Kazakhstan's and Denmark's first astronauts, Aidyn Aimbetov (right) and Andreas Morgensen (left), under Russian commander Sergey Volkov (centre). Citizens from 40 different nations have now flown in space.

2 March 1978
Vladimir Remek becomes the first non-Russian, non-American in space aboard Soyuz 28.

26 August 1978
East German Sigmund Jähn becomes the first German cosmonaut.

18 September 1980
Arnaldo Tamayo-Méndez becomes the first Cuban in space.

24 June 1982
Jean-Loup Chrétien of France becomes the first spationaut.

28 November 1983
Ulf Merbold becomes the first West German astronaut, aboard *Columbia*'s STS-9 flight.

3 April 1984
Rakesh Sharma becomes India's first cosmonaut.

5 October 1984
Marc Garneau becomes Canada's first astronaut with his flight aboard the Space Shuttle *Challenger* during STS-41G.

18 May 1991
Helen Sharman becomes the first Briton in space.

19 May 1996
Australian-born Andrew Thomas travels on the Space Shuttle *Endeavour*'s STS-77 mission.

CANADARM
The Canadian Space Agency's special relationship with NASA has allowed several Canadians to fly on the Space Shuttle. Here astronaut Chris Hadfield works in the Shuttle cargo bay during Endeavour's *STS-100 mission. The Canadian-built robot-arm (or Remote Manipulator System) is prominent in the foreground.*

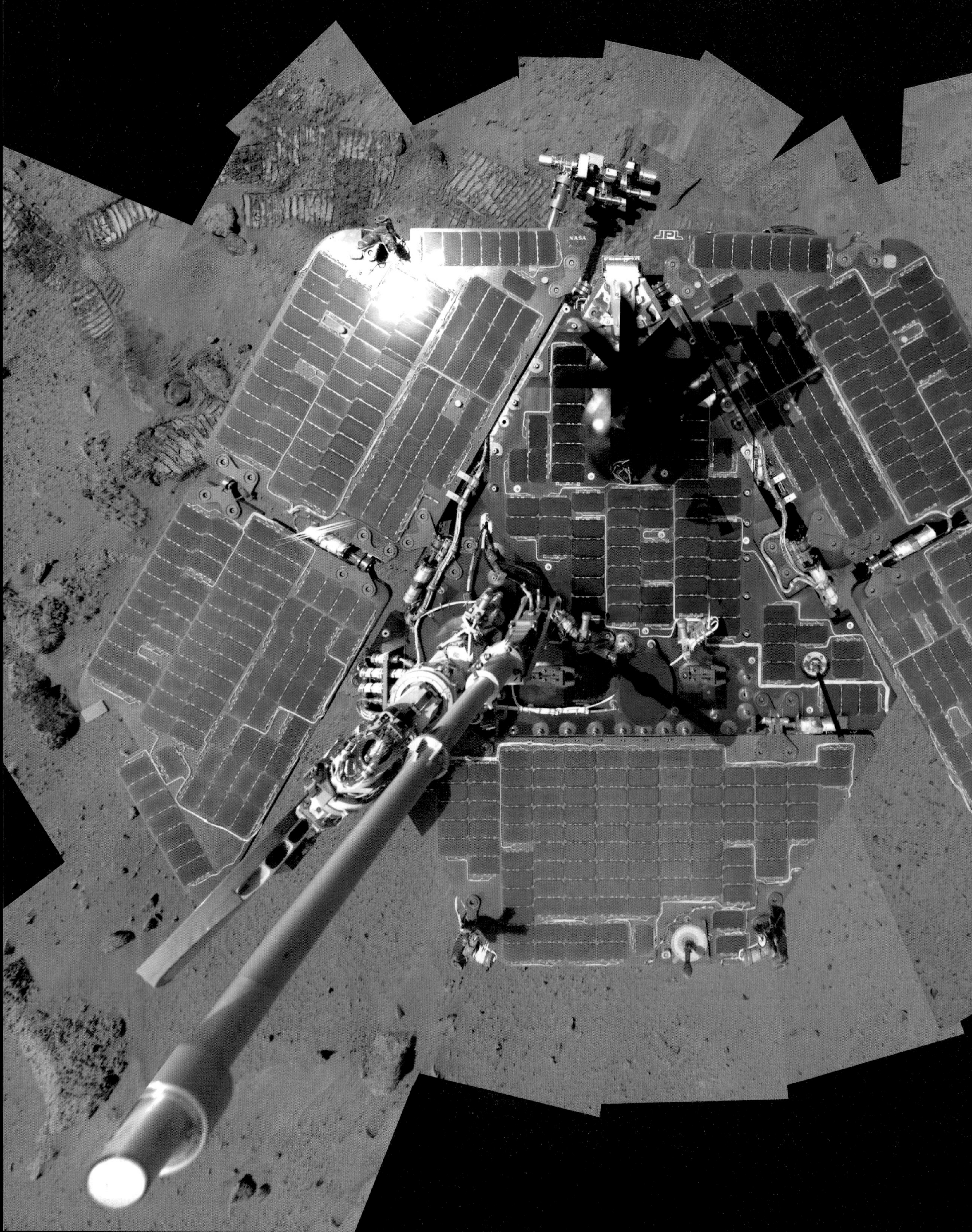
NASA
JPL

SATELLITES AND SPACEPROBES

THE SPACE AGE HAS TRANSFORMED our understanding of our own world and others. In 50 years, artificial satellites have gone from propaganda weapons to vital tools of humanity, watching over the Earth. They have gathered information on everything from hidden mineral deposits to long-term climate change, provided a new view of the Universe from beyond the atmosphere, and triggered a global revolution in communications that has helped to make the world a smaller place.

And while manned exploration of space has gone no further than our own Moon, our robot servants have gone much further. Spaceprobes have explored all the other major worlds that orbit the Sun, in addition to a number of the smaller objects that clutter the Solar System. They have ventured into environments that would quickly prove fatal to astronauts and to distant regions that would take decades for a manned mission to reach. The images and data they have sent back have not only revealed previously unseen worlds but also aided understanding of our own.

AUGUST 2005: ROVER ON A RED PLANET
Turning the gaze of its panoramic camera downwards, the Mars Exploration Rover Spirit photographs itself amid the dust of a Martian desert. After almost 20 months on the surface of Mars, the rover's solar arrays – providers of its energy and life – are still gleaming through only a thin veneer of dust.

4 October 1957
The satellite age begins with Sputnik 1's launch.

31 January 1958
Explorer 1 becomes the first satellite to return scientific data from orbit.

7 August 1959
Explorer 6 returns the first television pictures from orbit.

1 April 1960
NASA launches the first successful weather satellite, TIROS 1.

19 August 1960
The US retrieves useful reconnaissance images from orbit for the first time, returned from the Discoverer 14 satellite.

7 March 1962
NASA launches Orbiting Solar Observatory 1, arguably the first astronomical satellite.

16 August 1964
Syncom 3, the first successful geostationary comsat, is launched.

23 April 1965
The Soviet Union launches Molniya 1-01, the first comsat with a highly elliptical orbit.

23 July 1972
NASA launches ERTS-1, the first Earth resources satellite.

Orbiting the Earth

In the decades since Sputnik 1 first sent a simple radio signal back to Earth, artificial satellites have transformed not only our view of Earth and the wider Universe but also many aspects of everyday life.

As with many revolutionary technologies, it took some time for the full potential of the artificial satellite to be recognized. The wave of apprehension that greeted news of Sputnik 1 in the West was born of a fear that satellites might be used as weapons platforms to rain down missiles on a defenceless enemy. The military on both sides were also aware that satellites could act as spies in the sky, beyond the range of ground-based weapons. Indeed, many early US experimental satellites, such as Discoverer and CORONA (see p.250) and some of the Soviet Cosmos series, were in fact orbiting spy cameras. Another early application lay in monitoring large-scale weather patterns from above – the first NASA weather satellite, TIROS (Television Infrared Observation Satellite) 1, was launched in April 1960.

It took longer, however, for scientists to confirm the potential of satellite-borne cameras for studying the Earth in general. After Gordon Cooper returned pin-sharp photos and reported seeing individual buildings from his *Faith 7* Mercury capsule in 1963, NASA began to develop these applications in earnest. Today, remote-sensing satellites study many aspects of our planet's geology, oceanography, climate, and ecology. Astronomers have also made use of satellites – a location above the atmosphere holds obvious advantages for instruments requiring a clear view into deep space. Another predicted application was the communications satellite, or comsat – an orbital platform to bounce signals between distant places on Earth, overcoming the limitations of ground-based radio signals (see p.246). However, even enthusiasts could not foresee the revolution that comsats would trigger.

Orbital mechanics

The shape of a satellite's orbit depends on its function. The further from the Earth a satellite is, the longer it takes to complete an orbit – not only because it has further to travel, but also because it moves more slowly along its path through space, thanks to the Earth's weaker gravitational pull. The particularly useful geostationary orbit, first identified by Arthur C. Clarke (see p.246), is used by comsats and other craft that need to stay over a single spot on the Earth's surface. In this orbit, a satellite sits precisely 35,786km (22,227 miles) above the equator, orbiting the Earth in 23 hours 56 minutes – the same time the planet itself takes to rotate. The satellite therefore stays over a single point on the equator, providing a fixed location in the Earth's skies.

However, most satellites fly much closer to our planet, in low Earth orbit (LEO) a few hundred kilometres up, circling the world several times every day. This kind of altitude is close enough in for Earth-observing satellites to pick up fine details but far enough out to avoid drag from the upper atmosphere. For astronomy satellites, it is also high enough to avoid the atmospheric absorption of various types of radiation (see pp.250–57). Only space stations and large spacecraft typically orbit lower than this, skimming the upper atmosphere so that their orbits are unstable unless repeatedly boosted.

Very few orbits are circular – most are ellipses, dipping closer to the Earth on one side and rising higher above it on the other. Often the difference

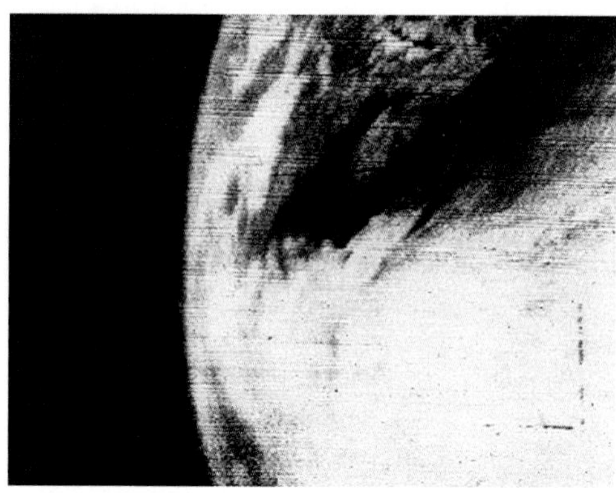

LIVE FROM ORBIT
After early experiments with the fax-like approach of scanning photographic film (see p.53), engineers soon realized that the most efficient way of returning moderate-resolution images from space was to use television camera tubes that transform images directly into electronic signals. This principle was trialled in early weather satellites such as TIROS 1.

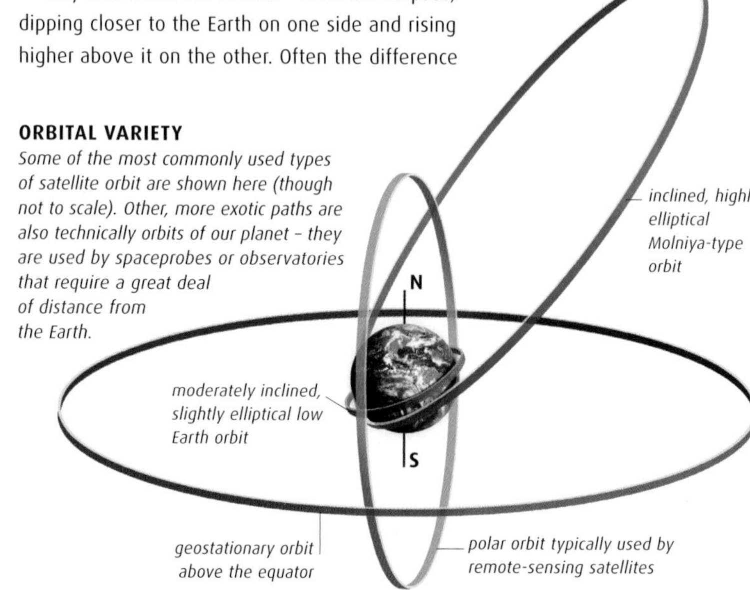

ORBITAL VARIETY
Some of the most commonly used types of satellite orbit are shown here (though not to scale). Other, more exotic paths are also technically orbits of our planet – they are used by spaceprobes or observatories that require a great deal of distance from the Earth.

is minor and has no operational effect, but some projects (such as the Soviet Molniya satellites) use highly elliptical orbits and deliberately take advantage of the satellite's slower speed when further from the Earth. In Molniya's case, the shape of the satellites' orbits ensures that they move very slowly across Russian skies, acting as easily tracked communications platforms in northern regions, where equatorial comsats may not be visible.

POLAR COSMODROME
Russia's northern cosmodrome at Plesetsk near Archangel is an ideal site for launching satellites into polar orbits and also into the highly inclined Molniya orbits used by Russian comsats.

BETWEEN ORBITS
Many satellites have their own integral rocket stages, called kick motors, to push them into their final orbit. Intelsat 603, launched into LEO by a Titan rocket in 1990, became stranded after its kick motor failed. It took a rescue mission from the Space Shuttle (see p.207) to finally get it to geostationary orbit.

Differing inclinations

Another important factor in a satellite's orbit is its inclination, or tilt relative to the equator. Although there are fuel advantages to launching satellites into orbits over the equator (see p.249), such orbits are useless for applications such as remote sensing, since the spacecraft will pass repeatedly over the same narrow strip of land and ocean. Instead, many types of satellite use inclined orbits that take them over high latitudes. Although the satellite only covers a narrow strip of land during each pass, the combined effect of repeated orbits and the Earth's daily rotation can gradually build up coverage of much of our planet's surface. The most extreme inclinations, known as polar orbits, allow a satellite to study the entire planet – but, just as some launch sites make it easier to reach equatorial and geostationary orbits, so only certain sites are suitable for polar launches.

TECHNOLOGY

DELTA LAUNCHERS

The Delta rocket originated as a three-stage vehicle built for NASA by the Douglas Aircraft Company (now part of Boeing), using elements of the Thor missile and the Vanguard rocket. Deltas rapidly became a mainstay of the US satellite launch programme, but after five decades of constant improvement the current variants bear little resemblance to the original. The highly successful Delta II, retired in 2018 after more than a hundred launches, was a three-stage rocket assisted by nine solid-fuelled boosters around its base, while the Delta IV is a modular system that may use a larger core assisted by two boosters, or even multiple rocket cores to launch heavier loads, as seen here.

Communication satellites

Orbiting communication relays have transformed our planet into a "global village" with near-instantaneous links between remote parts of the world.

10 July 1962
Telstar, the world's first fully functional comsat, is launched.

14 February 1963
Syncom 1 is the first attempt to launch a geostationary comsat.

19 August 1964
Syncom 3 becomes the world's first truly geostationary comsat.

20 August 1964
INTELSAT is established to set up a global comsat network.

23 April 1965
The Soviet Union launches the first of its Molniya comsats.

30 May 1974
NASA's Applications Technology Satellite (ATS) 6 is launched. As the first satellite to offer direct broadcast to relatively small receivers, it paves the way for satellite TV.

5 May 1997
The first five satellites in the Iridum telephone network are launched aboard a Delta II rocket.

16 August 2016
China launches the Micius satellite as part of its Quantum Experiments at Space Scale (QUESS) programme to test secure quantum cryptography.

Even before the dawn of the Space Age, it was clear that satellites had the potential to revolutionize communications, and the military were among the first to see the advantages. Telephone lines had limited capacity and were vulnerable to physical damage. Radio signals were fast but travelled in straight lines and so had limited range due to the curvature of the Earth (it was sometimes possible to bounce signals off the reflective ionosphere layer in the upper atmosphere, but results could be unpredictable). In contrast, a satellite high in the sky could be used as a reflector, allowing signals beamed from one part of the world to be received anywhere else that the satellite was visible above the horizon. NASA's Echo project (see p.51) tested the idea with a simple passive reflector in orbit, while Telstar (see panel, right) was the first satellite with the capability to receive, amplify, and re-transmit signals.

Geostationary satellites

For many applications, the ideal comsat orbit is the geostationary one noted by Arthur C. Clarke in 1945 (see panel, below, and p.245). In this orbit, a satellite maintains a steady position in the sky, and a receiver or transmitter aimed at it will not need to move to track it. NASA pioneered the use of this orbit with its experimental series of Syncom satellites. The first truly geostationary communications satellite, Syncom 3, was launched in August 1964 and had an immediate impact as it carried live pictures from the 1964 Tokyo Olympics to American broadcasters.

HISTORY FOCUS

TELSTAR

After various experiments with orbiting reflectors and taped transmissions from orbit, Telstar became the world's first true communications satellite when it was launched by a Delta rocket on 10 July 1962. Built at Bell Telephone Laboratories in the United States, the satellite was part of a joint programme developed by Britain, France, and the US. Telstar had the ability to receive signals from the ground, amplify them (using power from the solar cells on its spherical surface), and retransmit them through the horn antennae around its equator. It went into service as soon as it reached orbit, successfully transmitting television signals, phone calls, and even faxes across the Atlantic.

The benefits of comsats were so clear that, even as the technology was being proved, President Kennedy was calling for the establishment of an international organization to set up a global communications network. Nine months after Kennedy's death, the International Telecommunications Satellite Organization, or INTELSAT, came into being in August 1964 with 11 initial member states. It launched its first satellite, Early Bird, in April 1965, and its membership grew to more than 100 nations before it was privatized in 2001. INTELSAT's success paved the way for commercial companies, and the organization also formed a template for similar regional comsat networks and other satellite applications.

BIOGRAPHY

ARTHUR C. CLARKE

After serving in the RAF during the Second World War (working on the development of radar), Arthur C. Clarke (1917–2008) became an active member of the British Interplanetary Society. From the 1940s, he found fame as a science-fiction writer and shrewd prophet of future technology – his most famous book is probably *2001: A Space Odyssey*. Clarke suggested the use of geostationary satellites as signal relays in a 1945 article for *Wireless World* magazine, and although he was not the first to note the usefulness of geostationary orbits, his (independent) proposal was the first to be widely noticed. However, Clarke did fail to predict the rise of microelectronics, believing at the time that his relays would have to be manned.

GROUND STATIONS
Although two-way links to low Earth orbit (LEO) satellites can be established with smaller antennae, special equipment is needed to send signals to comsats in geostationary orbit. Satellite ground stations typically use movable parabolic antennae that can generate and send high-frequency radio signals in a focused beam. The size of antennae needed for most applications has shrunk considerably since the 1960s, when huge dishes were built at sites such as the UK's Goonhilly Earth Station (shown here).

Communications from lower orbits

While geostationary orbits are well suited for tasks such as broadcasting television and radio signals to fixed domestic antennae, the power requirements and cost of launch are significant barriers. Other comsat applications use a different approach, with satellites in low or intermediate orbits that have a smaller footprint (the area on the ground from which they can be seen at a given time). Medium-Earth orbit satellites are typically tracked using moving antennae, while those in low Earth orbit rely on the "constellation" principle, in which a swarm of satellites in orbit ensures regular passage over the regions where communication is required. One well-known example is the Iridium network, a series of 66 satellites that offers direct satellite communications to anyone with a suitable telephone. Although landline and mobile telephony between major landmasses today rely mostly on fibre-optic undersea cables, Iridium and similar systems provide a vital link to remote areas and regions hit by natural disasters that knock out fixed infrastructure. Increasingly, constellations of small, relatively cheap satellites are being used to offer other innovative communications services, such as satellite-based internet access and the automated tracking of assets such as shipping containers.

Mixed signals

A key challenge faced by all comsats lies in transmitting and receiving multiple signals at once and maintaining the security and integrity of data. Today, most signals are digital in nature, and complex "multiplexing" procedures are used to interweave different streams of data that can then be decoded by the intended recipient. All modern comsats transmit and receive signals using microwaves (short-wavelength radio waves), but the precise "band" in which they do so varies depending on their orbit, available power, and data requirements – in general, lower-frequency signals carry less data but require less power.

In 2016, China was the first to test a technique that promises to provide new security for sensitive communications in the future. Its Micius satellite uses the principles of "quantum cryptography" – an application of the strange behaviour of subatomic particles that allows information to be sent between two locations in a stream that can only ever be read by the intended recipient – any attempt to intercept the signal inevitably corrupts it.

GEOSTATIONARY COMSAT

As the efficiency of solar cells has improved, modern communications satellites have been able to broadcast increasingly powerful signals, allowing for easier reception by Earth-based receivers.

IRIDIUM FLARE

Thanks to the orientation of their antennae, the first generation of Iridium satellites produce bright, predictable flares in the night sky when they catch the sunlight in orbit.

17 September 1959
Experimental navigation satellite Transit 1A is launched, but it fails to reach orbit.

13 April 1960
Transit 1B successfully reaches orbit on a Thor launch vehicle.

18 December 1962
The first element of the operational Transit system is launched into orbit on a Scout rocket.

22 February 1978
Construction of the US Global Positioning System begins with the launch of NAVSTAR 1.

12 October 1982
The Soviet Union launches the first satellite in its GLONASS system.

30 October 2000
China launches BeiDou-1A, beginning construction of its first satellite navigation network.

1 May 2001
"Selective Availability" of GPS is switched off, allowing civilian access to high-quality position data for the first time.

21 October 2011
The decade-long launch programme for Europe's Galileo Positioning System begins with the first of 30 operational satellites.

Satellite navigation

Adapting techniques from ground-based radio navigation to work with orbiting satellites has led to a revolution in the precision and ease with which we can calculate and track our location on the Earth's surface.

The principle of using orbiting satellites as navigation beacons emerged soon after the 1957 launch of Sputnik 1. US physicists William Guier and George Weiffenbach, working at the Johns Hopkins University on ways to discover more about the mysterious new object, realized they could analyze its orbit by studying the Doppler shift in its signals – the slight change of frequency or pitch produced as Sputnik moved towards or away from an Earth-based receiver. The following spring, working with their colleague Frank McClure, they realized that the principle could be reversed – if the orbits of several satellites were precisely known, then Doppler shifts in their signals could be used to calculate the exact location of the receiving station on Earth. By late 1958, the US Navy and NASA's defence counterpart, the Advanced Research Projects Agency (ARPA) had commissioned Johns Hopkins University to develop the Transit programme – a series of satellites that tested this principle and put it into practice. Between 1959 and 1988, several dozen Transit satellites were launched, but the Doppler approach had several drawbacks – most importantly, its accuracy was no greater than around 200m (660 ft). As a result, the Department of Defense sponsored work on faster systems throughout the 1960s, testing the ability to broadcast highly accurate time signals from orbit. From the late 1970s, the US began assembly of the NAVSTAR Global Positioning System – the first network to use the principles of modern GPS.

HISTORY FOCUS

TRANSIT SATELLITES

The Transit system was initially developed to offer precise satellite navigation to US Navy nuclear submarines, but it soon found many other uses in both military and civilian spheres. The satellites followed polar orbits at relatively low altitude, orbiting Earth in around 106 minutes. Five satellites were required in order to provide basic coverage (an opportunity to fix location to within a few hundred metres every few hours), but in practice ten were usually maintained in orbit in case some failed. Earlier Transit satellites such as Transit 2B (the lower of two satellites in the picture below, with the GRAB 1 signals intelligence satellite sitting above it) weighed 136kg (300 lb), but this mass was later significantly reduced to enable its launch by the cheaper Scout rocket.

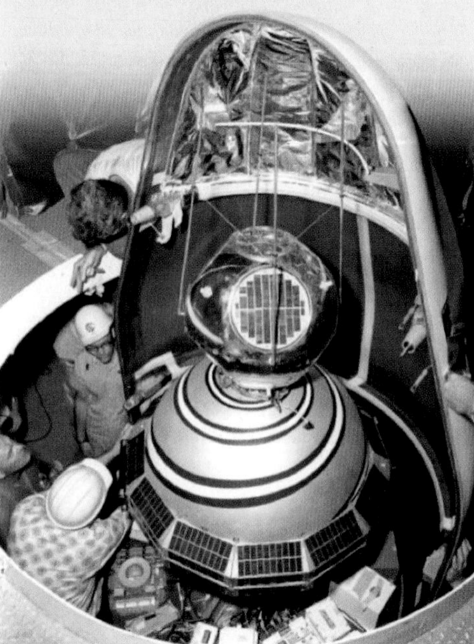

How GPS works

NAVSTAR (now simply known as the Global Positioning System) relies on principles of radio navigation, established long before the Space Age, coupled with precisely controlled satellite signals. Around 32 satellites follow intermediate orbits that circle the Earth twice per day, with the overall constellation arranged on varying orbital planes so that several satellites will be visible from different directions above the horizon from any location at a given time. Each satellite in this

FOUR-WAY LAUNCH
In November 2016, an ESA Ariane 5 rocket launched from French Guiana used a new payload-dispenser system to deploy no fewer than four Galileo GPS satellites into orbit at once – the first quadruple-satellite launch.

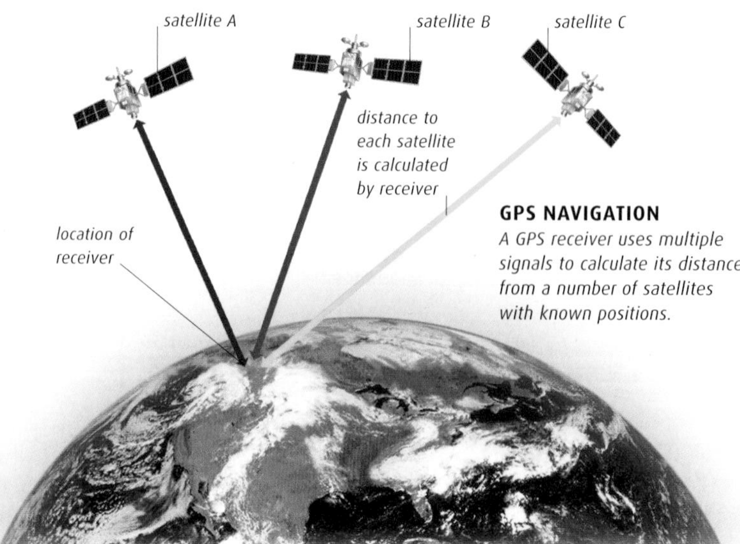

GPS NAVIGATION
A GPS receiver uses multiple signals to calculate its distance from a number of satellites with known positions.

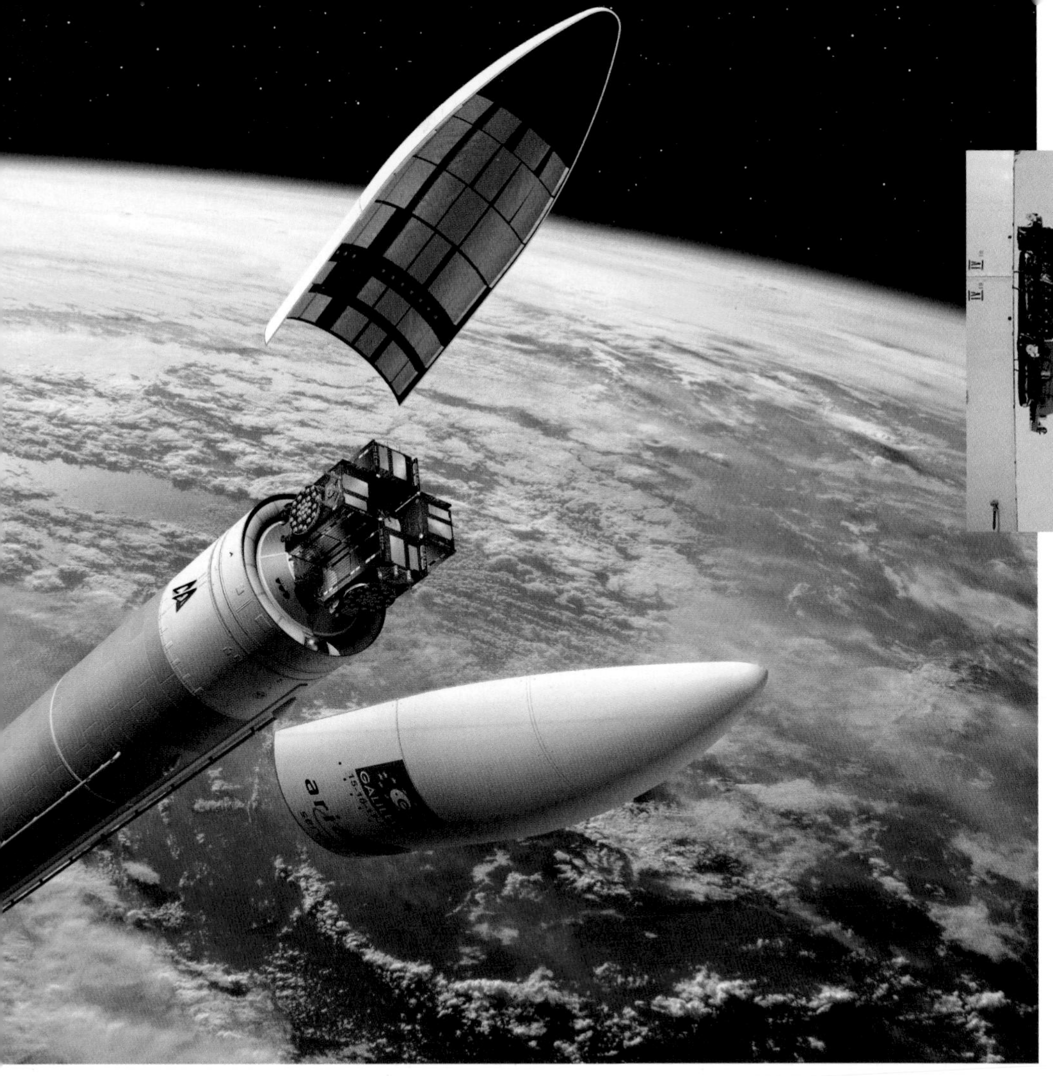

SATELLITE PREPARATIONS
A Galileo GPS satellite is loaded into its protective shroud prior to launch on a Soyuz rocket with a Fregat upper stage. The satellite's manoeuvring motors are wrapped in gold foil.

"space segment" sends out a digital signal whose components begin at precisely timed 30-second intervals. A receiver (the "user segment") scans for signals from different satellites and compares their time signatures to work out how long each signal has travelled for, and therefore the distance to each satellite. The receiver uses a precise ephemeris (orbital model) to calculate the satellite positions and therefore its own location – three satellites will suffice for a fix at sea level, but four are needed to provide altitude data and correct for errors in the receiver clocks.

Each satellite signal contains not only timing information, but also an up-to-date ephemeris (which the receiver periodically downloads in order to retain precision) and an almanac containing further information about the satellite network. A final element, the network of satellite ground stations known as the control segment, monitors the orbits of the satellites and the precision of their clocks, uploading a new ephemeris to each one every two hours and an updated almanac once a day.

Rival systems

President Reagan's 1983 decision to make NAVSTAR available for civilian use (albeit with some deliberate degradation of accuracy) ensured that the US system became synonymous with GPS navigation, allowing anyone to pinpoint their location on the Earth to within a few metres. Given the system's military applications, however, it is unsurprising that the Soviet Union chose to develop its own alternative. An early system (called Parus or Tsikada in its military and civilian guises) used the same principle as the US Transit satellites, while from the 1980s a GPS-like global network called GLONASS was deployed. China, meanwhile, developed coverage of Asia through limited "BeiDou" satellite constellations from around 2000 and is currently nearing completion of a network, known as BeiDou-3, with global coverage. Meanwhile, concerns about overdependence on a US military system, along with an eye to potential commercial benefits, have led the European Union to devise its own rival civilian system, called Galileo, which should also be complete by 2020.

AERIAL SURVEYOR
GPS-equipped drone aircraft can be programmed to overfly and survey specific regions and gather useful information. Here, a drone sweeps over fields in France, measuring indicators of crop health. This data can then be used by the farmers to treat specific areas and improve productivity.

The GPS revolution

While in-car "satnav" was introduced by Japanese manufacturers as early as 1990, and has since become ubiquitous, the true scope of applications for GPS is only now becoming clear, thanks to twin revolutions in robotics and consumer electronics. Many people now carry GPS receivers in their smartphones, which use navigational data for a variety of purposes, ranging from the obvious (map-reading applications) through the ingenious (various location-based gaming experiences) to the more opaque (pushing targeted, location-dependent advertisements on social media). Professional applications for the same technology include surveying, archaeology, and town planning.

It is in the field of automation, however, that the power of GPS is most widely felt. Satellite navigation lies at the heart of automated vehicles such as farming equipment, drone aircraft, and self-driving cars, while inexpensive transponders can allow precise tracking of containerized freight and even individual components for "just in time" delivery. "Big data" applications, meanwhile, can harvest apparently insignificant information from devices such as users' phones and match it to location to offer new insights.

1 April 1960
NASA launches TIROS 1, its first weather satellite.

12 August 1960
The re-entry capsule from US spy satellite Discoverer 13 is successfully recovered.

28 July 1962
The Soviet Union launches its first successful Zenit spy satellite under the codename Cosmos 7.

3 March 1969
Apollo 9 carries an experimental multi-spectral imager into orbit.

June 1971
The first civilian Soviet space station, Salyut 1, studies the potential of multispectral imaging.

23 July 1972
ERTS 1, the first remote-sensing satellite, is launched.

17 May 1974
NASA launches SMS 1 (Synchronous Meteorological Satellite), the first geostationary weather satellite.

25 June 1974
The first successful military space station, Salyut 3, carries a large reconnaissance camera onboard.

28 June 1978
NASA's Seasat carries the first synthetic aperture radar into orbit.

Watchers from above

Early satellites looked down at Earth to monitor the weather or collect military intelligence, while remote-sensing orbiters developed since the 1970s have transformed the way we look at and understand our world.

The ability to observe the Earth from space was an obvious early application of satellite technology, but initial thoughts in this direction were focused principally on the possibility of gaining an upper hand in terms of military intelligence. The US, in particular, recognized the potential of high-altitude aircraft to gather intelligence without the risk of interception and funded the secretive U-2 reconnaissance aircraft, which flew at altitudes of up to 21km (13 miles) over Soviet territory. Satellite surveillance became an extension to this application – all the more so for the US after a U-2 was unexpectedly downed over Soviet territory and its pilot captured in May 1960.

Rival programmes

The US Air Force began development of its CORONA programme using the "Keyhole" (KH) film recovery technique as early as 1956. By January 1959, now operating as a USAF/CIA "black programme", the first CORONA satellite, named Discoverer 1, was ready to fly. However, dependence on the still unreliable USAF Thor-Agena rocket for launches cost the US programme dearly – Discoverer 1 was destroyed in a launchpad explosion, and many other problems followed. It was not until August 1960 that Discoverer 13 successfully returned a film-return capsule from orbit.

As CORONA became an increasingly reliable tool for US intelligence, it disappeared behind a veil of secrecy in the National Reconnaissance Office. The mid-1960s saw development of the GAMBIT series, carrying larger, higher-resolution telescopes. From 1972, the CORONA design was replaced by the larger and more ambitious KH-9 Hexagon, while from 1976 digital-imaging satellites, capable of matching film resolution, were introduced. Initially known as the KH-11 Kennen, and later known as Crystal, these satellites used large telescopes to observe the Earth and are believed to share some design features with the Hubble Space Telescope.

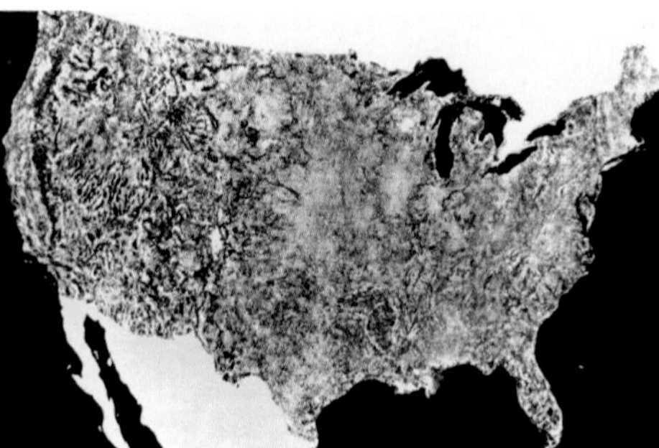

THE BIG PICTURE
An early display of remote sensing's potential was this giant, cloud-free photographic map of the contiguous United States, assembled in 1972 from 595 satellite images by the US Department of Agriculture. ERTS 1's orbit allowed it to take images from a constant altitude of 912km (560 miles) in the same lighting conditions.

Still less is known about the Soviet Union's spy satellites. When its Zenit series began in 1961, it used a Vostok-like design with a reusable camera unit replacing the cosmonaut in the Descent Module (see p.58), but as the series developed over an astonishing 33 years (mostly launched under the catch-all Cosmos codename), the Zenit design evolved considerably. Very little information has been made public about the later Yantar series, first deployed in 1974 and still launched by Russia under Cosmos designations.

BIG BIRD'S VIEW
This declassified 1979 image from a US National Reconnaissance Office KH-9 Hexagon satellite (nicknamed "Big Bird") reveals Soviet planes on the ground at an airfield near Moscow.

Earth remote sensing

Of course there is also another, far more open side to the observation of the Earth from space – the study of our geography and environment for scientific purposes. Meteorology was another easily foreseen use for early satellites, and NASA launched its first experimental weather satellite, TIROS 1,

COMMERCIAL SPY
Today, the use of satellite surveillance is no longer limited to governments – it is also available to paying customers. GeoEye's Ikonos-2 is the world's first commercially operated reconnaissance satellite, providing high-resolution, multispectral, and true-colour images such as this one of San Francisco.

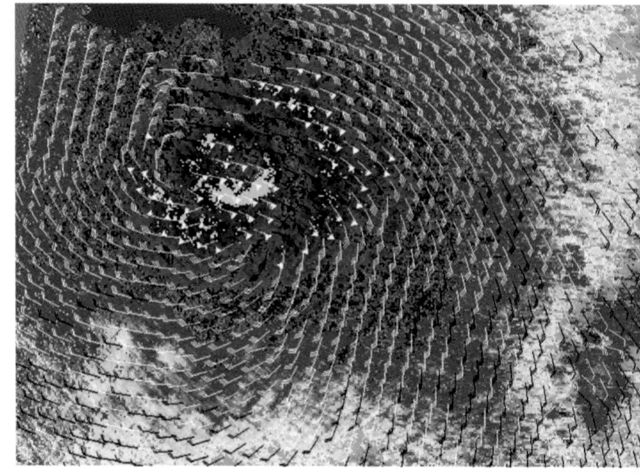

HURRICANE FORCE
Scatterometry enables satellites, such as NASA's QuikSCAT, to measure wind speeds at the ocean surface, giving advance warning of the growing strength of weather systems such as 2005's devastating Hurricane Katrina.

in 1960 (see p.244). Early weather satellites remained close to the Earth in low polar orbits, photographing one strip of the planet at a time. The first geostationary weather satellite, able to keep an entire hemisphere of the Earth in constant view, was not launched until 1974.

The true potential of satellites for Earth observation, however, only became clear when early astronauts reported seeing surprising detail from orbit. Subsequently, Apollo 9 carried experiments to study the Earth from space, including the first use of multispectral imaging (see panel, right). Soviet cosmonauts carried out similar experiments aboard Salyut space stations, as did the crew of Skylab (see pp.170–173).

NASA launched its first satellite dedicated to these new remote-sensing techniques in 1972. Its Earth Resources Technology Satellite (ERTS) series was renamed Landsat after the launch of Landsat 2 in 1975, and the programme continues today. The Landsats are among the most successful of remote-sensing satellites, but many others have been developed. ESA and the Soviet Union launched their own equivalents, and as scientists have found new ways to probe our planet's properties, many more specialized missions have been launched.

Monitoring our planet

Today's satellites monitor every aspect of the Earth, from air temperature to ocean circulation, and from wave height to wind speed. Spectrometers can analyze radiation emitted or reflected from the ground below, revealing everything from surface temperatures and crop usage to buried water and minerals. Synthetic aperture radar (SAR) can gather detailed information about the shape and composition of the landscape (see p.267), while microwave radars can penetrate the topsoil to reveal underground features. Another radar technique, called scatterometry, can measure wind speeds close to the sea surface, while microwave sounding units scan swathes of the atmosphere, building up temperature profiles from the Earth's surface to the upper stratosphere. Although the interrelationships between all these different properties are intricate, remote-sensing satellites are offering scientists the most detailed view yet of our planet and enriching our understanding of exactly how the Earth's complex climate operates.

TECHNOLOGY

MULTISPECTRAL IMAGING

The technique of multispectral imaging is simple but powerful – a camera fitted with a series of filters takes images of the same area in specific wavelengths (colours) of visible light and sometimes in infrared or ultraviolet light. When the images are compared or combined, features and properties that are normally invisible can be seen and analyzed – as in this map of the Malaspina Glacier in Alaska, which in visible light is simply lost in the snowy landscape. Often researchers will want to revisit the same area at intervals in order to reveal changes in the landscape. In this case, it is important that all the images are lit from the same angle, so remote-sensing satellites frequently occupy "Sun-synchronous" polar orbits: the satellite maintains its orientation to the Sun at all times (in other words, its orbit progresses slightly each day, circling the Earth once every year, while the Earth rotates daily beneath it).

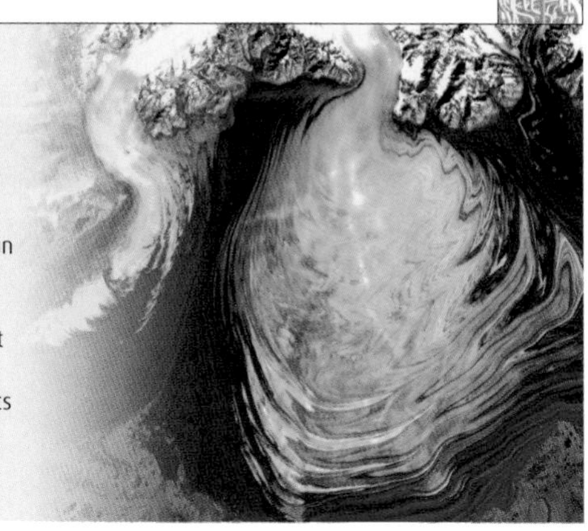

Astronomy in orbit

The arrival of the Space Age provided huge new opportunities for astronomers – finally they could send instruments above the Earth's atmosphere for a clear view of the Universe.

8 April 1966
NASA launches OAO-1, but it fails in orbit.

4 July 1968
NASA launches RAE-1, which deploys a cross-shaped 450m (1,500ft) antenna in orbit.

7 December 1968
The Orbiting Astronomical Observatory 2 is launched and begins ultraviolet observations.

July 1969
Gamma-ray bursts are first observed by US Vela satellites.

12 December 1970
SAS-1 (Uhuru), the first X-ray detector in space, is launched.

15 November 1972
NASA launches SAS-2, the first dedicated gamma-ray observatory.

12 November 1978
The Einstein satellite (HEAO-2) becomes the first imaging X-ray telescope in orbit.

25 January 1983
IRAS, the first infrared space telescope, is launched and operates for ten months.

Astronomers have always been frustrated by the Earth's atmosphere – even on a clear night, turbulent air currents can distort and blur telescope images. In the 19th and 20th centuries, their frustration grew as they learned that visible light was just a small part of a spectrum of electromagnetic waves ranging from very long radio waves to ultra-short, high-energy gamma rays – and that the atmosphere did a thorough job of blocking out nearly all of them.

So when captured V-2 rockets became available at the end of the Second World War, astronomers were eager to utilize them to take a look at the Universe from beyond the atmosphere. Rocket-borne detectors soon revealed that space was full of exotic radiations – ultraviolet radiation from the Sun was discovered in 1946, and solar X-rays in 1949. Radio waves from the Milky Way (the plane of our galaxy) had previously been discovered from the ground in 1932 by American engineer Karl Jansky.

Early satellites added to these discoveries, sometimes unexpectedly. Long-wavelength radio waves (thought to originate in clouds of cool dust and gas) were revealed in the early 1960s by satellites that were actually built to study the Earth's ionosphere. The first X-ray source beyond the Solar System (now suspected to be a black hole) was identified in 1962. Dedicated observatories soon followed, though the earliest, such as the Orbiting Solar Observatory (OSO) series, were intended to reveal more about the Sun rather than target more distant objects. The first successful satellites designed to look further afield were NASA's Radio Astronomy Explorer (RAE) missions, launched from 1968 onwards. The RAEs recorded radio waves from the Sun, Jupiter, and various sources elsewhere in our galaxy.

The first X-ray astronomy satellite was NASA's Small Astronomy Satellite (SAS) 1, also known as Uhuru, launched in 1970. This revealed X-ray sources scattered across the sky and led to a variety of later missions.

HISTORY FOCUS

OBSERVATORIES ON THE MOON

Before the microelectronics boom of the 1970s, it seemed as though many satellite instruments, including those for astronomy, would be impossible to automate. As a result, astronomy looked like being one of the main roles of any future moonbase. Operating under a long day-night cycle and in a thin atmosphere, a telescope on the Moon could see clearly into deep space – with protection from the surface glare, it could even operate when the Sun was above the horizon. A lunar observatory would also make it possible to construct larger telescopes than any that could be put in orbit. Today, advances in electronics and ingenious design have overcome many limitations of orbiting telescopes, but radio astronomers in particular still dream of building a telescope on the lunar far side.

OAO-1 SATELLITE
The Orbiting Astronomical Observatories were a series of ultraviolet telescopes launched from 1966 onwards.

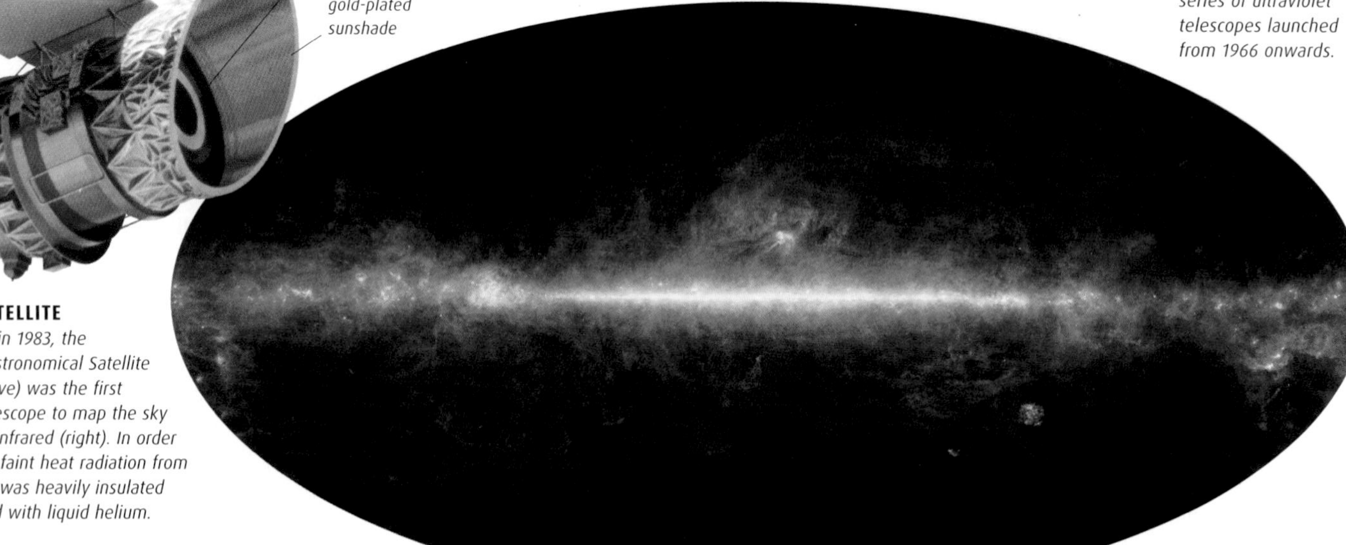

IRAS SATELLITE
Launched in 1983, the Infrared Astronomical Satellite (IRAS, above) was the first orbital telescope to map the sky in the far infrared (right). In order to pick up faint heat radiation from the sky, it was heavily insulated and cooled with liquid helium.

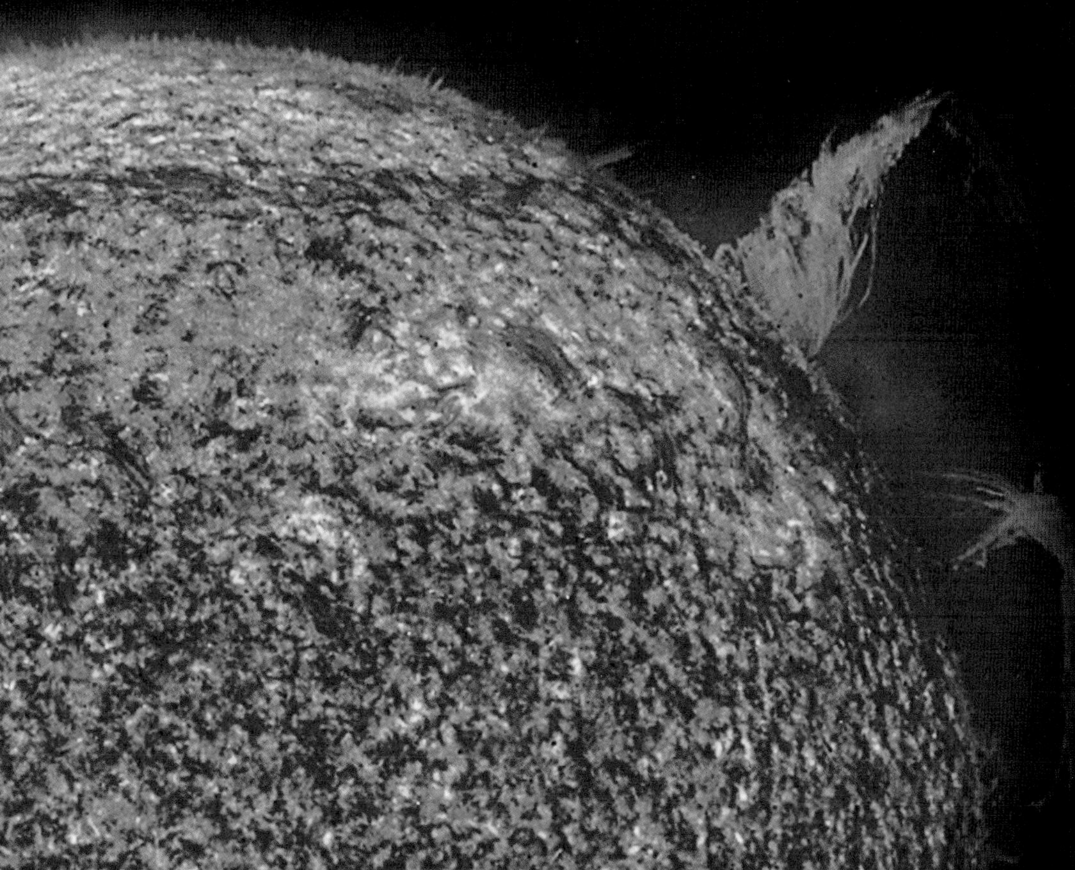

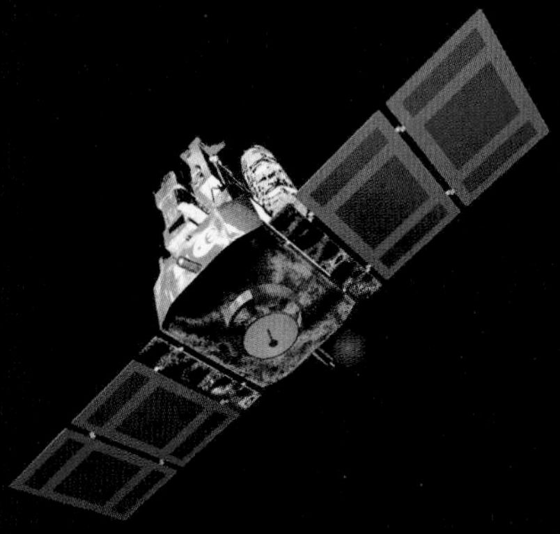

CLOSE TO THE SUN
While most astronomy satellites look into deep space, some spacecraft direct similar instruments towards the Sun. Launched in 1995, SOHO, the Solar and Heliospheric Observatory (above), is a joint project of NASA and ESA. It has studied the Sun in visible and ultraviolet light for more than two decades, revealing turbulent activity such as prominences (left) and enormous outbursts known as coronal mass ejections (below).

Another new field of astronomy opened when US Vela spy satellites, designed to look for evidence of nuclear testing on Earth, detected bursts of gamma rays from deep space. Their existence was confirmed in 1972 by SAS-2, which also found gamma-ray sources in the remnants of exploded stars. ESA's Cos-B gamma-ray satellite of 1975 was followed by several French experiments carried aboard Soviet spacecraft and space stations.

Beyond the blue, within the red

The first successful ultraviolet satellite was Orbiting Astronomical Observatory (OAO) 2. Launched in 1968, it studied the ultraviolet properties of thousands of stars, as well as objects such as comets and galaxies. Other ultraviolet observatories soon followed.

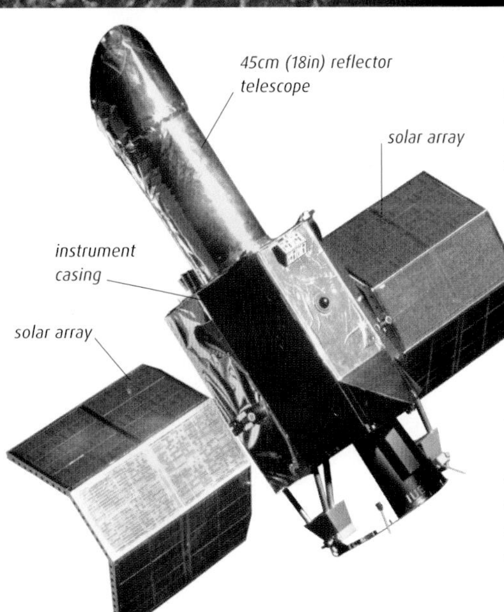

The last major area of the electromagnetic spectrum to be explored from space was the infrared. The nature of this heat radiation creates unique challenges – since the telescope itself is an infrared source, it must be cooled to very low temperatures to avoid swamping the weak infrared light from stars and other objects. The challenges were first overcome with the Infrared Astronomical Satellite (IRAS), a joint British, American, and Dutch mission launched in 1983.

ULTRAVIOLET OBSERVATORY
The highly productive International Ultraviolet Explorer (IUE), a joint venture between NASA, ESA, and the United Kingdom, operated for 18 years and was the first satellite that astronomers could operate from the ground in real time.

25 April 1990
The Space Shuttle *Discovery* deploys the Hubble Space Telescope into low Earth orbit.

December 1993
Endeavour recaptures the HST for a 10-day servicing mission as well as the installation of the COSTAR corrective optics system.

February 1997
In a second servicing mission, *Discovery* fits new instruments, including an infrared detector.

May 1999
The HST Key Project, an effort to measure the age and scale of the Universe from the brightness of distant stars, concludes.

December 1999
A third servicing mission by *Discovery* replaces the HST steering gyroscopes.

March 2002
Servicing mission 3B, by *Columbia*, installs a new camera, revives the infrared detector, and replaces the solar arrays for a second time.

May 2009
Servicing mission 4 sees the *Atlantis* crew give the HST a complete overhaul, installing new equipment to last another decade.

The Hubble Space Telescope

The first large orbiting telescope to study the Universe in visible light, the Hubble Space Telescope (HST) has revolutionized our view of the cosmos, answering some key questions in modern astronomy and raising new ones.

Astronomers have long been aware of the benefits of a telescope in space. Visible light is one of the few types of radiation that pass through our atmosphere relatively unscathed, but its passage through the air to even the highest mountaintop observatories is still affected by blurring and distortion. A telescope above the atmosphere would be immune to these problems. While Earthbound telescopes might have larger mirrors and therefore greater light grasp (the theoretical ability to see fainter objects), an orbiting observatory would see everything in pin-sharp detail.

The orbiting Hubble Space Telescope was championed by Lyman Spitzer (see panel, right). It finally received funding in 1977 and was scheduled for an October 1986 launch when tragedy struck the Space Shuttle *Challenger* (see pp.202–203). Although the remaining Shuttles returned to flight in late 1988, a slot could not be found to deploy the HST until *Discovery*'s STS-31 mission of April 1990.

BIOGRAPHY

LYMAN SPITZER

American astronomer and physicist Lyman Spitzer Jr. (1914–97) was the first person to suggest putting an optical telescope into Earth orbit in his 1946 paper *Astronomical Advantages of an Extra-Terrestrial Observatory*. Already a rising star of astrophysics before the Second World War, he became head of his department at Princeton University aged only 33. His interests ranged from interstellar gas and dust to plasma physics and nuclear fusion. After leading the design of NASA's Copernicus ultraviolet telescope (see p.250), he was asked in 1965 to chair a committee planning a "large space telescope". He did a great deal to convince Congress and sceptical scientists to back the project.

Teething troubles

The optical design of the HST was similar to many Earthbound telescopes, using a series of mirrors to bend light to a focus behind the large primary mirror, in any of four instrument modules. The initial instrument suite incorporated five instruments (two cameras, two spectrographs for analyzing light, and a photometer for measuring the precise brightness of objects). The HST was always designed for a long life, with occasional visits from the Shuttle to perform repairs and install new instruments. This was just as well, because when scientists at Baltimore's Space Telescope Science Institute began to put the telescope through its paces, they discovered that the primary mirror had a minute flaw in its shape – and as a result, all Hubble's images were blurred.

While the HST was not entirely crippled by the problem, it was a huge embarrassment. Fortunately, a way was found to compensate for Hubble's short sight (see panel, opposite). A rescue mission in December 1993 was an outstanding success, paving the way for three more servicing missions.

Since the repair, the HST has been one of the world's most productive scientific instruments, making countless observations and helping to solve many conundrums about the Universe. It has also been a source of good publicity for NASA, its beautiful, spectacular, and sometimes profound images frequently making the headlines. Such is the telescope's popularity that in 2006, NASA gave in to public pressure and lifted its own ban on non-ISS Shuttle flights, scheduling a final servicing mission for 2009 that allowed the telescope to continue operations for a further decade.

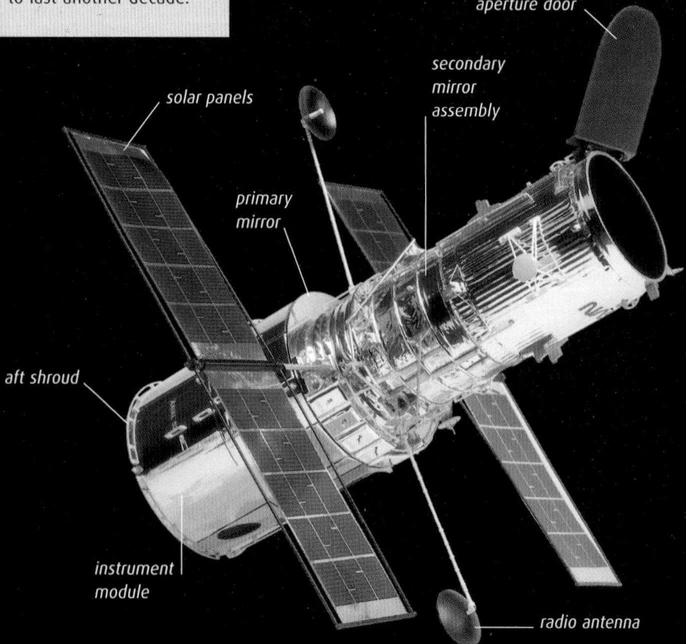

ANATOMY OF A SPACE TELESCOPE
Light entering the HST is reflected off a convex primary mirror, which bounces rays back on converging paths to a smaller secondary mirror. From here, it is folded back again, through a hole in the centre of the primary and brought to a focus at the instruments.

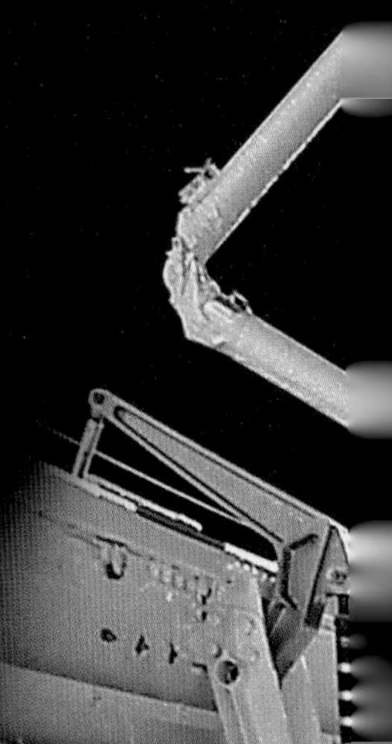

RELEASED IN ORBIT
Discovery*'s robot arm delicately places the HST in orbit 612km (380 miles) above the Earth during initial deployment in April 1990. Placing the telescope well above the atmosphere meant taking the Shuttle to an unusually high altitude.*

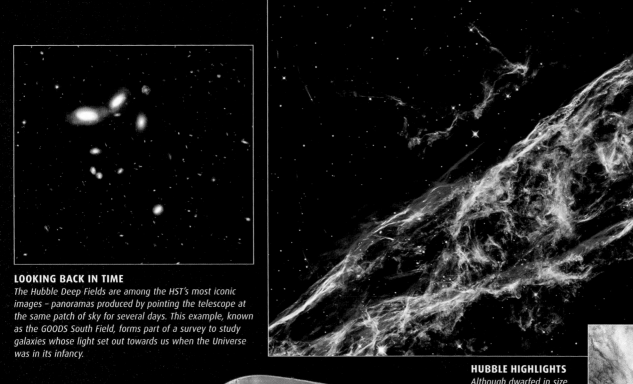

LOOKING BACK IN TIME
The Hubble Deep Fields are among the HST's most iconic images – panoramas produced by pointing the telescope at the same patch of sky for several days. This example, known as the GOODS South Field, forms part of a survey to study galaxies whose light set out towards us when the Universe was in its infancy.

HUBBLE HIGHLIGHTS
Although dwarfed in size by some Earth-based telescopes, Hubble's location in orbit offers unique clarity. False-colour images combine views taken with different filters to reveal intricacies in delicate gas clouds such as the Veil Nebula (above). Other subjects, such as spiral galaxy Messier 66 (right), need no such tricks to reveal detail in the light from their billions of stars.

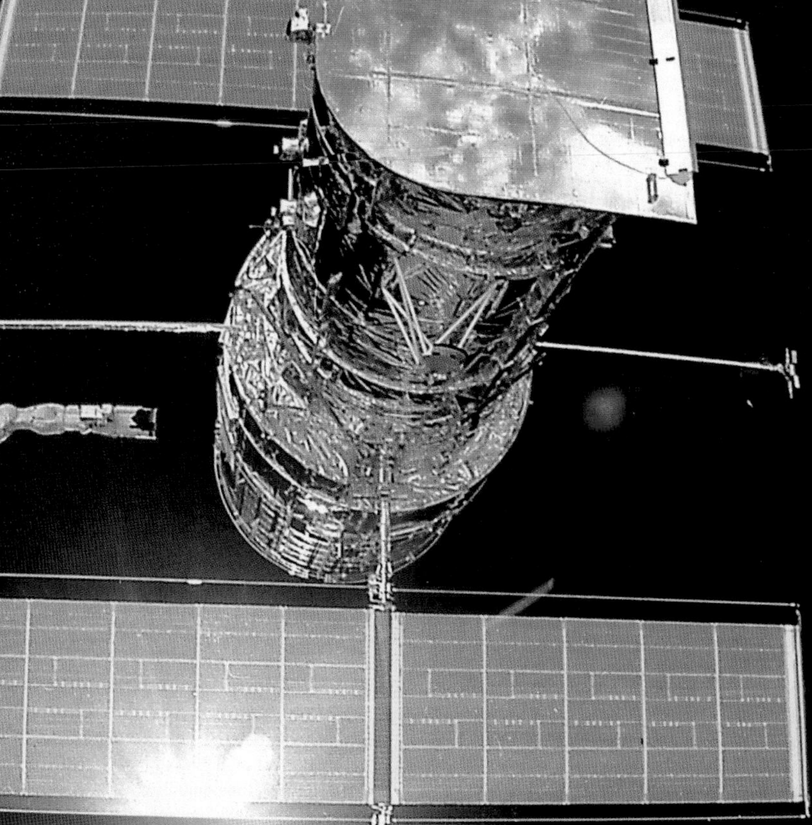

TECHNOLOGY

FIXING HUBBLE

Since there was no way to replace or repair Hubble's original primary mirror in orbit, engineers and scientists devised an optical box of tricks called the Corrective Optics Space Telescope Axial Replacement (COSTAR). This used ten precisely ground mirrors to adjust the path of light through the telescope and bring it into focus. It was built to slot into one of Hubble's existing instrument bays, diverting corrected light back into any of the other three. With COSTAR in place, the improvement in Hubble's images was spectacular.

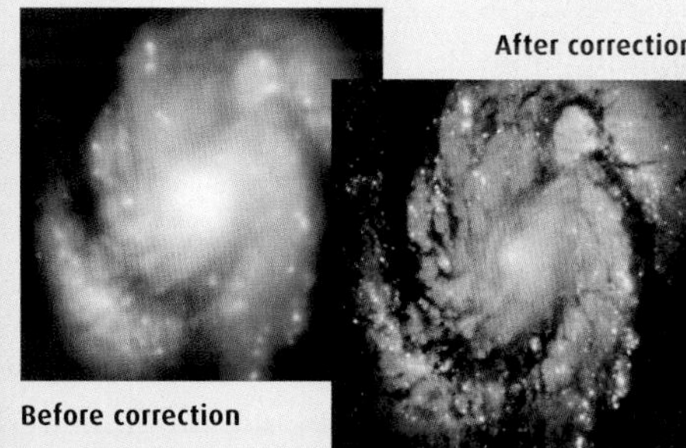

5 April 1991
The Compton Gamma-Ray Observatory (CGRO) is deployed into low Earth orbit from the Space Shuttle *Atlantis*.

23 July 1999
The Chandra X-ray Observatory is deployed from *Columbia*.

4 June 2000
For safety reasons, CGRO is deliberately taken out of orbit while NASA can still control its re-entry.

September 2001
Scientists announce Chandra's detection of X-ray flares from a giant black hole at the centre of the Milky Way.

25 August 2003
The Spitzer Space Telescope is launched on a Delta rocket.

22 March 2005
NASA announces Spitzer's detection of light from planets around other stars.

7 March 2009
NASA's Kepler spacecraft is launched on a hunt for distant exoplanets.

19 December 2013
ESA launches its Gaia satellite to build our most accurate map of the Milky Way so far.

Orbiting observatories

The success of the Hubble Space Telescope (HST) triggered a golden age of astronomy, with new satellites imaging the sky's invisible radiations or using ingenious new techniques to learn more about the Universe.

During the 1980s, many astronomers realized the benefits of a series of parallel observatories that could study the sky at different wavelengths. Optical (and near-ultraviolet) images from Hubble could then be complemented by simultaneous observations from other telescopes, revealing how the visible appearance of an object related to its changing properties in other wavelengths. This idea ultimately came to fruition in the Great Observatories programme, a series of four satellites (including the HST) launched to observe the Universe in visible and near-ultraviolet light, gamma rays, X-rays, and the infrared.

Great observatories

The second of the Great Observatory satellites, known as the Compton Gamma-Ray Observatory (CGRO), was launched in 1991. In order to detect the most energetic radiations in the Universe, it was equipped with four cumbersome instruments – not telescopes in a conventional sense, but instead detectors "tuned" to detect gamma rays of different energies and localize the directions from which they came. In nine years of operation, CGRO detected more than 400 new gamma-ray sources (10 times the number previously known), associated with some of the most bizarre objects in the Universe. However, with a weight of 17,000kg (37,400lb), much of CGRO could have struck the ground during an uncontrolled re-entry, so it was deliberately de-orbited under control once its guidance gyroscopes began to fail.

By the time CGRO's mission had been brought to an end, a third satellite in the programme had been launched. The Chandra X-Ray Observatory studies the Universe at slightly less energetic radiations, but it too raised challenges for telescope design, since X-rays approaching a mirror head-on can pass straight through it. Chandra therefore uses an ingenious series of curved metal sections nested one inside the other. X-rays strike the mirrors at shallow angles, ricocheting off each in turn and eventually coming to a focus on the X-ray detectors. Chandra has operated successfully for almost two decades and in 2012 was joined by NuSTAR (Nuclear Spectroscopic Telescope Array), a telescope for higher-energy X-rays based on the same principles.

The final Great Observatory, the Spitzer Space Telescope, targeted weak infrared radiation from the Universe's coolest objects. Launched in 2003, its instruments were chilled to −268°C (−450°F) using liquid helium coolant, which was exhausted over time. Nevertheless, the mission operated at full capacity until 2009, and it has continued to operate at shorter, warmer wavelengths while other satellites, such as ESA's Herschel Space Observatory, have carried on its work in the far infrared.

Planet-hunting telescopes

The 1990s saw a major astronomical breakthrough with the discovery of the first exoplanets (planets orbiting other stars). While early finds were made using ground-based telescopes, astronomers soon

radiator sheds excess heat into space

CHANDRA: X-RAYS
The Chandra X-ray Observatory detects high-energy radiation from objects such as supernova remnants and material being pulled into black holes.

SPITZER: INFRARED
The Spitzer Space Telescope studies infrared radiation, allowing it to peer through opaque gas and dust clouds and to detect relatively cold, dark material in our galaxy and beyond.

COMPTON: GAMMA RAYS
The CGRO measured radiation from some of the most violent events in the Universe, including supernova explosions and matter-antimatter collisions.

KEPLER BEFORE LAUNCH
NASA's Kepler spacecraft is inspected in a factory clean room prior to shipping and launch. Kepler is effectively a giant telescopic camera, designed with a field of view 4,000 times larger than that of the Hubble Space Telescope.

Ultraviolet radiation

X-ray image

Mid-infrared (Spitzer)

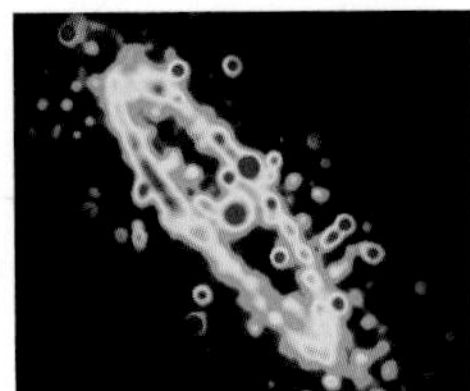

Ground-based radio image

SAME GALAXY, DIFFERENT WAVELENGTHS
The nearest major galaxy to our own, the Andromeda Galaxy (M31), has been studied at many different wavelengths, revealing an array of hidden features – many associated with ring-like dust clouds or the giant black hole at its core.

realized that one useful search technique could only work from orbit. The so-called transit method relies on observing a field of stars continuously over a prolonged period, looking for tiny dips in their brightness as orbiting planets pass in front of their parent stars. NASA's Kepler mission, launched in 2009, put the idea to work on a grand scale, staring at a distant cloud of stars continuously for months on end and discovering thousands of candidate exoplanets. The TESS mission, launched in 2018, is applying a similar principle to bright stars across the whole sky.

Mapping the Universe

In addition to general-purpose telescopes like the Great Observatories and their successors, some satellites can be tasked with studying the entire sky in a very specific way. For example, the European Space Agency's Gaia mission uses a relatively small but precise telescope to record tiny shifts in the apparent positions of stars as it is carried from one side of Earth's orbit to the other. This parallax effect, caused by our shifting point of view, is smaller for objects that are further away, so it can be used to measure stellar distances directly. The first satellite to carry out a parallax survey, Hipparcos, flew in the early 1990s, but Gaia offers far more accuracy and extends the technique across far greater distances, encompassing a billion stars across the Milky Way.

A series of highly specialized astronomical satellites has also revealed the secrets not of the present-day Universe but of its distant past. Microwave observatories are tuned to the weak radio waves emitted by the cosmic background radiation released in the aftermath of the Big Bang. Variations in the properties of these waves, such as tiny variations in temperature, reflect structures in the very early Universe. As early as 1992, NASA's Cosmic Background Explorer (COBE) satellite found ripples in the background radiation, but successor missions, such as the Wilkinson Microwave Anisotropy Probe (WMAP) and ESA's Planck, have revealed far more detail.

DEEP BACKGROUND
A map from ESA's Planck satellite shows the cosmic microwave background radiation from across the sky, revealing variations in the temperature of the Universe when it was just 380,000 years old.

TESS
NASA's Transiting Exoplanet Survey Satellite (TESS) scans broad swathes of the sky for 27 days at a time, looking for the telltale variations in brightness that could indicate a planet passing in front of a distant star. The mission is expected to discover some 2,000 new Earth-like exoplanets.

GAIA
The European Space Agency's Gaia spacecraft orbits in a gravitational "sweet spot" 1.5 million km (930,000 miles) from Earth, where it can avoid gravitational and thermal disturbances. Over five years, it is taking multiple precise measurements of the positions of a billion stars.

Later Moon probes

Through the early 1970s, the Soviet Union compensated for its failure to land men on the Moon with a series of complex unmanned probes. More recent missions have studied our satellite's mineral resources in detail.

12 September 1970
Luna 16, the first robot probe to return samples of Moon rock, launches.

17 November 1970
Luna 17 releases its Lunokhod rover onto the Moon.

19 August 1976
Luna 24's sample-return capsule blasts off from the Sea of Crises.

19 February 1994
Clementine enters orbit around the Moon.

6 January 1998
NASA launches its Lunar Prospector probe.

31 July 1999
Lunar Prospector is crashed into the Moon's south polar region.

27 September 2003
ESA launches its SMART-1 Moon probe.

3 October 2007
Japan's SELENE probe enters lunar orbit.

23 June 2009
NASA launches its Lunar Reconnaissance Orbiter.

14 December 2013
China's Chang'e 3 lander touches down on the Moon and releases a small rover, Yutu.

If it had successfully landed on the Moon in July 1969, Luna 15 might have given Soviet scientists some comfort even as Apollo 11 upstaged their efforts. It was the first in a new generation of Luna probes, equipped with an automatic drill and a small capsule that could bring a sample of lunar dust back to Earth orbit under remote control. Instead, the spacecraft retrorockets failed, sending it slamming into its target zone in the Sea of Crises.

Luna phase II

Despite this failure, unmanned craft clearly offered a cost-effective way for Soviet scientists to continue exploring the Moon. Luna 16 worked perfectly, landing in the Sea of Fertility in September 1970 and drilling 100g (3½oz) of rock that was returned to Earth over Soviet territory. Another new type of mission was pioneered by Luna 17 in November 1971. After landing on the Sea of Rains region, the probe released an automatic rover, Lunokhod 1. The size of a small car, it explored the lunar surface using solar power for some 321 days.

The Soviet programme continued until 1976, through seven more missions – Luna 21 carried another Lunokhod, while Lunas 18, 20, 23, and 24 were sample-return missions (though 18 and 23 failed). Lunas 19 and 22 were advanced orbiting surveyors, each operating for more than a year.

TECHNOLOGY

POLAR SNAPSHOT

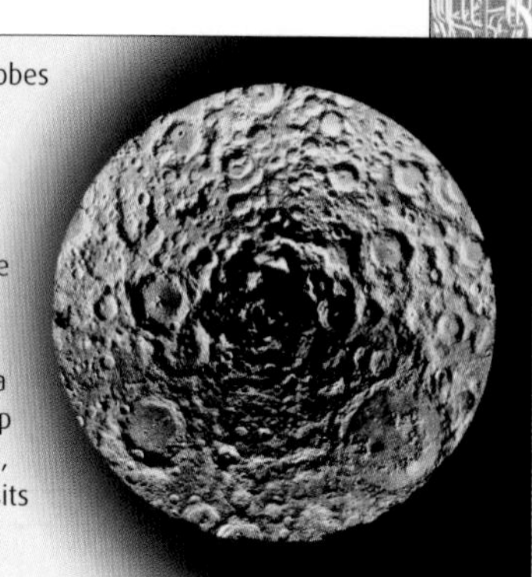

Because the Apollo missions and earlier lunar probes had only moderately inclined orbits, they did not get a good look at the lunar poles. However, their images hinted that there might be a truly enormous depression at the lunar south pole – most likely an ancient impact basin. Clementine confirmed the presence of this vast crater, the South Pole-Aitken Basin, some 2,500km (1,550 miles) across. Because the Sun only ever rises a few degrees into the polar skies, there are deep craters within the basin that never see sunlight, and some of these craters seem to shelter deposits of ice – perhaps deposited by comet impacts.

Back to the Moon

Following the end of the Luna programme, the Moon was ignored for almost two decades. It was only in 1994 that a small spacecraft called Clementine reignited interest in our satellite. A collaboration

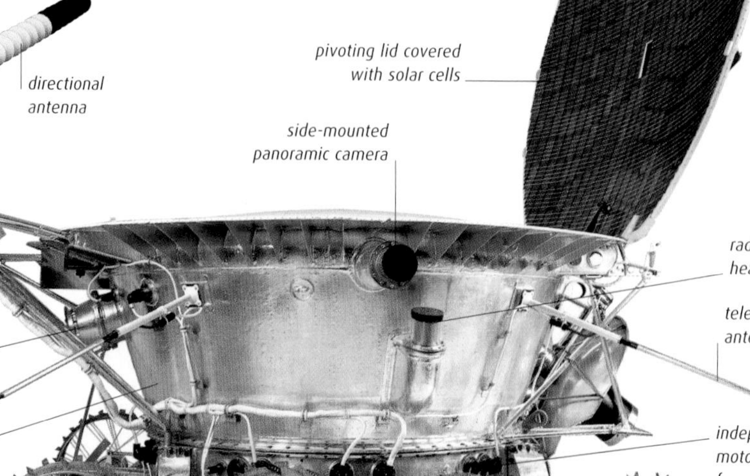

LUNOKHOD ROVER
Some 2.5m (8ft) long and 1.5m (5ft) wide, the Lunokhod had a strong resemblance to a bathtub on wheels. The rover's lid was covered in solar panels. It could tilt to face the Sun or close up to keep the probe warmer in the lunar night.

LUNA SAMPLER
Luna 16 and its successors used a two-stage spacecraft. A descent stage fitted with cameras and instruments acted as a launch pad for the cylindrical return capsule.

LUNAR PROSPECTOR
An engineer makes final adjustments to Lunar Prospector before its launch in 1998. The probe is already mounted on the upper stage that would send it from Earth orbit towards the Moon.

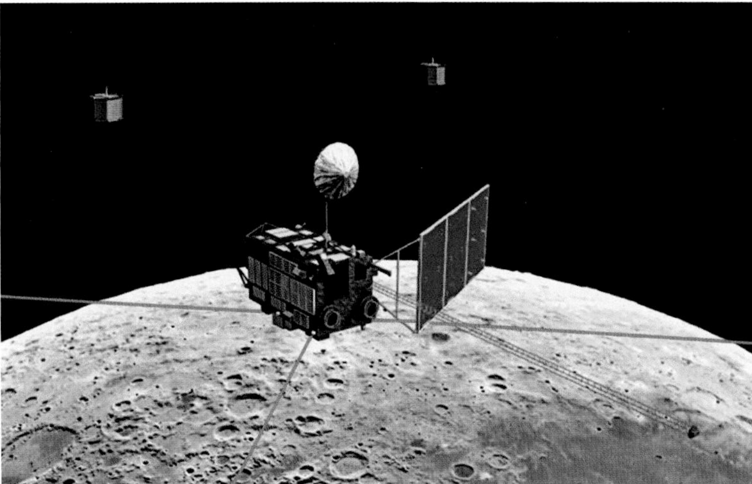

ORBITAL SURVEY
JAXA's SELENE (SELenological and ENgineering Explorer) mission (also known by its nickname "Kaguya"), orbited the Moon for 20 months, compiling detailed maps using HDTV cameras and measuring lunar gravity.

between NASA and the US Department of Defense, this was the first probe developed under a "faster, better, cheaper" philosophy introduced at NASA following a series of high-profile, expensive failures. It went from concept to launch in 22 months and delivered 1.8 million digital images in just two months of orbital operations. Clementine found the first hints of ice at the lunar poles (see panel, opposite) and also photographed lunar terrain at different wavelengths, revealing colour differences that hinted at various minerals in the lunar surface. Building on these discoveries, NASA launched the more ambitious Lunar Prospector in 1998. This used remote-sensing techniques to compile the first comprehensive mineral maps of the Moon, and reinforced evidence for ice on the surface. NASA hoped to clinch the case by crashing the probe into a suspect crater at the end of its mission, but the impact failed to produce the hoped-for spray of icy debris.

In the new millennium, other nations have taken a renewed interest in the Moon. In 2003, ESA launched a lunar probe of its own, SMART-1, which used an ion engine (see p.285) to reach the Moon for the first time, while 2007 saw the launch of lunar missions by both Japan and China (SELENE and Chang'e 1), both of which surveyed the Moon from orbit before making planned impacts. The following year, India followed suit with its own Chandrayaan-1 orbiter mission and a dedicated impact probe.

While NASA's Lunar Reconnaissance Orbiter has been mapping our satellite in unprecedented detail since 2009, more recent US missions have been aimed at answering specific questions. For instance, the twin spacecraft of 2012's Gravity Recovery and Interior Laboratory (GRAIL) mission mapped the Moon's internal structure, while the Lunar Atmosphere and Dust Environment Explorer (2013–14) studied particles in space surrounding the Moon.

THE VIEW FROM CLEMENTINE
Venus hangs just beyond the Moon in this photograph taken by Clementine above the night side of the Moon. The faint illumination on the right side of the Moon is caused by sunlight reflecting off the Earth.

Exploring Mercury

Reaching the innermost planet was a tough challenge for NASA, but techniques used to make flybys in the early 1970s paved the way for future missions elsewhere in the Solar System. However, it was not until a generation later that a spaceprobe returned to Mercury.

3 November 1973
Mariner 10 leaves Earth en route for Mercury.

5 February 1974
Mariner 10 flies past Venus, using a gravity-assist manoeuvre for the first time.

29 March 1974
Mariner 10 makes the first flyby of Mercury.

16 March 1975
Stabilized by "sailing" the solar wind, Mariner 10 makes its third and final Mercury rendezvous.

3 August 2004
MESSENGER is launched from Cape Canaveral on a Delta II rocket.

2 August 2005
MESSENGER flies past Earth in a gravity-assist manoeuvre that sends it first towards Venus and then on to Mercury.

14 January 2008
MESSENGER makes an initial flyby of Mercury – the first of three prior to entering orbit.

18 March 2011
The MESSENGER probe becomes the first spacecraft to orbit Mercury.

16 April 2012
Following completion of MESSENGER's primary mission and mapping of Mercury's entire surface, NASA engineers lower the probe into a closer eight-hour orbit.

30 April 2015
After exhausting its fuel in late 2014, MESSENGER crashes into Mercury's far side, marking the end of its mission.

The problems in reaching Mercury arise through the immutable laws of planetary motion, which dictate that planets orbiting closer to the Sun move more rapidly than those further out. Orbiting at an average of just 58 million km (36 million miles) from the Sun and with a year that lasts only 88 Earth days, tiny Mercury moves through space at around 48km (30 miles) per second. Earth, in contrast, moves along its orbit at a relatively sluggish 30km (19 miles) per second, and this dictates the speed "inherited" by spaceprobes departing Earth orbit for other worlds. A spacecraft launched from Earth to Mercury, therefore, has to pick up a huge amount of speed, and when NASA turned its attention towards Mercury in 1968, there was simply no way to launch even a lightweight probe with sufficient acceleration.

Chasing Mercury

Fortunately for planners of the Mariner 10 mission, there was an alternative – why not put the spacecraft in an elliptical orbit around the Sun, designed so that it would orbit once in every two Mercury years and come close to the planet at its perihelion point – not once but perhaps several times? Such an orbit would require far less energy and make the mission achievable, but even putting Mariner 10 on this elliptical course would be a challenge. The eventual solution called for the use of an untried theoretical technique known as a gravity assist, or "slingshot", at Venus (see panel, right). Launched in the early 1970s, Mariner 10's mission, therefore, not only gave scientists a first close-up look at Mercury, but also acted as a rehearsal for later missions that would use the slingshot technique to explore the outer planets, such as the TOPS probes (see p.264).

Mariner 10

The spacecraft made its first rendezvous with Mercury in March 1974 and sent back images of a cratered, baking world devoid of atmosphere but with many features suggesting that it has had an unusual history. Perhaps the most spectacular discovery was the enormous Caloris Basin, an impact scar some 1,300km (800 miles) across, but elsewhere huge cliffs cut across the landscape in many places, and density measurements suggested that Mercury has an unusually large metallic core. Mariner 10 made two further flybys of Mercury, in September 1974 and March 1975, but because the probe's orbit met the planet's every two Mercury years and this period is exactly three Mercury "days", the same areas were illuminated on each flyby, rendering it impossible to map more than 45 per cent of the surface.

In Mercury's orbit

Despite the success of Mariner 10, plans for any further exploration of Mercury languished for some time – interest in this small planet, it seemed, was outweighed by the cost and technological barriers to putting a probe in orbit. In the late 1980s, however, JPL scientist Chen-Wan Yen showed how a mission

SOLAR SAILOR
When Mariner 10 ran out of fuel and lost the ability to make controlled roll manoeuvres, ground controllers used the pressure of solar wind particles on its solar panels to steer the probe to its final, and closest, Mercury flyby.

TECHNOLOGY

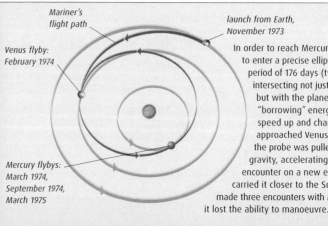

THE JOURNEY OF MARINER 10

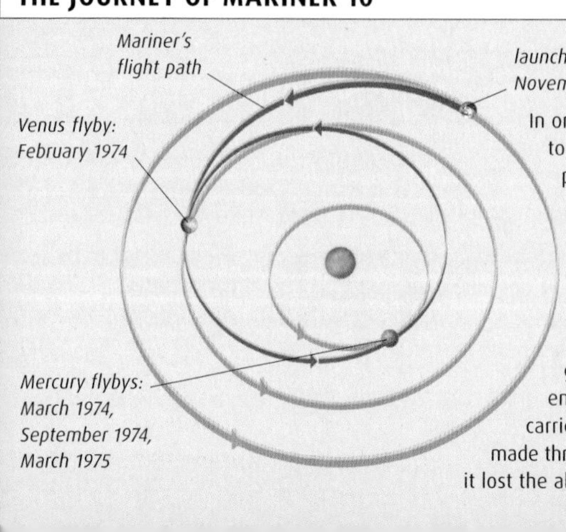

In order to reach Mercury, Mariner 10 had to enter a precise elliptical orbit with a period of 176 days (two Mercury years), intersecting not just with Mercury's orbit but with the planet itself. It did this by "borrowing" energy from Venus to speed up and change course. As it approached Venus in February 1974, the probe was pulled in by the planet's gravity, accelerating so that it left the encounter on a new elliptical path that carried it closer to the Sun. Mariner 10 made three encounters with Mercury before it lost the ability to manoeuvre.

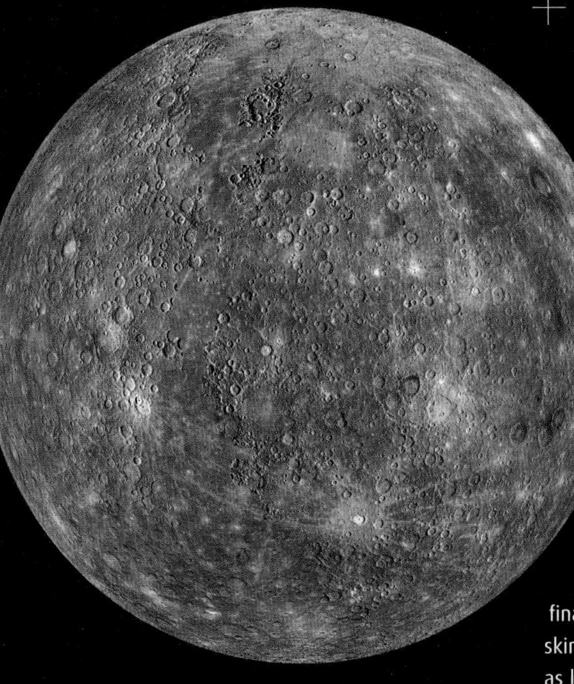

HOT WORLD
Despite its obvious resemblance to the Moon, the searing-hot planet Mercury seems to have a considerably more complex history.

MERCURY'S MINERALS
This false-colour mosaic made during MESSENGER's first flyby reveals differences in surface minerals linked to past impacts and volcanic eruptions.

could achieve Mercury orbit using multiple flybys of the inner planets. A decade later, NASA finally gave the go-ahead for a Mercury orbiter mission, eventually known as MESSENGER (Mercury Surface, Space Environment, Geochemistry, and Ranging).

The spacecraft began its lengthy journey in August 2004 and made no fewer than 15 circuits of the Sun during its inwards spiral to join Mercury. En route, it executed gravity-assist manoeuvres to pick up speed at Earth itself, Venus (twice), and three times during flybys of Mercury. Thus, by the time it entered orbit in March 2011, MESSENGER had already added considerably to knowledge of the innermost planet.

Alongside an array of cameras, the probe carried spectrometers for analyzing minerals and gases, a laser altimeter for measuring the height of the surface, and instruments to map Mercury's gravitational and magnetic fields. Temperature control so close to the Sun was an obvious concern throughout – a large sunshade protected the delicate electronics from solar radiation, but Mercury's surface is so hot that thermal radiation from the planet itself was also potentially harmful. Therefore, the probe's final 12-hour polar orbit was highly elliptical, skimming rapidly over the surface at altitudes as low as 200km (125 miles), before retreating to greater distances of up to 15,000km (9,300 miles).

MESSENGER operated at Mercury for more than four years, far exceeding initial expectations. Among other discoveries, its instruments revealed past volcanic activity associated with many impact craters and the curious orientation of the planet's magnetic field, which is distinctly offset from Mercury's core. Intriguingly, it also detected water ice and carbon-based organic compounds lurking in permanently shadowed craters at the planet's poles (probably dumped there by comet impacts).

From late 2014, its fuel exhausted, MESSENGER began a final inwards spiral. It made a new crater in Mercury's surface in late April 2015, sending back images until minutes before impact.

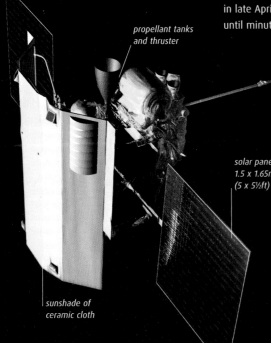

propellant tanks and thruster

magnetometer on 3.6-m (11½-ft) boom

solar panel "wings", 1.5 x 1.65m across (5 x 5½ft)

sunshade of ceramic cloth

LIGHTWEIGHT PROBE
Despite its use of gravity assists, the MESSENGER probe still needed to be relatively small in order to reach Mercury. Standing slightly taller than an average adult male, more than half of its 1,100kg (2,425lb) launch mass comprised propellant for manoeuvring in orbit.

1971

Early Mars missions

The first Mariner flybys of Mars had suggested that the planet was a barren, cratered ball of rock, but throughout the 1970s orbiters and landers revealed a far more complex, fascinating world.

9 May 1971
Mariner 8 is launched from Cape Canaveral, but crashes when its launch vehicle fails.

30 May 1971
Mariner 9 is launched successfully toward Mars.

14 November 1971
Mariner 9 enters Martian orbit during a dust storm.

December 1971
The first usable pictures are returned from Mariner 9.

27 October 1972
Mariner 9 is switched off after exhausting the fuel in its attitude control motors.

20 August 1975
Viking 1 launches from Cape Canaveral.

9 September 1975
Viking 2 launches successfully.

19 June 1976
Viking 1 enters orbit around Mars.

20 July 1976
After a hoped-for landing on 4 July to celebrate the US Bicentennial is postponed, the Viking 1 orbiter releases its lander, which parachutes into the atmosphere and lands successfully in the Chryse Planitia region.

7 August 1976
Viking 2 arrives at Mars.

3 September 1976
Viking 2's lander touches down in Utopia Planitia.

Although Mariners 4, 6, and 7 had all made successful flybys of Mars in the 1960s, no spaceprobe had yet gone into orbit around the planet – but this would change with the Mariner 9 mission. Launched by an Atlas-Centaur rocket, this probe (with its failed twin, Mariner 8) was the first to be equipped with a retrorocket that would allow it to lose excess speed and drop into orbit around Mars. When it arrived after a six-month journey in November 1971, the Red Planet was enveloped in one of its periodic global dust storms, but as the atmosphere began to clear over the following weeks, an unexpected landscape emerged. The northern hemisphere was dominated by huge volcanoes towering above smooth lowland plains. These included a peak which was later named Olympus Mons – the largest volcano in the Solar System, some 500km (300 miles) across and rising to 27km (17 miles) above the average Martian surface. Perhaps even more impressive was the deep canyon system called the Valles Marineris (Mariner Valleys). These huge scars in the Martian surface dwarf Earth's own Grand Canyon. They stretch for more than 4,000km (2,500 miles) around the Martian equator and are more than 10km (6 miles) deep and 600km (375 miles) across in places. Unlike their Earthly equivalent, it was clear that these valleys were produced by faults in the Martian crust, not a result of water erosion – but elsewhere, Mariner 9 photographed winding valleys, "islands", and outflow channels that looked very much like the result of water flowing in the planet's ancient past.

How had the three previous probes so misled the scientists? By a fluke of their flightpaths, each had photographed only the rocky, cratered highlands dominating the southern hemisphere – Mars is a world of two halves, and until 1971 its interesting side had been missed.

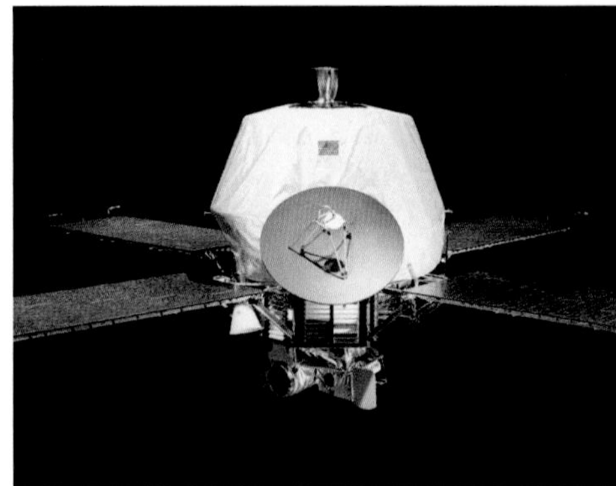

FIRST ORBITER
Mariner 9 was different from earlier Mariner probes in several respects – it had larger solar arrays to generate power and a retrorocket to slow it down as it reached Mars.

Outposts on Mars

In 349 days of operation, Mariner 9 transformed the scientists' image of Mars – now it seemed that the

VIKING'S VIEW
The Viking orbiters compiled a comprehensive photographic atlas of Mars, allowing the production of mosaics of entire hemispheres. This view is dominated by the Valles Marineris canyon system and huge volcanoes on the left-hand limb.

VIEWS FROM THE GROUND
The Viking landers found variations in the terrain around the planet. Viking Lander 1's location (below) had fewer rocks and more windblown sand dunes. Viking Lander 2's landing site (right) was strewn with many more rocks, resembling Earth's volcanic basalts.

TECHNOLOGY

THE VIKING ORBITERS AND LANDERS

The Viking orbiters were based on the successful Mariner template but considerably enlarged. Cameras and other instruments were mounted on a movable platform attached to an octagonal body 2.4m (100in) across. The body held most of the probe's electronics, while an antenna beamed data from the orbiter and lander back to Earth. On the back of this octagon was a rocket motor that slowed the probe down as it arrived at Mars, while the lander itself, sealed inside a sterile outer shell, was attached to the front. The landers were thoroughly sterilized before launch to prevent contamination of the Martian surface or the biology experiments with organic material or bacteria from Earth. Once at Mars, Viking's operators used images from the orbiter to study potential landing sites before releasing the lander.

possibilities of water and even life at some point in Martian history were back on the agenda. The advanced Viking missions, in development since 1969 and now scheduled for launch in 1975, suddenly became a focus of intense interest.

Twin Viking spacecraft were planned, each with two elements – an orbiter and a lander (see panel, above). Viking 1 arrived in orbit around Mars in June 1976, while Viking 2 arrived in August. The orbiters continued to photograph the entire planet at relatively high resolutions until 1980, but attention soon switched to the landers, which parachuted to the ground in July and September. The pictures they sent back recorded the landscapes of the Chryse Planitia and Utopia Planitia regions respectively, and in each case, the general impression was of a landscape of orange-red sand littered with rocks beneath a salmon-pink sky.

As well as cameras, the landers carried other equipment to study the surface. This included seismometers to detect any Martian earthquakes, a weather station that reported temperatures ranging from -120°C (-184°F) at night to -14°C (7°F) in the day, and a sampler arm to scoop up and study Martian soil.

Most attention, however, focused on the biology processor – a set of three experiments that looked for evidence of photosynthesis, bacteria, or just organic (carbon-based) matter in the soil samples. While there was no sign of photosynthesis at work, and the tests for organic matter also drew a blank, the soil did appear to react when "fed" with nutrients. Most scientists thought this was due to unusual chemistry rather than life, but the results were inconclusive. However, it would be a long time before another probe was able to continue Viking's work.

MARS VIKING LANDER
To slow its descent through the thin Martian atmosphere, each Viking lander used parachutes at first and then retrorockets. The lander's sterile shell also shielded it during atmospheric entry, before falling away during the final approach. Onboard antennae allowed for communications with Earth, either sending signals directly or relaying them through the orbiter.

Voyages among giants

The successful use of a gravity assist by Mariner 10 opened the way for ambitious missions to cross the vast distances between the giant outer planets of the Solar System.

2 March 1972
Pioneer 10 is launched on its 20-month long journey to Jupiter.

5 April 1973
Pioneer 11 sets out on its journey to Saturn via Jupiter.

3 December 1973
Pioneer 10 makes the first flyby of Jupiter.

5 March 1979
Voyager 1 makes its flyby of Jupiter. Voyager 2 follows four months later.

1 September 1979
Pioneer 11 makes the first flyby of Saturn.

12 November 1980
Voyager 1 flies through the Saturn system and makes a close approach to Titan.

24 January 1986
Voyager 2 makes the first flyby of Uranus.

25 August 1989
Voyager 2 makes the first flyby of Neptune.

25 August 2012
Voyager 1 becomes the first human object to enter interstellar space.

The space in which the inner, Earthlike worlds of the Solar System orbit is dwarfed by the orbits of the four outer planets. Even the closest of these giants, Jupiter, orbits 5 astronomical units (AU) from the Sun (an astronomical unit is the average distance between Earth and the Sun, roughly 150 million km or 93 million miles). The outermost giant, Neptune, is some 30 AU away from the Sun.

The giants are huge balls of gas surrounded by large families of moons and moonlets – complex systems that were an obvious scientific target for spaceprobes. But their great distance placed them beyond the range of easy targets for Space Race spectaculars – their exploration would be a long-term endeavour.

NASA launched two initial probes, Pioneers 10 and 11, towards Jupiter and Saturn in 1972 and 1973 respectively. These became the first objects to cross the asteroid belt between Mars and Jupiter, proving that it was largely empty space and the risk of a collision was small. Pioneer 10 reached its target in December 1973, and sent back pictures that improved on Earthbound telescopes, but still left many questions unanswered. Pioneer 11 swung past Jupiter in 1974 and used the giant planet's gravity to swing it towards Saturn, which it reached in 1979. By this time, a more sophisticated pair of probes was already following in its wake.

high-gain antenna
particle experiments
radioisotope thermal generator (RTG)
wide- and narrow-angle cameras
main system bus with propulsion system
infrared and ultraviolet instruments

VOYAGER
The twin Voyager probes bear a distinct resemblance to their Mariner ancestors. Stabilized by small thrusters, they orient themselves in space by tracking the Sun and the bright star Canopus.

Planning the Voyagers

During their initial studies of gravity assist in the 1960s, scientists had noticed that the giant planets would fall into a particularly neat arrangement in the late 1970s. A spacecraft launched during this alignment, which only occurs once every 176 years, would be able to make a "grand tour" of the outer Solar System, receiving from each giant planet in turn a gravity boost that would alter its course and accelerate it towards its next target. The opportunity was too good to miss, and NASA began to develop plans for a Thermoelectric Outer Planets Spacecraft (TOPS). However, as the mission grew more ambitious and complex, its budget spiralled, leading to its cancellation amid the cutbacks of 1972.

With just a few years until the launch window, NASA went back to the drawing board, designing a simpler probe based on Mariner technology. By the time the mission was ready in 1977, the twin spacecraft had developed so far from Mariner that they were given a new name – Voyagers.

Grand tourists

The Voyager mission design called for Voyager 1 to travel on a faster trajectory than its sibling, and so Voyager 2 was launched first, on 20 August 1977, and Voyager 1 took off 15 days later. Voyager 1

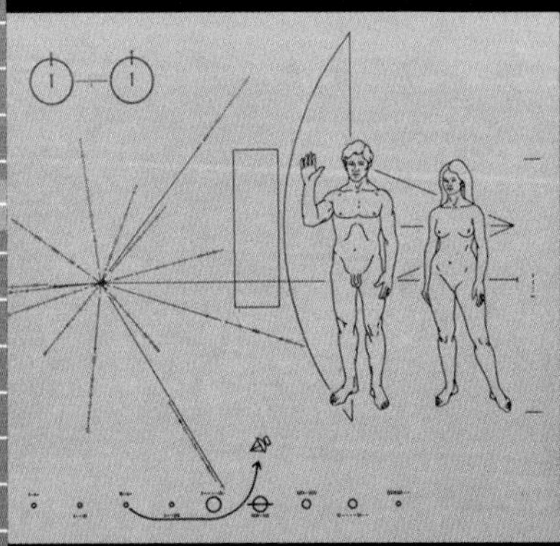

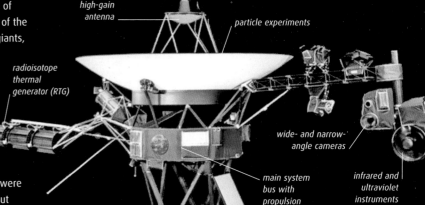

PIONEER AND ITS PLAQUE
The Pioneer probes (right) to Jupiter and Saturn were to be the first human objects sent to go beyond the Solar System, and the mission scientists felt it was important to send a greeting to any alien civilisation that might one day find them. To this end, they devised a plaque (above) that shows a man and a woman, along with basic directions to reach Earth.

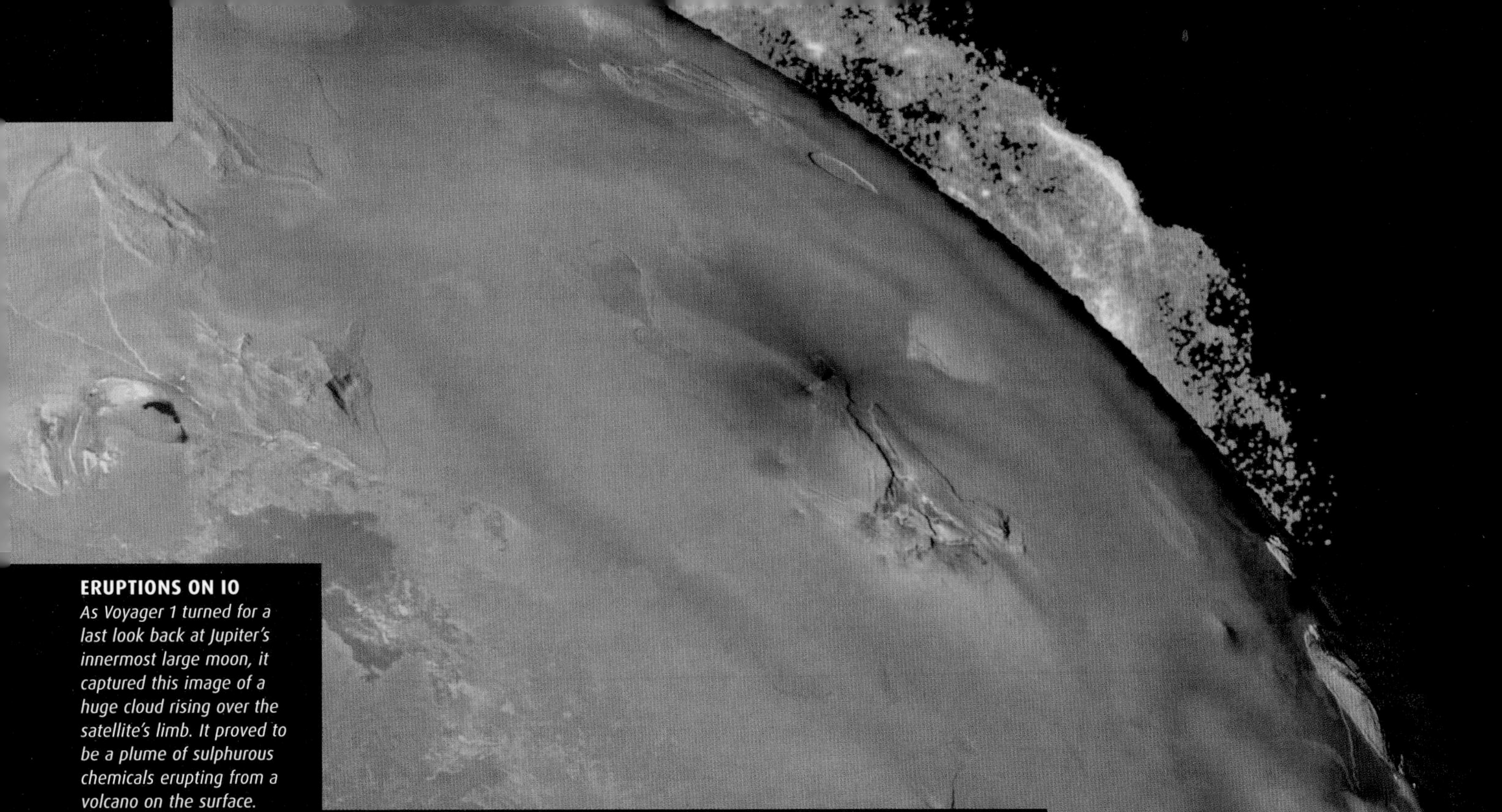

ERUPTIONS ON IO
As Voyager 1 turned for a last look back at Jupiter's innermost large moon, it captured this image of a huge cloud rising over the satellite's limb. It proved to be a plume of sulphurous chemicals erupting from a volcano on the surface.

swung past Jupiter on 5 March 1979, followed in July by its twin. The probes sent back the first close-up images of the large moons Io, Europa, Ganymede, and Callisto. These revealed active sulphur volcanoes on Io, an icy crust with hints of a hidden ocean on Europa, and a thin ring of dust around Jupiter itself.

At Saturn, the paths of the Voyagers diverged. Voyager 1 swung close to the planet, photographing its famous rings and flying past the giant moon Titan, which proved to be covered in a smoggy orange atmosphere. Voyager 2 flew past further out, using gravity assist to sling it on towards an encounter with Uranus in 1986. Here it found an eerily placid green world and an array of moons, including the bizarre Miranda with a surface that seems to be a jumble of different terrain types.

The final leg of the probe's marathon journey carried it on to Neptune in 1989. This proved to be a far more active world, with dark storms and high winds raging in its blue atmosphere. Its large satellite, Triton, was even stranger – with geysers of liquid nitrogen and a surface temperature of -238°C (-396°F). Leaving Neptune behind in its wake, Voyager 2 headed into the outer limits of the Solar System – like Voyager 1 and its Pioneer siblings, it is moving fast enough to leave the Solar System forever and wander among the stars.

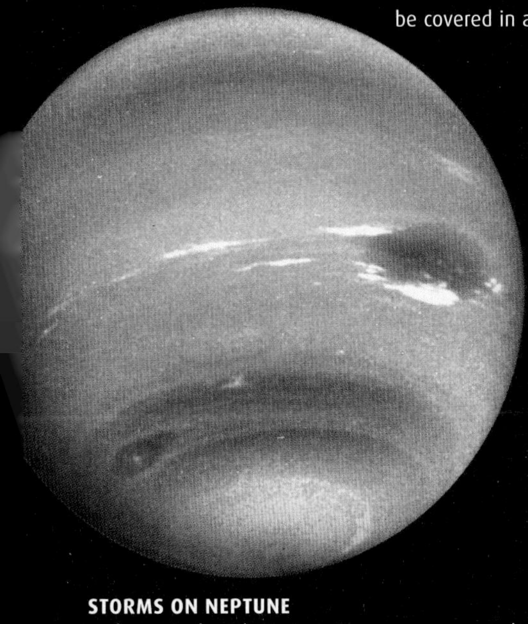

STORMS ON NEPTUNE
So far out in the Solar System, Neptune was expected to be a placid, deep-frozen world. Instead, Voyager 2 found huge storm systems and some of the highest wind speeds in the Solar System – all powered by an unknown internal energy source.

TECHNOLOGY

SLINGSHOTS ACROSS THE SOLAR SYSTEM

Gravity assist put strict constraints on the flightpaths of the Voyager probes. The main goals of the mission were to revisit Jupiter and Saturn and get a good look at Jupiter's four large satellites and Saturn's moon Titan. Uranus and Neptune were an optimistic afterthought, but a close Titan fly by would make it impossible for a probe to continue towards them, so mission designers came up with a contingency plan. Voyager 1 took a fast route to Saturn and Titan, while Voyager 2 followed a slower path that could be diverted to Titan if something went wrong. Fortunately nothing did, and the extension of Voyager 2's mission was authorized shortly after its Saturn encounter.

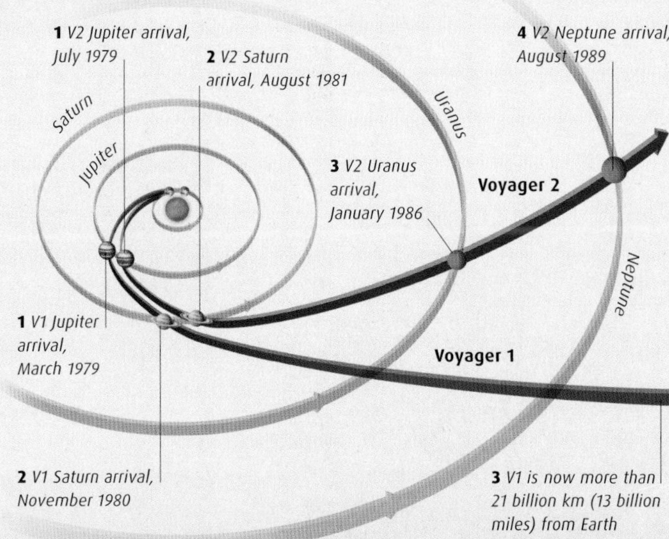

Venus explorers

Although early investigations of Venus had punctured the image of our nearest planetary neighbour as a potential tropical paradise, it remained a fascinating, if challenging, target for later missions.

Despite successful flybys of Venus in 1962 and 1967 (see p.55), NASA's efforts to explore the Solar System in the 1960s were largely focused on Mars and, of course, the Moon. This left the field clear for the Soviet Union, after many struggles, to take a lead in the exploration of Venus.

Flybys and first landings

Throughout the 1960s, a series of failed Soviet Venus missions was (mostly) covered up by the simple strategy of concealing their true destination until the spacecraft were well on their way. Veneras were intended to make close flybys but failed en route, while Veneras 3 through 6 were designed to parachute directly into the atmosphere on what were effectively suicide missions (the probes carried no modifications to cushion their landings). Venera 3 presumably hit the surface in March 1966 but failed to send back any data, while Venera 4 sent back brief signals from the atmosphere in October 1967, before it too lost contact. Veneras 5 and 6 fared little better in May 1969.

It was now clear that Venus's oven-like atmosphere was more than capable of destroying delicate spacecraft, so later probes were heavily armoured against the hostile conditions. Though it was initially thought to have failed during descent, analysis of radio signals from Venera 7's December 1970 mission showed air pressure levelling off at a mere 90 times that on Earth, and temperatures steadying at 475°C (887°F), confirming that the lander had in fact reached the ground and sent back the first signals from the surface of another planet. Later landers continued the path of improvement, with Venera 8 operating for almost an hour on the surface.

Soft landings and orbiters

Later Venera missions took a different approach, with a two-part design launched on the powerful Proton rocket. A relay spacecraft carried engines that allowed it to brake and enter orbit. This finally allowed the landing probes to return images from the Venusian surface (see panel, left).

VENUSIAN ATMOSPHERE
The Pioneer Venus Orbiter used its ultraviolet camera to capture this image of weather patterns in the dazzling clouds of Venus in February 1979, shortly after its arrival in orbit. The Multiprobe mission (right) carried four cone-like probes that sent back detailed infromation about Venusian weather conditions.

VENERA 10 LANDER
Heavy armour fitted to the later Venera landers raised their weight to an impressive 1,560kg (3,440lb). In addition to heat shields, circulating coolant redistributed heat, extending this lander's survival time to 65 minutes.

Meanwhile, after the brief Mariner 10 flyby of 1974 (see p.260), NASA finally turned its attention back to Venus in the late 1970s. Two Pioneer Venus missions were launched in May and August 1978, each with a unique objective. The Pioneer Venus Orbiter carried radar mapping equipment to study the surface beneath the clouds, while the Pioneer Venus Multiprobe deployed four small atmospheric probes that sent back information about weather conditions as they descended into the toxic clouds. The final pair of Venera probes, 15 and 16, also carried radar equipment, helping to pave the way for a mission that would finally uncover the Venusian landscape.

TECHNOLOGY

IMAGES FROM THE SURFACE

The thick Venusian atmosphere did such a good job of swamping radio signals from the planet's surface that the later Soviet probes used a different strategy to earlier missions. On Veneras 9 through 14, the landers were carried to the planet by a mothership that remained in orbit while they descended to the surface. The mother ship acted as a signal relay, picking up the data sent from the lander below and re-transmitting it to Earth through clear space. The monochrome and later colour images revealed a Venusian surface scattered with apparently volcanic rocks on a darker layer of "soil" – in fact it now seems that almost all of the surface material originated in volcanic eruptions.

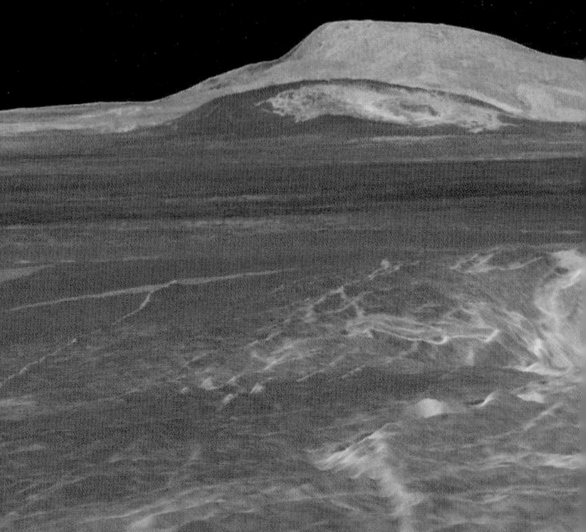

MAPPING VENUS
Data gathered by Magellan was used to produce global elevation maps of Venus such as this one (right). Highland areas are shown in cream and white, and lowlands in blue and purple.

Lifting the veil

Earth-based telescopes had used radar altimetry to identify broad features on the surface of Venus since 1961. The principle is simple – short "pings" of radio waves are fired at the target, and a receiver picks up their reflected echo. Because the waves travel at the fixed speed of light, the time they take to return can reveal the distance of the reflecting surface. The major challenges are a matter of resolution – how accurately is the radar beam directed? Pioneer Venus and the last Veneras offered a vast improvement on previous attempts, but by the 1980s, a new technique called synthetic aperture radar was being used in Earth remote sensing, and NASA was keen to send it to Venus.

Initially known simply as the Venus Radar Mapper, in 1986 this mission was renamed Magellan, in honour of the great Portuguese explorer. By sending out "chirps" of radar rather than mere "pulses", and measuring the echoes at various points as they moved along its orbit, Magellan would not only produce images with far higher resolution than before but could also reveal other surface properties, such as the slope, roughness, and electrical conductivity of the terrain. Launched from the Space Shuttle *Atlantis* in May 1989, the spacecraft entered a polar orbit that ensured it passed over almost every point on the surface as the planet slowly rotated below. By the end of three Venusian rotations (each equivalent to 243 Earth days), Magellan had mapped 98 per cent of the landscape to a resolution of 100m (330ft). In October 1994, after four years in orbit, it was sent on a final plunge into the Venusian atmosphere.

More recently, after another pause in exploration, the European Venus Express probe arrived in orbit in 2006, beginning a campaign of long-term atmospheric observations that have revealed secrets of the complex Venusian weather. Its mission concluded in 2014, just before the delayed arrival of Japan's Akatsuki probe on a search for signs of lightning and active volcanoes.

SIF AND GULA MONS
Altimetry data and other information from Magellan were combined to produce 3-D visualizations and flyovers of the Venusian terrain. This area of the Eistla Regio highlands is dominated by two huge volcanoes and a vast plain of solidified lava.

15 December 1970
Venera 7 makes the first successful landing on Venus.

22 October 1975
Venera 9 returns the first images from the Venusian surface.

16 November 1978
The Pioneer Venus Multiprobe releases the first of its probes into the planet's atmosphere.

4 December 1978
The Pioneer Venus Orbiter arrives to make the first radar maps of the planet.

10 August 1990
Magellan arrives at Venus and enters a near-polar orbit, circling the planet every 3 hours 15 minutes.

13 October 1994
Magellan plunges into the Venusian atmosphere and is destroyed at the end of its successful five-year mission.

11 April 2006
ESA's Venus Express arrives in orbit at the start of what will become a nine-year mission to study the atmosphere.

7 December 2015
Japan's Venus Climate Orbiter, Akatsuki, arrives at Venus after a five-year journey due to a missed orbital insertion.

BROKEN ANTENNA
Galileo's main high-gain "umbrella" antenna dish undergoes testing. When Galileo tried to unfold its high-gain antenna on the way to Jupiter, one of the spokes became trapped, leaving the craft's communications system crippled.

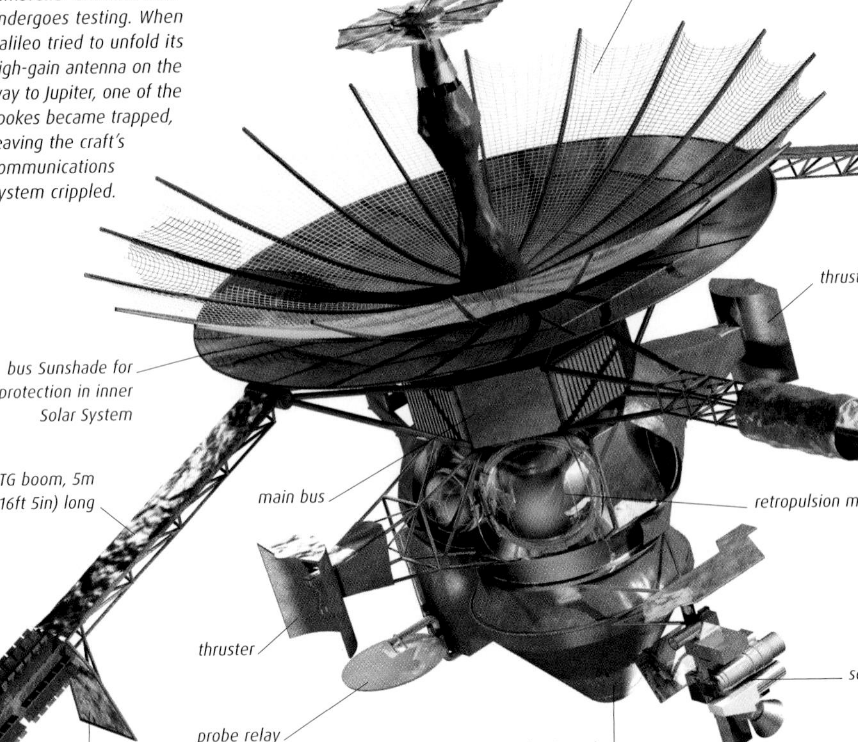

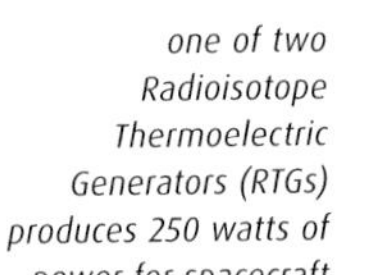

DEVELOPMENT TESTS
Before any spaceprobe is built, numerous mock-ups are produced to test how it might operate in outer space. Here, a model of Galileo is shown in its launch configuration at the JPL facility.

HEIGHT	5.3m (17ft 5in)
NUMBER OF INSTRUMENTS	10 on orbiter spacecraft
TOTAL MASS	2,223kg (4,900lb)
POWER SOURCE	2 x 7.8kg (17.2lb) plutonium RTGs
MANUFACTURER	Jet Propulsion Laboratory
LAUNCH DATE	18 October 1989
TERMINATION OF MISSION	21 September 2003

Galileo spacecraft

Originally known simply as the Jupiter Orbiter/Probe, Galileo, like many interplanetary orbiters, was spin-stabilized. When part of a spacecraft rotates, it acts like a gyroscope and helps to keep the craft in its desired attitude without the need for constant and wasteful corrections by the thrusters. In Galileo's case, the upper section, including antennae, field and particle experiments, and computers, rotated at three revolutions per minute. The lower section, containing cameras, other remote-sensing instruments, and the guidance sensors, kept a fixed orientation.

READY TO GO
Galileo stands in Kennedy Space Center's Vertical Processing Facility (VPF), ready for mating with its Inertial Upper Stage booster before its 1989 launch.

INSTALLATION

Galileo and its Inertial Upper Stage (IUS) were stowed in Atlantis's payload bay (left) prior to launch (below). The IUS attaches to a turntable that swivels upright in orbit, spins and then releases the probe.

JUPITER AT LAST

After years of delays to its proposed launch date and then a six-year journey through the Solar System, Galileo finally arrived at Jupiter on 7 December 1995 – where it would carry out eight years of work.

MONITORING GALILEO

Mission Control was at the Jet Propulsion Laboratory in Pasadena, California. During the long cruise to Jupiter, the controllers found a way to compensate for the crippled antenna on Galileo.

GIANT TO JUPITER

The Galileo probe was designed to operate for an extended period in one of the Solar System's most hostile environments – the radiation belts around Jupiter. Its computers were vulnerable to damage from radiation and strong electric fields, and they were designed and programmed with a high level of fault protection to reduce the risk of errors.

10.9m (35ft 9in) fibre-glass boom isolates sensitive instruments from probe's own electromagnetic fields

magnetometer sensor

plasma wave subsystem measures electric fields

THE JUPITER ATMOSPHERIC PROBE

Galileo's probe was designed to plunge into the giant planet's upper cloudtops, sending data back to the orbiter as it fell. Like the main spacecraft, it used spin to stabilize its motion through space – the spacecraft itself was deliberately spun up to a higher rate prior to the probe's release. After a journey of almost five months, the probe entered the Jovian atmosphere. A heat shield protected it for the first three minutes of entry, then fell away as the drogue and then the main parachute deployed. The probe transmitted data back to the orbiter for some 59 minutes before contact was lost.

DIAMETER	1.3m (4ft 3in)
NUMBER OF INSTRUMENTS	6
TOTAL MASS	339kg (746lb)
POWER SOURCE	Onboard battery
MANUFACTURER	German Space Agency
DEPLOYMENT DATE	13 July 1995
JUPITER ATMOSPHERE ENTRY	7 December 1995

DESCENT MODULE

Tested in a specially designed "Giant Planet Facility", the probe was built to withstand the forces of Jovian gravity, heat, and atmospheric pressure.

PARACHUTE TESTING

The braking characteristics of the probe's parachute were tested in the wind tunnels at NASA's Langley Center. However, it eventually deployed a minute later than planned. Afterwards, technicians diagnosed a wiring fault – it was fortunate that the parachute opened at all.

Galileo to Jupiter

The Voyager probes had provided a tantalising taste of Jupiter and its varied moons. The next challenge was to put a probe into orbit around the giant planet and conduct a long-term study of the Jovian system.

LAUNCHED BY SHUTTLE
The Shuttle Atlantis *blasts off from Cape Canaveral on 18 October 1989, carrying Galileo at the start of its long journey to Jupiter.*

13 July 1995
Galileo releases its atmospheric probe towards Jupiter.

7 December 1995
The atmospheric probe enters Jupiter's atmosphere, sending back data for 59 minutes.

7 December 1997
After two years, Galileo completes its primary mission and moves into an extended phase, making closer flybys of Io and Europa.

7 December 1999
The Galileo mission enters a new phase known as the Galileo Millennium Mission.

30 December 2000
The Saturn-bound Cassini probe makes its closest approach to Jupiter, and mission controllers use both probes to study Jupiter simultaneously.

15 October 2001
Galileo makes its closest flyby of Io, passing just 180km (112 miles) above the surface.

21 September 2003
Galileo plunges into the Jovian atmosphere and is destroyed.

Although it did not reach Jupiter until 1995, the Galileo probe had a long gestation. Development of a Jupiter orbiter spacecraft with an atmospheric probe began even before the launch of the Voyagers in 1977. Germany joined the project in the same year, providing Galileo's propulsion systems in return for participation in the scientific programme.

Even after it was ready for launch, Galileo suffered a series of delays that ultimately pushed the mission back by eight whole years. At first, it was to be deployed in 1981, but the Space Shuttle's teething troubles put paid to that plan. The launch was rescheduled for 1984, but debates about how to boost the probe from Earth orbit saw it miss that chance. Finally scheduled for a May 1986 launch, Galileo then fell victim to the delays following the *Challenger* disaster. It set off at last in October 1989.

Planetary pinball

Safety restrictions introduced after *Challenger* limited the size of Galileo's booster rocket, so in order to get up speed for the trip to Jupiter, it was sent through a tortuous series of gravity assists involving a flyby of Venus and two further flybys of Earth. This Venus-Earth-Earth Gravity Assist (VEEGA) put Galileo on course to reach its target six years after its launch.

VOLCANO WORLD
Galileo's pictures of Io revealed that it is peppered with countless volcanoes and coloured by sulphur compounds from their eruptions. All this volcanic activity is driven by tidal forces from nearby Jupiter.

Even while the probe was on its long flight to Jupiter, its controllers and technicians were kept busy. Following the first Earth flyby, Galileo's high-gain antenna was supposed to unfold in order to allow long-distance communication, but the mechanism jammed and it was rendered useless. For a while, it looked as if Galileo might be an expensive failure, but fortunately the engineers found a way to send data through the emergency low-gain antenna at a slower rate. Galileo's onboard computers and the Earth-based receivers had to be reprogrammed to handle the new procedures. With this problem solved, Galileo was directed towards close encounters with two asteroids as it passed through the asteroid belt between Mars and Jupiter (see p.272). Then, while the probe was

LAUNCH FROM *ATLANTIS*
Deployment from the Shuttle meant that Galileo could only use an Inertial Upper Stage booster rocket, instead of a powerful Centaur upper stage that would have allowed a more direct route to Jupiter.

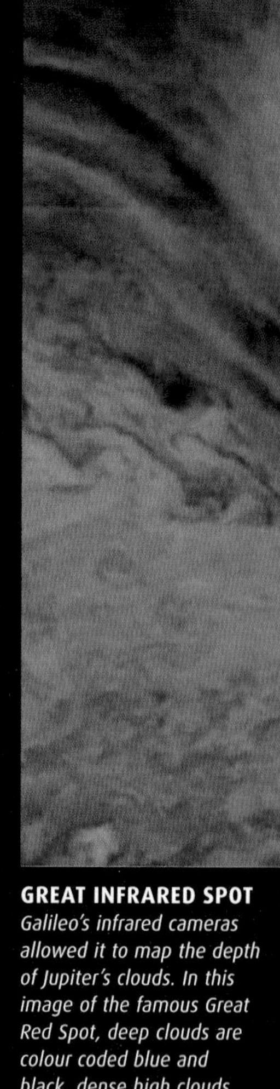

GREAT INFRARED SPOT
Galileo's infrared cameras allowed it to map the depth of Jupiter's clouds. In this image of the famous Great Red Spot, deep clouds are colour coded blue and black, dense high clouds are white, and high thin hazes are pink.

still a year from Jupiter, it had a spectacular view when Comet Shoemaker-Levy crashed into the giant planet in July 1994. Arriving at last, Galileo fired its main engine and dropped into orbit around Jupiter. During final approach, the atmospheric probe was released from the main spacecraft and plunged into the atmosphere (see panel, right).

Eight years of discovery

Now Galileo began its long scientific mission. It orbited Jupiter roughly once every two months, on long ellipses that brought it close to all of the major moons. The original plan for the mission assumed the probe might function for about two years – Jupiter is surrounded by radiation belts far fiercer than those around the Earth, which damage even the most robust electronics over time. As it was, Galileo beat even the most optimistic forecasts, surviving for more than eight years.

Over time, the probe was gradually brought closer to Jupiter itself, spending more time in the radiation zone. A final series of close encounters with Io saw its cameras damaged beyond repair in January 2002. During its time in orbit, Galileo had revolutionized ideas about the Jupiter system. In particular, it had revealed the huge extent of volcanic eruptions on Io (now known to be the most volcanic world in the Solar System) and bolstered the evidence that Europa's crust of scarred ice hides a global saltwater ocean kept warm by volcanic activity. Such an ocean might even harbour life, and in order to prevent contamination Galileo was steered to a fiery end in Jupiter's atmosphere on 21 September 2003.

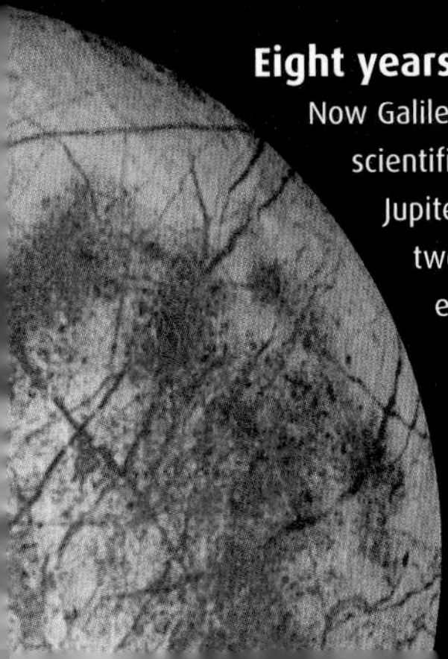

ICY EUROPA
Europa's brown stains are thought to show where its icy crust has cracked, allowing chemical-laden water to seep in and heal the scars.

TECHNOLOGY

THE PROBE'S STORY

Galileo's atmospheric probe entered the Jovian atmosphere at a speed of 47km (29 miles) per second. As it fell, it was slowed to below the speed of sound (0.35km/0.2 miles per second) in just two minutes. Friction with the atmosphere turned the probe into a flaming bullet and burned away more than half the mass of its 152kg (334lb) heat shield. Only then did the probe deploy its parachute and begin transmitting data. Fifty-nine minutes of temperature and pressure information were relayed via the main Galileo spacecraft to Earth, along with information about the chemical composition of the atmosphere, which turned out to be drier than expected, with less water vapour. The probe finally overheated and stopped transmitting at a depth of 146km (91 miles).

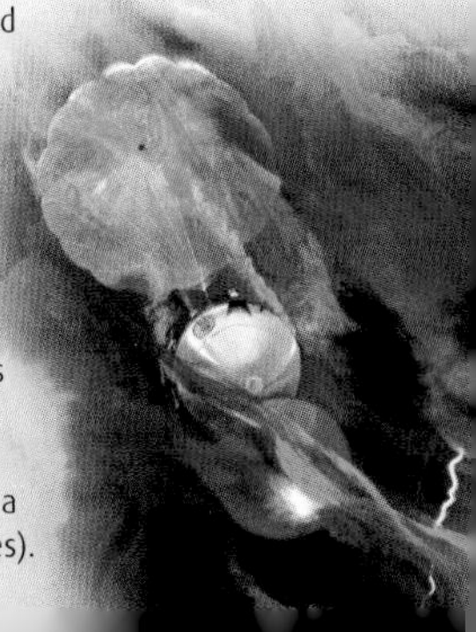

1986

Comets and asteroids

The Solar System's smaller bodies range from objects that have remained largely unaltered since its creation to worlds with complex geology to rival any planet. Since the 1980s, many of them have been visited.

March 1986
A series of five spaceprobes study Halley's Comet during its passage round the Sun.

29 October 1991
The Galileo Jupiter probe makes the first asteroid flyby, of 951 Gaspra.

14 February 2000
NEAR-Shoemaker arrives in orbit around asteroid 433 Eros.

2 January 2004
NASA's Stardust flies through the coma of Comet Wild 2 and collects samples of comet dust.

2 March 2004
ESA's Rosetta probe is launched on its 10-year journey to Comet 67P.

4 July 2005
Deep Impact launches its projectile into Comet Tempel 1 and studies the resulting explosion.

19 November 2005
Japan's Hayabusa probe touches down on asteroid 25143 Itokawa and collects dust from its surroundings.

16 July 2011
Dawn enters orbit around asteroid 4 Vesta.

6 August 2014
Rosetta takes up its orbit around Comet 67P.

6 March 2015
Dawn arrives at the largest asteroid, 1 Ceres, and enters orbit.

Between and beyond the orbits of the eight major planets there are countless small objects, ranging in size from mere boulders to minor worlds comparable to the Moon, and in composition from rocky asteroids close to the Sun to frozen comets and "ice dwarfs" in the space between and beyond the giant planets. Some comets follow long elliptical orbits that bring them close to the Sun at perihelion, when their surface ices can grow warmer and evaporate to form an extended atmosphere, or "coma", and tail. Most comets and asteroids are thought to preserve pristine material from the dawn of the Solar System, unaffected by the geological and chemical processes that have shaped the larger planets and moons.

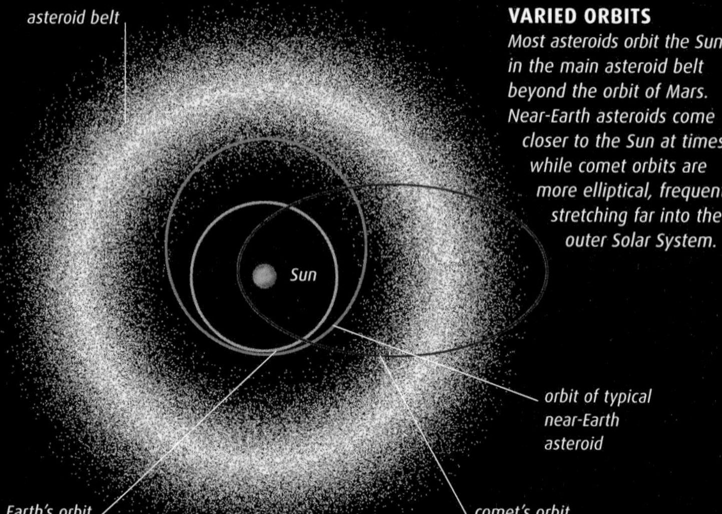

VARIED ORBITS
Most asteroids orbit the Sun in the main asteroid belt beyond the orbit of Mars. Near-Earth asteroids come closer to the Sun at times, while comet orbits are more elliptical, frequently stretching far into the outer Solar System.

Early investigators

In 1986, the famous and predictable Comet Halley became the first of these small worlds targeted for spaceprobe investigation. In an unprecedented collaboration, ESA, the Soviet Union, and Japan's ISAS agency launched an armada of probes to investigate different aspects of the comet. NASA did not participate, though it had laid its claim to the first comet probe by diverting a pre-existing mission to intercept Comet Giacobini-Zinner in 1985.

The armada of probes sent to Halley – the Soviet Vegas 1 and 2, the Japanese Sakigake and Susei, and Europe's Giotto – all flew past the comet at the peak of its activity in March 1986. Together, they built a comprehensive picture of the comet, including Giotto's images of its "dirty snowball" nucleus taken from just 600km (370 miles) away.

The first close-up images of asteroids came a few years later as the Galileo probe crossed the main asteroid belt on its way to Jupiter, and in 1996, NASA launched a spaceprobe dedicated to asteroid research. This was NEAR-Shoemaker, the Near Earth Asteroid Rendezvous mission. After a long journey, it slipped into orbit around asteroid 433 Eros on Valentine's Day, 14 February 2000. Over the following months, the probe mapped Eros with an array of instruments, before finally descending to a gentle touchdown on the surface a year after arrival.

Since the success of NEAR-Shoemaker, studies of comets and asteroids have accelerated, with a wide range of missions to study their properties. NASA's Stardust, launched in

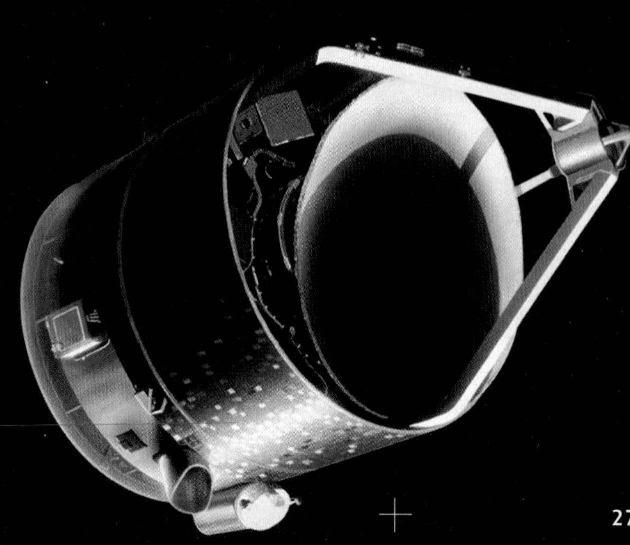

GIOTTO
The European Space Agency's first mission beyond Earth orbit, Giotto was based on a GEOS research satellite bus. An aluminium and kevlar dust shield offered protection as it flew into the heart of Comet Halley.

February 1999, rendezvoused with Comet Wild 2, collecting a sample of particles from its coma on a lightweight material called an aerogel before looping back to Earth in 2006 and ejecting its cargo in a re-entry capsule.

Another impressive NASA mission was the appropriately named Deep Impact, which fired a barrel-like 370kg (814lb) projectile into the nucleus of Comet Tempel 1 in July 2005 in order to study the material ejected into space. The results caught many scientists by surprise and led to a rethink of previous comet models – Tempel 1 turned out to be far more dusty, and less icy, than expected.

Japan's JAXA launched its own Hayabusa probe in May 2003. This aimed to touch down on near-Earth asteroid 25143 Itokawa and collect samples before returning to Earth. Although the mechanism intended to blast material from the surface of Itokawa failed, the collection system still swept up some floating dust, which was returned to Earth in a re-entry capsule as Hayabusa flew past in June 2010.

Perhaps the most ambitious comet probe so far, however, is ESA's Rosetta. Launched in 2004, it spent a decade travelling to its rendezvous with Comet Cheryumov-Gerasimenko (also known as 67P). Entering a finely balanced orbit between May and August 2014, it then flew alongside the comet for two years as it approached the Sun and became active. Although a small lander called Philae failed after landing in a permanently shadowed area, the mission finale offered some compensation as Rosetta itself made a gentle landing on the surface.

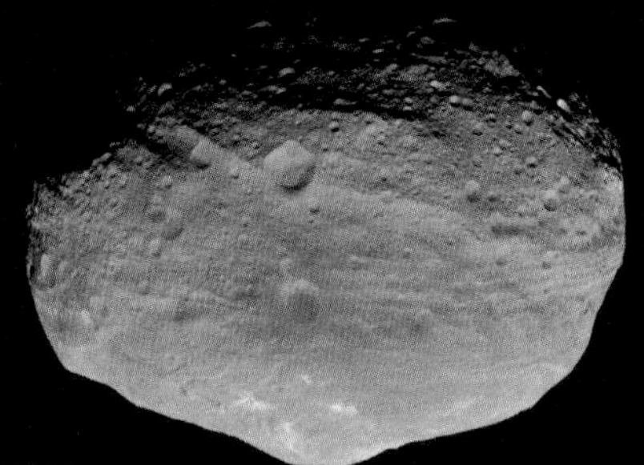

VESTA
The shape of the asteroid belt's second-largest body is distorted by two enormous craters called Rheasilvia and Veneneia. Dawn revealed a contrasting mix of ancient terrains and younger ones formed by volcanic activity.

Orbiting asteroids

Not all asteroids and comets, however, are small and primitive bodies – some of the bigger asteroids display signs of prolonged geological activity. Two of the biggest, 1 Ceres and 4 Vesta, were investigated by NASA's Dawn mission, launched in 2007. Powered by an ion thruster (see p.285), Dawn reached Vesta, the Solar System's second-largest asteroid, in July 2011 and remained in orbit for 14 months. It then flew on to the even bigger Ceres, entering orbit in March 2015. The spacecraft carried visible-light and infrared cameras, and spectrometers for imaging the asteroids' surfaces, alongside a Gamma-Ray and Neutron Detector (GRaND) used to identify individual elements in the surface rocks. At Ceres, it identified hints that a liquid-water ocean lies hidden beneath the crust.

UPPER STAGE ATTACHMENT
Dawn is secured to the upper stage of its Delta II launch vehicle in a Cape Canaveral clean room. The mission had a bumpy road to launch, cancelled in 2003, reinstated in 2004, and then put on hold for several months in 2005–06.

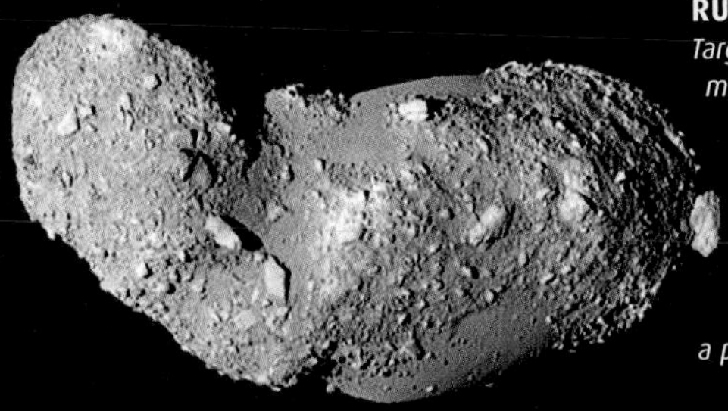

RUBBLE-PILE ASTEROID
Targeted by JAXA's Hayabusa mission, the near-Earth asteroid 25143 Itokawa is surprisingly lacking in surface craters. It is thought to be little more than a cavity-riddled collection of fragments, loosely reassembled after a past collision.

67P/CHURYUMOV-GERASIMENKO
In two years orbiting the 4.3-km- (2.7-mile-) long nucleus of "Comet 67P", Rosetta mapped a bizarre and changing landscape of jagged cliffs, rubble fields, and sunken pits. As the comet rounded the Sun, violent eruptions of ice from beneath the surface (inset) sent boulders flying for hundreds of metres and created large areas of subsidence.

1997

Return to Mars

After a break in Martian exploration of almost two decades, the late 1990s saw the beginning of a new wave of more sophisticated probes – orbiters, landers, and rovers that have transformed our view of the Red Planet.

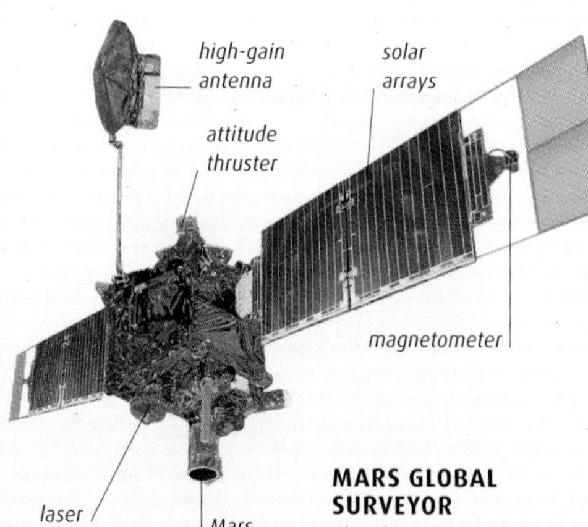

MARS GLOBAL SURVEYOR
The thrusters on MGS were able to tilt the probe 30° in any direction to photograph the Martian surface at oblique angles, and even look at objects close to Mars such as its two small moons and other orbiting spacecraft.

4 July 1997
Mars Pathfinder and its Sojourner rover touch down on Mars.

12 September 1997
Mars Global Surveyor (MGS) enters orbit and begins aerobraking.

23 September 1999
Mars Climate Orbiter crashes into the planet during orbit insertion.

3 December 1999
Mars Polar Lander is lost shortly before landing on the planet.

24 October 2001
Mars Odyssey enters orbit and begins aerobraking.

25 December 2003
Mars Express enters orbit. Beagle 2 drops into the Martian atmosphere but fails during landing.

4 January 2004
The *Spirit* Mars Exploration Rover lands in Gusev Crater.

25 January 2004
The *Opportunity* Mars Exploration Rover lands in Meridiani Planum.

22 March 2010
The grounded *Spirit* rover loses contact with Earth.

The 1980s and early 1990s were a bad time for Mars-bound spaceprobes. NASA's immediate plans to follow up on the success of Viking were shelved as the rising cost of the Space Shuttle programme forced cutbacks elsewhere, and a series of Soviet probes fell victim to a variety of accidents and mishaps. For example, two sophisticated Phobos probes were lost in the late 1980s – the first when a faulty signal from Earth accidentally ordered the probe to shut down, the second for unexplained reasons shortly after Phobos 2 had entered orbit.

Even when NASA did finally launch a new mission to Mars, it proved to be a false dawn, as Mars Observer, sent on its way in September 1992, mysteriously lost contact shortly before entering orbit. With such bad luck, some people even started to joke about a "curse of Mars".

The long wait ends

It was not until 1997 that a relatively small spacecraft, Mars Pathfinder, finally touched down in the Ares Vallis region. As the probe unfolded its triangular, petal-like solar panels, it released a small rover called Sojourner, which trundled onto the Martian soil on 4 July and immediately began to explore the rocky landscape around it. Sojourner operated for 83 Martian days, well beyond its expected lifetime, and even while it was still roaming the surface, another mission slipped into orbit above it. Mars Global Surveyor (MGS) brought with it sophisticated, high-resolution cameras that could distinguish objects just a few metres across. In order to slow itself into a lower orbit, however, the probe used the aerobraking technique rehearsed by the Magellan Venus mission (see p.267), slowly spiralling inwards until it was ready to begin its serious work in April 1999. MGS provided a stunning new view of Mars, sending back detailed pictures for some seven-and-a-half years – six years beyond its primary mission. Perhaps its most important revelation was the presence of eroded gullies on some canyon slopes and crater walls, which many experts believe are evidence of liquid flowing on Mars in the very recent past. Elsewhere, the presence of near-pristine lava flows suggested that Mars may still have volcanic activity too.

SOJOURNER ROVER
The Sojourner rover was a small but robust vehicle, just 65cm (26in) long and weighing 10.6kg (23.3lb). With a maximum speed of 60cm (24in) per minute, it could range up to 500m (1,650ft) from the Pathfinder lander.

PRESIDENTIAL PANORAMA
A mosaic of photographs from the Mars Pathfinder lander, this so-called Presidential Panorama includes ten separate images of the Sojourner Rover investigating the nearby landscape. The lander itself is surrounded by the deflated airbags that cushioned its landing.

CRATER IN THAUMASIA
This view from the Mars Express High Resolution Stereo Camera (HRSC) shows a 20-km (12½-mile) wide crater in the Thaumasia mountain range in the Martian southern highlands. Stereo images like this can offer planetary scientists better insight into true shape of landscape features, and a new understanding of how they formed.

TECHNOLOGY

THE ILL-FATED BEAGLE

Beagle 2 (named after the ship on which naturalist Charles Darwin travelled) was a small probe designed specifically to look for signs of life on Mars. The probe separated from the Mars Express orbiter on 19 December 2003 and entered the atmosphere on Christmas Day. It was supposed to touch down in the equatorial region of Isidis Planitia, open up, and use a robot arm to collect rock samples for analysis in various instruments. But nothing was heard from the Martian surface – it seems likely that Beagle 2 was disabled by hitting a crater wall as it landed.

Following the failure of two spacecraft – Mars Polar Lander and Mars Climate Orbiter, both lost near their final destinations during the launch opportunity of 1999 – NASA toasted success once again two years later, when the 2001 Mars Odyssey mission reached orbit. Designed to complement MGS, Odyssey included imagers and spectrometers to study Mars at different wavelengths and probe the chemicals in its rocks. Its most important finding was probably the presence of huge amounts of hydrogen (later confirmed as icy permafrost) around both poles.

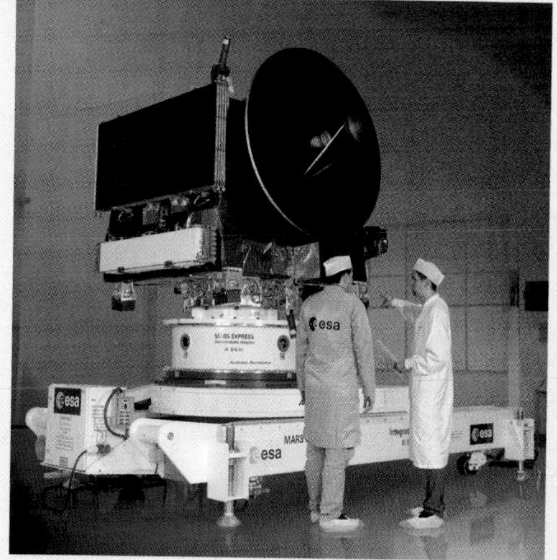

Europe goes to Mars

ESA's first interplanetary probe, Mars Express, arrived at the Red Planet on Christmas Day 2003. The probe had two parts – an orbiter and a small, and ultimately doomed, British-built lander called Beagle 2 (see panel, above). The orbiter carried spectrometers to investigate the chemistry of the surface and atmosphere, ground-penetrating radar to study buried features, and a camera that produced three-dimensional views by photographing the surface from two slightly different angles. This hugely successful mission has lasted more than 15 years, making intriguing discoveries such as subterranean lakes and methane in the atmosphere (see pp.282–83).

In 2004, two new NASA rovers arrived on Mars. The Mars Exploration Rovers *Spirit* and *Opportunity* (see over) were larger and more robust than Sojourner. *Spirit*'s landing site in Gusev Crater had a broad, river-like channel flowing into it, where scientists hoped to discover traces of an ancient lake. Surprisingly, however, rocks in the region turned out to be mostly volcanic. Later in its mission, *Spirit* photographed dust devils (and even benefitted from their winds cleaning its solar panels). It also found bright silica sand just beneath the surface, potentially formed by ancient hot springs. After covering some 7.7km (4⅘ miles) of terrain, the rover became stuck in soft soil in May 2009 and eventually lost contact with Earth in March 2010.

Opportunity's landing site in Meridiani Planum, meanwhile, was thought to lie near the shore of an ancient shallow sea, and when the probe visited a small nearby crater its operators were delighted to find rocks and minerals that probably formed under water. The rover has continued to explore the Martian surface over more than 15 years, travelling across some 45km (28 miles) before a seemingly permanent loss of contact during a 2018 dust storm.

TECHNOLOGY

NASA'S *SPIRIT* AND *OPPORTUNITY*

Mars Exploration Rover

NASA's second generation of Mars rovers built on the success of 1997's Mars Pathfinder mission. But the two Mars Exploration Rovers (MERs) were to be much larger and more robust vehicles, capable of direct communication with Earth or the various Martian orbiters without using a base-station relay. A unique suspension and drive system, powered by a large array of solar panels on the MER's upper surface, allowed it to negotiate almost any hazard on the Martian surface with ease. Cameras and other scientific instruments were mounted on a raised mast and on a small robot arm - the Instrument Deployment Device (IDD).

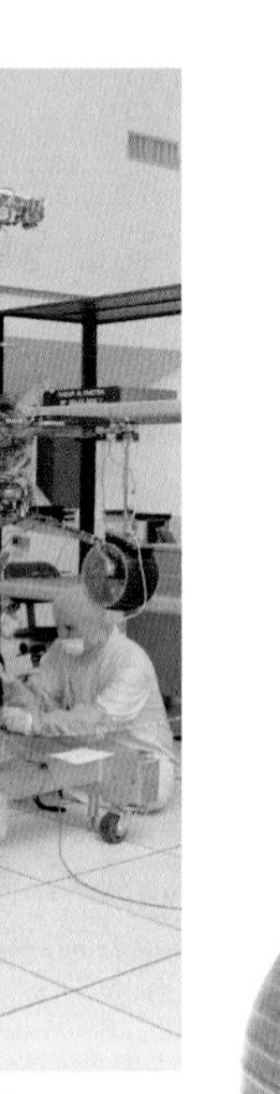

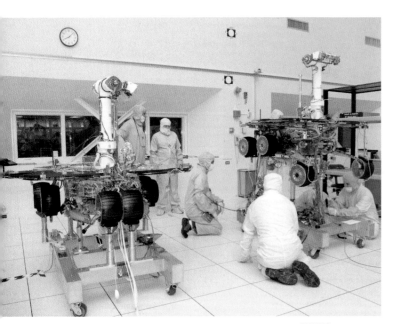

TWINS

The MER rovers were built side-by-side at the Jet Propulsion Laboratory. Developing a pair of probes is always a good policy as it is a cost-effective way of doubling the science yield from a project and also helps insure against failures.

TWO GENERATIONS OF ROVER

MER meets Sojourner in the lab at JPL. The larger rover borrowed the rocker-bogie suspension system developed for Sojourner - each wheel has independent suspension and keeps contact with the ground at all times, reducing the risk of sudden bumps and shocks. The MER has rolled over large rocks and rough terrain with few problems.

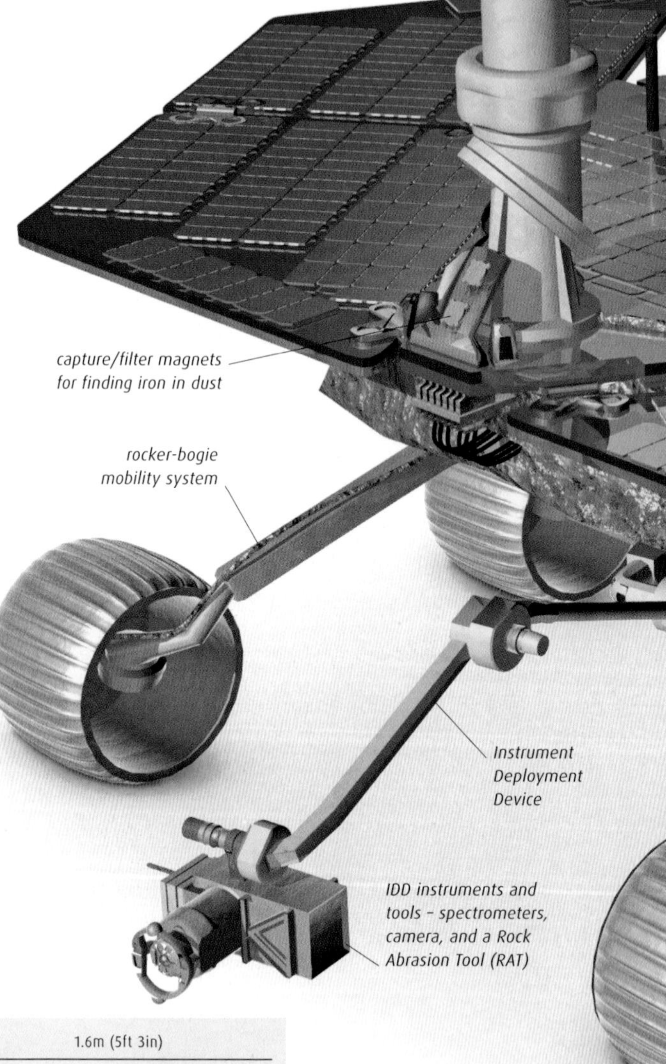

LENGTH	1.6m (5ft 3in)
HEIGHT	1.5m (4ft 11in)
WEIGHT	185kg (407lb)
NUMBER OF INSTRUMENTS	7 including PanCam
POWER SOURCE	Top-mounted solar arrays
MANUFACTURER	JPL/Caltech
LAUNCH DATES	10 June 2003 (*Spirit*), 8 July 2003 (*Opportunity*)

SLOW BUT SURE

The MER is a high-performance vehicle: independent motors on each wheel allow it to negotiate most obstacles, and it can turn on the tightest of angles. However, it is not the fastest vehicle - its top speed on Mars is about 5cm (2in) per second, and it stops every few seconds to allow its computer to study the route ahead.

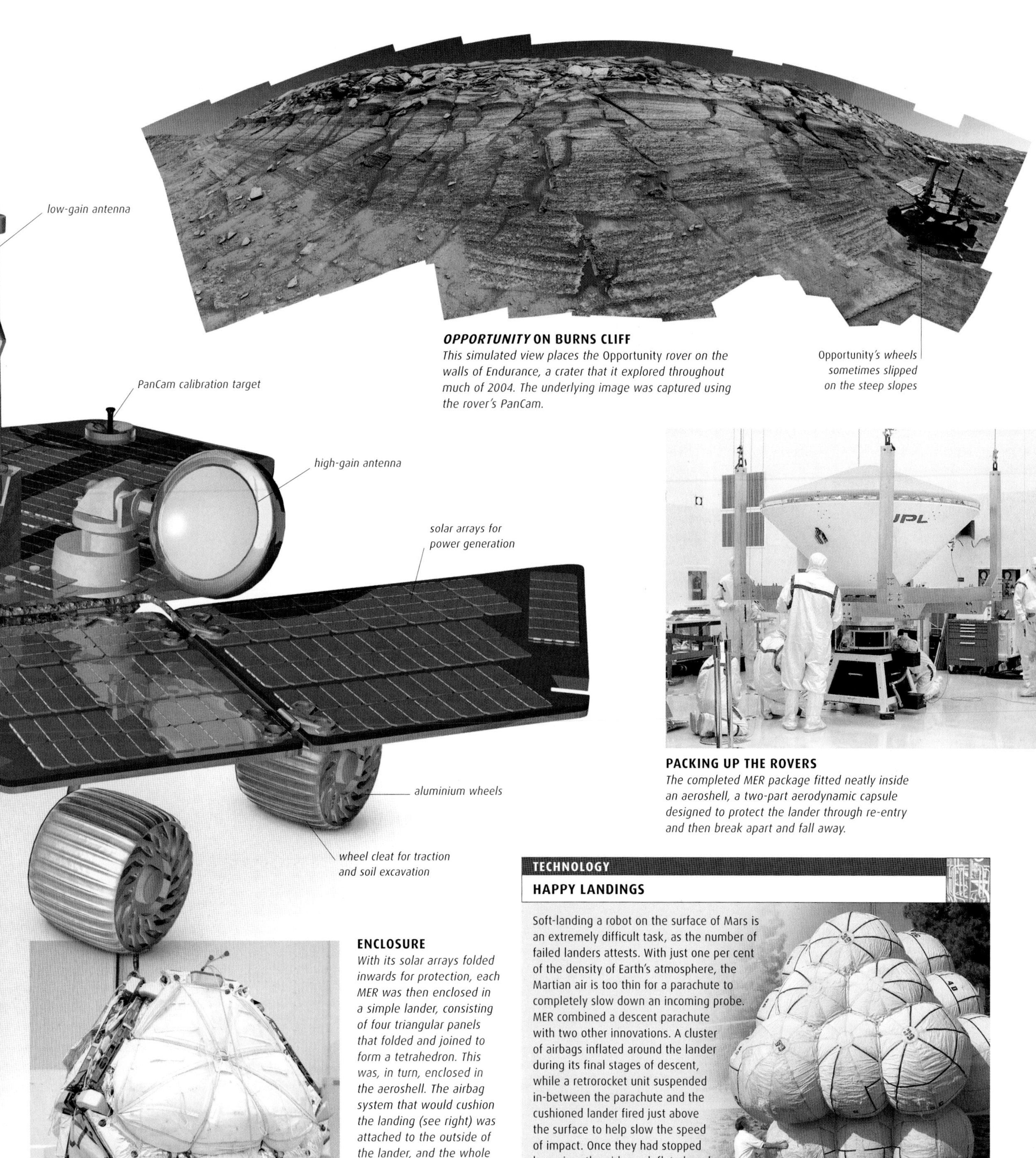

***OPPORTUNITY* ON BURNS CLIFF**
This simulated view places the Opportunity rover on the walls of Endurance, a crater that it explored throughout much of 2004. The underlying image was captured using the rover's PanCam.

PACKING UP THE ROVERS
The completed MER package fitted neatly inside an aeroshell, a two-part aerodynamic capsule designed to protect the lander through re-entry and then break apart and fall away.

ENCLOSURE
With its solar arrays folded inwards for protection, each MER was then enclosed in a simple lander, consisting of four triangular panels that folded and joined to form a tetrahedron. This was, in turn, enclosed in the aeroshell. The airbag system that would cushion the landing (see right) was attached to the outside of the lander, and the whole MER assembly attached by tether to the retrorocket and braking parachute unit.

TECHNOLOGY

HAPPY LANDINGS

Soft-landing a robot on the surface of Mars is an extremely difficult task, as the number of failed landers attests. With just one per cent of the density of Earth's atmosphere, the Martian air is too thin for a parachute to completely slow down an incoming probe. MER combined a descent parachute with two other innovations. A cluster of airbags inflated around the lander during its final stages of descent, while a retrorocket unit suspended in-between the parachute and the cushioned lander fired just above the surface to help slow the speed of impact. Once they had stopped bouncing, the airbags deflated, and powerful motors in the lander unit's "petals" pushed the vehicle into a horizontal position.

Cassini–Huygens

The Cassini mission was the most sophisticated spaceprobe so far, and it successfully orbited among the moons and rings of Saturn for over a decade.

11m (36ft) magnetometer boom

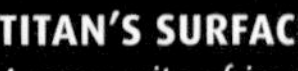

TITAN'S SURFACE
A composite of images taken at three wavelengths by Cassini's VIMS instrument peers through Titan's hazy atmosphere to reveal surface features.

Following the Voyager encounters with Saturn in the early 1980s, it was clear that the ringed planet and its moons still held enough secrets to justify an orbiter mission similar to the Galileo one already planned for Jupiter. The project became an international effort after a committee formed to look at future collaboration between NASA and the European Space Agency ESA recommended an orbiter-and-probe mission similar to what NASA had in mind. It was named Cassini after Gian Domenico Cassini, the 17th-century Italian-French astronomer who discovered the main division within Saturn's rings.

Although it subsequently evolved into a much larger, heavier spacecraft, Cassini's original design was similar to Galileo's – both were to be "Mariner Mark II" craft, as was a third, cancelled mission, the Comet Rendezvous Asteroid Flyby (CRAF).

While Galileo carried an atmospheric probe to descend into Jupiter's clouds, Cassini's cargo was even more ambitious – a lander that would parachute through the murky haze of Titan and send back data and pictures from the surface. This lander, Huygens, was ESA's main contribution to the project.

In order to power its array of scientific instruments, Cassini carried a large payload of plutonium in its radioisotope thermal generators. The possibility of an accident created much controversy before launch and during Cassini's flyby of Earth in 1999 (although smaller quantities of plutonium had already been used by previous spaceprobes). Fortunately, the launch, by a Titan IVB/Centaur rocket on 15 October 1997, went flawlessly. With a total mass at launch of 5,655kg (12,441lb), the probe was so big that it needed a complex series of gravity assists to get up to a reasonable speed. For this reason, the flight to Saturn took more than six years, with flybys of Venus (twice), Earth, and Jupiter. Cassini reached Jupiter in late 2000 and produced a unique scientific opportunity – the Saturn-bound probe was able to make long-range observations of the giant planet and its moons at the same time as Galileo.

Arrival at Saturn

Cassini's main engine executed a 96-minute burn to drop the spacecraft into orbit around Saturn on 1 July 2004. On its way into the Saturnian system, the probe had already provided the first close-up photographs of the outer moon Phoebe. Now it could begin its work in earnest, with cameras,

BUILDING HUYGENS
Engineers fit the back cover to the Huygens probe. The gold tiles offered protection against the extreme temperatures encountered during entry into Titan's atmosphere.

BACKLIT RINGS
On 15 September 2006, Cassini passed into Saturn's shadow. Over three hours it took 165 images of the still-sunlit rings, resulting in this magnificent panorama. The box marks the distant glimmer of the Earth.

11 June 2004
Cassini flies by Saturn's frozen moon Phoebe.

1 July 2004
After a journey of over six years, Cassini enters orbit around Saturn.

25 December 2004
The Huygens lander separates from Cassini, following its own trajectory towards Titan.

14 January 2005
Huygens enters Titan's atmosphere, sending back pictures and data throughout its descent and landing.

21 July 2006
Using radar imaging, Cassini detects what appear to be lakes on the surface of Titan.

12 March 2008
Cassini flies through one of the Enceladus ice plumes, just 50km (30 miles) from the surface.

5 December 2010
Cassini observes the beginnings of a great Saturnian storm that recurs every 30 years.

15 September 2017
Cassini's mission ends with a deliberate crash into Saturn's atmosphere.

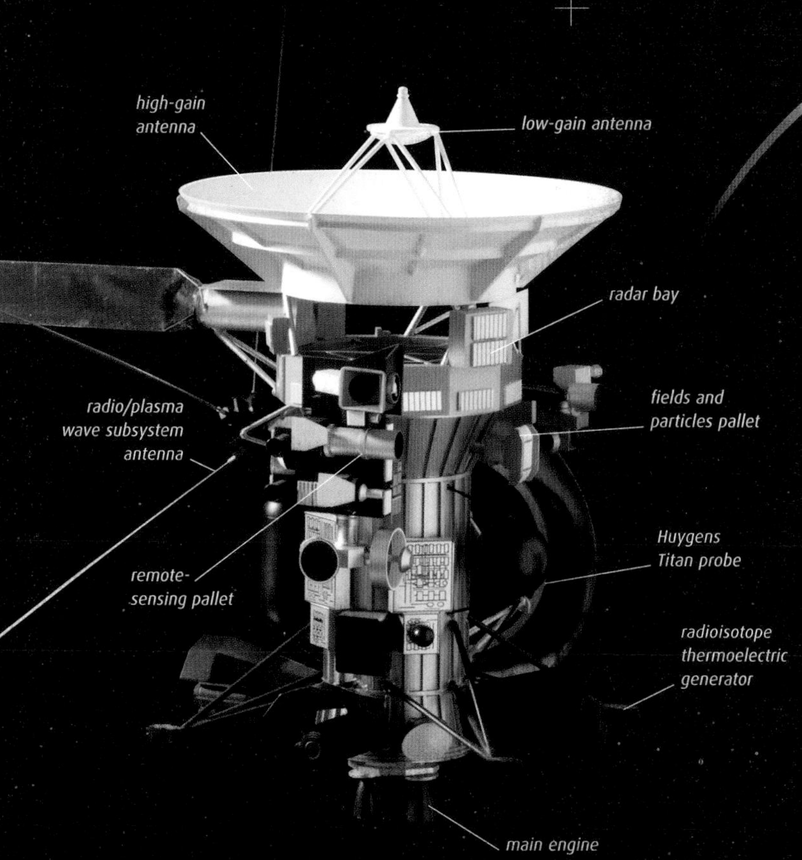

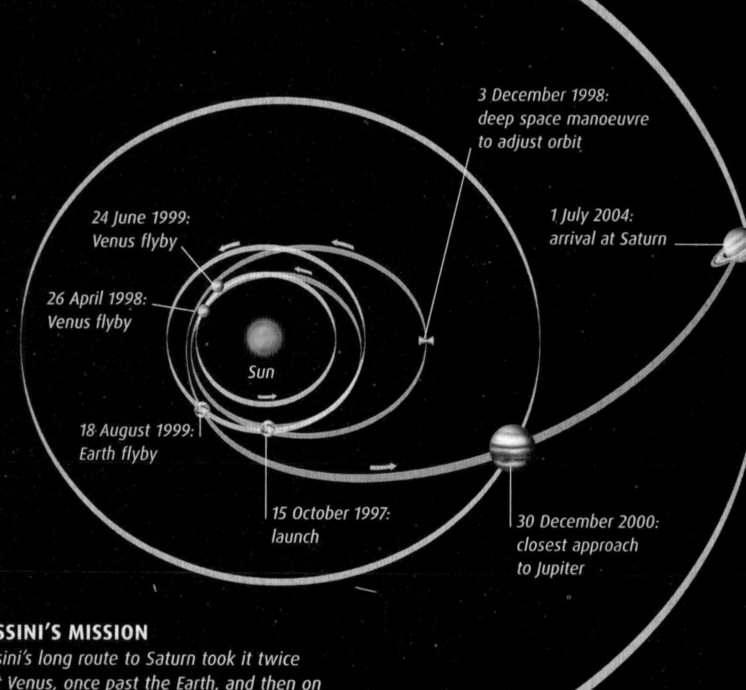

CASSINI'S MISSION
Cassini's long route to Saturn took it twice past Venus, once past the Earth, and then on to Jupiter and finally Saturn. Once there, the probe (left) was built to operate for at least four years and some 80 orbits of the planet.

spectrometers, magnetometers, and a wide variety of other instruments recording every aspect of the planet and its rings and satellites. The lander separated from the main spacecraft on Christmas Day, and Huygens dropped into Titan's atmosphere on 14 January 2005.

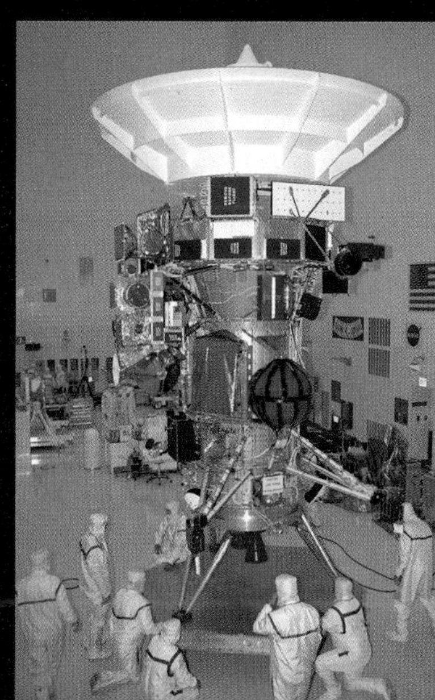

GIANT PROBE
The enormous Cassini dwarfs technicians fitting instruments to it in Kennedy Space Center's Payload Hazardous Servicing Facility prior to launch.

As the probe fell towards Titan, aerial photographs revealed an eerily Earth-like eroded shoreline, and Huygens landed on what seemed to be a pebble-strewn river delta. This confirmed suspicions that Titan is a world where methane (which freezes at -182°C/-296°F) plays a similar role to water on Earth, occurring as ice, liquid, and vapour.

Discoveries in orbit

Following this early highlight, the Cassini orbiter began its work in earnest. The primary mission was originally planned to last for just three years, but Cassini's large engine and plentiful propellant ultimately allowed it to continue its looping orbit around the planet and its rings and moons for much longer. One of the orbiter's first and most spectacular discoveries was the huge plumes of ice crystals erupting from fractures in the surface of Saturn's small inner moon, Enceladus. In March 2008, Cassini was directed to make a daring flyby that came within 50km (30 miles) of the airless moon and took it straight through one of these plumes, allowing its instruments to confirm the presence of water. It is now clear that Enceladus has a deep ocean hidden beneath its surface, making it a prime candidate for possible alien life in the Solar System.

Cassini also went equipped to tackle Titan's hazy atmosphere, with a near-infrared camera that can see through the orange smog and map the terrain below. The probe's radar, meanwhile, confirmed the existence of lakes of seasonal liquid methane and ethane that form around the moon's winter pole.

Other moons proved less active but still revealed fascinating histories. Dione has towering ice cliffs, Hyperion appears to be the broken-up remnant of a much larger moon, and Iapetus has a dark coating of "soot" on one hemisphere and a bizarre ridge running around much of its equator. Saturn itself, though outwardly placid, has revealed storms just as violent as those on Jupiter, while its rings seem to be in a constant state of flux and change as they are twisted and distorted by the gravity of the nearby moons.

The discovery of a potentially habitable environment on Enceladus made safe disposal of Cassini, and avoidance of contamination, a priority. Consequently in September 2017, the probe ended its mission in style with a spectacular "Grand Finale" dive to its destruction in Saturn's atmosphere.

ICY ENCELADUS
About 500km (300 miles) in diameter, Enceladus is made mostly of ice, with some rock. Enhanced-colour images from Cassini reveal the blue "tiger stripes" associated with the eruption of water-ice plumes.

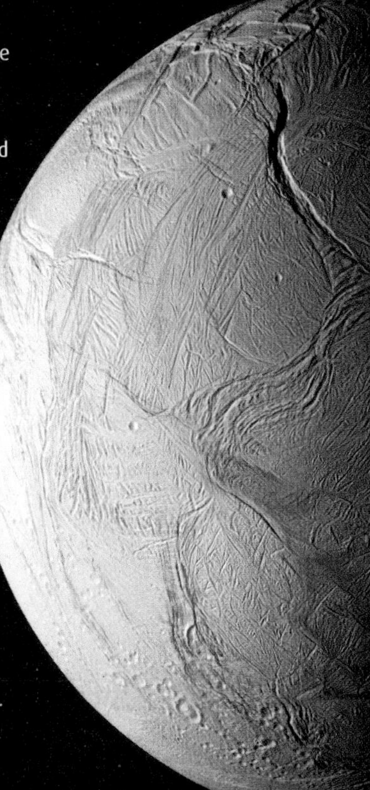

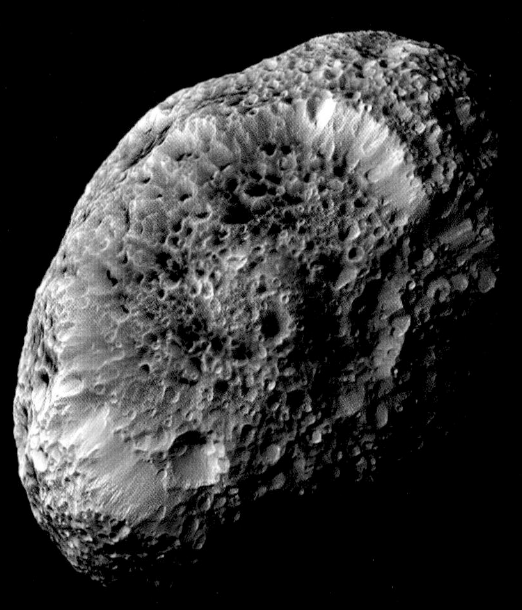

CASSINI'S TOUR OF SATURN

All the giant planets have rings around them, but Saturn's are by far the most spectacular – vast, bright planes of orbiting ice chunks, arranged in countless ringlets. Cassini has kept well clear of them after an initial manoeuvre took it through the outer limits of the ringplane in order to get into orbit, but its spectacular images (right) have revealed fine structures and the presence of short-lived twists and knots of material where Saturn's family of moons and moonlets exert their gravitational influence. The moons themselves have proved equally impressive – Enceladus (top) shoots enormous plumes of water into the sky from geysers powered by an underlying liquid ocean. While most of the water falls back onto the moon as bright snow, some is swept up into orbit around Saturn, forming a broad, faint ring beyond the central system. Meanwhile Hyperion (above) is surely the strangest-looking large moon in the Solar System. Its spongy texture seems to be the result of the Sun's heat acting on dark "pits" in the surface and evaporating buried ice.

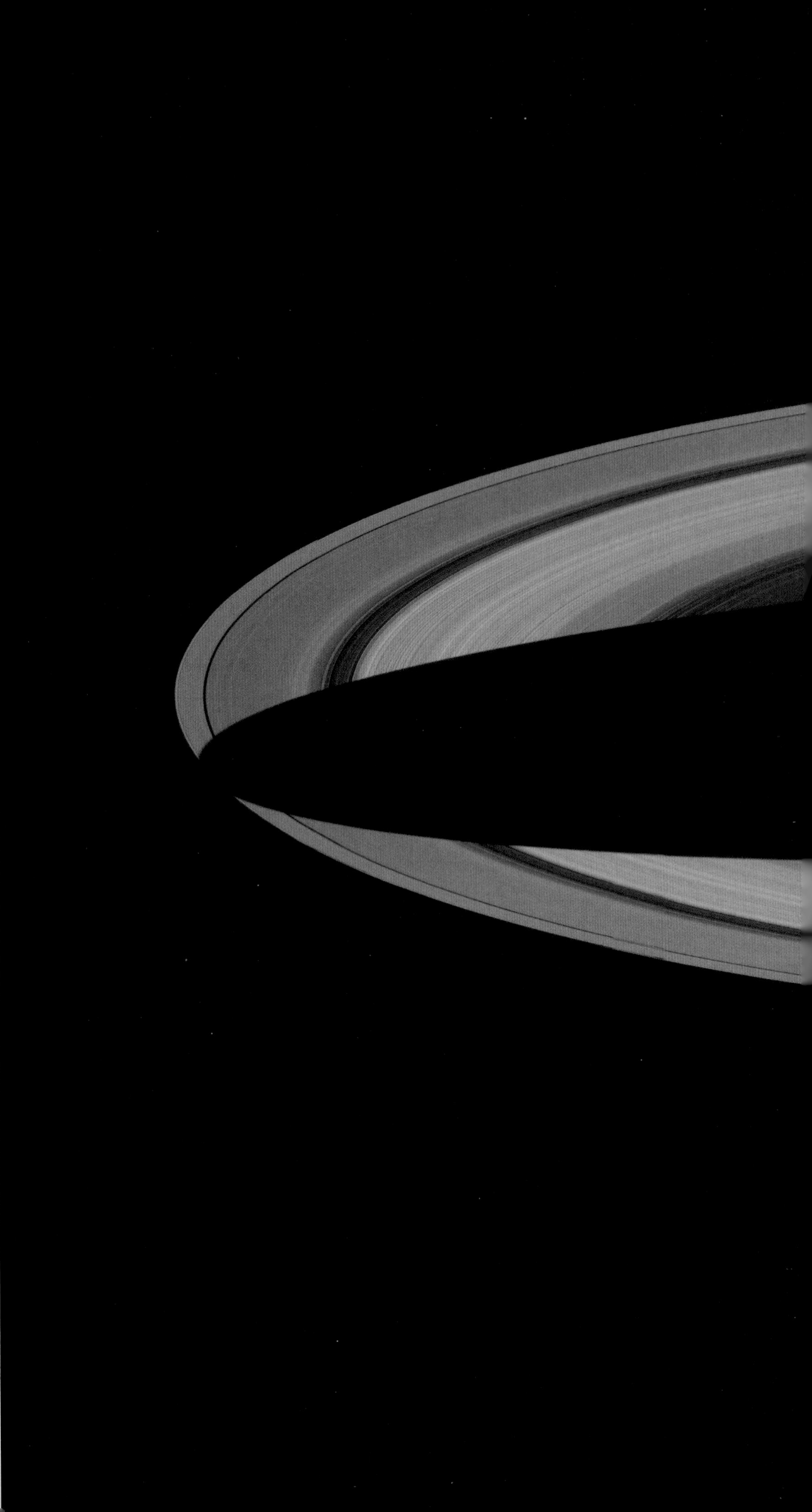

10 March 2006
NASA's Mars Reconnaissance Orbiter arrives at the Red Planet.

25 May 2008
NASA's Phoenix lander touches down in the Vastitas Borealis region.

2 November 2008
Phoenix is shut down as it begins to lose power.

8 November 2011
Russia's Fobos-Grunt and the Chinese Yinghuo-1 probe are destroyed in a launch failure.

26 November 2011
NASA's *Curiosity* rover (formally known as Mars Science Laboratory) is launched on an Atlas V rocket.

6 August 2012
Curiosity executes a complex landing procedure to begin its work on Mars.

22 September 2014
NASA's MAVEN enters Martian orbit on a mission to analyse the atmosphere.

24 September 2014
India's Mangalyaan probe arrives at Mars.

19 October 2016
ESA's ExoMars Trace Gas Orbiter arrives in orbit. Its Schiaparelli lander crashes due to an engine misfire.

Mars transformed

As exploration of the Red Planet continues at an ever-accelerating rate, new orbiting probes and surface explorers are revealing ever more Earth-like aspects of the Martian past.

In 2006, NASA's Mars Reconnaissance Orbiter (MRO) reached Mars to replace the ageing Mars Global Surveyor (MGS). The mission arrived with immaculate timing, entering its final orbit two months before contact with MGS was finally lost. The orbiter carries a much-improved camera called HiRISE that, at maximum resolution, can take recognizable photographs of objects as small as 1m (3ft) across. MRO's main role was to take detailed images of craters, canyon walls, and sediment layers, looking for more evidence of water and trying to estimate just when it disappeared from the surface (see opposite). The probe also carries spectrometers to study the rock chemistry and radar to look beneath the Martian surface, and can act as a communications relay for surface missions.

New views from orbit

With MRO and ESA's continuing Mars Express mission still delivering photography of the Martian surface, more recent Mars orbiters have focused on specialized scientific studies. NASA's MAVEN (Mars Atmosphere and Volatile Evolution) mission uses technology from Earth remote-sensing satellites to measure the chemistry and large-scale properties of the Martian atmosphere, while the ExoMars Trace Gas Orbiter focuses on minor gases. In particular, it hopes to discover more about the seasonal "burps" of methane detected by Earth-based telescopes and the *Curiosity* rover (see opposite). These intriguing bursts of gas could be associated either with rare geological processes or some form of microbial life.

EXOMARS TGO
The first phase of ESA's ambitious ExoMars mission, the Trace Gas Orbiter (TGO) arrived in orbit in October 2016. Unfortunately, a stationary lander called Schiaparelli, which had travelled alongside it, crashed during descent.

Sadly, an ambitious Russian mission called Fobos-Grunt, which aimed to return rock samples from the Martian moon Phobos and put a Chinese probe in Martian orbit, failed before escaping Earth's orbit in 2011, but India's ISRO space agency was more successful, with its Mars Orbiter Mission, known as Mangalyaan, reaching Mars in 2014. And older missions continue to deliver stunning results – in 2018, Mars Express scientists unveiled evidence, gathered from years of radar measurements, for a huge lake of briny water beneath the south polar cap.

Landing on ice

May 2008 saw the arrival on Mars of a static lander called Phoenix. Reusing instruments designed for the failed 1999 Mars Polar Lander and another cancelled mission, it was sent to study conditions around the north polar cap. In order to cope with the cold conditions and weak polar sunlight, Phoenix carried a number of additional electrical heaters and a pair of large, fan-like solar panels to help power them. Phoenix touched down on the Vastitas Borealis,

PHOENIX LANDER
A mosaic self-portrait of NASA's polar lander Phoenix shows one of the two fan-shaped solar arrays used to gather weak high-latitude sunlight.

a great plain surrounding the polar ice cap. As its retrorockets blew away the dust, they exposed bright underlying material that Phoenix soon confirmed to be water ice. As well as cameras, the lander carried a weather station, a robot arm with close-up imager and sample collector, and onboard laboratories for studying the soil's chemical makeup and physical properties.

It was known from the outset that the mission would last for a matter of months before it succumbed to the onset of the Martian winter, but Phoenix outperformed expectations – the primary mission was designed to run for 90 Martian days around northern midsummer, but in the end it operated for 157 days before being overcome by the autumn frosts.

Curious explorer

NASA's *Curiosity* rover was a major advance on its predecessors *Spirit* and *Opportunity* in both size and ambition. Touching down in August 2012 near Aeolis Mons, the central peak of the 154-km (96-mile) Gale impact crater, the rover's primary aim was to assess whether the ancient Martian environment could have supported life. Gale, a 3-billion-year-old lake bed now lined with sedimentary rocks, was an ideal target.

Curiosity's instruments include panoramic and close-up cameras, an X-ray spectrometer for analyzing rock chemistry, a robot arm equipped (for the first time on Mars) with a drill, and two laboratories for sample analysis. Initial studies showed that the rover had landed in what was once a fast-flowing streambed, and as it set out on a path towards Aeolis Mons, it continued to take rock and atmosphere samples, finding plentiful evidence for a habitable past environment and abundant ice mixed with the soil. *Curiosity*'s atmospheric instruments found that methane levels vary dramatically depending on the seasons, and in 2018, its drill found complex carbon-based "organic" chemicals in the sedimentary rocks. Such chemicals are not necessarily evidence of life, but they are the sort of remnants it might leave behind.

WATER ON THE SURFACE?
In 2011, MRO photographed dark streaks that change with the seasons on the Sun-facing slopes of Newton crater. They may be created by water seeping from the crater walls.

CURIOSITY ON MARS
A 3-D model (above) of Gale Crater combines data from several orbiting spaceprobes and shows the landing location of the Mars Science Laboratory Curiosity *(left). In more than five years on the surface, the rover has explored around the central peak of Aeolis Mons, covering more than 18km (11 miles) of Martian terrain.*

24 October 1998
NASA launches its Deep Space 1 technology test bed mission, trialling the use of ion thrusters for space propulsion.

29 July 1999
Deep Space 1 flies past asteroid 9969 Braille.

22 September 2001
Deep Space 1 flies past Comet 19P/Borrelly.

19 January 2006
NASA's New Horizons probe blasts off on an Atlas V launch vehicle.

28 February 2007
New Horizons observes Jupiter during a gravity-assist flyby.

21 May 2010
An H-IIA rocket launches JAXA's prototype solar-sail probe IKAROS.

5 August 2011
Juno is launched from Cape Canaveral.

14 July 2015
New Horizons flies within 12,500km (7,800 miles) of Pluto.

5 July 2016
Juno arrives in orbit at Jupiter.

1 January 2019
New Horizons passes 2014 MU69 on its voyage through the Kuiper Belt.

Exploring at the limits

In recent years, a new generation of spaceprobes has pushed the boundaries of technology, delivering stunning new images and scientific data from remote and inaccessible parts of the Solar System.

Following an official reclassification of the planets in 2006, NASA could legitimately claim to have sent probes to all the major worlds of the Solar System – but what of the former ninth planet, Pluto? Astronomers now classify it as a dwarf planet, and part of the Kuiper Belt, a ring of icy objects beyond Neptune predicted in 1951 by Gerard Kuiper (see panel, below). In January 2006, NASA had launched a probe on a speedy journey to investigate both Pluto and its companions in the Solar System's uncharted outer reaches.

Racing to Pluto

New Horizons was the fastest spacecraft ever to leave Earth, travelling at a speed of 58,536kph (36,373mph). A gravity assist at Jupiter in 2007 boosted its speed still further, sending it beyond the orbit of Neptune and into the Kuiper Belt within just eight years. One reason for setting such a blistering pace was Pluto's relative proximity – the planet is currently retreating from its closest approach to the Sun in 1989.

The mission made its short but eventful passage through the Pluto system in January 2015, not only photographing the planet and its giant satellite Charon but also identifying the chemical composition of their surfaces, measuring Pluto's atmosphere, and studying the duo's interaction with the weakening solar wind. The first close-up images of Pluto were a huge surprise to most scientists – they revealed not a deep-frozen relic unchanged from the dawn of the Solar System but instead a world with hugely varied landscapes that has clearly been geologically active in the relatively recent past.

Following this success, the probe flew on to a 2019 rendezvous with a much smaller Kuiper Belt Object (KBO) called 2014 MU69. As New Horizons hurtles towards interstellar space, it will continue to monitor other KBOs that come within range.

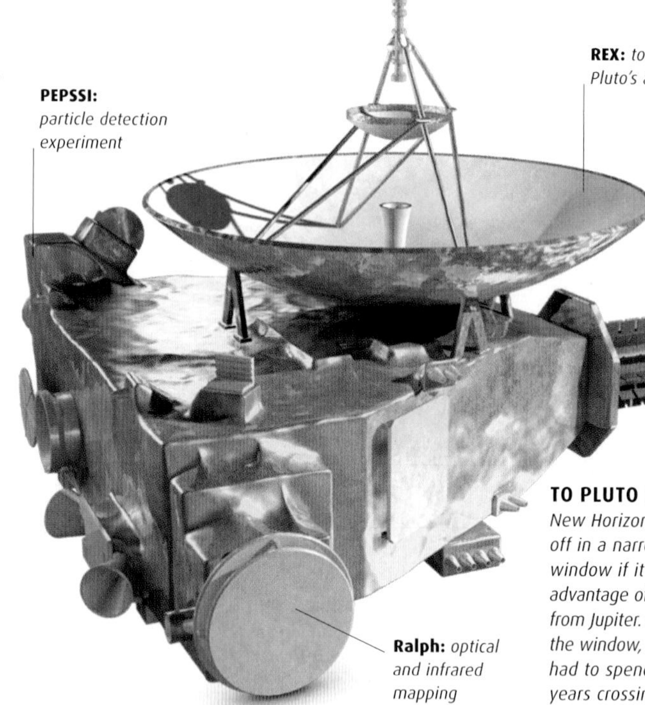

TO PLUTO AND BEYOND
New Horizons had to take off in a narrow launch window if it was to take advantage of a gravity-assist from Jupiter. If it had missed the window, it would have had to spend three more years crossing space.

Jupiter up close

NASA's Juno spacecraft might be targeting somewhat more familiar territory, but it is still pushing the limits of space technology in other ways. Launched from Earth in August 2011, it is the most distant space mission yet to rely entirely on solar power, with three huge "wings" designed to generate 435 watts of electricity in a region where sunlight has just 4 per cent of the strength it has at Earth. Arriving at Jupiter in July 2016, Juno entered a unique and highly elliptical 53-day orbit that takes it over Jupiter's poorly studied polar regions, speeding above the clouds at an altitude of just 4,200km (2,600 miles) during "perijove" before retreating to a maximum distance of 8.1 million km (5 million miles). This

FROZEN SURFACE
New Horizons' images of Pluto revealed a world of complex geology, with large areas where the icy surface is being continually renewed as fresh ice pushes up from beneath.

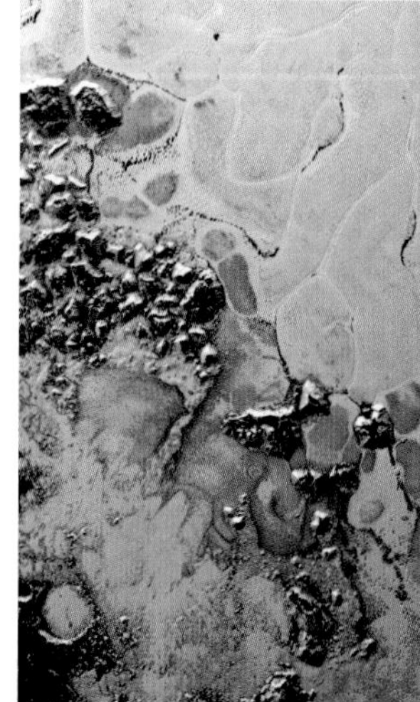

BIOGRAPHY
GERARD KUIPER

Dutch-born astronomer Gerard Peter Kuiper (1905–73) had a career that ranged across the Solar System. After studying at the University of Leiden, Kuiper left to work in the United States, joining Chicago University's Yerkes Observatory in 1937. While working here, he discovered Uranus's moon Miranda and Neptune's moon Nereid, found that Titan has an atmosphere, and predicted the existence of what would become known as the Kuiper Belt. In 1960, he moved to Tucson to become the first director of the University of Arizona's new Lunar and Planetary Laboratory. He also served as chief scientist on NASA's series of Ranger lunar crash-landers.

SWIRLING CLOUDSCAPE
This enhanced-colour Juno view of Jupiter's turbulent clouds spans an area larger than Earth, as seen from an altitude of 13,345km (8,292 miles) above the cloud tops. The mission's JunoCam imager has an exceptionally wide field of view, generating images with a peak resolution of 15km (9⅓ miles) per pixel at its closest to the planet.

IKAROS
JAXA's propulsion test bed consists of a spinning satellite rigged to a reflective solar sail with a surface area of 400sq m (4,300sq ft). During the spacecraft's cruise towards Venus, it was successfully accelerated and steered using the radiation pressure of sunlight.

keeps the spacecraft safely outside of the giant planet's fierce radiation belts for most of its orbit, while providing detailed close-ups of Jupiter's intricate and ever-changing clouds.

The spacecraft instruments include a microwave radiometer for detecting short-wavelength radio waves that pass through Jupiter's upper cloud layers and can reveal atmospheric structure and chemistry down to around 600km (370 miles). An ingenious tracking experiment, meanwhile, probes the planet's deeper structure and internal density by measuring the slight shifts in Juno's radio signals as its orbit is modified by the shape of Jupiter's gravitational field. A final key area for Juno's investigations is Jupiter's extensive and powerful magnetic field.

Plans to modify Juno's orbit to a mere 14 days, allowing for more science passes within the original two-year mission plan, were abandoned in early 2017 due to the discovery of an engine valve problem – if the engines had misfired, they could have flung the probe into an unusable orbit. However, in June 2018, with Juno in good health, NASA announced that its mission would be extended for three more years before a planned de-orbit into Jupiter's atmosphere.

New propulsion

Some missions drive science forwards in other ways, testing innovative propulsion systems that will later be adopted by other probes. The success of NASA's 1998 Deep Space 1 mission, for example, has had a lasting influence. Rather than use a traditional chemical rocket, it was the first spacecraft to test an ion drive (see panel, right). This highly efficient system generated a tiny amount of force over many months, gradually accelerating the spacecraft by an additional 4.3km/s (2⁷⁄₁₀ miles per second) and sending it to rendezvous with both an asteroid and a comet. Ion engines have since been used on JAXA's Hayabusa comet probe, NASA's Dawn asteroid orbiter (see p.272–73), and the ESA/JAXA BepiColombo mission (see p.306).

Japan's IKAROS, launched in 2010, could prove equally influential for future spaceflight, since it is the first mission to test the long-discussed concept of the solar sail. IKAROS (Interplanetary Kite-craft Accelerated by Radiation Of the Sun) accelerates without propellant by harnessing the tiny but persistent force from sunlight and other radiations bouncing off a huge reflective sail. The same technology may be used in the late 2020s for a JAXA mission to the Trojan asteroids near Jupiter.

TECHNOLOGY

THE ION DRIVE

Solar electric propulsion (often known as the ion drive) is a highly efficient alternative to chemical rockets – though unfortunately it is not capable of generating the huge thrusts needed to launch spacecraft out of a strong gravitational field. Electricity from solar panels or another power source is used to create a high voltage across an ionization chamber. When atoms of an inert gas such as xenon are fed into the chamber, they are ionized, breaking apart into electrically charged particles called ions. The xenon ions are repelled by a charged plate in the ionization chamber and pushed out of the back of the engine, generating thrust. Ion engines can propel spacecraft to very high speeds, but run for months not minutes.

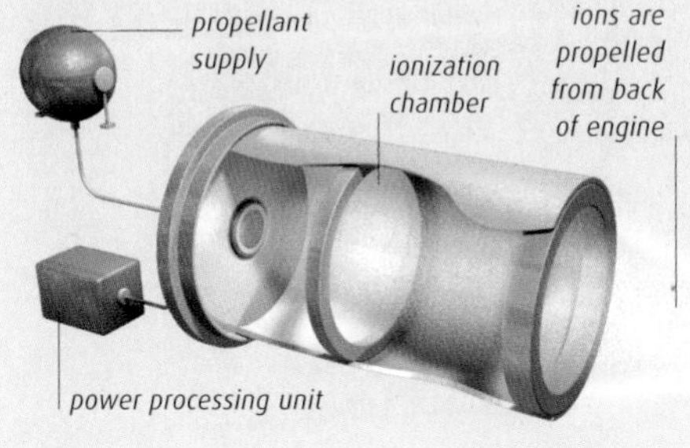

ORION OVER MARS
An artist's impression depicts a potential mission for NASA's new Orion Multi-Purpose Crew Vehicle to the Martian moon Phobos. Exploring the moons of Mars, perhaps in the late 2020s, would provide valuable experience before landing a crew on the Red Planet itself.

INTO THE FUTURE

WHEN FUTURE HISTORIANS write the story of mankind's early steps into space from a more distant perspective, 30 October 2000 may be a date of special significance – the last day on which humanity was entirely confined to Earth. The following afternoon, a Soyuz rocket lifted the first crew of the International Space Station into orbit, ready to occupy what should be our first permanent outpost in space. It seems fitting that the station is a collaborative effort between once-hostile nations.

But the ISS is a mere staging post to the Solar System – where next? NASA has lately refocused its human spaceflight efforts towards the Solar System beyond Earth orbit, with the Moon and nearby asteroids as potential targets. Meanwhile a younger space power, China, has launched its first manned spacecraft, including space stations, and has longer-term plans that also involve lunar bases. With these goals fulfilled, it will be time for mankind to look further afield – first to Mars, then deeper into the Solar System, and perhaps, one day, to the stars.

The ISS concept

Conceived in the 1980s as the West's answer to Salyut and Mir, the US space station Freedom eventually evolved into a truly global project – the International Space Station.

25 January 1984
US President Ronald Reagan announces the development of a new US space station to be called Freedom.

23 June 1993
After nine years of planning, constant revisions, and huge budget reductions, Freedom barely survives a Congressional vote on its proposed cancellation.

1 November 1993
NASA announces a partnership with Russia to develop a truly international space station.

29 June 1995
The Shuttle *Atlantis* docks with the Russian space station Mir, beginning a new phase of cooperation.

20 November 1998
The Russian Zarya Functional Cargo Block, the first element of the ISS, is launched.

7 December 1998
The Space Shuttle *Endeavour* links the US Unity module with Zarya.

26 July 2000
The arrival of the Russian Zvezda module makes the station ready for occupation.

In the early 1970s, budget cuts had forced NASA to choose between the Space Shuttle and a large space station, but in the early 1980s the political winds changed again. Shuttle flights were finally becoming routine, and relations between the superpowers were cooling into a new phase of the Cold War that might again extend into orbit (see panel, opposite). In 1984, President Reagan announced that the United States was at last going to build a permanent space station – with the politically loaded name Freedom.

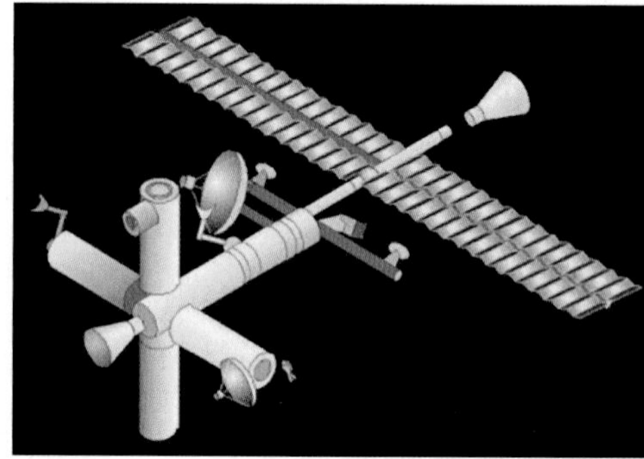

MIR 2
Authorized in 1976 as an eventual successor to Mir, the Mir 2 design went through many changes until, in response to the US Star Wars project, it threatened to become an orbiting battle platform.

From competition to collaboration

The new station's development was long and tortuous. Initial plans called for a truly huge outpost with a crew of 12, and the European, Japanese, and Canadian space agencies soon joined the project, agreeing to provide their own laboratory modules and other elements. Various designs were proposed, and projected costs spiralled, even as each successive redesign reduced the station's capability. Meanwhile, Shuttle flights carried various experiments to test techniques and technologies that might be used on the station. But the loss of the *Challenger* was a blow to US confidence in human spaceflight, while the liberalization and eventual disintegration of the Soviet Union brought an end to the Cold War rivalry that helped to justify Freedom. In 1993 the project survived a call for cancellation in the US Congress by a single vote.

"We can **follow our dreams** to distant stars, **living and working in space** ..."

US President Ronald Reagan, 25 January 1984

The writing was on the wall for Freedom, but the improved relationship with Russia led to a new way forward. Despite the poor state of their economy, the Russians had space-station experience that could help NASA make its station a reality. In 1993, officials from the US and Russian space agencies met to agree on a joint enterprise: the resulting station would be a hybrid of elements from Freedom and Russia's own stalled Mir 2. At first called Space Station Alpha, before long it became simply the International Space Station (ISS).

POWER TETHER
In the early 1990s, two Shuttle missions flew an experiment called the Tethered Satellite System (TSS). Linked to the Shuttle by a long conducting wire, its movement through the Earth's magnetic field created electricity. At the time, this technology was considered for use on future space stations.

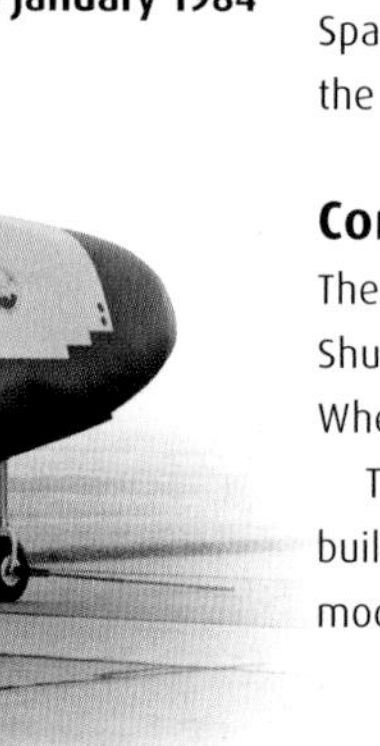

SPACE TAXI
NASA considered various options for crew transport to the space station – one was the HL-20 lifting body, which got as far as this engineering mock-up.

Construction begins

The new collaborative spirit was tested by the Shuttle-Mir missions of the mid-1990s (see p.216). When these came to an end, it was time to build.

The first element put into orbit was the Russian-built Functional Cargo Block. Known as Zarya, this module had a dual purpose. During the early stages

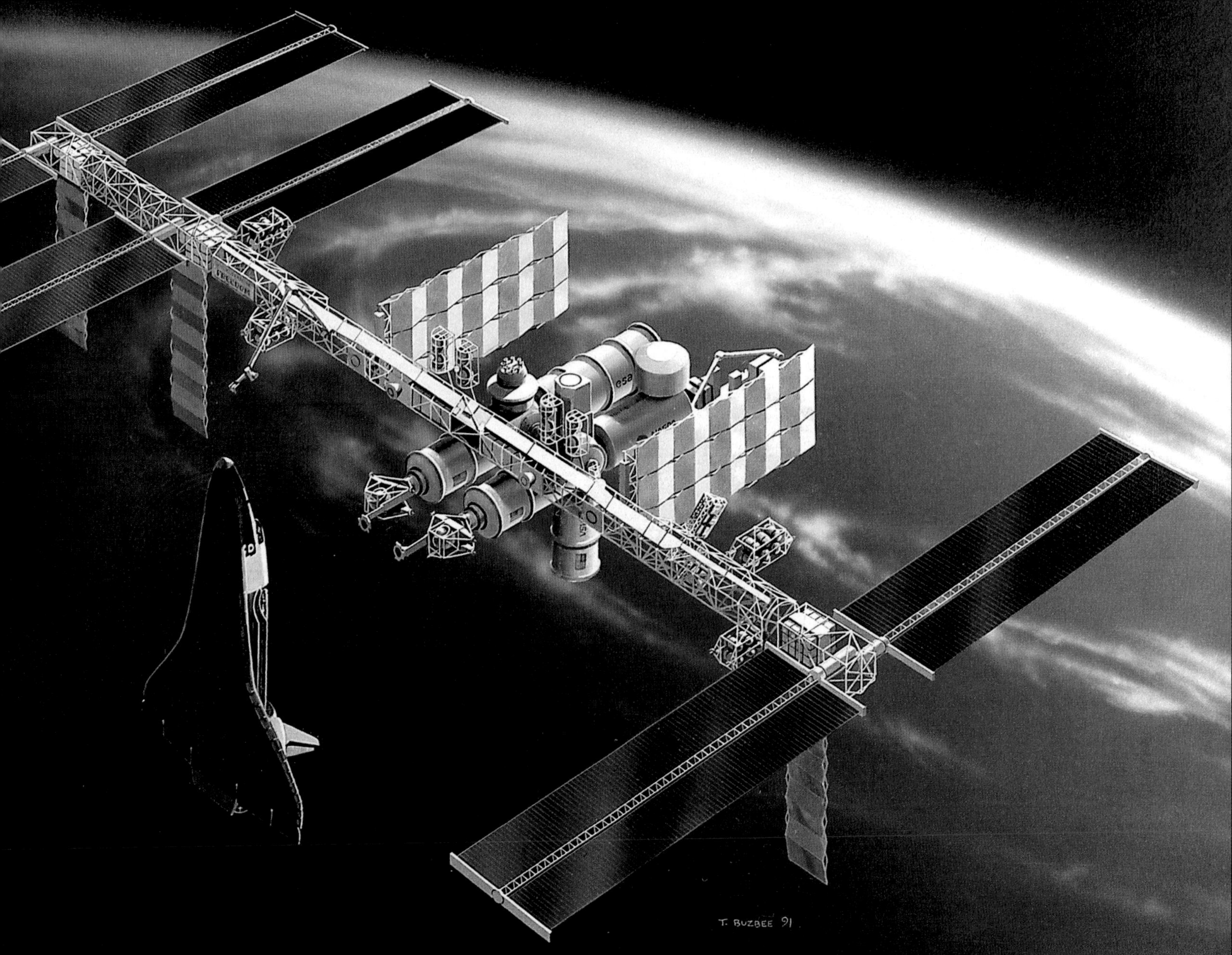

SPACE STATION FREEDOM

By the early 1990s, constant budget cuts and redesigns had seen the Freedom station evolve into a smaller configuration that bears some resemblance to the ISS. The main similarity is the use of the horizontal truss with solar arrays at each end and pressurized modules in the centre.

of construction, it would act as the station's heart, generating power and providing propulsion. As the ISS grew larger and these functions moved elsewhere, it would become a storage facility.

Endeavour rendezvoused with Zarya in December 1998, bringing with it NASA's Unity module, the first of three connecting nodes that would join the station's various elements together. However, there was a long delay before the next crucial element, Russia's Zvezda service module arrived. This provided living accommodation, life support, and environmental controls. Once it had docked in July 2000, the ISS was at last ready for its first crew.

TECHNOLOGY

STAR WARS

One of President Reagan's schemes to keep the upper hand in the new Cold War was the Strategic Defense Initiative (SDI), better known as the Star Wars programme. Reagan's advisors persuaded him that the United States, and perhaps its allies, could be protected from nuclear attack by a network of missile-detecting satellites and an array of science-fiction weapons. The proposed SDI arsenal included satellite-based interceptor missiles, energy weapons such as lasers, and ground-based missile-defence systems. Development stalled due to engineering problems and budget restrictions, although some of its technologies are still being pursued. Had SDI succeeded, it would have given the US a decisive lead in the Cold War arms race – but the effects of that could have been unpredictable and dangerous.

TECHNOLOGY

A PERMANENT OUTPOST IN SPACE

The International Space Station

The design of the ISS has changed repeatedly, even after its construction finally got underway. The design set in place by early in the new millennium was a hybrid of elements from NASA's original Freedom station and the Russian "Mir 2", with international contributions from Europe, Japan, and Canada. The station is dominated by the central truss that runs out from Node 1 along the entire length of the station. Sections of truss on the station's starboard (right) side as seen from Zvezda are given the designations S1 to S6 as they run out from the central Z0 and S0 trusses. The solar-array wings take the name of the truss to which they are attached, and the system is mirrored by port sections P1 to P6 on the left side.

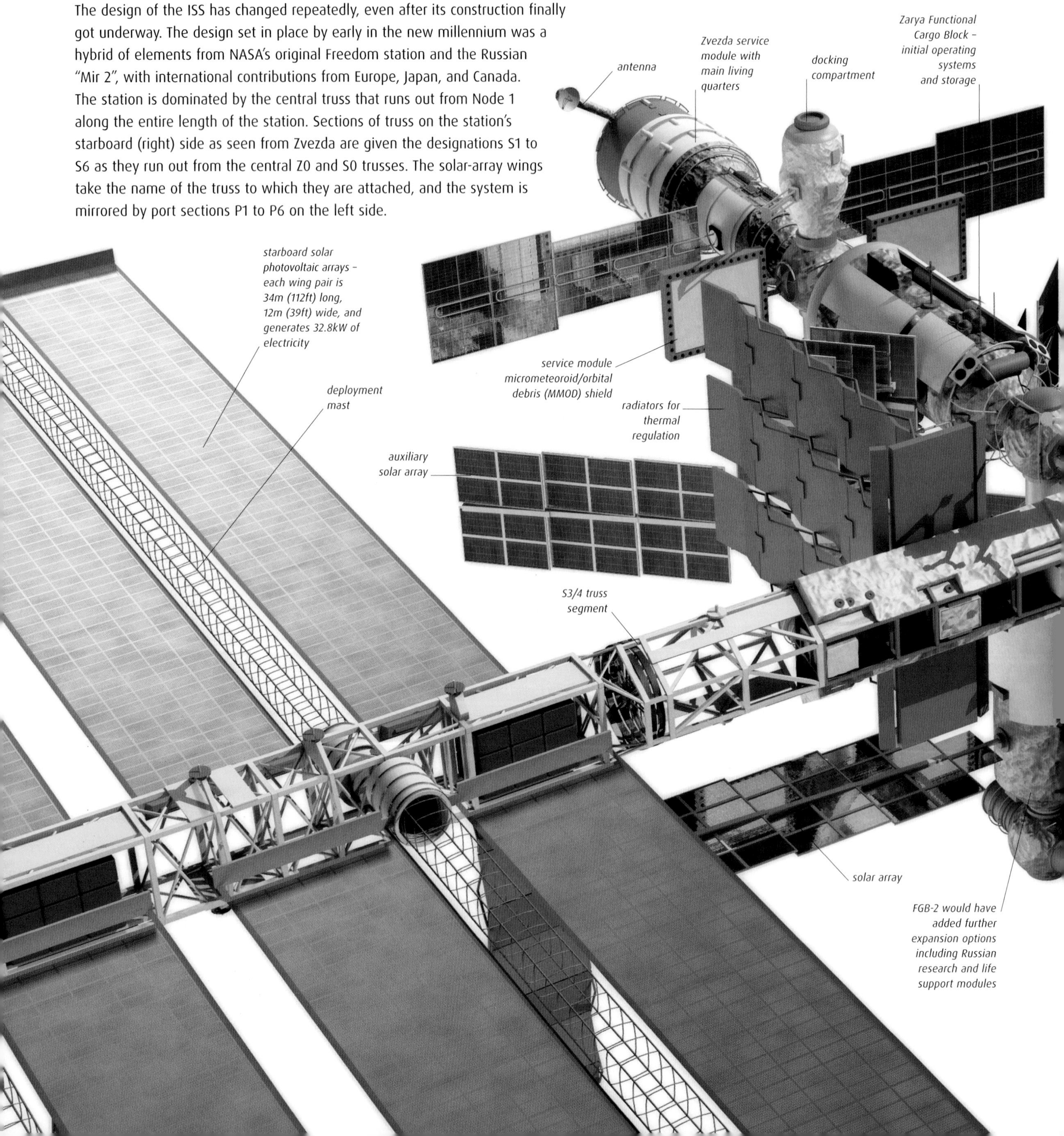

antenna

Zvezda service module with main living quarters

docking compartment

Zarya Functional Cargo Block – initial operating systems and storage

starboard solar photovoltaic arrays – each wing pair is 34m (112ft) long, 12m (39ft) wide, and generates 32.8kW of electricity

service module micrometeoroid/orbital debris (MMOD) shield

deployment mast

radiators for thermal regulation

auxiliary solar array

S3/4 truss segment

solar array

FGB-2 would have added further expansion options including Russian research and life support modules

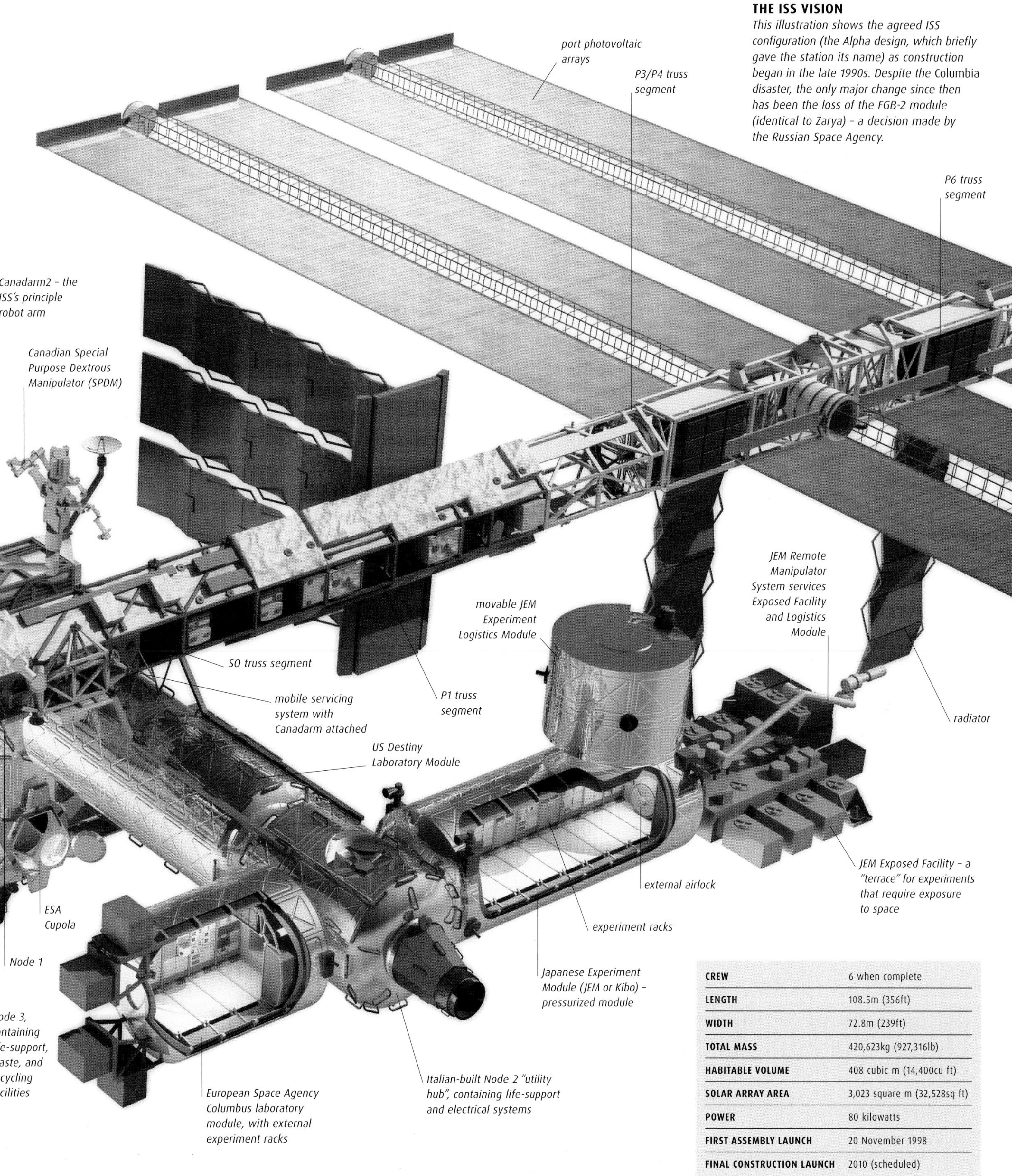

THE ISS VISION

This illustration shows the agreed ISS configuration (the Alpha design, which briefly gave the station its name) as construction began in the late 1990s. Despite the Columbia *disaster, the only major change since then has been the loss of the FGB-2 module (identical to Zarya) – a decision made by the Russian Space Agency.*

CREW	6 when complete
LENGTH	108.5m (356ft)
WIDTH	72.8m (239ft)
TOTAL MASS	420,623kg (927,316lb)
HABITABLE VOLUME	408 cubic m (14,400cu ft)
SOLAR ARRAY AREA	3,023 square m (32,528sq ft)
POWER	80 kilowatts
FIRST ASSEMBLY LAUNCH	20 November 1998
FINAL CONSTRUCTION LAUNCH	2010 (scheduled)

EARLY STAGES
(Top) From 1998 to 2000, the initial building blocks of the ISS, Unity and Zarya, remained docked in orbit for 18 months awaiting the next stage of construction.

GROWING STATION
(Above) A burst of construction work began in late 2000 and saw the addition of the Z1 truss (a temporary mounting point for the P6 truss and the first large solar array) as well as the Destiny laboratory and the Canadarm2 remote manipulator.

BACK TO WORK
(Right) During an EVA to install solar panel equipment, Shuttle astronauts Daniel Burbank and Steven MacLean hitch a lift on the Mobile Base System that allows Canadarm2 to move along the station's central truss. Atlantis's 2006 STS-115 mission saw ISS construction resume in earnest, following delays due to the loss of Columbia.

Early expeditions

The International Space Station has been continuously manned since the first crew arrived in October 2000. In its first decade, crews worked on construction and fitting-out alongside various scientific experiments.

MIR LIVES ON
A view into the Zvezda Service Module on the ISS shows the stations's main crew quarters. Zvezda began life as the core of the Mir 2 station and follows a design that dates back to the Salyuts.

Long-duration stays aboard the ISS are known as expeditions. Each typically lasts around six months, although since they are usually defined by a particular launch or return to Earth of personnel, there is often overlap. In the early years of ISS construction, expeditions were limited to three people (scaled back to just two after the *Columbia* disaster – see p.224), with frequent visitors on either the Space Shuttle or Russian Soyuz spacecraft. From 2007, as more room became available, the typical expedition crew rose to six, with an overlapping roster that usually sees three astronauts return to Earth midway through the expedition and three newcomers take their place on their own six-month flight.

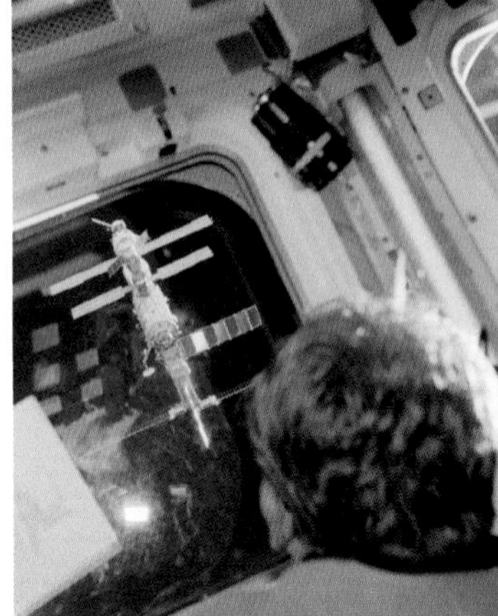

FIRST HANDOVER
Shuttle astronaut Andrew Thomas enjoys the view of ISS from Discovery *during its approach to the station in March 2001. This STS-102 mission saw the handover between the crews of Expeditions 1 and 2.*

Early expeditions

Expedition 1 set a pattern for other early crews. The commander was NASA astronaut William Shepherd, while the other two crew members were Russian – Sergei Krikalev and Yuri Gidzenko. Arriving on Soyuz TM-31, they spent most of their four-month flight getting the ISS fully operational and carrying out assembly tasks (see pp.296–97). In February 2001, they welcomed the crew of *Atlantis* and the US Destiny laboratory module. March saw *Discovery* arrive, bringing with it the Expedition 2 crew of Yuri Usachev, Susan Helms, and James Voss (one cosmonaut and two astronauts). In April, they were joined by the crew of Soyuz TM-32, including Dennis Tito, the first space tourist (see p.308), while in August, *Discovery* brought a new crew who spent four months concentrating largely on science. Expeditions 4 and 5 (spanning December 2001 to November 2002) doubled the number of experiments carried out onboard and received some of the first scientific payloads from private companies, while the launch of Expedition 6 saw the debut of the new "TMA" class of Russian Soyuz spacecraft, upgraded largely to accommodate taller passengers.

Throughout this period, the station continued to grow, with the addition of airlock modules, elements of the station's main framework, and its main robot arm, Canadarm2.

Back to basics

This phase of expansion came to an abrupt halt two months into Expedition 6, as the loss of *Columbia* deprived the station of its main supply vehicle and construction tool. With the station cut back to a crew of two and forced to rely on Russian spacecraft alone, Expeditions 7 to 11 were dominated by the need to keep the ISS operational, with science taking second place to routine maintenance and repair. The Space Shuttle finally returned to the ISS with *Discovery*'s STS-114 mission in late July 2005, delivering supplies halfway through Sergei Krikalev and John Phillips's Expedition 11.

2 November 2000
The first crew of the International Space Station arrives aboard Soyuz TM-31.

10 March 2001
The crew of Expedition 2 arrives aboard the Space Shuttle *Discovery*. The station's first handover is accomplished.

30 April 2001
The ISS welcomes Dennis Tito, the first of several space tourists on visiting Soyuz missions.

8 October 2001
The first spacewalk by ISS astronauts takes place to attach the Russian Pirs docking module.

26 April 2003
Following the loss of *Columbia*, Soyuz TMA-2 brings the Expedition 7 crew to the ISS, and the station is temporarily reduced to two crew members.

29 May 2009
The docking of Soyuz TMA-15 sees the official start of Expedition 20, commanded by Gennady Padalka and the first to achieve the full ISS complement of six.

BIOGRAPHY
SCOTT KELLY

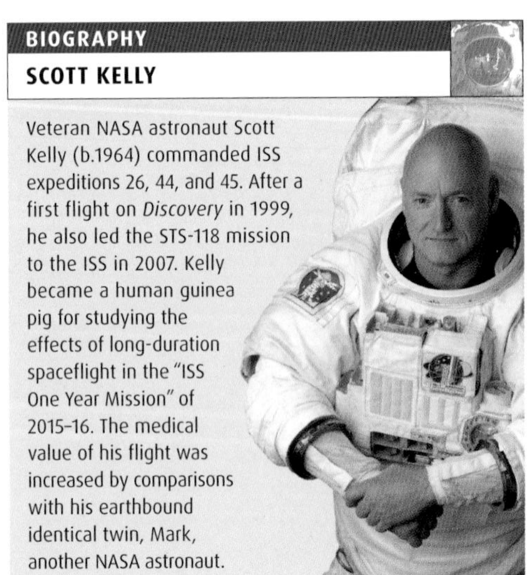

Veteran NASA astronaut Scott Kelly (b.1964) commanded ISS expeditions 26, 44, and 45. After a first flight on *Discovery* in 1999, he also led the STS-118 mission to the ISS in 2007. Kelly became a human guinea pig for studying the effects of long-duration spaceflight in the "ISS One Year Mission" of 2015–16. The medical value of his flight was increased by comparisons with his earthbound identical twin, Mark, another NASA astronaut.

WORKING OVER EARTH
ISS commander Peggy Whitson hangs on the side of the station during a November 2007 EVA. Construction of the station involved some 973 hours of spacewalks, most of them by ISS crewmembers.

DESTINY LABORATORY
Astronaut James Voss of Expedition 2 works in the US laboratory module, as visiting astronaut Scott Horowitz floats through the hatchway from the Unity module.

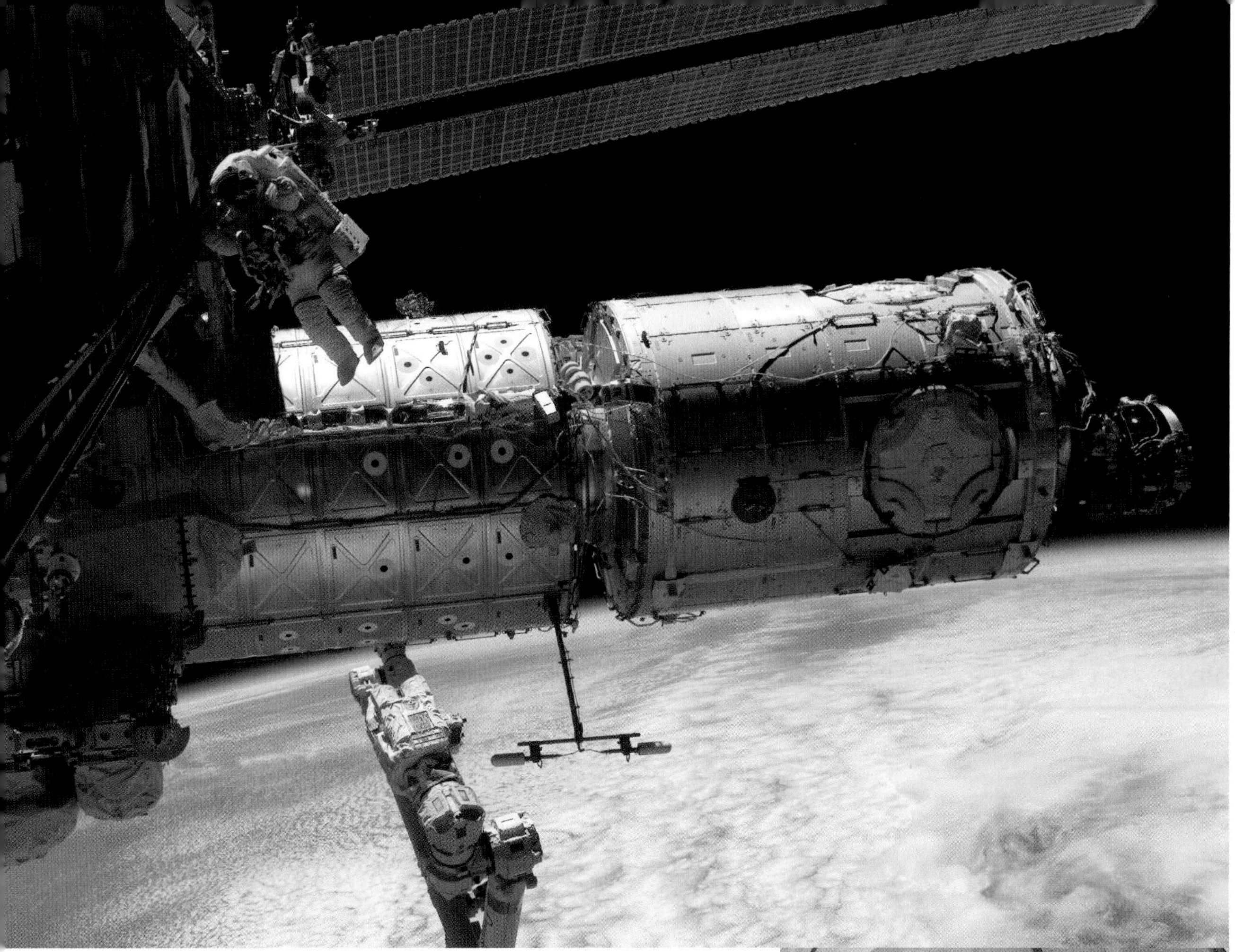

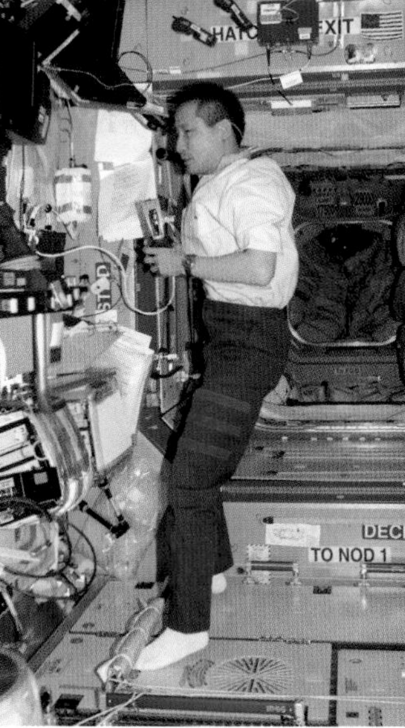

ORBITING EDUCATORS
Ed Lu of Expedition 7 presented a series of Saturday Morning Science *shows from the ISS that were shown back on Earth, demonstrating the effects of microgravity in the US Destiny laboratory.*

Continuing problems with the Shuttle led to another year-long grounding, and the crew of Expedition 12 also had to rely on Russian Soyuz and Progress spacecraft.

A growing crew

Discovery's successful second return in July 2006 saw normality restored. German astronaut Thomas Reiter, Europe's first ISS crewmember, arrived on the Shuttle to join Pavel Vinogradov and Jeffrey Williams and bring the Expedition 13 crew back up to three. Over the following years, a more complex, staggered schedule interwove Shuttle construction missions with Soyuz flights to ferry crew. Expedition 16, coinciding with the 50th anniversary of Sputnik 1 in 2007, saw the first female ISS commander, Peggy Whitson, and marked the beginning of a further expansion in crew numbers. The year 2008 saw the arrival of Europe's Columbus module and the first elements of Japan's complex Kibo laboratory, and by 2009's Expedition 20, improvements to the ISS environmental systems, the attachment of a second Soyuz "lifeboat", and an expansion of the Progress supply schedule allowed the station to operate with a full crew of six for the first time. With the arrival of final components over the next two years, the last departure of the Space Shuttle *Atlantis* on 19 July 2011 drew a line under the first phase in the station's history.

IN THE QUEST AIRLOCK
(Right) The Quest Joint Airlock arrived on Atlantis *in July 2001 and is a doorway to space for crew members in both Russian and US spacesuits. Expedition 4's Daniel Bursch (left) and Carl Walz are shown testing the airlock.*

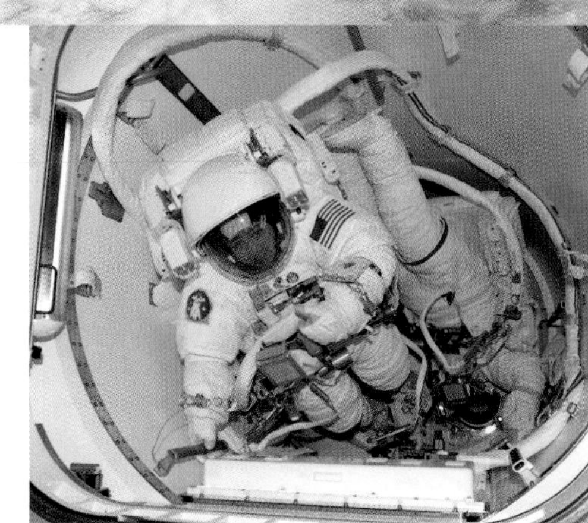

TECHNOLOGY

CUPOLA MODULE

February 2010 saw the arrival of an innovative ISS module featuring the largest windows ever used in space. Supplied by ESA and attached to the station's Tranquility node, the Cupola (from an Italian word for dome) is an "observatory" with seven windows that can be used for Earth observation, experiments, and to track docking operations. Here, Expedition 24's Tracy Caldwell Dyson simply takes a moment to enjoy the view.

CARGO FERRIES
A variety of cargo vehicles are used to supply the station, ranging from pressurized modules carried in the Shuttle cargo hold to autonomous spacecraft. Here, Canadarm2 snags the Japanese HTV-6 vehicle.

EXPERIENCE

MANNING THE ISS

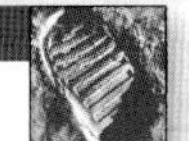

Expedition to "Alpha"

OFFICIAL PORTRAIT
William M. Shepherd, first commander of the International Space Station, and his crewmates Sergei Krikalev (left) and Yuri Gidzenko (right) pose in their Sokol spacesuits prior to launch.

The first crew of the International Space Station had an arduous list of duties, firstly getting the linked Zarya, Zvezda, and Unity modules fully operational, and then preparing the station for scientific work.

Launched from Baikonur aboard Soyuz TM-31 on 31 October 2000, Shepherd, Krikalev and Gidzenko docked with the station on 2 November. A Shuttle mission had visited in the previous month and installed a variety of external fittings, including communications antennae and the Z1 truss. Life-support systems had also been activated, so the crew were able to open the hatch from their Soyuz, enter the station, and get straight to work. Expedition commander Shepherd, a naval man by training, was bothered by the ISS's lack of a name, so he requested during the first radio conversation that, at least for this mission, it should revert to its 1990s name of Alpha. NASA administrator Daniel Goldin acquiesced, to applause in Mission Control.

"I think the **air quality** has been very good. I am surprised at the amount of air that gets moved through filters up here – it is pretty substantial. Anything we have that gets loose – food, things like that – they are in a filter in matter of **a minute or two** – it is pretty surprising. Odours? Non-existent. I am very surprised at how well all the **environmental equipment** has been working so far."

William Shepherd, interviewed by CNN, 18 December 2000

ERUPTION FROM SPACE
When the Mexican volcano Popocatapetl began to belch smoke in late-January 2001, the Expedition One crew managed to snap this stunning photo as the space station's orbit carried them northeast of the mountain.

AT WORK IN ORBIT
After launch from Baikonur on a Soyuz rocket, the Expedition One crew settled in and began to get the ISS systems up and running. Communications were an early priority – the crew had to rely initially on Russian communications equipment, with corded headphones that restricted movement. By the time Atlantis arrived with the Destiny module in February 2001, things were starting to feel more homely.

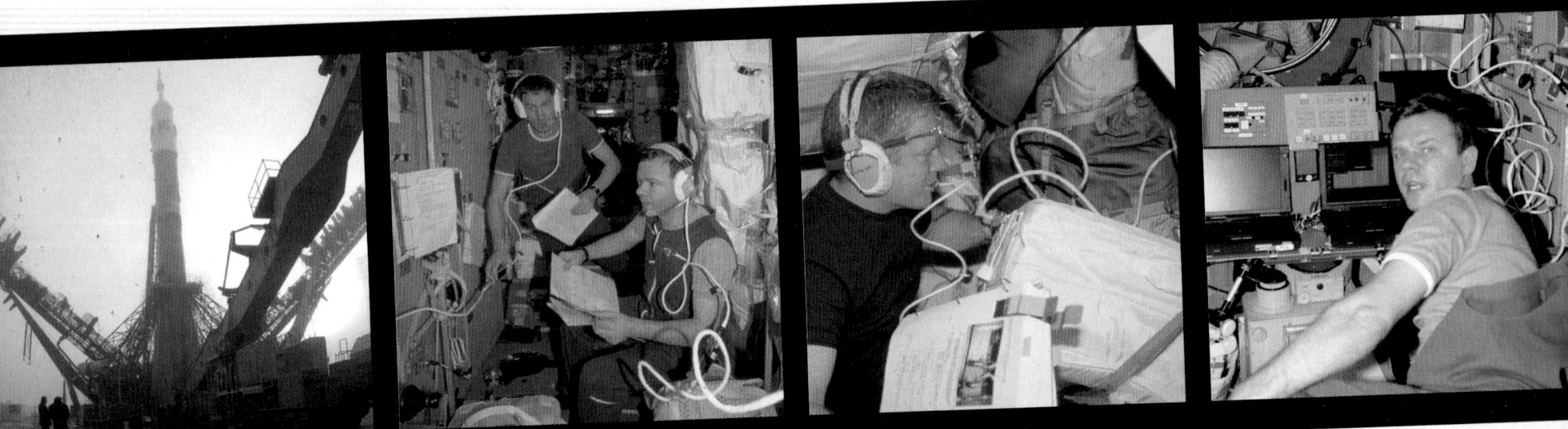

RELIEVED
This scene of Discovery's tail section against clouds on Earth was photographed from aboard the station after STS-102 had docked in March 2001, bringing with it a new crew for the ISS.

"We opened the hatches … and it was very pleasant to find ourselves in a place … with **good, clean air.**"

Sergei Krikalev, 2 November 2000

For the first month, the crew were restricted to the two Russian modules – they did not have sufficient power to run the Unity module, and so it remained sealed up until *Endeavour* arrived in December, bringing with it the first of the huge solar arrays. However, there was plenty to do unpacking the large amounts of equipment, ranging from clothes to laptop computers, left behind by visiting Shuttle missions prior to their arrival – and more arrived on two Progress ferries during the expedition. Unity contained a similar stash of equipment when finally opened, and by the time this was all installed, the next Shuttle mission was approaching. The main aim of *Atlantis*'s STS-98 mission in mid-February 2001 was to install the US Destiny module. There was plenty to do setting up the new laboratory, although it arrived without experiments aboard – the first of these arrived with *Discovery* on 10 March 2001. But that Shuttle also brought a new crew, and it would be their job to begin the station's scientific programme.

"… from all the people on the ground here in Houston and in Moscow and around the world [who] have supported you through this flight, it has been an **honor and privilege**, and you have our sincere thanks for your **outstanding duty** on Alpha. Your accomplishments are impressive. We witness your departure with both regret and happiness: regret because we will miss working with you; happiness because you are speeding home towards a reunion with family and friends and a **well-deserved rest.**"

Cady Coleman, Capcom, 18 March 2001

RETURN CEREMONY
(Left to right) Krikalev, Shepherd, and Gidzenko are feted upon their return to Earth at a reception at the Ellington Field Air Force Base near Houston.

The ISS in operation

Vith the departure of the last Space Shuttle mission during 2011's Expedition 28, the ISS was functionally complete. In the years since, it has continued to carry out groundbreaking space research.

Vith the Shuttle's retirement, the ISS was once again eliant entirely on Russian Soyuz capsules to bring up new crew members and return others to Earth at the ends of their missions. Progress cargo ferries would also continue to form the backbone of ISS logistics.

Going it alone

Somewhat embarrassingly, however, the first Progress launch of the new era (barely a month after the last Shuttle's departure), failed to reach orbit after its Soyuz-U third stage shut down prematurely. Fortunately, the station was well stocked at the time, but there were knock-on effects for the launch schedule as the next Progress mission was hurriedly brought orward and a Soyuz crew transfer launch pushed back.

Of course, Progress was no longer the only option available. An unprecedented variety of unmanned vehicles docked with the station in 2012 – an ESA Autonomous Transfer Vehicle, a Japanese HTV, and, for the first time, two reusable SpaceX Dragon modules (see pp.308–09). The following year saw a further supply option join the fray in the form of the American Orbital ATK Cygnus. Another 2013 landmark saw Chris Hadfield take command of Expedition 35 – the first Canadian ISS commander (and only the second non-US or non-Russian, after Belgian ESA astronaut Frank De Winne in 2009).

Fully operational

Although regular maintenance of the station continued and occasional new components had to be installed, most of the ISS crew's time could now be devoted to research. Station crew have carried out a huge range of experiments, in areas from Earth observation and materials science to medicine design and animal behaviour research. In 2014 a versatile 3-D printer was added to the station's laboratory, and a new electromagnetic levitator furnace, capable of melting material samples in weightless conditions, was added to ESA's Columbus module. At the other extreme, 2018 saw the installation of NASA's Cold Atom Laboratory, a cutting-edge physics facility that uses magnetic fields to cool individual atoms, slowing them to a halt in microgravity conditions so that their properties can be studied.

GOING VIRAL
ISS Commander Chris Hadfield became a social media celebrity during Expedition 35, both for his explanatory YouTube videos and for his weightless guitar-playing. His version of David Bowie's Space Oddity has been viewed more than 40 million times.

CUBESAT RELEASE
Deployed using the robot arm on Japan's Kibo laboratory in October 2012, three tiny CubeSats float free of the ISS. These small, low-cost devices were designed for use in radio communication experiments by schools and amateur radio enthusiasts.

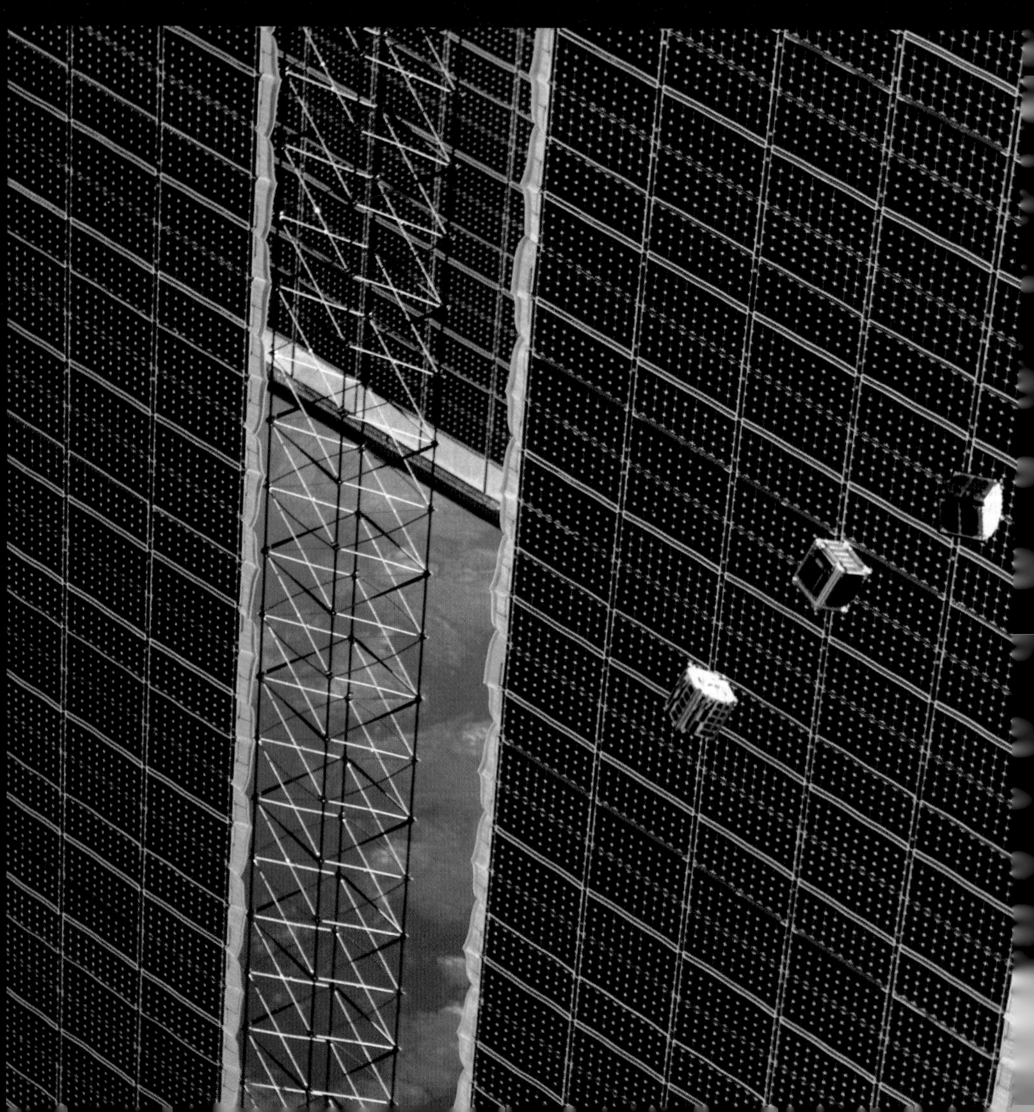

TECHNOLOGY

ROBOT HELPERS

Robots have been used in several ISS experiments. A prototype humanoid machine called Robonaut 2 was delivered to the ISS during one of the final Shuttle missions and subjected to various performance tests – it's hoped that remote-controlled "astronaut substitutes" could be used to carry out dangerous tasks such as EVAs during future deep-space missions. In 2018, German astronaut Alexander Gerst was accompanied into orbit by a floating "astronaut assistance system" called CIMON. Built by Airbus and the German DLR space agency, it can float freely in the ISS, using its artificial-intelligence programming to assist the crew.

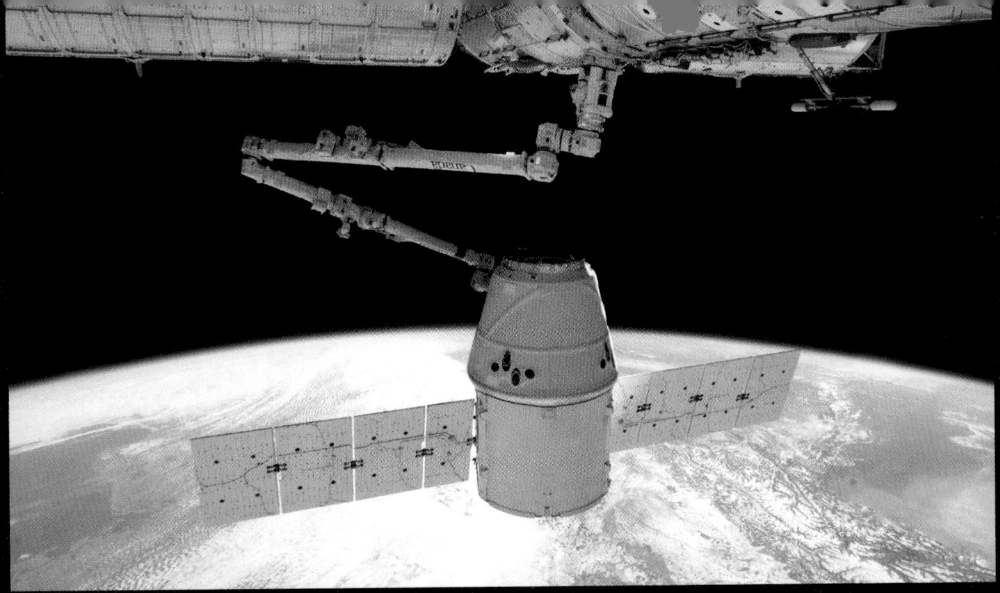

SPACEX DRAGON
Canadarm2 grapples a SpaceX Dragon capsule during mission SpX-2 in March 2013. The docking procedure involves the robot arm bringing Dragon into position next to the station's US segment before docking.

Towards the high frontier

As space agencies have increasingly shifted their direct priorities from low Earth orbit to exploration further afield in the Solar System, the ISS has hosted research to pave the way. Experiments have tested ways of promoting the growth of plants in space to ensure a supply of fresh foods, investigated the potential for robots to assist astronauts (see panel, opposite), and even installed a potentially revolutionary new space habitat (see below). Furthermore, from March 2015 to March 2016, NASA's Scott Kelly and Russia's Mikhail Kornienko became human guinea pigs for the "Year in Space" project, an exhaustive series of tests that allowed space-medicine researchers to improve their understanding of how prolonged space travel affects the human body.

SPACE SCIENCE
German ESA astronaut Alexander Gerst works on the Electromagnetic Levitator experiment during his 2014 "Blue Dot" mission. In 2018, during Expedition 57, Gerst became the second European ISS commander.

ORBITING GREENHOUSE
NASA astronaut Shane Kimbrough collects fresh red romaine lettuce grown in the International Space Station as part of the Veg-03 experiment in 2018. The lettuce is grown on nutrient-dispensing "plant pillows" and incubated in a special light box.

BEAM
In 2016, the Bigelow Expandable Activity Module (BEAM), a prototype space habitat that can be launched flat and inflated in orbit, was added to the station.

BIOGRAPHY

TIM PEAKE

Britain's first ESA astronaut, Major Tim Peake was actually the second person to fly with a UK flag (after the privately sponsored flight of Helen Sharman, see p.217). Launching on Soyuz TMA-19M in December 2015, he spent six months aboard the station during Expeditions 46 and 47. As well as a complex EVA, he carried out a wide variety of experiments and publicity works, including teleconferencing with schoolchildren and running the London Marathon on the ISS treadmill.

19 July 2011
With the departure of the Space Shuttle *Atlantis*, ISS crews are once again dependent on Russia's Soyuz spacecraft.

25 May 2012
An unmanned SpaceX Dragon capsule is docked to the station for the first time, remaining attached for five days.

29 September 2013
An unmanned Orbital ATK Cygnus capsule docks with the ISS for the first time.

27 March 2015
Soyuz TMA-16M blasts off, carrying Scott Kelly and Mikhail Kornienko to the ISS at the start of their "year-long" mission.

30 September 2015
NASA extends its contract with prime ISS contractor Boeing to prolong the station's operational lifetime until 2028.

2 March 2016
Kelly and Kornienko return to Earth after 342 days in space.

16 April 2016
The Bigelow Expandable Activity Module is berthed to an unused station airlock and inflated at the beginning of a year-long performance test.

1986

4 February 1986
During his State of the Union address, President Reagan calls for a hypersonic spaceplane called the NASP.

5 February 1986
The UK announces its HOTOL spaceplane, later cancelled after funding issues in 1988.

May 1993
The NASP project is cancelled.

18 August 1993
The DC-X Delta Clipper makes its first test flight.

31 July 1996
A launch pad fire destroys the DC-XA, leading to the project's cancellation.

November 1999
A prototype X-33 fuel tank fails in testing, leading NASA to cancel the project.

4 October 2004
During its 17th flight, SpaceShipOne claims the Ansari X Prize.

13 November 2006
A prototype SSTO vehicle from Blue Origin makes its first short flight.

22 April 2010
The Boeing X-37B is launched into orbit on a 224-day mission.

Successors to the Shuttle

Following the retirement of the Space Shuttle, NASA's new Orion spacecraft will see a return to the Apollo-era "conical capsule" template. But the idea of a winged spacecraft, it seems, is too good to disappear completely.

Reusability is a key part of a winged spacecraft's appeal – while the Space Shuttle system was only partially reusable and blighted with difficulties due to its overwhelming complexity, a vehicle that can launch, return to Earth intact, and be ready for another flight after minimal servicing could dramatically cut the costs of reaching orbit. Countless solutions to the challenge have been proposed over the years, and now it seems the idea is finally coming closer to reality.

SÄNGER'S *AMERIKA*
Spaceplanes were first thought of by Austrian engineer Eugen Sänger in the 1930s. He envisioned an orbital bomber launched from a rail track and accelerated by a rocket before takeoff.

Launch alternatives

The main challenge for a truly reusable spaceplane is how to cope with the huge fuel burden needed to reach space from the ground. Some systems use the same approach taken by earlier experimental US aircraft such as the Bell X-1 and the X-15 (see p.34 and p.188) – a large carrier aircraft does much of the hard work of reaching high altitude, before the much smaller vehicle is released and fires its rockets to carry it the rest of the way into space.

This principle was demonstrated successfully in 2004, when the winged SpaceShipOne, designed by aero engineer Burt Rutan (b.1943), made two suborbital hops into space within a fortnight to scoop the $10m Ansari X-prize. Rutan subsequently partnered with Virgin Galactic to scale up his innovative design, which also takes a unique approach to the problem of re-entry (see panel, opposite), with ambitious plans to carry paying space tourists aboard the larger SpaceShipTwo within the next few years. For the moment, however, this particular design is limited to suborbital spaceflight.

DAY TRIP TO SPACE?
Virgin Galactic's suborbital SpaceShipTwo, shown here docked to its carrier aircraft White Knight Two, can achieve speeds of 4,200kph (2,600mph) using a single hybrid rocket engine. In operation, it will carry six passengers and two pilots.

An alternative means of cutting down on the fuel needed to reach space is the hybrid "aerospaceplane" – a spacecraft that behaves like a normal aircraft in the atmosphere, accelerating to hypersonic speeds by using air to oxidize its fuel, before transforming into a rocket at high altitudes. When the concept was first suggested in the 1980s, in forms such as Britain's HOTOL and the US National Aerospace Plane (NASP), the technology required to make such vehicles a reality was still in its early stages of development. Now that ramjets have matured (see panel, below), the idea is being revisited in earnest. One promising scheme, Skylon, uses the SABRE engine from British firm Reaction Engines Limited, which liquefies atmospheric air for use as rocket oxidizer.

TECHNOLOGY

SPACESHIPONE

Ansari X-prize winner SpaceShipOne has a totally unique flight profile. Taken to high altitude by the White Knight carrier aircraft, it fires its rockets to enter a suborbital trajectory, re-enters the atmosphere like a shuttlecock, and glides back to Earth like a Space Shuttle. The spacecraft itself is just 8.5m (28ft) long, while the carrier has a wingspan of 28m (92ft). Launch from altitude helps cut down its fuel weight, while slow re-entry speed reduces the need for thermal shielding.

1 *powered by twin jet engines, White Knight carries SpaceShipOne to 15km (50,000ft)*

2 *released from White Knight, SpaceShipOne's unique rocket motor ignites, powering the spacecraft to 100km (60 miles) above the Earth at speeds of up to Mach 3*

3 *as SpaceShipOne drops back into the atmosphere, its wings "feather", rotating into a configuration that slows re-entry*

4 *after re-entry, the wings unfold and SpaceShipOne glides home*

Mini-shuttles

In addition to full-sized spaceplanes, a wide variety of proposals for compact, Shuttle-like spacecraft have come and gone. Most of these were intended to be carried into space on a vertical rocket, using "lifting body" aerodynamics (similar to those seen in the Dyna-Soar project of the 1960s, see p.188) on their descent back through Earth's atmosphere. Examples of such "mini-shuttles" include the French CNES space agency's Hermès project of the 1980s, Japan's HOPE (H-II Orbiting Plane) in the 1990s, and a proposed Crew Return Vehicle considered in the planning stages of the ISS. Russian space contractor Energia floated a similar idea for a spacecraft called Kliper in the early 2000s, with ESA and JAXA briefly involved, but the project was suspended indefinitely after 2006 for budgetary reasons.

Amid all these cancellations, the most successful example of the approach is the unmanned Boeing X-37. This small spaceplane, launched on both Atlas V and SpaceX Falcon rockets, is used by the US Air Force and Defense Advanced Research Projects Agency (DARPA) for unspecified military space operations that probably include propulsion and surveillance system tests. With no crew to support, it can spend months at a time in orbit.

Single stage to orbit

One final radical approach to the problem of reusable launch vehicles is to integrate spacecraft and rocket, so that the entire vehicle travels into space and returns to Earth intact. This single-stage-to-orbit (SSTO) concept was tested in the 1990s with a scale prototype called the McDonnell Douglas DC-X, or Delta Clipper, and was also applied to a planned NASA launch vehicle, the X-33. These projects faltered after a 1996 fire destroyed the DC-X prototype, but the concept remains sound, and the potential cost savings of a fully reusable rocket are impressive. Recently, the private sector has risen to the challenge, although rocket firms such as SpaceX and Blue Origin (see pp.308–09) are developing fully reusable multi-stage rockets rather than an all-in-one vehicle.

DARPA XS-1
Currently under development, the XS-1 project is a hybrid of the spaceplane and SSTO concepts – a reusable, plane-like first stage that can deploy an upper stage and payload to orbit before returning to base, refuelling, and flying again.

TECHNOLOGY

RAMJETS

Many spaceplane concepts rely on ramjet technology to help them fly through the atmosphere. A ramjet is a jet engine with few moving parts – rather than using turbines to draw in air for combustion, it relies on its own forward motion to force air into it at high speed. The shape of the engine then compresses and heats the air until fuel is injected, and the combustion produces thrust. Ramjets are often incorporated as a scoop-shaped inlet within the airframe of an aircraft. But they are only effective at speeds of 1,600kph (1,000mph) or more, so another engine is needed to reach these speeds. Fortunately, the rocket engines needed to operate in space can also be used to gain the required speed.

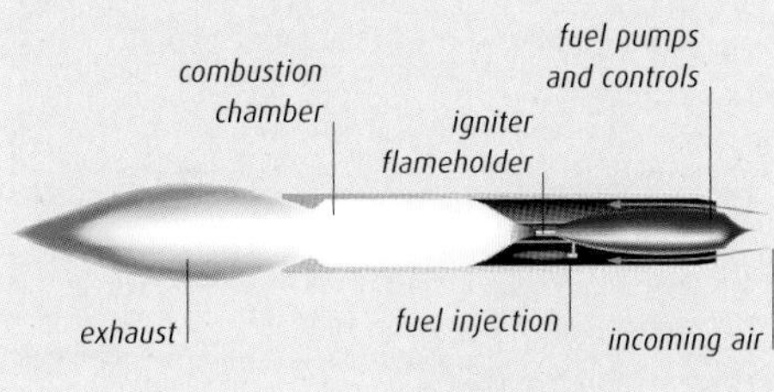

Chinese manned spaceflight

In the 21st century, China has embarked on an ambitious project to equal the achievements of the 20th-century superpowers, sending men into orbit and perhaps beyond.

DIVINE VESSEL
The Shenzhou spacecraft shares its three-part layout with the Russian Soyuz capsules, but is somewhat larger and has an extra pair of solar arrays on the orbital module.

20 November 1999
The unmanned Shenzhou 1 test rocket is successfully launched.

5 January 2003
The safe return of Shenzhou 4 after six days in orbit opens the way for a Chinese manned spaceflight.

15 October 2003
Yang Liwei becomes the first Chinese taikonaut, aboard Shenzhou 5.

16 October 2003
Shenzhou 5's re-entry module returns Yang to Earth after 14 orbits and 21 hours.

12 October 2005
Shenzhou 6 carries China's first two-man crew into orbit.

27 September 2008
The crew of Shenzhou 7 carry out an EVA during China's first three-man mission.

29 September 2011
The unmanned Tiangong-1 laboratory is launched into orbit.

18 June 2012
Shenzhou 9 docks with Tiangong-1 in orbit, allowing three taikonauts to carry out experiments on board.

Although China planned a manned spaceflight programme as early as the 1960s, its chief supporters fell victim to one of the government's periodic purges in the 1970s, and the plan seemed to be forgotten. There was some talk of Chinese cosmonauts flying aboard Mir in the 1980s, or even joining a Space Shuttle crew, but the world was taken by surprise when the Chinese government approved a new manned space programme, initially known as Project 921, in 1992.

Sino-Russian origins

The newly formed China National Space Administration (CNSA) benefited from an agreement signed with Russia in 1994, which gave them access to Soyuz capsules, blueprints, and Russian expertise, but despite an overall resemblance to the reliable Soviet spacecraft, CNSA's manned vehicle, called Shenzhou, is completely Chinese in design and manufacture. Like Soyuz, Shenzhou (meaning "divine vessel") has three separate elements – an orbital module, a re-entry module, and a service module. However, it is significantly larger than Soyuz and can be flown with two pairs of solar arrays – one pair on the service module and the other on the orbital module. This allows the orbital module to continue powered operation even after it has been jettisoned from the rest of the spacecraft and the crew have returned to Earth.

Shenzhou 1, an unmanned development test, was launched in 1999 by a CZ-2F Long March rocket. It orbited Earth 14 times, tracked by China's network of shipborne monitoring stations, before the re-entry sequence was triggered and the central module returned safely to Inner Mongolia.

Pioneering taikonauts

Three more Shenzhou test flights followed, using dummy astronauts and animal passengers to study conditions and the performance of life-support

EN ROUTE TO TIANGONG-2
A Long March 2F carrier rocket carrying the manned spacecraft Shenzhou 11 blasts off from the launch pad at the Jiuquan Satellite Launch Center in October 2016.

1957 1958 1959 1960 1961 1962 1963 1964 1965 1966 1967 1968 1969 1970 1971 1972 1973 1974 1975 1976 1977 1978 1979 1980 1981 1982 1983 1984 1985 1986 1987 1988 1989 1990 1991 1992 1993 1994 1995 1996 1997 1998 1999 2000 2001 2002 2003 2004 2005 2006 2007 2008 2009 2010 2011 2012 2013 2014 2015 2016 2017 2018 2019 2020

RETRIEVING CAPSULE
Engineers retrieve the Shenzhou 6 re-entry capsule from its landing site in Inner Mongolia's Siziwang Banner region. In contrast to early US and Soviet flights, all the Chinese Shenzhou missions so far have come down on target within a relatively small landing zone.

systems. Each of these flights carried experiments onboard, some of which remained in space with the orbital module for extended missions.

The launch of Shenzhou 5 saw the first flight of a Chinese spacefarer, or "taikonaut". Although the Chinese authorities announced the flight in advance, they did not permit a live broadcast of the launch, presumably in case something went wrong. Yang Liwei (see panel, below) blasted off from Jiuquan at 09:00 local time on 15 October 2003. He completed 14 orbits of the Earth before landing in Inner Mongolia in a virtual repeat of the Shenzhou 1 mission. Yang remained in the re-entry module throughout the flight, but experiments in the orbital module continued to function for five months after the craft had been abandoned in space.

BIOGRAPHY

YANG LIWEI

Born in China's Liaoning province, the first taikonaut Yang Liwei (b.1965) was fascinated by aviation from his youth and, after joining the People's Liberation Army in 1983, enrolled in its Aviation College No. 8. After graduation, he became a fighter pilot, accumulating more than 1,300 flying hours in his Air Force career. Selected as a candidate for manned spaceflight in 1998, he went through five years of intensive study and simulation at the Astronaut Training Base in Beijing. This led to his selection as the pilot for the Shenzhou 5 mission in October 2003.

READY FOR TAKEOFF
The three taikonauts of the first flight to a Chinese space station – from left to right: Liu Yang, Liu Wang, and mission commander Jing Haipeng (on his second Shenzhou mission) – prepare for launch aboard Shenzhou 9. As with all taikonauts so far selected, all three were experienced military pilots.

In marked contrast to the one-upmanship and rapid launch cycles of the Space Race era, China seems to have adopted a more measured strategy, with a gap of some two years before the launch of Shenzhou 6. This time, two taikonauts remained in orbit for almost five days, entering the orbital module for the first time to work and carrying out tests on the spacecraft's systems. Shenzhou 7, launched in September 2008, saw another step forward, with three taikonauts aboard. Although the mission lasted just three days, it included China's first spacewalk, a 22-minute EVA by Zhai Zhigang and Liu Boming.

Heavenly palaces

In September 2011, the Chinese space programme stepped up a gear with the launch of an unmanned module called Tiangong-1 (meaning "heavenly palace"). Intended as proof-of-concept for a Chinese space station, Tiangong was joined in orbit a month later by the unmanned Shenzhou 8, which carried out two successful docking operations with the module under remote control from the ground. A crewed mission, Shenzhou 9, followed in June 2012. Its three taikonauts (including China's first female spacefarer, Liu Yang) spent six days on the station, and a year later, an additional mission doubled the time spent docked to Tiangong-1. The station's mission officially ended in March 2016, but it remained in orbit for a further two years before re-entering the atmosphere in April 2018.

A second Tiangong module was lofted to orbit in September 2016 and visited the following month by the crew of Shenzhou 11. Despite speculation, a second manned expedition to Tiangong-2 has not yet taken place, but the station has not been idle – from April to September 2017, it participated in a docking exercise, as the unmanned cargo ferry Tianzhou-1 rendezvoused with it under remote control on three occasions, each time transferring fuel to maintain the station's orbit.

The final stage of Project 921, confirmed after much speculation in 2017, will involve the launch of a larger modular space station, with an elongated Tiangong-3 module at its core and several other modules. To be built between 2019 and 2022, the station will mark China's coming of age as a space power, potentially paving the way for international guest astronauts and a manned lunar mission.

DOCKING WITH TIANGONG-1
This artist's impression shows a Shenzhou spacecraft (front) docking to the Tiangong-1 laboratory in orbit. Tiangong is divided into a pressurized experiment module and a resource module fitted with engines and solar panels.

Exploring the Solar System

With operations in low Earth orbit increasingly handled by commercial spaceflight companies, space agencies such as NASA have refocused on long-term goals for human exploration further from our home planet.

The loss of *Columbia* in 2003 and the decision to wind down the Space Shuttle programme led to a fundamental reassessment of space priorities for the United States. The country was committed to its role in completing and operating the International Space Station for the foreseeable future, but the general recognition that the Shuttle design had been fatally risky from the outset highlighted questions about America's future in space that had gone unaddressed for a generation.

ORION TAKES SHAPE
An artist's impression shows the Orion MPCV, consisting of a conical crew module and cylindrical service module equipped with solar panels for power generation.

Where next?

Even in the early 2000s, the potential for other agencies and commercial operators to take over relatively routine space tasks such as launching goods (and even astronauts) to the ISS was becoming clear. Since that time, the commercial operators have certainly proved their reliability (see p.308), and the ambition of retargeting NASA's manned spaceflight priorities beyond Earth orbit has remained a constant. The political priorities directing efforts towards specific goals and targets, however, have proved somewhat less reliable.

In 2004, President George W. Bush laid the groundwork for the new era with a speech in which he announced America's "New Vision for Space Exploration" – an ambitious programme calling for a return to the Moon by the 50th anniversary of Apollo 11 in 2019, the establishment of a permanent Moon base in the following decade, and the long-awaited mission to Mars perhaps in the 2030s.

Galvanized by its new mission, NASA quickly went to work on the details of how it would return to the Moon with larger spacecraft and heavier cargoes. The plan unveiled over the following year, known as the Constellation programme, was complex but versatile, requiring both Earth-orbit and lunar-orbit rendezvous, and called for a new reusable spacecraft named Orion (see below) and two new "Ares" launch vehicles derived from reliable elements of Shuttle-era technology. Ares I was to be an adapted solid rocket booster (SRB) with a new liquid-fuelled upper stage for manned launches, while Ares V was liquid fuelled and at first intended to feature a cluster of five Space Shuttle Main Engines (SSMEs) for launching heavy cargoes.

However, spiralling cost projections, lagging schedules, and a global economic slowdown dealt a series of blows to Constellation. President Barack Obama ordered a commission to review US space activity in 2009 and announced a major policy shift a year later. The Ares launchers, and with them the Constellation programme and plans for a return to the Moon, were abandoned. Orion survived, with some modifications, while a new heavy-lift launch vehicle, the Space Launch System (see panel, opposite), would replace the Ares V. NASA astronauts would use new commercially developed spacecraft and launch vehicles for access to Earth orbit, while Orion's exploration goals would immediately look further afield, targeting a rendezvous with a near-Earth asteroid in the mid-2020s and a mission to Mars orbit around 2035. The arrival of the Trump administration in 2017 saw new discussions about partnering with commercial firms to return astronauts to the Moon ahead of the ambitious Orion goals, but NASA's current major programmes for spacecraft and rocket development have remained largely unaltered.

FUELLING SLS
The SLS liquid hydrogen tank stands 40m (130ft) tall and is the largest cryogenic fuel tank ever built. Combined with a smaller liquid oxygen tank and four RS-25 rocket engines, it forms the core stage of the SLS launch vehicle.

NASA's new spacecraft

The final iteration of Orion, known as the Multi-Purpose Crew Vehicle (MPCV), replaces three proposed variants under the Constellation programme. Resembling the Apollo spacecraft in basic profile, but equipped with technology that would have been unimaginable to the Apollo astronauts, Orion consists of a crew module (CM) capable of carrying four people on extended missions of several weeks (extending to months with the use of an additional habitat module) and a European Service Module (ESM) adapted from

LAUNCH ABORT SYSTEM
During launch, the Orion spacecraft is docked to a launch abort system – a small but powerful rocket that can carry the crew module safely away from the main launch vehicle in the event of problems.

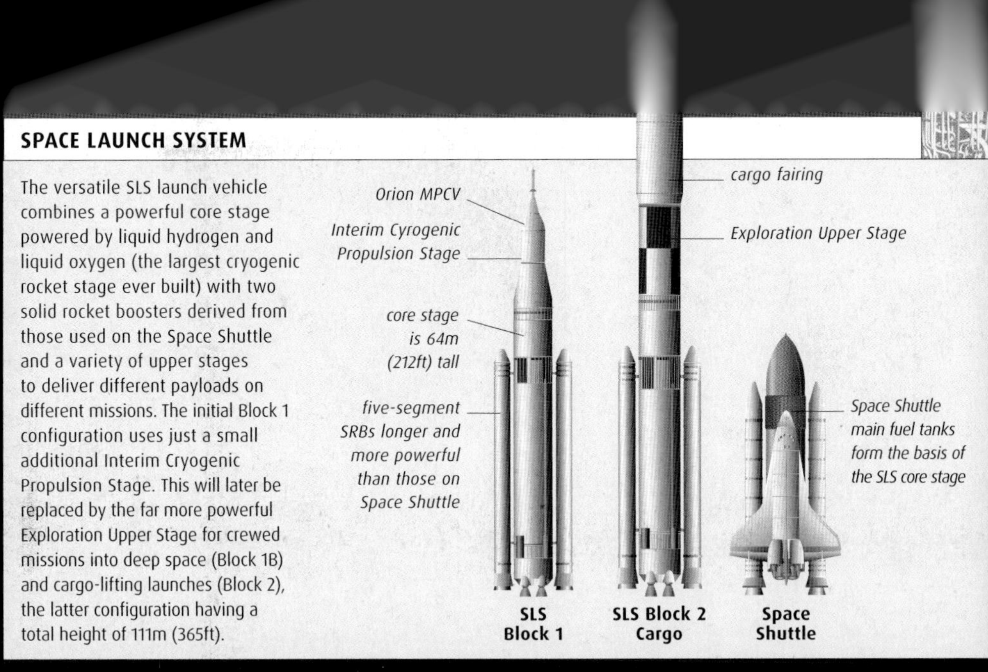

SPACE LAUNCH SYSTEM

The versatile SLS launch vehicle combines a powerful core stage powered by liquid hydrogen and liquid oxygen (the largest cryogenic rocket stage ever built) with two solid rocket boosters derived from those used on the Space Shuttle and a variety of upper stages to deliver different payloads on different missions. The initial Block 1 configuration uses just a small additional Interim Cryogenic Propulsion Stage. This will later be replaced by the far more powerful Exploration Upper Stage for crewed missions into deep space (Block 1B) and cargo-lifting launches (Block 2), the latter configuration having a total height of 111m (365ft).

ESA's Automated Transfer Vehicle (ATV, see p.231). After extensive testing, an unmanned Orion crew module first reached space on 5 December 2014, launched by a Delta IV Heavy rocket on a two-orbit flight around the Earth. The main crew module is intended to be reused up to 10 times, with plans for an unmanned loop around the Moon in 2019 as the Space Launch System becomes available, and a manned mission to establish a space station in lunar orbit in 2023.

Exploration options

Looking further ahead, where will human exploration of the Solar System take us next? Mars seems an obvious target, and it is increasingly likely that exploration of our neighbouring planet will involve a return flight to orbit and perhaps investigation of the Martian moons before a landing is attempted.

US spaceflight advocate Robert Zubrin outlined an influential profile for a return mission to the planet's surface, called Mars Direct, in the early resources such as water in the Martian soil has strengthened his argument that landing an automated factory to produce fuel for the return journey (reducing the risk of landing a fuel-laden spacecraft on the surface) could be a key element of such a mission. Meanwhile, a fringe of "Mars to Stay" supporters argue that the entire process can be simplified by sending astronaut volunteers on a one-way journey and ignoring the challenges of bringing them back.

And what of other space powers? ESA's plans for human exploration are wedded to the Orion project, but Russia is developing a new spacecraft called Federatsiya, with the aim of returning to the Moon in 2028. China, though keeping its plans typically close to its chest, almost certainly has its own lunar ambitions. In rare public statements, its leaders have even spoken of exploring Mars and Saturn.

MODULES FOR MARS
An artist's impression shows two Orion spacecraft docked to one end of a Mars transfer vehicle. Such a vehicle, consisting of a pressurized habitat, life support, and a (still-hypothetical) nuclear-powered rocket, would be assembled in Earth orbit through a series of unmanned launches.

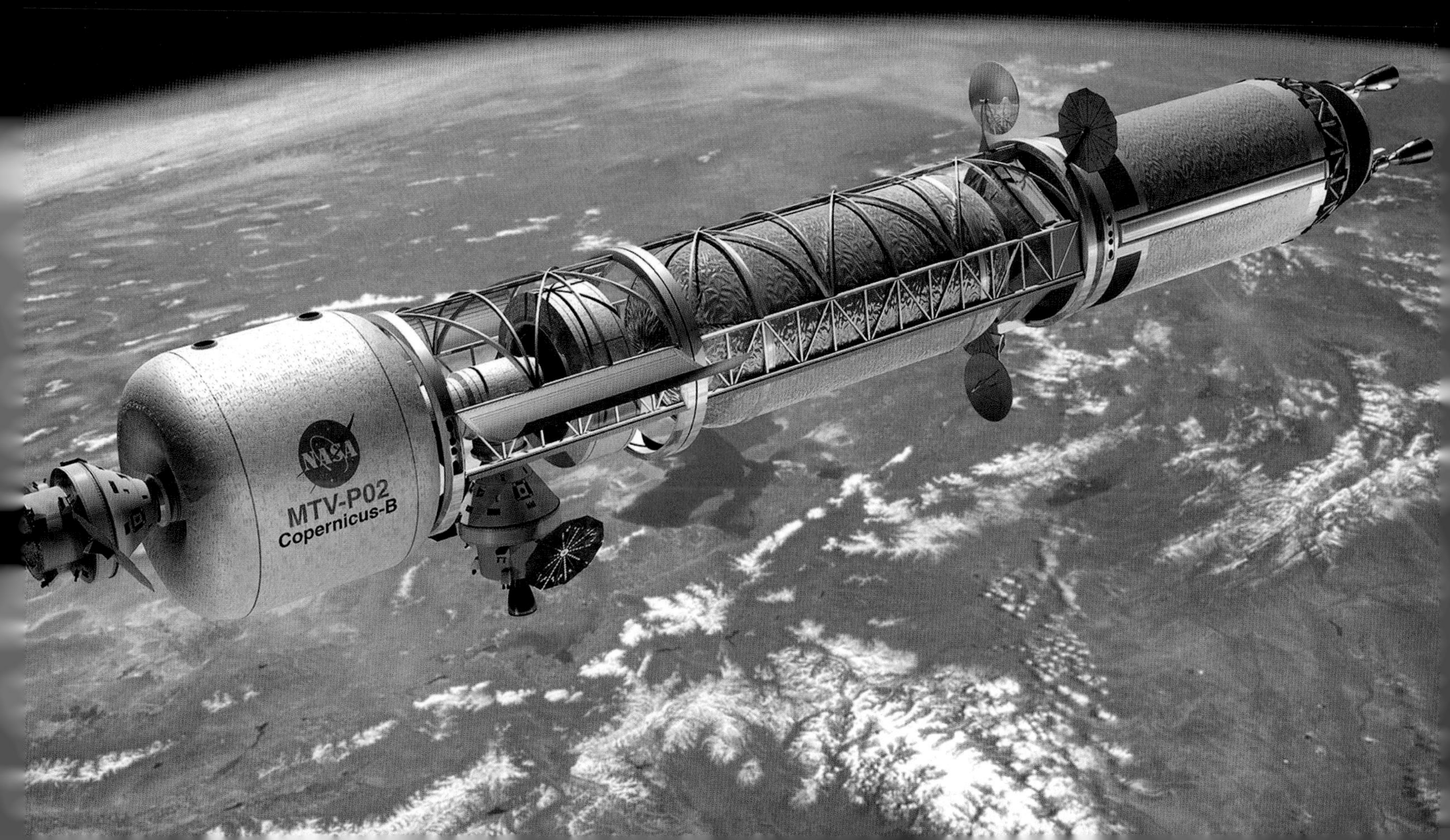

Future spaceprobes

In the next decade, new missions will explore uncharted corners of the Solar System and new aspects of familiar planets, from the interior of Mars to the lonely Trojan asteroids and beyond.

Plans for future unmanned spaceprobes are notoriously prone to change – priorities alter, budgets spiral, and of course things go wrong. However, some future plans are more certain than others – and barring accidents, probes currently in the late stages of development or en route to their destinations will be surprising us with new discoveries before long.

Towards the Sun

Late 2018 saw the launch of two missions towards our local star. The Parker Solar Probe will brave the searing temperatures of the Sun's outer corona on a seven-year mission that will bring it closer to the Sun than any previous artificial object. Meanwhile, the ESA/JAXA BepiColombo mission will be on a long journey to Mercury, involving a complex sequence of gravity-assist flybys of Earth, Venus, and Mercury itself, before finally entering orbit around the innermost planet in 2025. BepiColombo uses ion thrusters (see p.285) as its main propulsion system, with only small chemical rockets for final adjustments in orbit. Once safely at Mercury, two separate spacecraft will be deployed from the Mercury Transfer Module engine unit to begin around 30 months of data collection in Mercury orbit. The mostly European-built Mercury Planetary Orbiter (MPO) carriers a range of cameras, spectrometers to analyze the surface composition, a laser altimeter, and other instruments. The mostly Japanese Mercury Magnetospheric Orbiter (MPO) carries instruments to study the planet's magnetic field and the particle environment created as the solar wind streams past the planet. A small surface lander was cancelled in 2003 for budgetary reasons.

New Martian insights

Mars will inevitably remain at the heart of our exploration of the Solar System, with efforts redoubling in the light of exciting results from the *Curiosity* rover (see p.283) and still-uncertain plans for manned missions. By 2019, the first results are expected from InSight, a NASA lander that carries a drill and seismometer to investigate the planet's internal structure. Beyond this, the 2020 launch window may see an armada of missions. NASA's Mars 2020 rover mission will be based on *Curiosity*, but its onboard experiments will focus specifically on

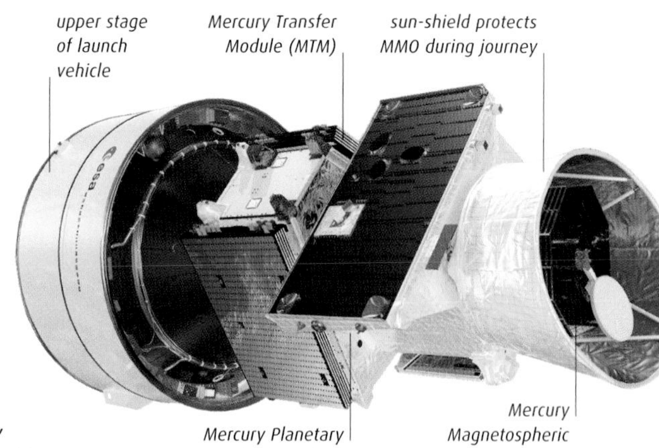

BEPICOLOMBO
This illustration shows the various elements of the BepiColombo spacecraft in transit mode. After arrival at Mercury, the MMO and MPO will be deployed to separate orbits.

READYING INSIGHT
The InSight lander's protective backshell is fitted into place in a clean room at Lockheed Martin Space Systems in Denver, Colorado. Problems with a seismometer instrument led to the spacecraft missing its original 2016 launch date.

TECHNOLOGY

LUCY'S ORBITAL PATH

NASA's Lucy mission will use a gravity assist from Earth to send it on a 12-year looping flight path that takes it through both the "leading" and "trailing" Trojan swarms (orbiting at gravitational sweet spots called L4 and L5). After encountering a small main-belt asteroid in 2024, the mission will journey first to the leading, L4 group, flying past four asteroids in 2027–28 as it slows on the outer edge of its orbit. It will then swing back through the inner Solar System, passing Earth en route to an encounter with one of the largest Trojans, a binary system called Patroclus-Menoetius, in the trailing L5 group.

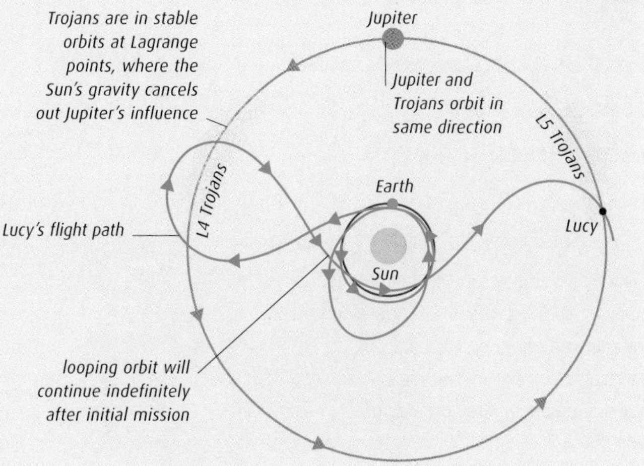
Trojans are in stable orbits at Lagrange points, where the Sun's gravity cancels out Jupiter's influence
Jupiter
Jupiter and Trojans orbit in same direction
L5 Trojans
L4 Trojans
Earth
Lucy's flight path
Lucy
Sun
looping orbit will continue indefinitely after initial mission

tilting solar arrays

3m (10ft) high-gain antenna

antennae for REASON ice-penetrating radar

EUROPA CLIPPER
NASA's planned probe to Europa will use large solar panels to generate electricity rather than radioisotopes that might contaminate the moon.

the search for Martian life. It will also carry with it a drone called the Mars Helicopter Scout.

The ExoMars Rover, a joint ESA/Russian project, is also planned to launch in this window, and China is planning a Martian debut that will comprise an orbiter, a lander, and a small rover, launched on a single rocket. Furthermore, the United Arab Emirates may become the first Arab country to send a mission to Mars with Hope, an orbiter planned to collect more data about the Martian atmosphere.

Looking further ahead, various other missions are already being discussed, with ISRO actively working on a follow-up to India's Mars Orbiter Mission, and various countries looking to be the first to return a sample from the Martian surface to Earth.

INSIGHT LANDER ON MARS
An artist's impression shows InSight after landing in the Elysium Planitia region of Mars. The spacecraft borrows design elements from the Phoenix mission, and will send a self-burrowing heat probe 5m (16ft) into the Martian crust to explore its properties.

Outer realms

Further out in the Solar System, both NASA and JAXA have plans for missions to the mysterious realm of the Trojan asteroids. The Trojans are a curious family of asteroids that share Jupiter's orbit but avoid disruption from the giant planet's gravity by gathering in two clouds at the Lagrange points, gravitational neutral zones some 60 degrees behind and ahead of the planet itself. Understanding the composition and distribution of the Trojans will tell us more about the Solar System's early evolution; one theory is that they were ejected into their present locations during an early reshuffling and migration of the giant planets. The two groups of Trojans will both be investigated by NASA's ambitious Lucy mission (see panel, above) scheduled for launch in 2021, while JAXA has plans to send OKEANOS, a solar-sailing follow-up to IKAROS (see p.285) in the same direction later in the decade.

Meanwhile, plans are also being made for two new missions to Jupiter's giant Galilean moons. ESA's Jupiter Icy Moons Explorer (JUICE) will investigate the three outer moons – Europa, Ganymede, and Callisto – all of which are now thought to have hidden oceans beneath their surfaces. Launching in 2022, JUICE should arrive at Jupiter around 2030 and, after a series of encounters with Europa and Callisto, eventually enter orbit around Ganymede, the largest moon in the Solar System, by 2033. A small Ganymede lander is also being considered for the mission.

NASA's Europa Clipper, meanwhile, will focus on the innermost of this icy trio, whose ocean lies closest to the surface and which is thought to be one of the most likely places in the Solar System for alien life to have evolved. Launching around 2022–25, it will enter an elliptical orbit around Jupiter that permits at least 45 close flybys of Europa.

EXOMARS TESTING
A prototype of ESA's robotic ExoMars rover explores the arid Chilean desert during testing in 2013. Conducted remotely with a time delay as if the rover were on Mars, such missions help scientists learn how to "drive" and collect scientific data on the Red Planet.

A commercial revolution

The 21st century has seen a major shift in the availability of access to space, both for private enterprise and potential space tourists. No longer the preserve of superpowers, spaceflight today features startups, entrepreneurs, and even holiday brochures.

The idea that one day we might holiday in space would have seemed quite feasible to Wernher von Braun and his colleagues when they wrote their *Collier's* articles in the early 1950s. But as Cold War politics and constant budget cuts got in the way, the idea that space would ever be opened up to the masses faded away, or at least retreated into an unimaginably distant future. Space travel, it seemed, would only ever be open to hardened professionals – pilots, engineers, and the occasional lucky scientist. It's all the more surprising, then, that today the pendulum has swung the other way, with space tourism a reality and companies developing plans to make it cheaper and more accessible. Meanwhile, an entire "new space" sector of entrepreneurial industry has sprung into existence, making use of cheaper launch vehicles and smaller, off-the-shelf satellites for a variety of applications that would have seemed unimaginable two decades ago.

Open for business

The first person widely recognized as a space tourist, US businessman Dennis Tito, paid $20 million for his seat on Soyuz TM-32 in 2001. Several similar flights helped boost the Russian Space Agency coffers before the programme petered out in 2009. However, one of the early Soyuz space tourists, Iranian-American software entrepreneur Anousheh Ansari, had a more lasting impact through her co-funding of the Ansari X-Prize, a $10 million competition to develop a reusable, piloted, suborbital spacecraft. The prize, eventually claimed by Burt Rutan's SpaceShipOne in 2004 (see p.301), garnered 26 separate entries and effectively kick-started a commercial spaceflight industry that has blossomed in the years since.

FEMALE SPACE TOURIST
Anousheh Ansari during training for her trip to the ISS. Ansari launched aboard Soyuz TMA-9 on 18 September 2006, and spent nine days aboard the station before returning to Earth on 29 September.

CUBESATS
Designed on a 10-cm (4-in) cubic framework, NASA's PhoneSat 2.4 is a typical single-unit CubeSat. These small devices offer a cheap means of testing innovative technologies – in this case, using an Android smartphone to provide key satellite functions.

Although the relatively cheap space tourism promised by the X-Prize itself is still on the cusp of becoming reality, the competition demonstrated that it was possible to "do" space in faster, cheaper, and more adaptable ways than the bureaucratic space agencies would permit. In the satellite industry, for instance, entrepreneurs rapidly adopted the innovative CubeSat – a template for simple and cheap construction of satellites using small cubic units weighing no more than 1.33kg (3lb). Faster and cheaper microelectronics allow even such small satellites to carry a variety of different experiments, and their ability to "piggyback" on the launch of larger payloads allows them to be launched very cheaply, opening up access to space for academic institutions, small businesses, and even crowdfunded projects.

SpaceX and its rivals

Meanwhile, the launch industry too has been shaken up by the arrival of newcomers. Several Internet entrepreneurs, including Paypal co-founder Elon Musk (see panel, right) and Amazon's Jeff Bezos, have used their wealth to fund development of private space enterprises. Musk's SpaceX in particular has set out to challenge the traditional aerospace companies in competition for commercial and government launch contracts. Boosted by the award of a delivery contract to the ISS, its Falcon series of rockets have grown steadily in capacity and sophistication. They have also tackled the holy grail of launch systems – reusability.

The principle behind SpaceX's remarkable achievements in this field is to retain fuel in the rocket's lower stage after separation, and then use onboard guidance to return for a vertical landing, either at the home spaceport or on a remote-controlled drone barge. After several tests, the first successful landing and recovery came in December 2015, with the first launch of a recycled Falcon 9 first stage in March 2017. SpaceX has also been developing spacecraft – the unmanned Dragon capsules, which have delivered cargo to the ISS since 2012, and the human-rated Dragon 2, designed to carry people, which should start taking US astronauts to the station in 2019 and will see a return of orbital space tourism.

Other private launch companies have, so far, been somewhat less centre-stage. Jeff Bezos's Blue Origin has been developing rocket engines for use in other companies' launch vehicles, while also working on its

BIOGRAPHY

ELON MUSK

South African-born US citizen Elon Musk (b.1971) dropped out of a PhD at Stanford University, California, USA, to pursue a business career and made billions as a software and Internet pioneer in the late 1990s and early 2000s. He returned to his passion for engineering with firms such as Tesla, SpaceX, and the Boring Company. Widely acclaimed as a technological innovator, he has cultivated a public image through appearances in popular media and stunts, such as launching his own Tesla car into space, while also proving to be a master of politics to win US government contracts and support.

own reusable spaceflight systems – the suborbital New Shepard and the orbital New Glenn. Despite successful tests, however, schedules have slipped and plans for manned flights have been delayed.

While Bezos has the capital and commitment to ride out the inevitable uncertainties of the space business, some "new space" enterprises are less well placed, and reliance on venture capital in particular has led to many promising projects falling by the wayside. Others, however, have flourished – for example, Rocket Lab (see p.239) began to offer its own commercial launches from New Zealand with its Electron rocket in June 2018, while Sierra Nevada Corporation is pursuing development of a small lifting-body spaceplane called Dream Chaser for missions to low Earth orbit.

Falcon Heavy and beyond

One intriguing question, of course, is how commercial spaceflight will extend beyond Earth orbit. Private companies have long considered exploiting the potential metallic wealth of asteroids (which have been largely unaffected by the geological processing that makes Earth's precious elements so hard to extract). Meanwhile, the spectacular 2018 debut of the SpaceX Falcon Heavy rocket showed for the first time that a commercial enterprise could launch payloads to explore the wider Solar System. SpaceX's current ambitions seem almost boundless, with development of the even larger BFR (Big Falcon Rocket) proceeding apace. This huge launch system comprises a 58-m (190-ft) reusable first stage and a second stage integrated to an interplanetary spaceship. Whether SpaceX will fulfil its extraordinary plan for a human colony on Mars by the late 2020s, however, remains to be seen.

FALCON HEAVY RETURN
Just as impressive as Falcon Heavy's launch was the simultaneous return of the two Falcon 9 "booster" rockets. Only the central core rocket was lost, due to an engine problem as it approached its drone barge.

FALCON HEAVY LAUNCH
The maiden voyage of SpaceX's Falcon Heavy launch vehicle took place from Kennedy Space Center's Pad 39A on 6 February 2018. Three identical Falcon 9 first stages and a small upper stage combined to create the most powerful rocket seen since the Soviet Energia.

Out of the cradle

What is the future of humanity in space? Will mass space travel and the colonization of other worlds ever become a reality, or will space travel remain forever the preserve of an elite few?

More than a century ago, Konstantin Tsiolkovskii was already certain of the answer – he spoke eloquently of mankind's need to spread its wings and journey to other worlds. Oberth, Goddard, von Braun, and Korolev were among the many who agreed with him and strove to fulfil his original dreams. And yet, just as technology reached a point where such dreams could be a reality, political realities intervened and the future was stolen away.

Today, there are many who argue that the manned exploration of space is an expensive diversion from problems on Earth – wars, poverty, and environmental crisis. Ironically, the revolution in ecological awareness that created a generation of critics can be traced in part to 1968, when those first famous pictures of the Earth alone in space were sent back by the astronauts of Apollo 8.

By the 1990s, manned spaceflight risked becoming an irrelevance. Even many enthusiasts felt that the retreat to Earth orbit was a diversion from the real aims of exploring the Solar System and settling other worlds. Robert Zubrin (see p.305) argued that such an approach was like abandoning exploration of the Americas after Columbus.

However, the sudden appearance of space tourism and the redirection of NASA to manned exploration beyond Earth orbit have seen the start of what may be a rapid change. Continuing Zubrin's analogy, enthusiasts for this new age of democratic space exploration have pointed out that it also took several decades after those first pioneering voyages for the settlement and exploration of the New World to begin in earnest. Today, NASA has plans for its first moonbase and a future directive to continue to Mars. China is only taking its first steps into space, yet it is already looking as far afield as Mars and Saturn. And the enterprising space tourism community plans to follow its first wave of suborbital vehicles with spacecraft capable of reaching orbit and docking with existing or future space stations.

LIVING IN ORBIT
In one futuristic space habitat design, a pair of rotating cylinders 22km (13.7 miles) long and 6.2km (3.9 miles) across could support up to 20 million people continuing an independent civilization in space.

AMONG THE STARS
Physicist, historian of science, and one of the first to propose large-scale space habitats, Irishman John Desmond Bernal (1901–71) predicted that one day the human race might split irrevocably into two species: Earthkind and Spacekind.

Visions of the future

By the centenary of Sputnik 1, there may well be orbiting hotels and commercial flights to the Moon. The national space agencies may have established a base on Mars, and manned expeditions may be venturing further afield. Commercial exploration of the Moon and asteroids is already being planned – our satellite is rich in mineral resources, and any major space construction project would almost certainly use materials mined there instead of those launched at great expense from Earth. The asteroids, while harder to reach, are an even richer potential source of valuable minerals and metals.

We may never see space stations as ambitious as the space habitats suggested by the American physicist Gerard O'Neill (see panel, left), but self-sustaining colonies on other worlds seem increasingly plausible as we learn more about the resources

BIOGRAPHY

GERARD O'NEILL

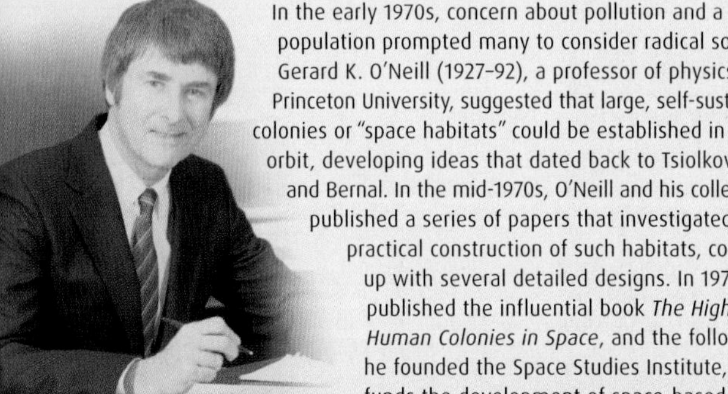

In the early 1970s, concern about pollution and a growing population prompted many to consider radical solutions. Gerard K. O'Neill (1927–92), a professor of physics at Princeton University, suggested that large, self-sustaining colonies or "space habitats" could be established in Earth orbit, developing ideas that dated back to Tsiolkovskii and Bernal. In the mid-1970s, O'Neill and his colleagues published a series of papers that investigated the practical construction of such habitats, coming up with several detailed designs. In 1976 he published the influential book *The High Frontier: Human Colonies in Space*, and the following year he founded the Space Studies Institute, which funds the development of space-based industry.

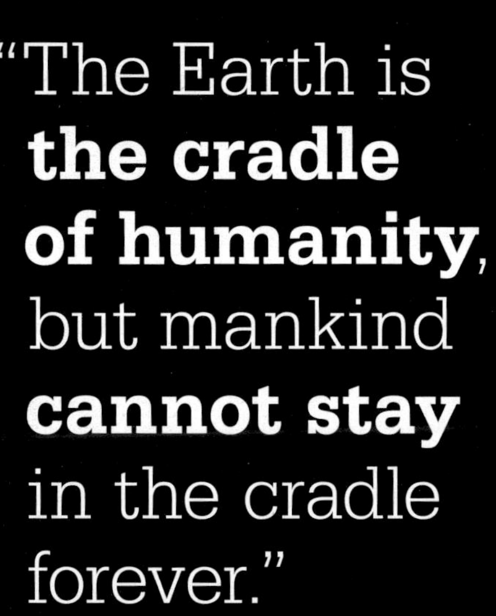

"The Earth is **the cradle of humanity**, but mankind **cannot stay** in the cradle forever."

Konstantin Tsiolkovskii, 1903

scattered across the Solar System. It may even be possible one day to terraform other worlds – seeding them with bacteria and gases that turn them eventually into hospitable environments, just as once happened on Earth itself.

Perhaps the ultimate dream is to venture across the stars and become truly independent of Earth and the Solar System. The distances involved are astronomical in the truest sense of the word – even the nearest star is 8,900 times more distant than Neptune. Expeditions to the stars may not happen for centuries and, if and when they do, they will almost certainly involve technologies that are currently in their infancy (such as nuclear propulsion – see panel, right) or not yet thought of. But as renowned cosmologist Stephen Hawking has said, the colonization of space may ultimately be the only way to ensure the long-term future of the human race.

TECHNOLOGY

NUCLEAR-POWERED SPACE TRAVEL

A journey to the stars in any reasonable timescale would require a new propulsion system – neither chemical rockets nor ion engines are up to the job. In the 1950s, US physicists Theodore Taylor and Freeman Dyson showed how a spacecraft might be powered instead by a series of explosions from small nuclear bombs detonating in its engine. A prototype of their Project Orion was tested with chemical explosives, but the use of nuclear material was so controversial that the concept was shelved. In the 1970s, the idea was revived by the British Interplanetary Society in Project Daedalus, the first detailed study of a practical starship design.

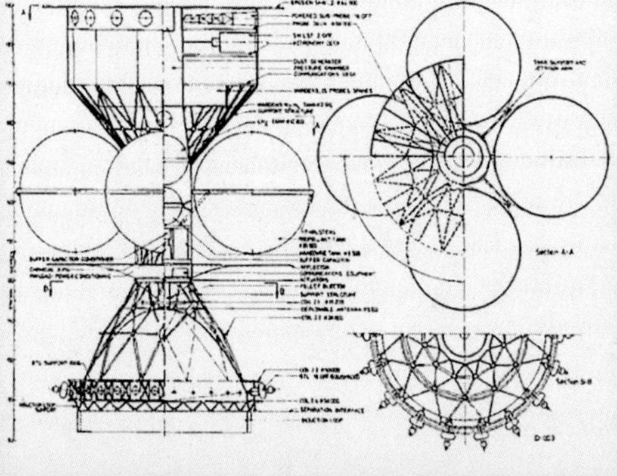

GLOSSARY

A4 The original designation of the early German rocket that flew as the V-2 missile.

ablative heat shield *see* **heat shield**

aerospaceplane A spaceplane designed to operate in the Earth's atmosphere using an alternative to rocket propulsion – usually a ramjet or scramjet.

Aerozine A rocket fuel consisting of a mixture of hydrazine and unsymmetrical dimethylhydrazine (UDMH).

Agena An upper rocket stage used on Thor, Atlas, and Titan launch vehicles and also as a docking target for several Gemini missions.

Almaz A Soviet military space-station design, flown as Salyuts 2, 3, and 5.

apogee The point in the orbit of a satellite or spacecraft where it is furthest from the Earth.

Ariane A series of European launch vehicles, operating since 1979 and widely used for commercial and scientific satellite launches.

Atlas A long-running US launch-vehicle series, originating from the first US Air Force Intercontinental Ballistic Missile.

attitude The orientation of a spacecraft or space station in space. Attitude adjustments can be made in roll, pitch, or yaw axes.

Baikonur Cosmodrome The main launch centre for the Soviet and Russian space programmes, located in Kazakhstan near the town of Tyuratam (originally named to deceive intelligence agencies into believing it was near the town of Baikonur itself).

ballistic A term used to describe a missile or spacecraft that makes its descent through the atmosphere under the influence of gravity and atmospheric drag alone, with no aerodynamic lift; the term also describes the behaviour of projectiles such as cannonballs.

boilerplate capsule A full-sized but not fully equipped replica of a finished spacecraft design, used in the early stages of testing for studying aerodynamic characteristics and other properties.

booster A small rocket attached to the side of a larger rocket stage to give extra thrust during launch.

Capcom An abbreviation for Capsule Communicator, normally the only person at NASA Mission Control who communicates with astronauts in space. Typically, the role is taken by a trained astronaut.

Centaur A type of upper rocket stage used to launch spaceprobes and satellites. The Centaur was the first rocket to successfully use high-energy cryogenic propellants.

Chang Zheng *see* **Long March rocket**

CM Abbreviation for the Command Module of the US Apollo spacecraft.

combustion chamber The part of a rocket engine where the fuel and the oxidant mix and combust, generating thrust against the forward-facing part of the chamber as the exhaust escapes from the nozzle at the rear.

comsat An abbreviation of communications satellite – a satellite used for receiving and re-transmitting signals to and from ground stations.

Cosmos A long-running series of Soviet and Russian satellites, comprising several different concealed programmes and often used to hide failed missions in other series.

cryogenic propellant A rocket propellant (fuel or oxidant) that must be stored at extremely low temperatures and which usually requires ignition in order to start a chemical reaction. Despite the problems in handling them, cryogenic propellants can be extremely powerful.

CSM Abbreviation for the combined Command and Service Module of the US Apollo spacecraft.

CZ *see* **Long March rocket**

Delta A long-running and highly successful series of US launch vehicles.

DOS A Russian acronym for Permanent Orbital Station, the Soviet space-station design developed in the late 1960s by the Korolev design bureau as an adaptation of the military Almaz station.

drogue parachute A small parachute used to slow a spacecraft down, usually directly after re-entry to the atmosphere and before the main parachute opens.

elliptical orbit An orbit with the shape of an ellipse (a "stretched circle"). As well as a centre, an ellipse has two foci, with the centre of mass being orbited at one focus. Because an orbiting object's speed is dependent on its distance from the mass that it orbits, it moves more slowly at one end of the ellipse than at the other.

Energia The Soviet/Russian space company formed from the former OKB-1 design bureau of Sergei Korolev. Also a heavy-lift rocket produced by the company for launching the Soviet Buran space shuttle.

equatorial orbit An orbit directly above the Earth's equator. Equatorial orbits are comparatively easy to reach because the Earth's rotation gives rockets an immediate boost if they are launched eastwards from on or close to the Earth's equator.

ESA The European Space Agency, formed from the merger of the European Launcher Development Organization (ELDO) and the European Space Research Organization (ESRO) in 1975.

escape velocity The speed at which a spacecraft must travel if it is to escape the Earth's gravitational field – 11.2km (7 miles) per second. It is not necessary to reach escape velocity in order to orbit the Earth.

ET Abbreviation for the large external fuel tank of the Space Shuttle.

flight deck In the Space Shuttle orbiter, the upper deck containing flight controls and seating for most of the crew during launch.

fuel One element of a spacecraft's propellant. The fuel mixes with an oxidant and combusts to create exhaust that pushes the spacecraft forward.

g force A measure of acceleration forces. 1*g* is typical Earth gravity, but during launch and re-entry spacefarers experience accelerations of several *g*.

geostationary orbit Also known as geosynchronous orbit. An orbit in which a satellite above the Earth's equator moves around the planet in the same direction as the Earth's rotation and with the same period (roughly 23 hours, 56 minutes). This means that the satellite remains over the same point on the equator and occupies a fixed point in the sky as seen from Earth. Geostationary satellites are ideal for weather-observation and comsats.

GPS An abbreviation for Global Positioning System – a network of satellites that allows a computerized Earthbound receiver to work out its position by receiving signals from three or more satellites in orbit. The original GPS system uses US NAVSTAR military satellites, but rival systems, including the Russian GLONASS and the European Galileo, are also often known simply as GPS.

gravitational slingshot *see* **gravity assist**

gravity An attractive force generated by a massive object, which pulls other objects towards it or holds them in orbit.

gravity assist A technique used to speed up and change the direction of a spaceprobe without burning fuel, by flying close to and "borrowing" a small amount of energy from a planet or moon.

ground station A radio receiving dish for communication with spacecraft, satellites, and spaceprobes.

Guiana Space Centre The launch site for the European Space Agency's Ariane rockets, at Kourou, French Guiana.

heat shield A protective layer that shields a spacecraft from the heat of re-entry. Most heat shields are ablative – they burn away during re-entry, carrying the heat away from the spacecraft. Other systems, such as the Space Shuttle orbiter's ceramic tiles, must absorb heat without transmitting it to the hull.

H-series rocket Japan's most widely used launch vehicle.

hydrazine A toxic chemical, commonly used as a rocket fuel because of its violent and spontaneous chemical reaction with many oxidants. It is used in the Space Shuttle's Auxiliary Power Units.

hypergolic propellant A rocket fuel that reacts spontaneously with its oxidant (avoiding the need for an ignition system), and which can usually be stored at relatively normal temperatures.

inertial guidance A guidance system that uses gyroscopes and accelerometers to calculate a vehicle's position and motion by dead reckoning (a method of navigation in which position is determined relative to a known point of departure using measurements of speed, heading, and time).

Inertial Upper Stage A large independent rocket stage used for putting satellites or other payloads into their final orbit or escape trajectory after they have been deployed to low Earth orbit by the Space Shuttle or another launch vehicle.

ion engine A propulsion system that uses the ionization of a chemical propellant in a strong electric field in order to generate thrust. Ion engines are very efficient but produce very small amounts of thrust for very long periods, contrasted with chemical rockets that produce large amounts of thrust for brief periods. They are usually powered by solar arrays.

JAXA The Japan Aerospace Exploration Agency, Japan's space agency, founded in 2003 from the merger of the Institute of Space and Astronautical Science (ISAS), the National Aerospace Laboratory of Japan (NAL), and the National Space Development Agency (NASDA).

Johnson Space Center (JSC) The site of NASA's main Mission Control and many other elements of its manned spaceflight programme, at Houston, Texas.

Juno An adapted version of the Jupiter-C used to launch some of the first US satellites.

Jupiter-C A modified Redstone missile used to carry the warhead of a Jupiter missile into space for re-entry tests.

Kennedy Space Center (KSC) The main US launch complex at Cape Canaveral in Florida. The Cape itself was known as Cape Kennedy between 1963 and 1973 in memory of the assassinated US President.

kick motor A small rocket motor built into a satellite and used to move it from low Earth orbit to its final location.

Korabl Sputnik Any of the later Soviet Sputnik satellite launches (4 onwards) that were in fact unmanned tests of Vostok spacecraft.

launch vehicle A complete vehicle, usually consisting of several rocket stages, boosters, and perhaps other components, used to launch payloads into space (often simply referred to as a rocket).

LH2 An abbreviation for liquid hydrogen, a powerful cryogenic fuel.

lifting body An aircraft or spaceplane that has only small wings, if it has wings at all. Lifting bodies rely on the shape of the fuselage to generate aerodynamic lift – they are typically triangular, with convex upper or lower hulls.

liquid-fuelled rocket A rocket in which fuel and oxidant are mixed together and react explosively, creating an expanding mixture of exhaust gases that escape through an exhaust nozzle. The reaction against the escaping gases pushes the rocket forwards. Liquid-fuelled rockets are more complex than solid-fuelled ones, but they are also more versatile, since the flow of fuel can be throttled, stopped, and restarted.

LM An abbreviation for the Apollo spacecraft Lunar Excursion Module (also LEM) – the spiderlike lander that actually put astronauts on the Moon.

Long March rocket A series of Chinese launch vehicles, used in manned and unmanned space programmes.

low Earth orbit An orbit a few hundred kilometres above the Earth, often abbreviated to LEO. Low Earth orbits are typically used by manned spacecraft and space stations, Earth-observing satellites, and as a temporary orbit for satellites later launched into higher orbits by an Inertial Upper Stage, a Payload Assist Module, or a kick motor.

LOX An abbreviation for liquid oxygen, a powerful cryogenic oxidant.

Marshall Spaceflight Center (MSFC) The principle US centre for launch-vehicle development and testing, developed from the US Army's former Redstone Arsenal facility at Huntsville, Alabama.

mass A property of the amount of material present in an object. Mass is unaffected by a gravitational field, unlike weight.

microgravity The term for conditions experienced in orbit – although the effects of gravity are much reduced, they are almost never completely absent.

mid-deck The lower habitable deck of the Space Shuttle orbiter, where equipment and, sometimes, experiments are stored.

MKS A Russian abbreviation for Reusable Space System, the Soviet attempt to develop a reusable space shuttle, also known as Buran.

Molniya orbit A highly elliptical, inclined orbit typically used by communications satellites for countries at high latitudes and named after the Soviet Molniya comsat system. A Molniya orbit sees a satellite spend a large amount of time visible in the skies of a particular part of the Earth, so that it can be easily tracked by ground stations.

monopropellant A class of rocket propellants that can act as both fuel and oxidant in the right conditions – one example is hydrogen peroxide.

multispectral imaging A technique used by remote-sensing satellites and other spacecraft that involves photographing areas at different wavelengths of light (different colours) and analyzing the images to bring out hidden features and reveal surface composition.

N_2O_4 Dinitrogen tetroxide, a commonly used hypergolic propellant that functions as an oxidant.

NASA The National Aeronautics and Space Administration, the US space agency, established in 1958 as successor to NACA, the National Advisory Committee for Aeronautics.

nozzle The exhaust outlet from the combustion chamber of a rocket. Exhaust gases typically escape from the combustion chamber at high temperatures through a narrow opening – the bell-shape of the nozzle forces the gases to expand rapidly, cooling them but increasing their speed so that they leave the rocket at up to ten times the speed of sound.

nuclear propulsion A theoretical propulsion system that would use the explosions of countless small nuclear devices to push a large spacecraft forwards. Nuclear propulsion is one potential way of accelerating a future starship to very high speeds.

OKB-1 The design bureau run by Sergei Korolev from 1946 until his death in 1966. OKB-1 (sometimes known simply as Korolev) was responsible for much of the Soviet space effort. Since 1974 it has been known as Energia.

OMS The Orbital Maneuvering System, a pair of medium-sized rocket engines at the rear of the Space Shuttle orbiter that are used for adjusting the spaceplane's orbit and as retrorockets for re-entry.

orbit A path that one object follows around another, more massive one due to the force of gravity. An orbit traces a path through space along which the tendency of the object to fly off in a straight line is precisely balanced by the inward gravitational pull of the more massive object.

orbiter The proper name for the spaceplane element of the Space Shuttle system – the orbiter is the vehicle that reaches space, carries out its mission, and then returns to Earth.

oxidant A chemical, used as a rocket propellant, that undergoes a chemical reaction (combustion) with a fuel to generate exhaust gases and push a rocket forwards. Unlike other types of engine, rockets require an oxidant as well as fuel because they must operate in a vacuum – other engines use oxygen from the atmosphere to burn their fuels.

payload The cargo that a launch vehicle delivers into orbit.

Payload Assist Module An independent rocket engine (smaller than an Inertial Upper Stage) attached to the base of a satellite released from the Space Shuttle, which is used to put the satellite into its final orbit.

perigee The point in the orbit of a satellite or spacecraft at which it comes closest to the Earth.

perihelion The point on its orbit when a planet or other body comes closest to the Sun.

pitch The rotation of a spacecraft about its lateral (side-side) axis – for example, the angle from nose to tail of a Space Shuttle.

Plesetsk Cosmodrome The northern launch site for Soviet and Russian rockets, located close to the Arctic Circle and ideal for launching rockets into polar and high-inclination orbits.

polar orbit An orbit around the Earth that passes over (or very close to) the planet's poles. Typically used by Earth-observing satellites, a polar orbit allows the satellite to fly over most of the Earth's surface as our planet rotates beneath it each day.

Proton rocket A Soviet heavy-lift rocket used for launching heavy unmanned payloads such as space-station components.

R-7 A Soviet ballistic missile developed by Sergei Korolev that forms the basis of Soviet launch vehicles such as the Sputnik, Vostok, and Soyuz rockets.

ramjet A jet engine with few moving parts and no turbine, in which the speed of the aircraft through the atmosphere forces air into the engine at high pressure. Fuel is then added, with the combustion producing forward thrust. Ramjets are a key element of many aerospaceplane concepts, but they only function efficiently at supersonic speeds.

RCS An abbreviation of Reaction Control System, a series of small rocket engines (thrusters) scattered over the surface of a spacecraft and used for adjusting its attitude in yaw, roll, and pitch axes.

Redstone An American ballistic missile, developed by Wernher von Braun at Redstone Arsenal, which formed the basis for many early US launch vehicles.

re-entry The return of a spacecraft or other object into the Earth's atmosphere, during which it may be heated to extreme temperatures by friction with air molecules.

remote sensing The scientific study of the Earth from space.

retropack A set of discardable retrorockets strapped over the heat shield of the Mercury space capsule in order to slow it for re-entry.

retrorocket A rocket system used for slowing a spacecraft down rather than accelerating it. Retrorockets are used to begin re-entry to the Earth's atmosphere, or to slow spaceprobes down when they arrive at their destination.

rocket A propulsion system that drives a vehicle forwards through the principle of action and reaction and is capable of working in a vacuum. The term is also used casually to refer to entire launch vehicles.

roll The rotation of a spacecraft about its longitudinal (front-back) axis – for example, the tilt of the Space Shuttle's wings.

RP-1 A form of kerosone used as fuel in US rockets.

RSA An abbreviation for the Russian Federal Space Agency, formed in the early 1990s to manage various aspects of the former Soviet Union space programme. It is also known as Roskosmos.

RTG An abbreviation for Radioisotope Thermoelectric (or Thermal) Generator. An RTG uses heat produced by a sample of radioactive material to generate electricity for spaceprobes travelling in distant parts of the Solar System.

Salyut A series of Soviet space stations, incorporating both DOS (Salyuts 1 and 4) and Almaz (Salyuts 2, 3, and 5) stations, along with more advanced designs (Salyuts 6 and 7).

satellite Any object that moves around a more massive one due to the effect of its gravitational attraction. Satellites may be either natural (moons of the various planets) or artificial.

Saturn rockets A series of US heavy-lift launchers developed by Wernher von Braun in the early 1960s. Saturn I was based on a cluster of Redstone-type rockets, while the Saturn V used massive new engines and high-energy cryogenic propellants. Saturn IB was a hybrid, based on Saturn I but with an upper stage borrowed from the Saturn V.

scramjet A modified ramjet design in which combustion happens while the fuel and air are moving at supersonic speed.

shaft and trunnion NASA's term for the two axes in which the telescope and sextant of the Apollo guidance computer could be moved for targeting the Sun, stars, and other astronomical objects.

solar array A panel-like or wing-like arrangement of solar cells that converts sunlight into electricity for use by spacecraft.

solar sail An experimental propulsion method that uses the pressure of radiation from the Sun to push a spacecraft forwards. Solar sails are only capable of low acceleration but can reach very high speeds.

solid-fuelled rocket A rocket in which fuel and oxidant are mixed (usually with other chemicals) and stored in a solid state. When ignited, the rocket burns like a firework and gases escape through an exhaust nozzle, pushing it forwards. A solid-fuelled rocket can only be ignited once.

Soyuz A long-running Soviet spacecraft series, first launched in 1967 and later upgraded to Soyuz-T (1980), Soyuz-TM (1987), and Soyuz-TMA (2002).

Soyuz rocket A reliable Soviet/Russian rocket, derived from the R-7 missile and used to launch Soyuz spacecraft.

spaceplane A spacecraft with aerodynamic properties that allow it fly like an aircraft or glider for at least part of its time in the atmosphere.

spaceprobe An automatic vehicle sent to explore the Solar System away from Earth. Spaceprobes can include flyby missions, orbiters, and landers.

Sputnik rocket A Soviet rocket developed by Sergei Korolev from the R-7 missile and used to launch the first satellites.

SRB An abbreviation of Solid Rocket Booster, the rockets that assist the Space Shuttle during launch.

SSME An abbreviation of Space Shuttle Main Engine, the engines on the back of the Shuttle orbiter that burn fuel from the External Tank during launch.

stage A section of a launch vehicle that burns its fuel and then separates and falls away from the rest of the vehicle.

steering vane A movable deflector that can affect the path of exhaust from a rocket engine, controlling the direction in which the vehicle moves.

STS An abbreviation for Space Transportation System, the official name of the US Space Shuttle. Each Shuttle mission is given an STS designation followed by a number (though the numbers do not necessarily indicate launch order).

Sun-synchronous orbit A polar orbit that also circles the Earth's equator once a year, keeping pace with the angle of the Sun so that the ground below is illuminated at a constant angle.

telemetry A stream of data sent automatically from a spacecraft to Earth, containing information about the status of its onboard systems.

thrust The forward force generated by a rocket engine, often measured in kilograms-force (kgf). One kilogram-force is the force exerted by a weight of one kilogram in Earth gravity, equivalent to 9.81 newtons (the official SI unit of force).

thruster A small rocket engine used, for example, in attitude adjustments, such as in an RCS.

thrust structure A structure above a rocket engine that takes the brunt of the engine's forward thrust, thereby preventing it from pushing into its own fuel tanks.

Titan A long-running series of US launch vehicles, originating in the US ballistic missile programme.

TKS A Soviet ferry spacecraft that was developed for use with the Almaz space stations but was eventually used as the basis for several of the modules on the Mir station.

translunar Literally "moon-crossing" – the path taken by a spacecraft from the Earth to the Moon.

turbopump A high-speed pump that supplies fuel and propellant from the tanks of a liquid-fuelled rocket to the combustion chamber.

UDMH Unsymmetrical dimethylhydrazine, a widely used hypergolic rocket fuel.

V-2 The first large liquid-fuelled rocket, designed by Wernher von Braun and used as a missile by Germany during the Second World War.

Vernier engine A small rocket engine on a movable gimbal, mounted away from the main engines and used to steer a launch vehicle.

VfR An abbreviation of *Verein für Raumschiffahrt*, the German rocketry society of the 1930s.

Vostok rocket A Soviet launch vehicle, derived from the R-7 missile, used to launch the first manned spacecraft.

weight The force that acts on an object with mass in a gravitational field – while an object's mass remains constant, its weight may vary depending on the strength of local gravity.

weightlessness The condition of "free fall" experienced by people and objects when the effects of gravity are cancelled out in orbit.

yaw The rotation of a spacecraft around its vertical axis, the "crossways" orientation of a vehicle such as the Space Shuttle.

zero gravity *see* **microgravity**

INDEX

Page numbers in italic refer to illustrations.

A

B

C

D

E

M

N

O

P

R

S

ACKNOWLEDGMENTS

Author's acknowledgments
I'd like to thank everyone at DK, MP3, and elsewhere who helped turn this book from my personal hobby-horse into a reality. If I try to list them all I'll be sure to miss someone out, but particular thanks are due to Liz Wheeler and Peter Frances at DK and David Preston at MP3, without whom it really wouldn't have happened. I should give a special mention to everyone at South Florida Science Museum, Kennedy Space Center, and the US Space and Rocket Center, for their help and hospitality during our mad photographic dash around the United States.

Thanks also to friends and family for their interest, encouragement, and patience when I dropped off the radar during long periods of writing.

And I'd like to dedicate this book to my Mum and Dad – my never-failing sources of support and encouragement.

Publisher's acknowledgments
DK would like to thank Tamlyn Calitz for her editorial contribution. Miezan van Zyl, Rebecca Warren, Rob Houston, Ed Wilson, and Manisha Thakkar also gave editorial assistance. Initial presentation design work was done by Mark Lloyd at On Fire and Peter Laws. Jim Jackson did additional design work, and John Goldsmid provided DTP design help. For their work on the second edition, DK would like to thank Vanessa Bird for revising the index, Jamie Ambrose for proof-reading, and Bharti Bedi for editorial assistance.

MP3 acknowledgments
Many thanks to Rick Newman of HighTechScience.org for the kind loan of his space artefacts, and Elizabeth Dashiell and all the staff at the South Florida Science Museum, West Palm Beach for their assistance in photographing the collection; the Kennedy Space Center and the US Space and Rocket Center, Huntsville for granting access to their exhibits; Anatoly Zak and James Oberg for granting access to their private photography collections; Jody Russell at the Johnson Space Center and Houston One Great Photo Lab for processing help; Dave Shayler at Astro Info Service Ltd for supplying mission patches; INP Media Ltd for producing video grabs; Carole Ramsey and Sally Wortley for editorial assistance; Bob Bousfield and Jeff Carroll for DTP help.

Smithsonian Enterprises
Carol LeBlanc Senior Vice President, Consumer and Education Products
Brigid Ferraro Vice President, Consumer and Education Products
Ellen Nanny Senior Manager Licensed Publishing
Kealy Gordon Product Development Manager

Picture credits
(a=above, b=below/bottom, c=centre, f=far, fp=full page, l=left, r=right, t=top)
2–3, NASA/JPL-Caltech; **4tl**, Getty Images; **4bl**, NASA; **4bc**, NASA; **5tl**, Science Museum/Science & Society Picture Library; **5cl**, NASA; **5c**, NASA; **5**, John Frassanito & Associates Inc; **5cra**, NASA; **6–7**, NASA; **8tc**, NASA; **8bc**, Bettmann/Corbis; **9c**, NASA; **10–11**, Getty Images; **12tl**, NASA; **12bl**, NASA; **12tr**, Science Photo Library; **12cr**, Science Museum/Science & Society Picture Library; **13t**, Western Reserve Historical Society; **13bl**, NASA; **13br**, Bettmann/Corbis; **14tr**, NASA; **14bl**, Mary Evans Picture Library; **14bc**, Mary Evans Picture Library; **14br**, Mary Evans Picture Library; **14–15c**, Mary Evans Picture Library; **15tr**, NASA; **15cl**, Mary Evans Picture Library; **16**, RIA Novosti/Science Photo Library; **17tr**, RIA Novosti/Science Photo Library; **17cr**, NASA; **17br**, Hugh Johnson; **18cl**, Bettmann/Corbis; **18bc**, NASA; **18–19c**, NASA; **19t**, Bettmann/Corbis; **19c**, NASA/Science Photo Library; **19b**, Hugh Johnson; **20tl**, Austrian Archives/Corbis; **20bl**, NASA; **20bc**, Hulton Archive/Getty Images; **20br**, UFA/The Kobal Collection; **20–21t**, Akg-Images, London; **21tr**, Bettmann/Corbis; **21br**, Hulton-Deutsch Collection/Corbis; **22l**, RIA Novosti/Science Photo Library; **22tr**, Science Museum/Science & Society Picture Library; **22br**, V2rocket.com; **23bc**, RIA Novosti/Science Photo Library; **23r**, Hugh Johnson; **23tl**, NASA; **24tl**, Detlev van Ravenswaay/Science Photo Library; **24bl**, V2rocket.com; **25–25c**, NASA; **25c**, Science Museum/Science & Society Picture Library; **25bc**, V2rocket.com; **25tr**, Julian Baum; **25br**, Science Museum/Science & Society Picture Library; **26tl**, Hulton-Deutsch Collection/Corbis; **26b**, NASA; **27tl**, FoxPhotos/Getty Images; **27r**, Images de Dora, Le Maner & Sellier; **27b**, Images de Dora, Le Maner & Sellier; **26–27c** V2rocket.com; **28tl**, Nina Leen/Pix Inc/Time Life Pictures/Getty Images; **28l**, V2rocket.com; **28bl**, V2rocket.com; **28bc**, V2rocket.com; **28br**, V2rocket.com; **28–29t**, Corbis; **29bl**, Imperial War Museum; **29br**, V2rocket.com; **29tr**, Bettmann/Corbis; **30tl**, NOAA; **30b**, NASA; **30–31t**, Akg-Images, London; **31tr**, NASA; **31cr**, Bonestall Space/NASA; **31bl**, NASA; **32tl**, RIA Novosti/Science Photo Library; **32b**, RIA Novosti/Science Photo Library; **32br**, RIA Novosti/Science Photo Library; **33l**, NASA; **33tl**, RIA Novosti/Science Photo Library; **33tr**, RIA Novosti/Science Photo Library; **33br**, Anatoly Zak; **34tl**, USAF; **34br**, Bettman/Corbis; **35r**, Anatoly Zak; **35r**, NASA; **36tl**, USAF; **36bl**, USAF; **36t**, USAF; **36tc**, USAF; **36c**, USAF; **36cb**, USAF; **36b**, USAF; **37t**, USAF; **37cr**, USAF/Air Force Flight Test Center History Office; **37br**, USAF; **38tl**, Bettman/Corbis; **38bc**, NASA; **38tr**, US Army; **38–39c**, NASA; **39tr**, RIA Novosti/Science Photo Library; **39b**, Terry Pastor (The Art Agency); **40–41**, NASA; **42bl**, Novosti; **42br**, Bettmann/Corbis; **42tr**, Hank Walker/Time Life Pictures/Getty Images; **43bc**, RIA Novosti/Science Photo Library; **43cr**, Nat Farbman/Time Life Pictures/Getty Images; **43bc**, Dmitri Kessel/Life Magazine/Copyright Time Life Pictures/Getty Images; **44bl**, NASA; **44cb**, Bettmann/Getty Images; **44–45c**, NASA; **45tc**, NASA; **45br**, Hugh Johnson; **46bl**, Novosti/Science Photo Library; **46br**, Bill Bridges/Time Life Pictures/Getty Images; **46–47t**, Bettmann/Corbis; **47tc**, Blank Archives/Getty Images; **47tr**, Fox Photos/Getty Images; **47bc**, RIA Novosti/Science Photo Library; **48bc**, NASA; **48tr**, IMP BV; **49tl**, NASA; **49c**, NASA; **49tr**, NASA; **49bl**, NASA/Science Photo Library; **49b**, Hugh Johnson; **49br**, NASA; **50bl**, Keystone/Getty Images; **50br**, Hulton-Deutsch Collection/Corbis; **50–51t**, NASA; **51br**, Hugh Johnson; **52b**, Hugh Johnson (background)/Terry, Pastor (The Art Agency); **53t**, NASA; **53bl**, Novosti; **53br**, mentallandscape.com; **54bl**, Bonestell Space Art; **54br**, Andy Salmon; **54–55t**, Galaxy Picture Library; **55tc**, Galaxy Picture Library; **55tr**, Galaxy Picture Library; **55c**, Galaxy Picture Library; **55br**, NASA/Science Photo Library; **56b**, Genesis Picture Library; **56tr**, RAF; **57tc**, CNES; **57c**, SIRPA-ECPA 1995/ESA; **57bl**, Science Museum/Science & Society Picture Library; **57tr**, Anatoly Zak; **57br**, NASA/Science Photo Library; **58tr**, Bettmann/Corbis; **58bl**, Novosti; **58br**, Novosti; **59**, VideoCosmos; **60bl**, Novosti; **60bc**, Novosti; **60br**, Bettmann/Corbis; **60–61c**, Novosti; **61tc**, Bettmann/Corbis; **61br**, Novosti; **62cl**, Anatoly Zak; **62bl**, Anatoly Zak; **62c**, Anatoly Zak; **62tr**, RIA Novosti/Science Photo Library; **62br**, Hugh Johnson; **63l**, Hugh Johnson; **63tr**, Novosti-Kosmonautiki; **63c**, Anatoly Zak; **63cr**, Novosti-Kosmonautiki; **64tl**, Anatoly Zak; **64b**, VideoCosmos; **64–65b**, Anatoly Zak; **64tr**, Anatoly Zak; **65tl**, VideoCosmos; **65tc**, Anatoly Zak; **65tr**, Anatoly Zak; **65br**, Anatoly Zak; **66bl**, NASA; **66br**, NASA; **67**, NASA; **68l**, NASA; **68br**, NASA; **69t**, USAF; **69c**, NASA; **69bl**, NASA; **70tl**, NASA; **70l**, NASA; **70c**, NASA; **70bc**, NASA; **70br**, NASA; **70–71**, Julian Baum; **71tc**, NASA; **71tr**, NASA; **71br**, Hugh Johnson; **72–73**, NASA; **74tr**, NASA; **74bc**, NASA; **74–75c**, NASA; **75tr**, NASA; **75cr**, National Geographic Image Collection; **75br**, NASA; **76**, Novosti/Science Photo Library; **77l**, RIA Novosti/Science Photo Library; **77r**, Russian State Archive; **78tl**, Novosti; **78bl**, NASA; **78bc**, Roger Ressmeyer/Corbis; **78–79t**, Novosti; **78–79c**, Novosti; **78–79bc**, Novosti; **78–79b**, Novosti; **79bc**, Science Museum/Science and Society Picture Library; **79cr**, Novosti; **80bl**, Novosti; **80bc**, IMP BV; **80br**, Novosti; **80cr**, Novosti/Science Photo Library; **80–81tc**, Hulton-Deutsch, Collection/Corbis; **81bl**, Novosti; **81bc**, Novosti; **81br**, Bettmann/Corbis; **81tr**, Novosti; NASA; **82tr**, Ralph Morse/Time Life Pictures/Getty Images; **83**, Ralph Morse/Time, Life Pictures/Getty Images; **84tl**, Dean Conger/Corbis; **84l**, Ralph Morse/Time Life Pictures/Getty Images; **84cl**, Ralph Morse/Time Life Pictures/Getty Images; **84cr**, NASA; **84r**, NASA; **84–85tc**, NASA/Getty Images; **85bl**, NASA; **85br**, MPI/Getty Images; **86bl**, NASA; **86bc**, NASA; **86br**, NASA; **86tr**, NASA; **86–87t**, NASA; **87tr**, NASA; **87cr**, NASA; **87br**, NASA; **88**, NASA; **89tl**, NASA; **89bl**, NASA; **89tl**, NASA; **90–91**, NASA; **92bl**, Galaxy Picture Library; **92bc**, Galaxy Picture Library; **92tr**, NASA; **92–93t**, NASA; **93br**, NASA; **94tr**, NASA; **94cl**, NASA; **94br**, NASA; **94–95c**, Julian Baum; **95bl**, NASA; **95tr**, NASA; **95bc**, NASA; **95br**, NASA; **96br**, RIA Novosti/Science Photo Library; **96–97t**, RIA Novosti/Science Photo Library; **97cr**, Russian Space Archive; **97bc**, Novosti; **98tl**, NASA; **98bl**, Bettman/Corbis; **98tr**, NASA; **98br**, NASA; **99**, NASA; **100–101**, NASA; **102–103c**, NASA; **103tc**, Bettmann/Corbis; **103bl**, Bettmann/Corbis; **104tl**, AP/Empics, **104tc**, Novosti; **104–105c**, Bettmann/Corbis; **104tr**, AP/Empics; **105tl**, Novosti; **105tc**, Empics; **105tr**, Empics; **105cr**, Novosti; **105bl**, Time Life, Pictures/Getty Images; **105bc**, Rykoff Collection/Corbis; **106–107**, Bettmann/Corbis; **107t**, Bettmann/Corbis; **107bl**, Lynn Pelham/Time Life Pictures/Getty Images; **108–109**, NASA; **108bl**, NASA; **108tr**, Hugh Johnson; **109tl**, NASA; **109tc**, NASA; **109tr**, NASA; **110bl**, Time Life Pictures/NASA/Time Life Pictures/Getty Images; **110r**, NASA; **111t**, NASA; **111bl**, NASA; **111bc**, NASA; **111br**, NASA; **112tr**, NASA; **112bl**, NASA; **112bc**, NASA; **112br**, NASA; **113t**, NASA; **113bl**, NASA; **113cr**, NASA; **113br**, NASA; **116tr**, NASA; **116b**, NASA; **117t**, NASA; **117b**, Peter Bull; **118l**, NASA; **118tr**, NASA; **118br**, NASA; **119tr**, NASA; **119bl**, NASA; **119br**, NASA; **120cl**, NASA; **120tc**, NASA; **120tr**, NASA; **120c**, NASA; **120cb**, NASA; **120br**, NASA; **121tl**, NASA; **121c**, NASA; **121br**, Galaxy Picture Library; **122–123t**, Terry Pastor (The Art Agency); **122bc**, NASA; **122br**, Genesis Picture Library; **123cl**, NASA; **123bl**, NASA; **124tl**, NASA; **124tr**, NASA; **124c**, NASA; **124bl**, NASA; **124br**, NASA; **124–125c**, Julian Baum; **125tl**, NASA; **125tr**, NASA; **125bl**, NASA; **125bc**, NASA; **126bl**, Dean Conger/Corbis; **126r**, Edwin Cameron/Energia; **127tl**, Hugh Johnson; **127tr**, Novosti/Science Photo Library; **128l**, Novosti; **128tr**, Anatoly Zak; **128br**, Anatoly Zak; **129tl**, Novosti; **129tr**, Novosti; **129bl**, NASA; **129br**, Novosti; **130tc**, Anatoly Zak; **130tl**, Anatoly Zak; **130br**, Anatoly Zak; **130–131**, Chris Taylor; **131cl**, VideoCosmos; **131bl**, Anatoly Zak; **131tr**, VideoCosmos; **131cr**, VideoCosmos; **132bl**, NASA; **132cl**, NASA; **132br**, NASA; **133tr**, NASA; **133tl**, NASA; **133tr**, NASA; **133bl**, Ralph Morse/Time Life Pictures/Getty Images; **133c**, NASA; **133br**, NASA; **134tl**, NASA; **134bc**, NASA; **135bl**, NASA; **135bc**, NASA; **135br**, NASA; **136bl**, NASA; **136tr**, NASA; **137t**, NASA; **137bl**, NASA; **137bc**, NASA; **137br**, NASA; **139–139**, Time Life Pictures/NASA/Getty Images; **140b**, Galaxy Picture Library; **140bc**, NASA; **140rt**, NASA; **140rtc**, NASA; **140rc**, NASA; **140rcb**, Getty Images; **140rb**, NASA; **141tc**, NASA; **141cr**, NASA; **141br**, NASA; **142tr**, NASA; **142bl**, NASA/Galaxy Picture Library; **142bc**, NASA; **142br**, NASA; **143t**, NASA; **143br**, NASA; **144tl**, NASA; **144bl**, Science & Society, Picture Library; **144tr**, NASA; **145tl**, NASA; **145bl**, NASA; **145cr**, NASA; **145br**, NASA; **146br**, NASA; **148tc**, NASA; **148bl**, NASA; **149tl**, NASA; **149tr**, NASA; **149cr**, NASA; **149b**, Chris Taylor; **150–151t**, NASA; **150bl**, NASA; **150cl**, NASA; **150bc**, NASA; **150cr**, NASA; **150bl**, NASA; **151tr**, NASA; **151cr**, Ronald Grant Archive; **152–153**, NASA; **153c**, NASA; **153t**, NASA; **153b**, NASA; **154tl**, NASA; **154bl**, NASA; **154br**, NASA; **155**, NASA/Science Photo Library; **156tr**, NASA; **156cl**, NASA; **156c**, NASA; **156–157b**, NASA; **157tr**, NASA; **158tl**, NASA; **158cl**, NASA; **158–159**, Julian Baum, **159tl**, NASA; **159tc**, NASA; **159bc**, NASA; **159tr**, NASA; **159c**, Julian Baum; **159bc**, Julian Baum; **159br**, Julian Baum; **160–161**, NASA; **162–163**, Corbis; **162tr**, NASA; **162bc**, NASA; **163t**, NASA; **163bl**, NASA; **163cr**, NASA; **164bl**, NASA; **164–165**, NASA; **165tr**, NASA; **165cr**, NASA; **165br**, Galaxy Picture Library; **166tl**, NASA; **166tc**, NASA; **166tr**, NASA; **166bl**, NASA; **166bc**, NASA; **166br**, NASA; **167tl**, NASA; **167tc**, NASA; **167tr**, NASA; **167bl**, NASA; **167bc**, NASA; **167br**, NASA; **166–167b**, NASA; **168–169**, Science Museum/Science & Society Picture Library; **170tr**, NASA; **170c**, NASA; **170bl**, AP/Empics; **170br**, NASA; **171l**, NASA; **171tr**, NASA; **171cr**, NASA; **171br**, NASA; **172tc**, NASA; **172cl**, NASA; **172–173**, Hugh Johnson; **172br**, NASA; **173tc**, NASA; **173cr**, NASA; **173bl**, NASA; **173bc**, NASA; **173br**, Science & Society Picture, Library; **174tr**, NASA; **174c**, NASA; **175tl**, NASA; **175cl**, NASA; **175tr**, Novosti; **175br**, RIA Novosti/Science Photo Library; **176tl**, NASA; **176bl**, Arthur Rothstein/Corbis; **176tr**, AFP/Getty Science & Society Picture Library; **176cr**, NASA; **176c**, Bettmann/Corbis; **176bc**, NASA; 176br, Bettmann/Corbis; **177t**, NASA; **177br**, NASA; **178tl**, Bettmann/Corbis; **178bl** Eric Preau/Sygma/Corbis; **178–179b**, Novosti; **179tl**, VideoCosmos; **179tr**, Novosti; **179br**, James Oberg; **180bl**, Anatoly Zak; **180br**, Anatoly Zak; **180tr**, Anatoly Zak; **180–181b**, VideoCosmos; **182bl**, James Oberg; **182bc**, James Oberg; **182tr**, Novosti; **183t**, Novosti; **183c**, Corbis; **183br**, Novosti; **186–187**, NASA; **188cl**, Molniya; **188tr**, NASA; **188bl**, BAE Systems; **188–189b**, NASA; **189fp**, USAF; **189br**, NASA; **190bl**, NASA; **190b**, NASA; **190r**, Julian Baum; **190–191t**, Adrian Thompson; **191bl**, NASA; **191bc**, NASA; **191br**, NASA; **192bl**, NASA; **192b**, NASA; **192tr**, NASA; **192–193c**, NASA; **193tr**, NASA; **193br**, NASA; **194bl**, Science & Society Picture Library; **194tr**, NASA; **194–195**, NASA; **195tr**, NASA/Science Photo Library; **196cr**, Corbis; **196bl**, NASA; **196bc**, NASA; **196–197c**, Julian Baum; **197tc**, NASA; **197cr**, NASA; **198bl**, AP/Empics, **198br**, ESA; **198tr**, NASA; **199tl**, NASA/Science Photo Library; **199tr**, Bettmann/Corbis; **199bl**, AP/Empics; **199br**, Ed Kolenovsky/AP/Empics; **200bl**, NASA; **200tr**, Science & Society Picture Library; **201fp**, NASA; **201bl**, NASA/Science Photo Library; **202–203**, NASA; **202bl**, NASA/Getty Science & Society Picture Library; **203cl**, NASA/Science Photo Library, **203bl**, NASA; **203tr**, Diane Walker/Time Life Pictures/Getty Science & Society Picture Library, **204tl**, NASA; **204bl**, Bettmann/Corbis; **204bc**, Bettmann/Corbis; **204cb**, Time Life Pictures/NASA/Getty Science & Society Picture Library; **204br**, NASA; **204tr**, Bettmann/Corbis; **204–205c**, Corbis; **205bl**, Space Frontiers/Hulton Archive/Getty Science & Society Picture Library; **205cr**, Corbis; **206tl**, NASA; **206–207b**, NASA; **207tr**, NASA; **207tl**, Time Life Pictures/NASA/Getty Science & Society Picture Library; **207br**, NASA; **208–209**, NASA; **210bl**, VideoCosmos; **210tr**, Detlev van Ravenswaay/Science Photo Library; **210br**, VideoCosmos; **211**, VideoCosmos; **212cl**, Novosti; **212–213**, Ian Palmer; **213tr**, NASA/Time Life Pictures/Getty Science & Society Picture Library; **213cl**, VideoCosmos; **214–215**, Molniya; **215tl**, Molniya; **215c**, Novosti; **215br**, Molniya; **215bl**, Molniya; **216tl**, NASA; **216tr**, NASA; **216cl**, NASA; **216bl**, NASA; **216br**, NASA; **217tl**, NASA; **217bl**, NASA; **217tr**, Novositi; **218–219**, NASA; **220tl**, Reuters/Corbis; **220tr**, Empics; **220br**, Reuters/Corbis; **221**, NASA/Science Photo Library; **221br**, NASA; **222tl**, NASA; **222bl**, NASA; **222bc**, NASA; **222tl**, NASA; **222cl**, NASA; **222c**, NASA; **222bc**, NASA; **222br**, NASA; **223t**, NASA; **223br**, NASA; **224tl**, NASA; **224tr**, NASA/Science Photo Library; **224bc**, NASA; **224br**, Robert, McCullough/Dallas Morning News/Corbis; **224–225t**, NASA; **225bc**, NASA/Sandra Joseph and Kevin O'Connell; **226–227**, NASA; **228c**, ESA; **228–229b**, ESA; **228–229c**, CNES; **229cl**, ESA; **229tr**, ESA; **230bl**, ESA; **230bc**, ESA; **230tr**, Tony Ranze/Getty Science & Society Picture Library; **231l**, ESA; **231crb**, ESA-David Ducros, 2015; **232tl**, ESA; **232tc**, Safran; **232–233c**, Adrian Thompson; **232bc**, Ariane Space; **233bl**, Ariane Space; **233tr**, ESA-D. Ducros, 2014; **234bl**, VideoCosmos; **234tr**, Courtesy of JAXA; **234cr**, Courtesy of JAXA; **234–235b**, Courtesy of JAXA; **235tl**, Courtesy of JAXA; **235tr**, Japan Aerospace Exploration Agency (JAXA): © JAXA; **236**, Roger Ressmeyer/Corbis; **236tl**, Stefan, R. Landsberger/International Institute of Social History, Amsterdam; **237tr**, Bettmann/Corbis; **237bl**, Reuters/Corbis; **238tc**, Canadian Space Agency; **238bl**, Art Directors/Dinodia; **238br,** Indian Space Research Organisation; **238–239c**, Indian Space Research Organisation; **239tr**, Reuters/Eric, Sultan/Corbis; **239crb**, Rocket Lab; **240tl**, Empics; **240cl**, NASA; **240bl**, NASA; **240tr**, Hulton-Deutsch, Collection/Corbis; **240br**, Stanislav Krasilnikov/TASS/Alamy Stock Photo; **241**, Canadian Space Agency; **242–243**, Reuters/Corbis; **244br**, Peter Bull; **244cra**, Science History Images/Alamy Stock Photo; **245t**, NASA/Science Photo Library; **245bl**, Camera Press/ITAR-TASS; **245crb**, Joe Davila/U.S. Air Force photo; **246bl**, Luis Enrique Ascui/Getty Science & Society Picture Library; **246br**, Rolls Press/Popperfoto/Getty Images; **246cra,** Keystone/Getty Science & Society Picture Library; **247cr**, Panther Media GmbH/Alamy Stock Photo; **247b**, Tommy Eliassen/Science Photo Library; **248br**, Peter Bull; **248bc**, National Reconnaissance Office; **248–249t**, ESA-P. Carril; **249br**, Denis Charlet/Getty Images; **249cra**, ESA; **250bc**, National Reconnaissance Office; **250cra**, NASA; **251cra**, NASA/JPL; **251br,** NASA; **251t**, Courtesy of GeoEye; **252b,** IRAS; **252bl**, Corbis; **252tr**, NASA; **252cr**, NASA; **253t**, NASA; **253tr**, ESA/NASA; **253c**, NASA; **253bc**, ESA; **254bl**, NASA; **254tr**, Al Fenn/Time Life Pictures/Getty Science & Society Picture Library; **255cr**, ESA/Hubble: http://creativecommons.org/licenses/by/3.0/NASA, ESA and the Hubble Heritage (STScI/AURA)-ESA/Hubble Collaboration. Acknowledgement: Davide De Martin and Robert Gendler; **255tr**, NASA/ESA/Hubble Heritage Team; **255cla**, NASA, ESA, the GOODS Team, and M. Giavalisco (University of Massachusetts, Amherst); **255br**, NASA; **256bl**, NASA; **256bc**, NASA/JPL-Caltech; **256br**, NASA; **256–257c**, NASA: NASA/JPL-Caltech/Ball Aerospace; **257crb**, ESA/ATG medialab; background: ESO/S. Brunier; **257cr**, ESA, Planck Collaboration; **257cr**, DSS/Galaxy Picture Library; **257cb**, NASA/GSFC; **257tl**, MPE; **257tc**, GALEX; **257tr,** Philip, Perkins/Galaxy Picture Library; **257cl**, IRAS; **257c**, NASA/JPL-Caltech; **257r**, Rainer Beck/Philipp Hoernes/MPIfR; **258bl**, RIA Novosti/Science Photo Library; **258tr**, BMDO/NRL/LLNL/Science Photo Library; **259ca**, Japan Aerospace Exploration Agency (JAXA); **259tl**, NASA; **259b**, NASA; **260tr**, Science & Society Picture Library; **260bc**, Peter Bull; **261tl**, NASA/Johns Hopkins University Applied Physics Laboratory/Carnegie Institution of Washington; **261bl**, NASA/Johns Hopkins University Applied Physics Laboratory/Carnegie Institution of Washington; **261r**, NASA/Johns Hopkins University Applied Physics Laboratory/Carnegie Institution of Washington; **262c**, NASA; **262–262b**, NASA; **262cr**, NASA; **262–263t**, NASA; **263tr**, NASA/Science & Society Picture Library; **264tr**, NASA; **264bl**, NASA; **264bc**, NASA; **264–265t**, NASA; **265bl**, NASA; **265br**, Peter Bull; **266tl**, NASA; **266cl**, NASA; **266clb**, Novosti/Science Photo Library; **266cra**, ITAR-TASS News Agency/Alamy Stock Photo; **266–267b**, BP/NRSC/Science Photo Library; **267ca**, NASA/JPL/USGS; **268tl**, US Naval Observatory/Science Photo Library; **268cl**, NASA/Science Photo Library; **268–269c**, Hugh Johnson; **268br**, NASA; **269tl**, TLP/NASA/Getty Science & Society Picture Library; **269tc**, NASA; **269tr**, NASA; **269r**, Dan Groshong/AFP/Getty Science & Society Picture Library; **269bl**, IMP BV; **269c**, NASA; **269br**, NASA; **270tl**, NASA; **270tr**, NASA/Galaxy Picture Library; **270bl**, NASA; **270–271b**, NASA/Galaxy Picture Library; **271t**, NASA; **271br**, NASA; **272cr**, Peter Bull; **272bl**, Detlev van Ravenswaay/Science Photo Library; **273b**, ESA/Rosetta/NAVCAM, CC BY-SA IGO 3.0; **273cr**, Japan Aerospace Exploration Agency (JAXA); **273crb**, NASA: ESA/Rosetta/MPS/UPD/LAM/IAA/SSO/INTA/UPM/DASP/IDA; **273tr**, NASA/JPL; **273tc**, NASA/JPL/MPS/DLR/IDA/Björn Jónsson; **274tl**, NASA; **274bc**, Science Museum/Science & Society Picture Library; **274–275b**, NASA; **275tl**, ESA/DLR/FU Berlin; **275tr**, ESA; **275c**, ESA; **276cl**, NASA; **276bl**, NASA; **276–277c**, Julian Baum; **277t**, NASA/JPL-Solar System Visualization Team; **277cr**, NASA; **277br**, NASA; **277bl**, NASA; **278tl**, NASA/JPL/University of Arizona/University of Idaho; **278b**, NASA; **278tr**, ESA; **279tl**, Science & Society Picture Library; **279tr**, Peter Bull; **279bl**, NASA; **279br**, NASA/JPL/Space Science Institute; **280tl**, NASA/JPL-Caltech/Space Science Institute; **280cl**, NASA/JPL/Space Science Institute; **280–281**, NASA; **282cra**, ESA-D. Ducros; **282b**, NASA/JPL-Caltech/University of Arizona/Texas A&M University; **283tr**, NASA/JPL-Caltech/Univ. of Arizona; **283c**, NASA/JPL-Caltech/ESA/DLR/FU Berlin/MSSS; **283bl**, NASA/JPL-Caltech/MSSS; **284br**, NASA/Johns Hopkins University Applied Physics Laboratory/Southwest Research Institute; **284bc**, Bettmann/Corbis; **284cr**, NASA; **285cr**, JAXA/Military Collection/Alamy Stock Phoo; **285tl**, NASA/JPL-Caltech/SwRI/MSSS/Gerald Eichstadt/Sean Doran; **285br**, Hugh Johnson; **286–287**, NASA; **288bl**, NASA; **288tr**, NASA; **288cr**, NASA; **289t**, NASA; **289br**, RKK Energia; **290–291**, Hugh Johnson; **292tl**, NASA; **292cl**, NASA; **292–293**, NASA; **294bc**, NASA/Robert Markowitz; **294tl**, NASA; **294ca**, NASA; **294crb**, NASA; **294–295t**, NASA; **295bc**, NASA; **295br**, NASA/Johnson; **295c**, NASA; **295cra**, NASA; **296tl**, NASA; **296cr**, NASA; **296bl**, NASA; **296bcl**, NASA; **296bc**, NASA; **296bl**, NASA; **296–297t**, NASA; **297tr**, NASA; **297br**, NASA; **298bl**, Copyright Airbus 2018; **298tr**, NASA; **298–299b**, NASA; **299cra**, ESA/NASA; **299cb**, NASA/Robert Markowitz; **299tl**, NASA; **299c**, NASA; **299cr**, NASA; **300cl**, Joshua Hildwine, through Dan Johnson, Luft46.com; **300–301b**, Bloomberg/Getty Images; **301cb**, DARPA Outreach; **301br**, NASA; **301tr**, Hugh Johnson; **302tl**, China, Photos/Getty Science & Society Picture Library; **302–303c**, STR/AFP/Getty Images; **303br**, Alejandro Miranda/Alamy Stock Photo; **303cra**, AFP/Getty Images; **303bl**, Li Gang/China Features/CORBIS; **304br**, NASA Image Collection/Alamy Stock Photo; **304ca**, ESA: NASA; **304cb**, NASA/MAF/Steven Seipel; **305**, NASA Image Collection/Alamy Stock Photo; **306tr**, ESA/ATG medialab; **306bl**, NASA/JPL-Caltech/Lockheed Martin; **306–307b**, NASA/JPL-Caltech; **307cb**, ESO/G. Hüdepohl (atacamaphoto.com); **307tr**, NASA/JPL-Caltech; **308crb**, ROBYN BECK/AFP/Getty Images; **308ca**, NASA; **308bl**, Sergei Ilnitsky/Corbis; **309cr**, Science Collection/Alamy Stock Photo; **309b**, Joe Raedle/Getty Images; **310–311tr**, NASA; **310tr**, NASA/Ames Research Center; **310bl**, Space Studies Institute; **311br**, British Interplanetary Society. **Endpaper images:** *Front and Back:* **Alamy Stock Photo:** Dennis Hallinan.